Autodesk
Inventor® 2013

L. Scott Hansen

Southern Utah University

Connect
Learn
Succeed™

AUTODESK® INVENTOR® 2013

Published by McGraw-Hill, a business unit of The McGraw-Hill Companies, Inc., 1221 Avenue of the Americas, New York, NY 10020. Copyright © 2013 by The McGraw-Hill Companies, Inc. All rights reserved. Printed in the United States of America. No part of this publication may be reproduced or distributed in any form or by any means, or stored in a database or retrieval system, without the prior written consent of The McGraw-Hill Companies, Inc., including, but not limited to, in any network or other electronic storage or transmission, or broadcast for distance learning.

Some ancillaries, including electronic and print components, may not be available to customers outside the United States.

 This book is printed on recycled, acid-free paper containing 10% postconsumer waste.

1 2 3 4 5 6 7 8 9 0 QDB/QDB 1 0 9 8 7 6 5 4 3 2

ISBN 978–0–07–3522708
MHID 0–07–3522708

Vice President & Editor-in-Chief: *Marty Lange*
Editorial Director: *Michael Lange*
Global Publisher: *Raghothaman Srinivasan*
Executive Editor: *Bill Stenquist*
Marketing Manager: *Curt Reynolds*
Development Editor: *Lorraine Buczek*
Project Manager: *Melissa M. Leick*
Buyer: *Sandy Ludovissy*
Media Project Manager: *Prashanthi Nadipalli*
Cover Designer: *Studio Montage, St. Louis, Missouri*
Cover Image: *[Inset photo] Mark Gundersen/ Copyright 2006 by Marine Advanced Research, Inc.*
[main photo] Marine Advanced Research, Inc and Autodesk, Inc.
Compositor: *Laura Hunter, Visual Q*
Typeface: *10.5 /12.5 Palatino*
Printer: *Quad/Graphics Dubuque*

All credits appearing on page or at the end of the book are considered to be an extension of the copyright page.

Library of Congress Cataloging-in-Publication Data

Hansen, L. Scott.
    Autodesk Inventor 2013 / L. Scott Hansen. — 1st ed.
        p. cm.
    ISBN 978–0–07–352270–8 (alk. paper)
    1. Autodesk Inventor (Electronic resource) 2. Engineering graphics. 3. Engineering models—Data processing. I. Title.

    T353.H2454   2013
    620'.0042028553—dc23

                                                                            2012013855

www.mhhe.com

ABOUT THE AUTHOR

L. Scott Hansen received his A.A.S degree in Electro-Mechanical CAD from Pima Community College in Tucson, Arizona. He received his B.S. and M.S. degrees in Vocational Education from Northern Arizona University and also received a Ph.D. in Applied Science and Technology from the University of Wyoming. He is currently the Department Chair of Engineering Technology and Construction Management and Associate Professor of Engineering Technology at Southern Utah University. He teaches freshman through senior-level courses in the CAD/CAM Engineering Technology program. Hansen's software application experience includes IBM Fastdraft, VersaCAD, AutoCAD, Inventor, SolidWorks, Solid Edge, CATIA V5, and Mastercam. In his spare time, he has performed extensive design and fabrication work to build and modify automobiles, dune buggies, boats and motorcycles along with residential construction projects.

TABLE OF CONTENTS

TABLE OF CONTENTS

PREFACE

ABOUT THIS BOOK

The philosophy behind this book is that learning computerized drafting programs is best accomplished by emphasizing the application of the tools rather than spending time on the theoretical principles underpinning engineering graphics and computer-aided design. Students also seem to learn more quickly and retain information and skills better if they are actually creating something with the software program. The driving force behind the entire presentation in this book is "learning by doing."

The instructional format of this book centers on making sure that students learn by doing and that students can learn from this book on their own. In fact, this is one thing that differentiates my book from others: the emphasis on being able to use the book for self-study. The presentation of Inventor is structured so that the no previous knowledge of using any CAD program is required. My belief is that Inventor is mastered best by concentrating on applying the program to create different types of solid models, starting simply and then using the power of the program to progressively create more complex solid models. The Drawing Activities at the end of each chapter are more complex iterations of the part developed by each chapter's objectives.

Because CAD programs are highly visual, there are more graphical illustrations showing how to use the program than there are verbal instructions. This reinforces the "learn by doing" philosophy since a student can see exactly what the program shows, and then step through progressive commands to implement the required operations. Rather than using a verbal description of the command, a screen capture of each command is replicated. This book's emphasis is on providing the screen captures of the program's commands to develop the skills needed for the creation of models and parts that are illustrated in the book. The screen captures have also been enhanced with small arrows to point out exactly what is most important in each screen capture.

Learning by doing requires Drawing Activities that reinforce each chapter's objectives. All Part/Model exercises that form the core of each chapter begin simply and build up in level of complexity throughout the text as the student's skills with the program improve. Those skills are reinforced and improved through the end of chapter problems. The Chapter Problems have been carefully developed to compliment the skills taught in the part/model detailed in the chapter objectives.

It has been my experience that the "learn by doing" model produces faster mastery of computer-aided drafting programs than comparable lecture modes where the commands are explained and then the students are expected to implement those commands. I have found that students are eager to take over "driving the program" and if they have the correct roadmap, they quickly pick up how to operate the program and progress quickly. My hope is that you will find equal success in using this book in your Autodesk Inventor courses. I welcome your feedback and if you need specific on-site Inventor training, please contact me at hansens@suu.edu.

ACKNOWLEDGEMENTS

I would like to thank my wife Linda, my son Jordan and daughter Morgan for their constant support and understanding while I was writing this book. I spent quite a few evenings away from them working late into the night to complete it. I also want to thank my father, Leo Hansen and mother Sherrell Hansen for their support and encouragement. I also want to thank Dr. Clair Hill for giving me my first teaching experience, which has opened many doors for me and enhanced my life and career.

I would like to acknowledge the valuable comments and suggestions of the following individuals who reviewed the manuscript for this text: Eduardo Chan, San Jose State University; William A. Ross, Purdue University; and U. Sunday Tim, Iowa State University.

INSTRUCTOR RESOURCES

Instructors will find solutions to the Chapter Problems at the textbook website: www.mhhe.com/hansen

Contact your local McGraw-Hill representative for password access.

CHAPTER 1

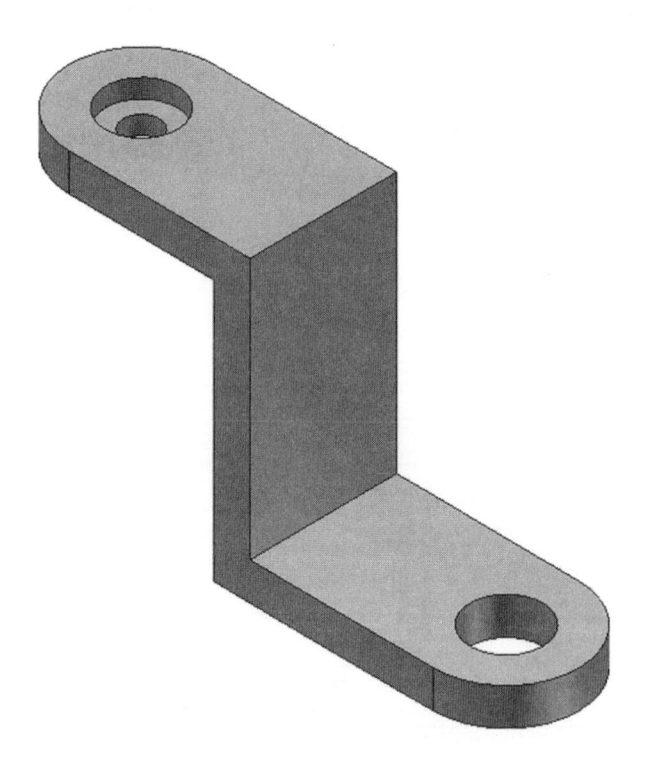

Chapter 1 includes instructions on how to design the part shown above.

GETTING STARTED

OBJECTIVES

1. Create a simple sketch using the Sketch Panel

2. Dimension a sketch using the Dimension command

3. Extrude a sketch in the Model/Part Features Panel using the Extrude command

4. Create a fillet in the Model/Part Features Panel using the Fillet command

5. Create a hole in the Model/Part Features Panel using the Extrude command

6. Create a counter bore in the Model/Part Features Panel using the Hole command

1. Start Inventor by moving the cursor to the **start** button in the lower left corner of the screen. Click the left mouse button once.

2. A pop-up menu of the programs that are installed on the computer will appear. Scroll through the list of programs until you find Autodesk Inventor Professional 2013.

3. Move the cursor over **Autodesk Inventor Professional 2013** and left-click once.

FIGURE 1-1

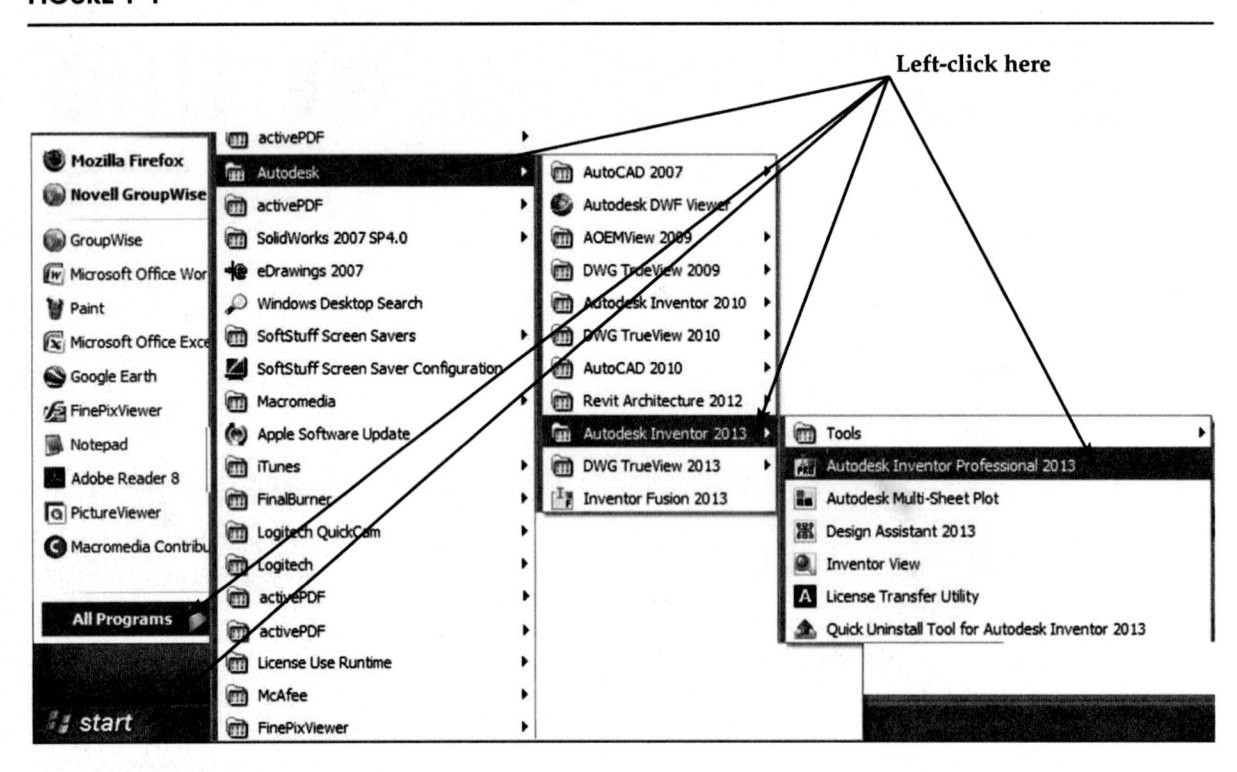

4. Autodesk Inventor Professional 2013 will open (load up and begin running).

5. The Autodesk Product "for placement only" banner will appear. Left-click on the **Close** icon as shown in Figure 1-2.

FIGURE 1-2

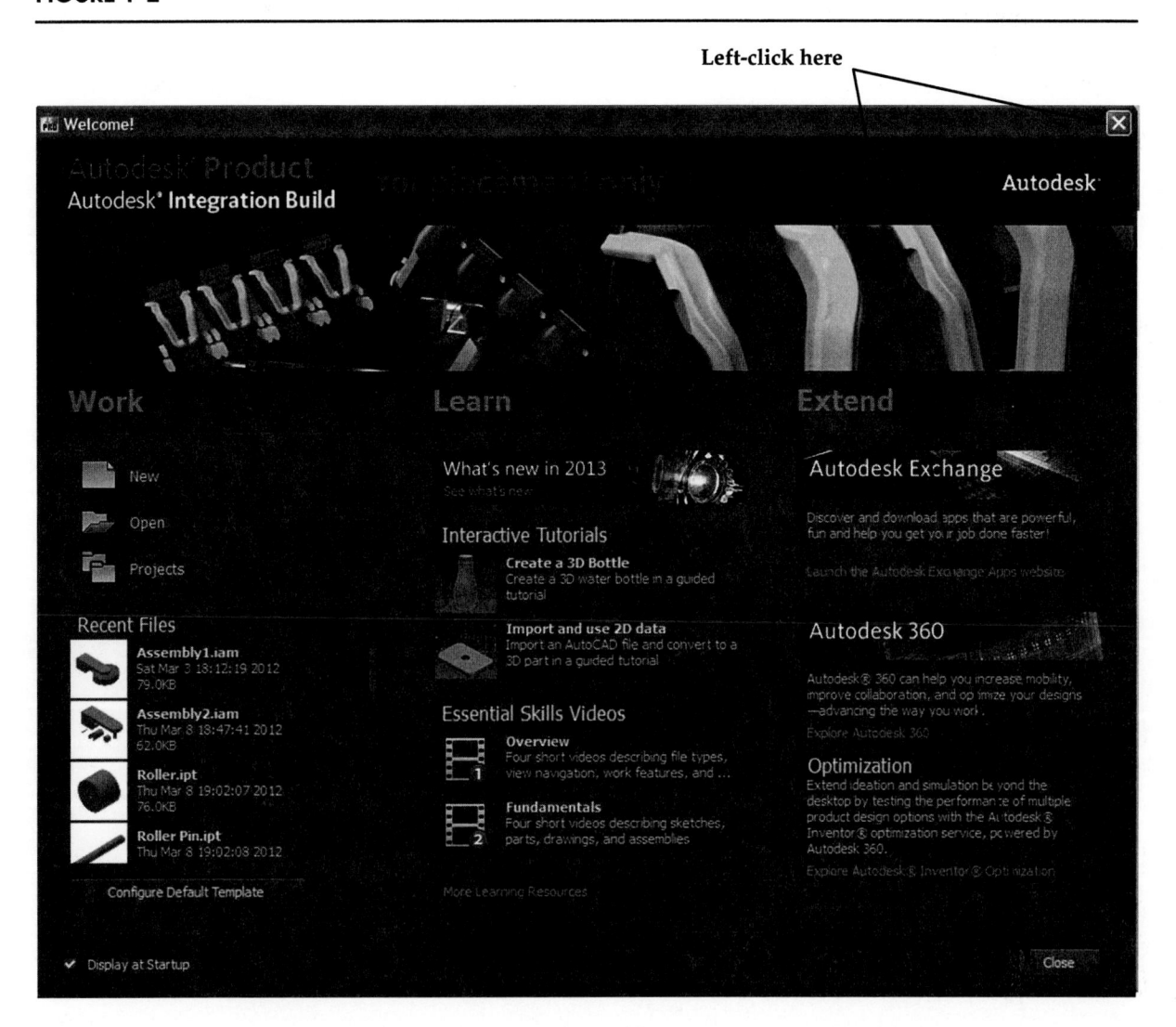

6.   Left-click on the **New** icon (white piece of paper) located in the upper left portion of the screen as shown in Figure 1-3.

FIGURE 1-3

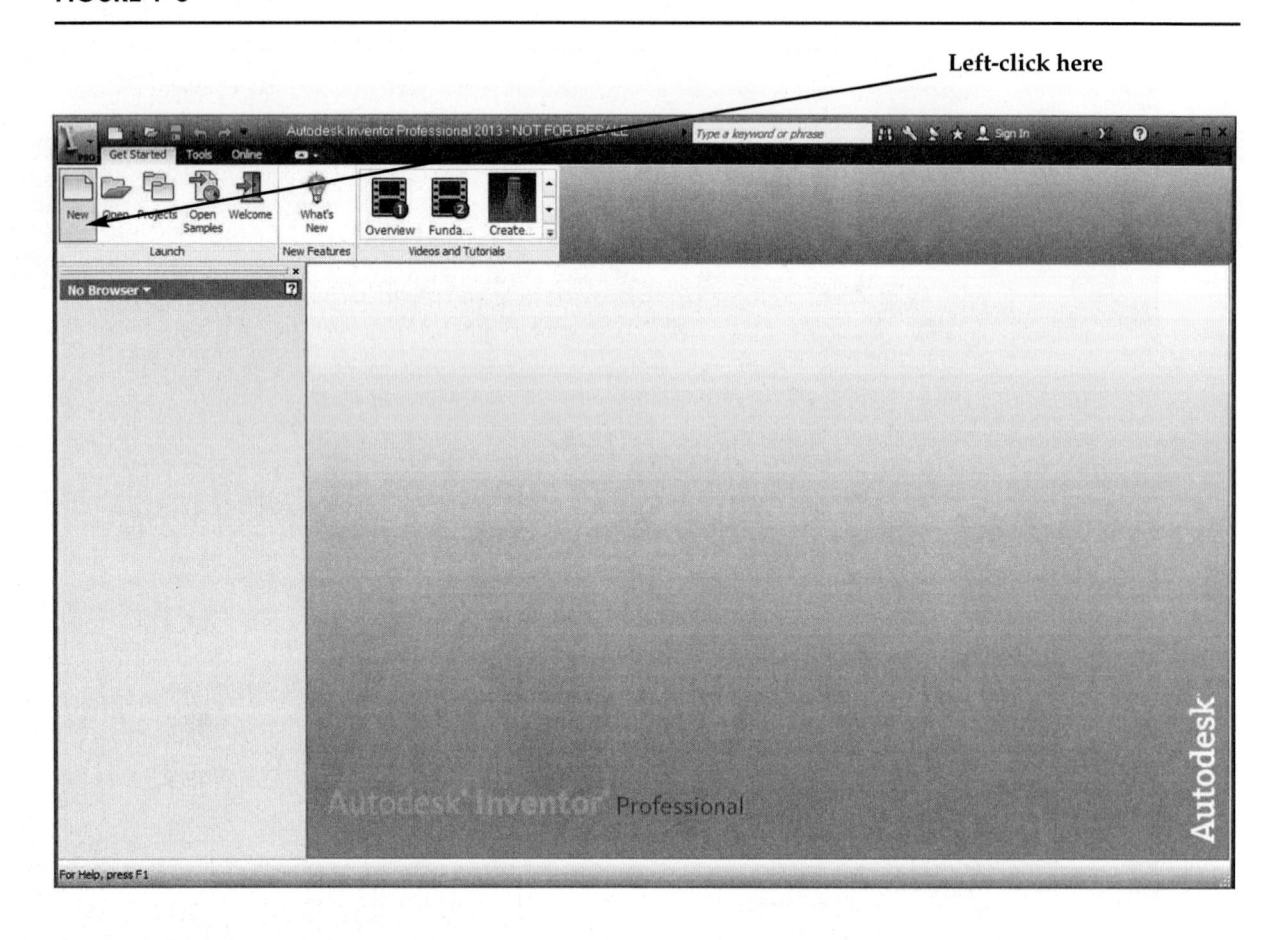

7. The Create New File dialog box will appear. Left-click on the **English** folder at the upper left corner of the dialog box. After left-clicking on the **English** folder, left-click on **Standard (in).ipt** as shown in Figure 1-4.

FIGURE 1-4

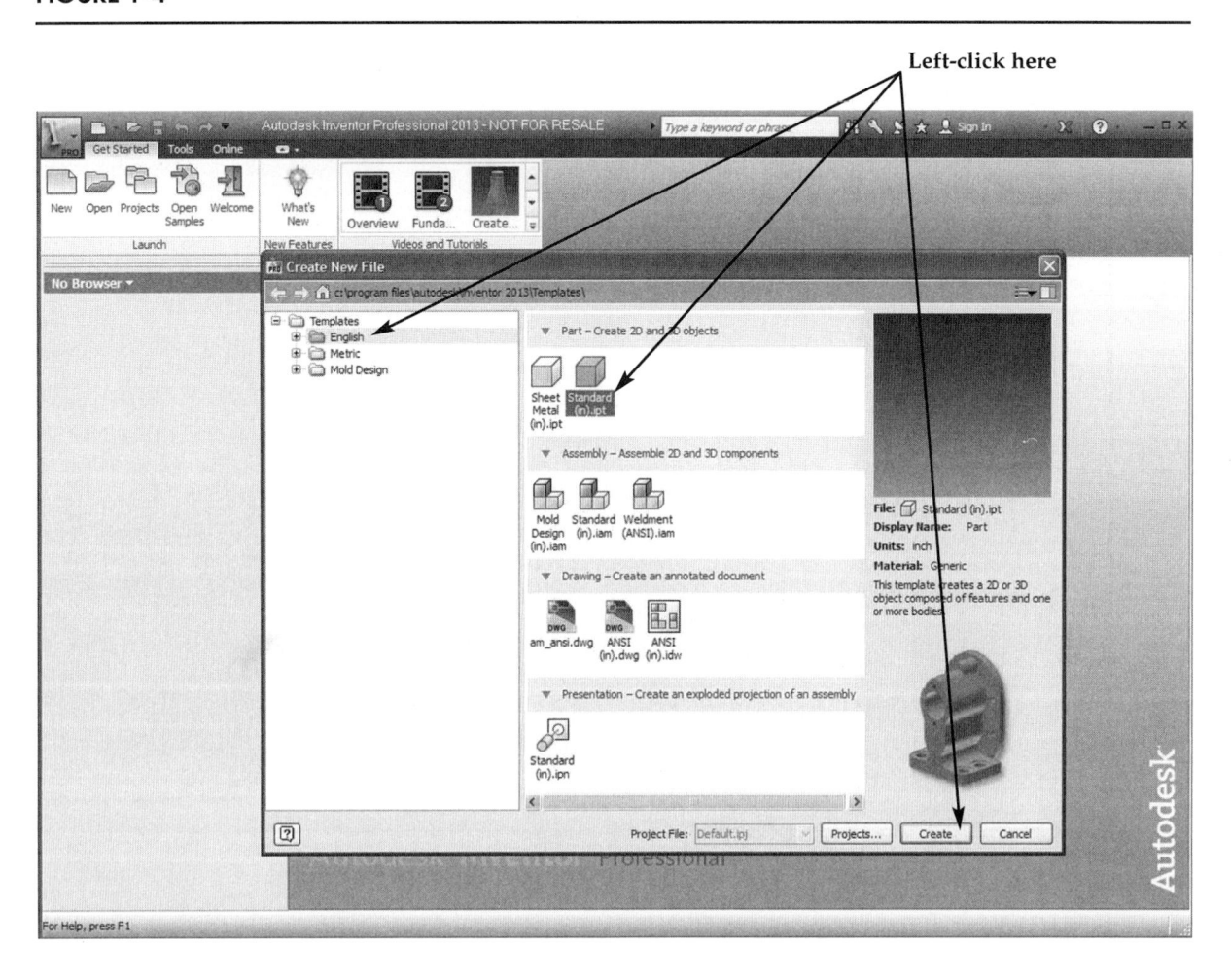

8. Left-click on **Create**.

9.   Inventor is now ready for use. The screen should look similar to Figure 1-5.

FIGURE 1-5

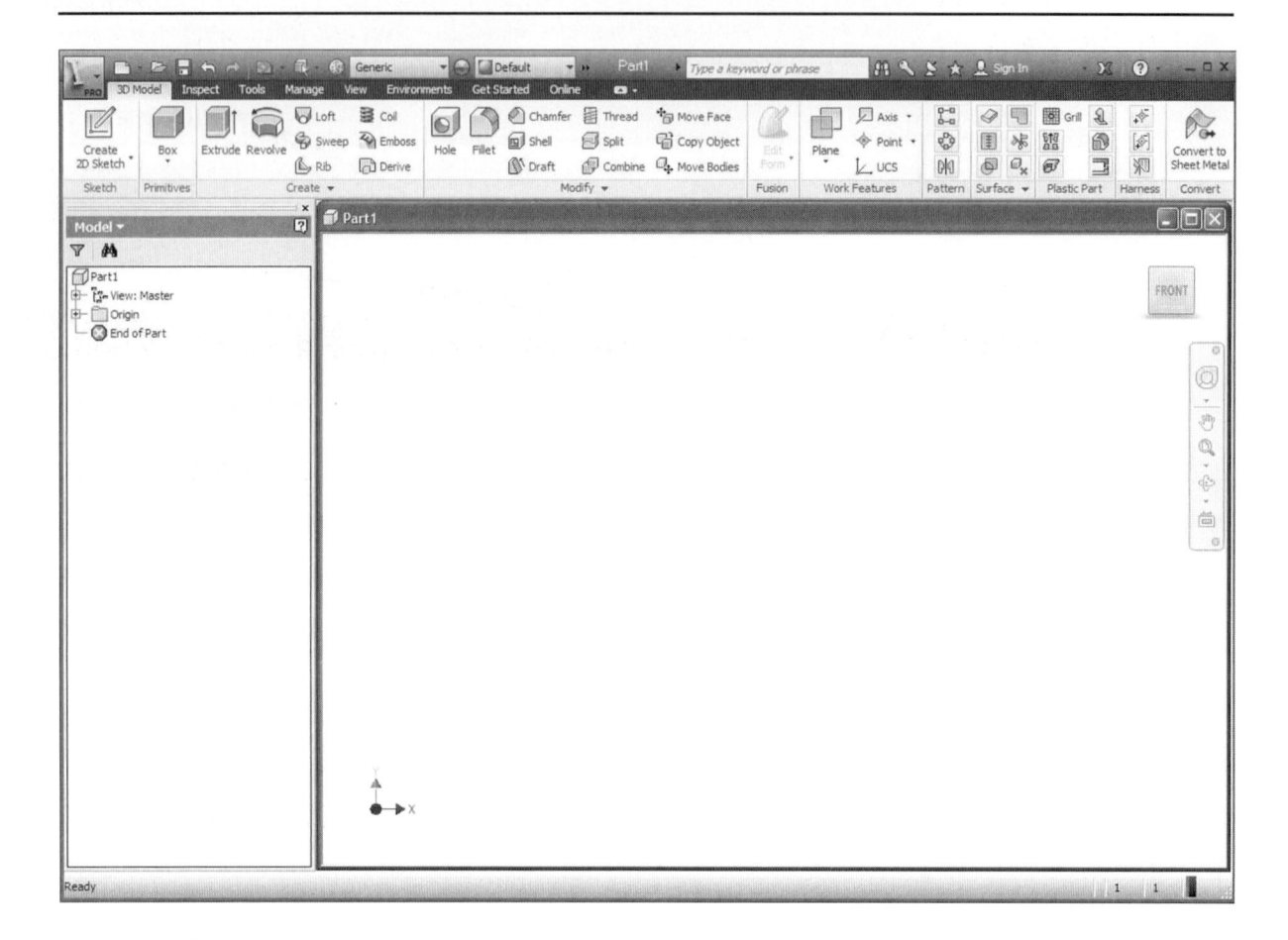

CREATE A SIMPLE SKETCH USING THE SKETCH PANEL

10.   Begin a drawing by first constructing a "sketch." Move the cursor to the upper left portion of the screen and left-click on the **3D Model** tab (if not already selected). Left-click on **Create 2D Sketch**. Select the "Front Plane" as shown. To know what any icon or command will do, move the cursor over the icon or command and wait a few seconds. A yellow banner will appear describing the icons or commands function.

FIGURE 1-6

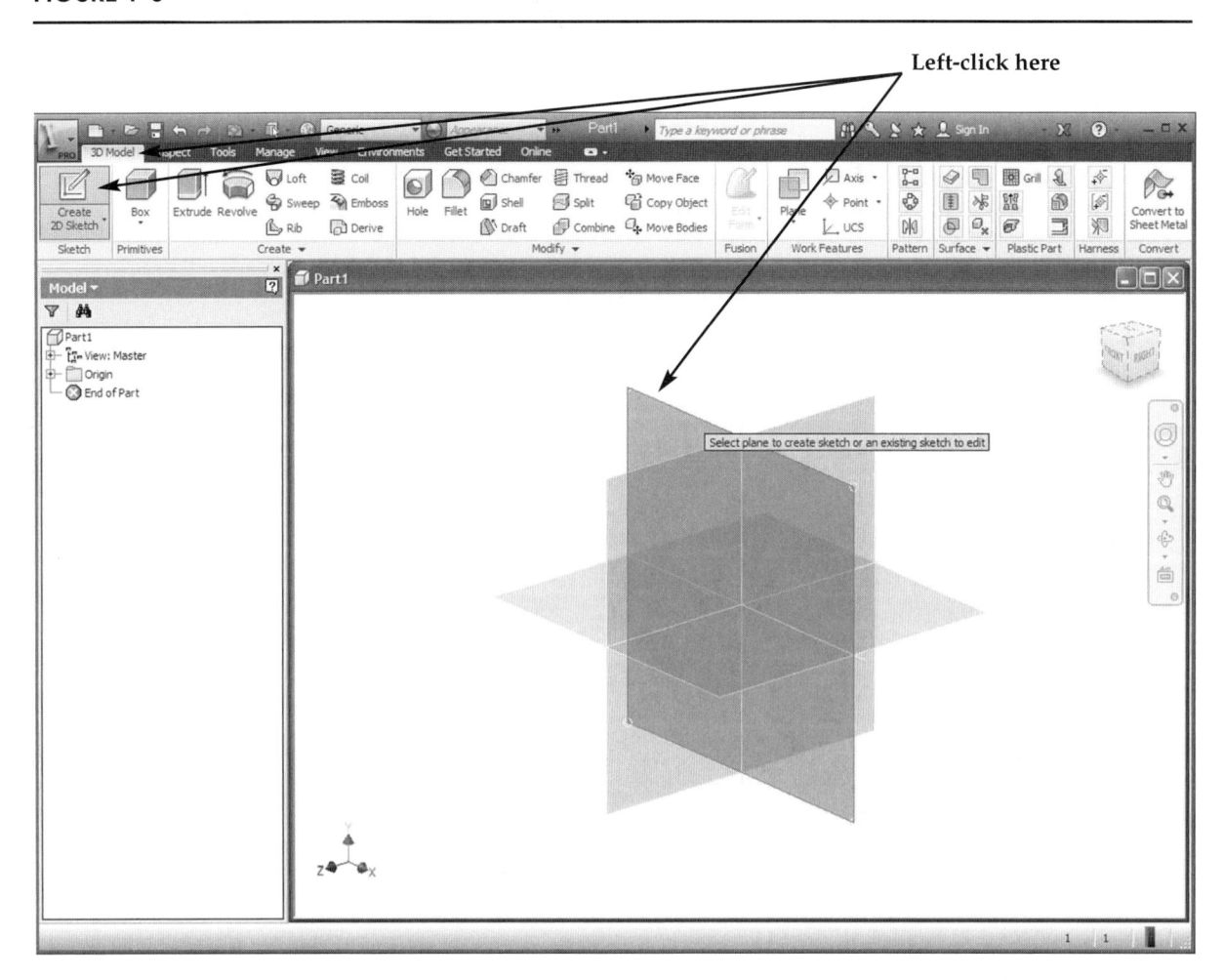

11. Your screen should look similar to Figure 1-7. Left-click on the **Sketch** tab if needed. Move the cursor to the upper left portion of the screen and left-click on **Line** as shown in Figure 1-7.

FIGURE 1-7

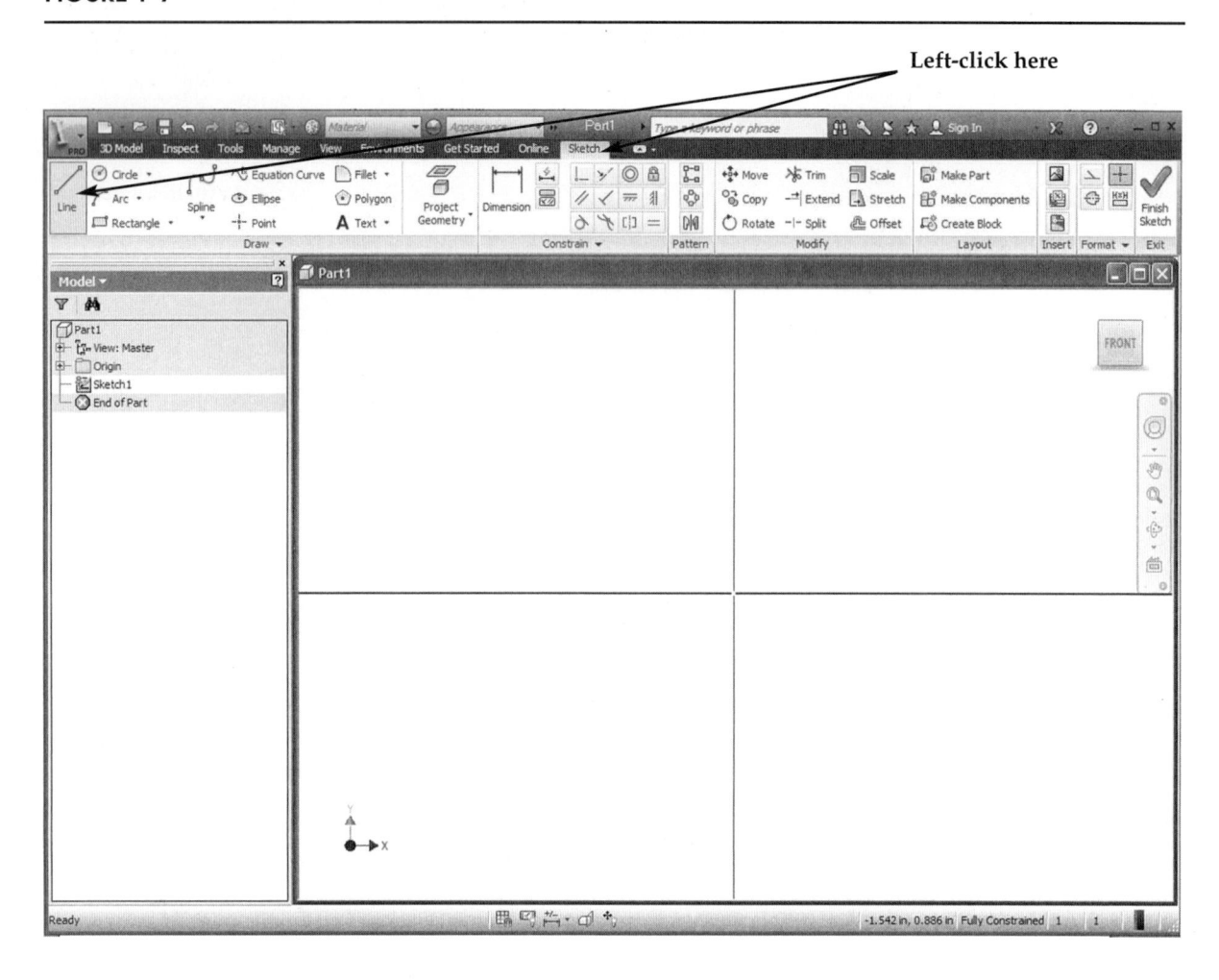

12. Move the cursor somewhere in the lower left portion of the screen and left-click once. This will be the beginning end point of a line as shown in Figure 1-8.

FIGURE 1-8

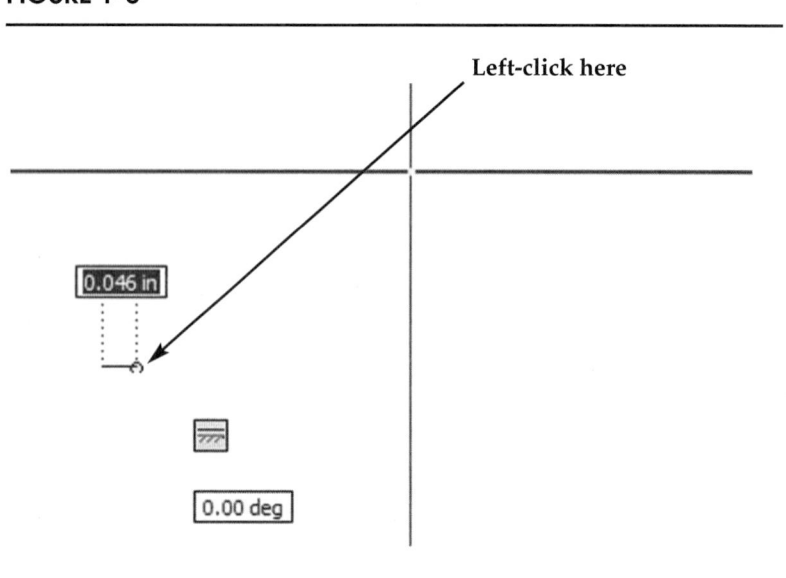

13.   Move the cursor towards the lower right portion of the screen and left-click once as shown in Figure 1-9.

FIGURE 1-9

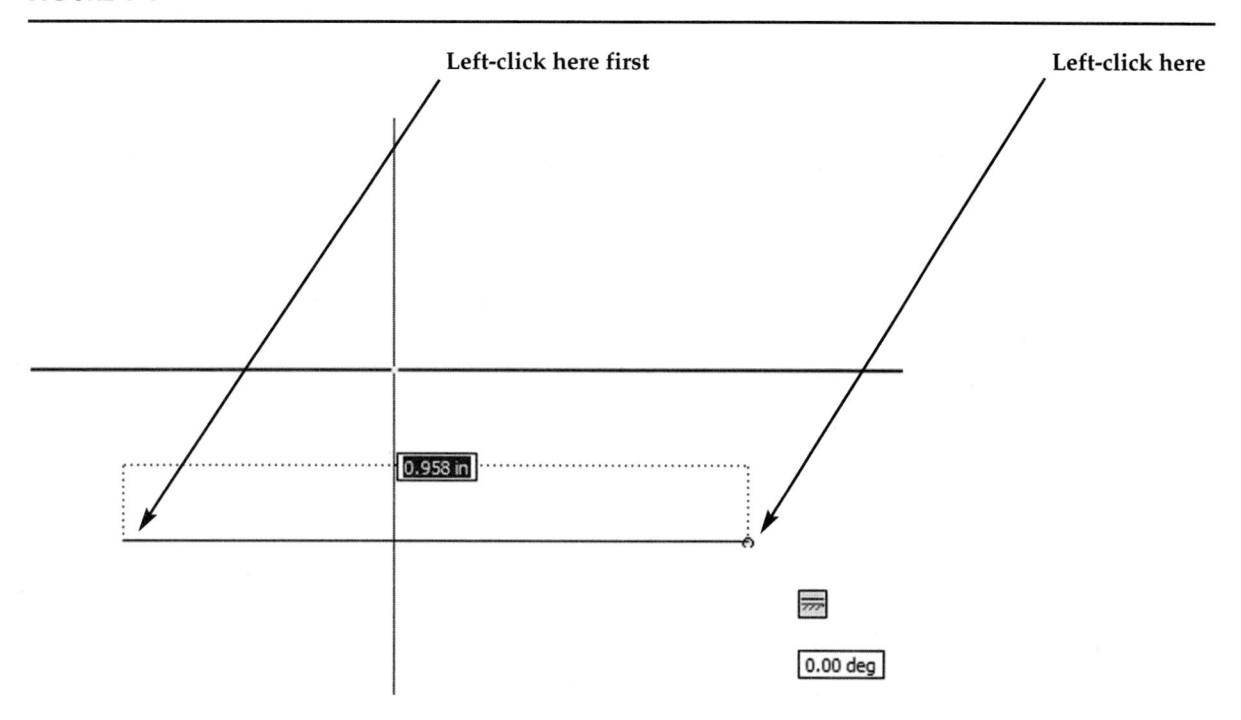

14.   While the line is still attached to the cursor, move the cursor towards the top of the screen and left-click once as shown in Figure 1-10.

FIGURE 1-10

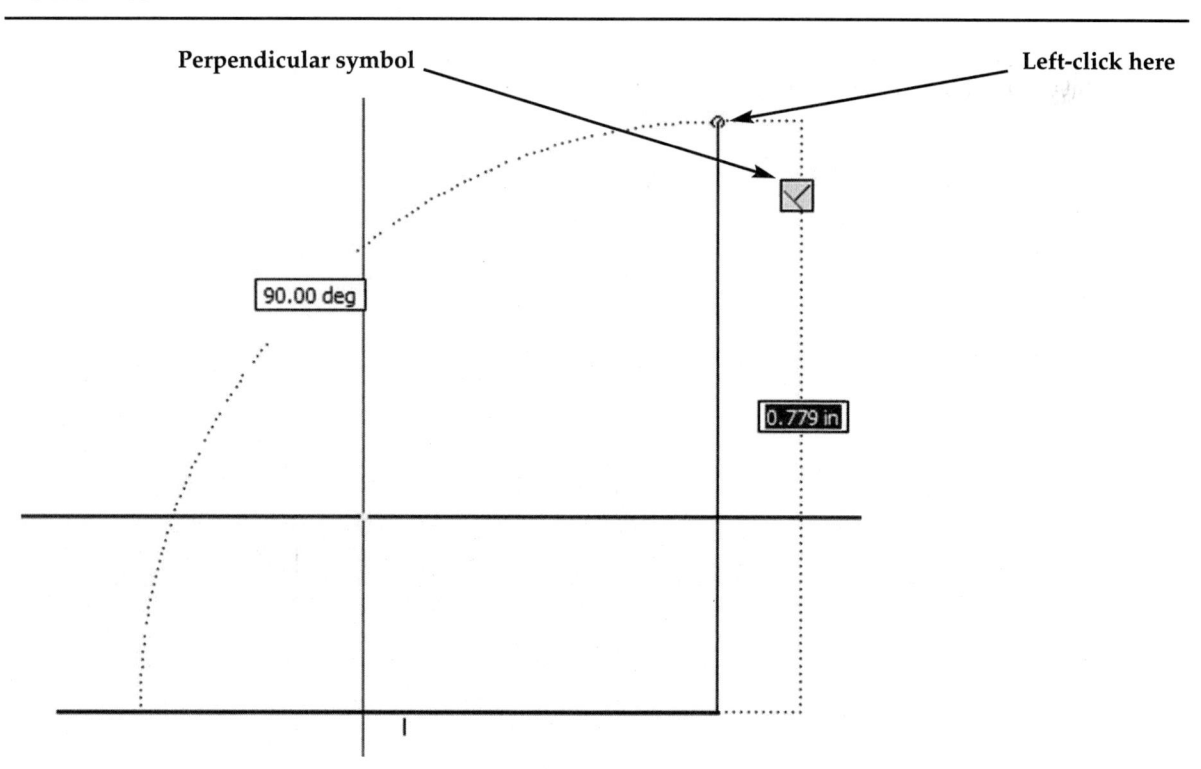

15.   This signifies that the vertical line is exactly 90 degrees (perpendicular) to the horizontal line.

16.   With the line still attached to the cursor, move the cursor towards the left side of the screen as shown in Figure 1-11.

FIGURE 1-11

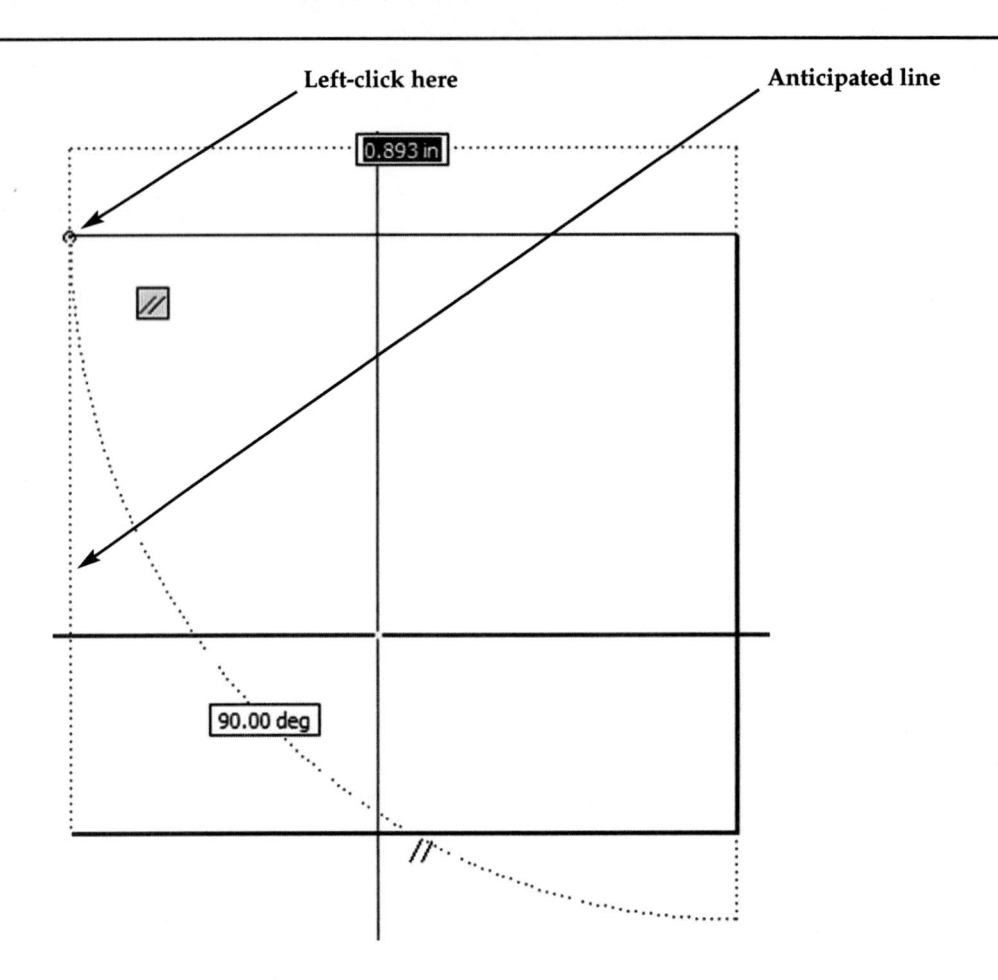

17.   Notice the line of small dots connecting the first and last points together. Left-click once when the small dots appear as shown in Figure 1-11.

18.   This will form a 90-degree box. Move the cursor down towards the original starting point. Ensure that a green dot appears (as shown in Figure 1-12) at the intersection of the two lines. This indicates that Inventor has "snapped" to the intersection of the lines. After the green dot appears, left-click once as shown in Figure 1-12.

FIGURE 1-12

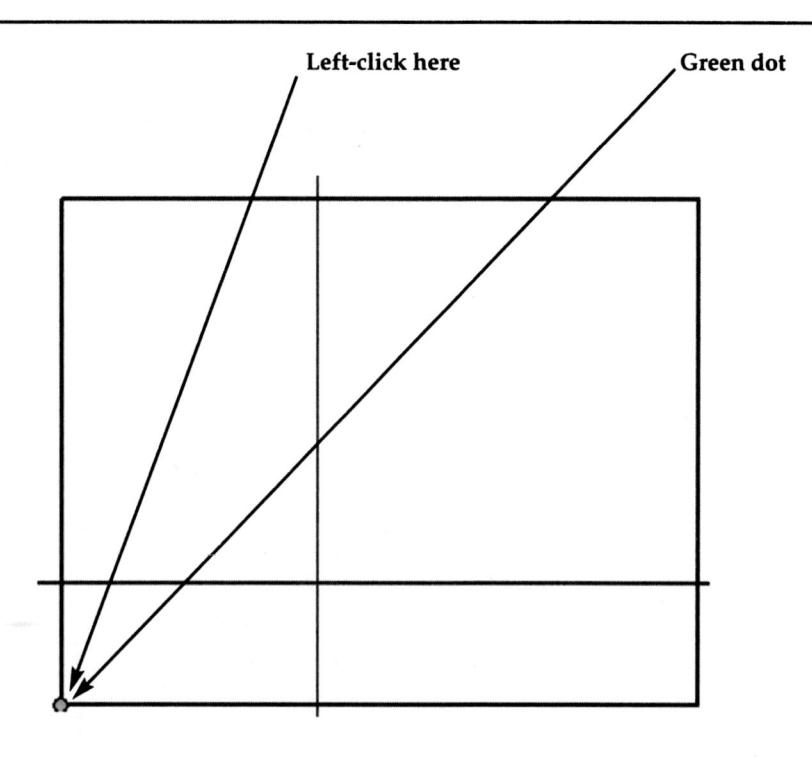

19.   Your screen should look similar to Figure 1-13.

FIGURE 1-13

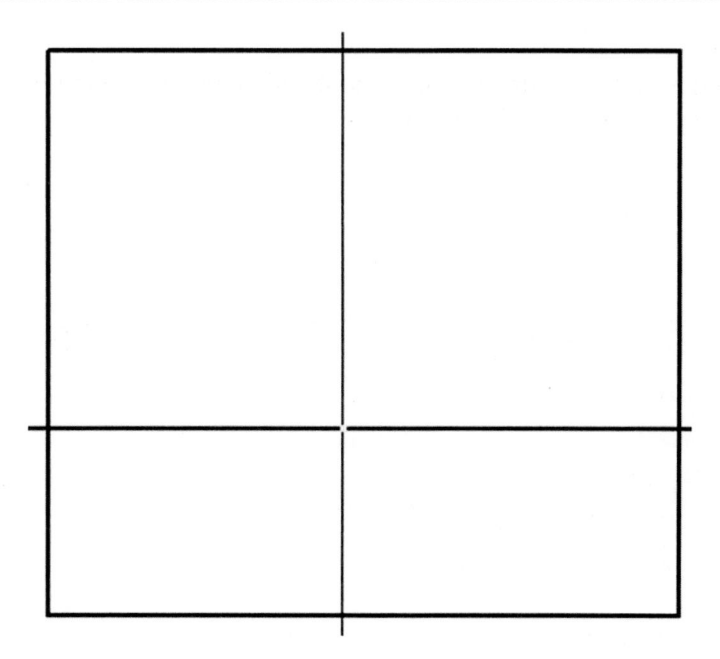

DIMENSION A SKETCH USING THE DIMENSION COMMAND

20.   Right click anywhere around the sketch. A pop-up menu will appear. Left-click on **OK** as shown in Figure 1-14. Pressing the **Esc** key will also "get out" of the Line command.

FIGURE 1-14

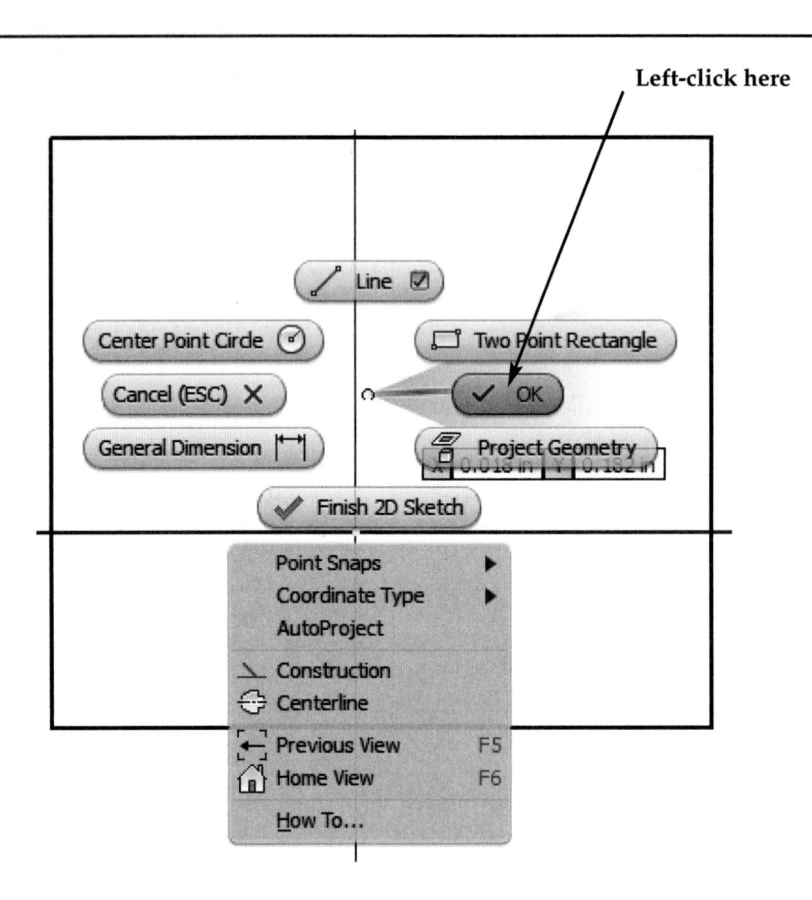

21.   Move the cursor to the upper middle portion of the screen and left-click on **Dimension** as shown in Figure 1-15.

FIGURE 1-15

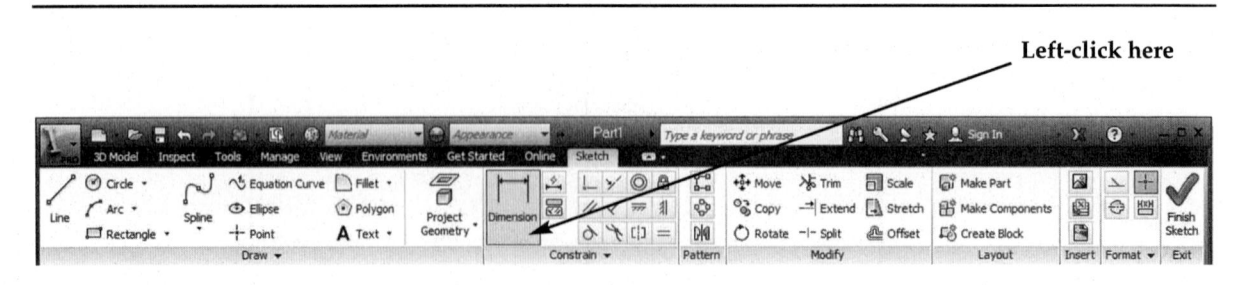

22.   After selecting **Dimension**, move the cursor over the bottom horizontal line. The line will turn red as shown in Figure 1-16. Select the line by left-clicking anywhere on the line <u>or</u> on each of the end points. To use the end points of the line, move the cursor over one of the end points. A small red square will appear. Left-click once and move the cursor to the other end point. Another red square will appear. Left-click once. The dimension will now be attached to the cursor. Move the cursor up and down to verify it is attached.

FIGURE 1-16

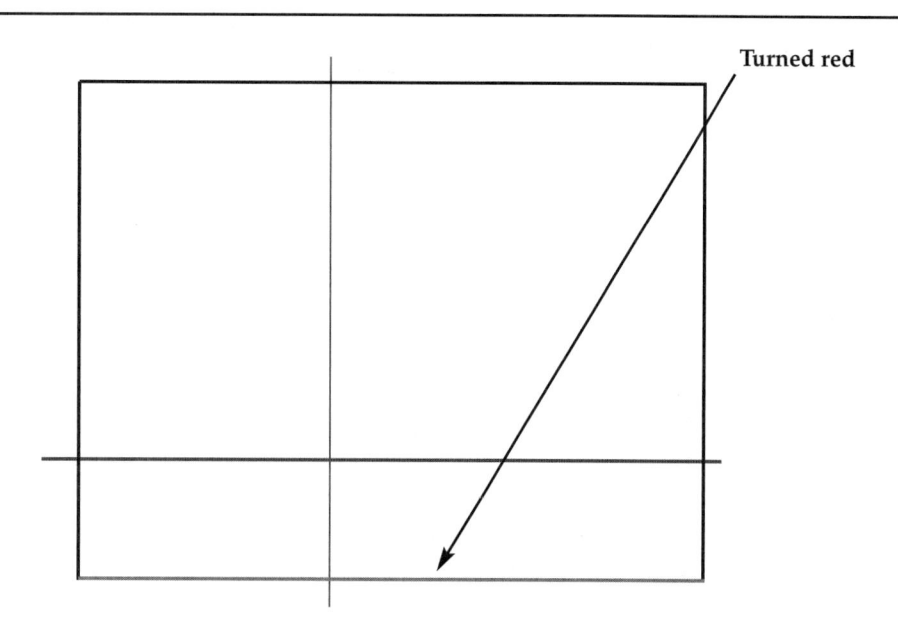

Turned red

23.   Move the cursor down. The actual dimension of the line will appear as shown in Figure 1-17.

FIGURE 1-17

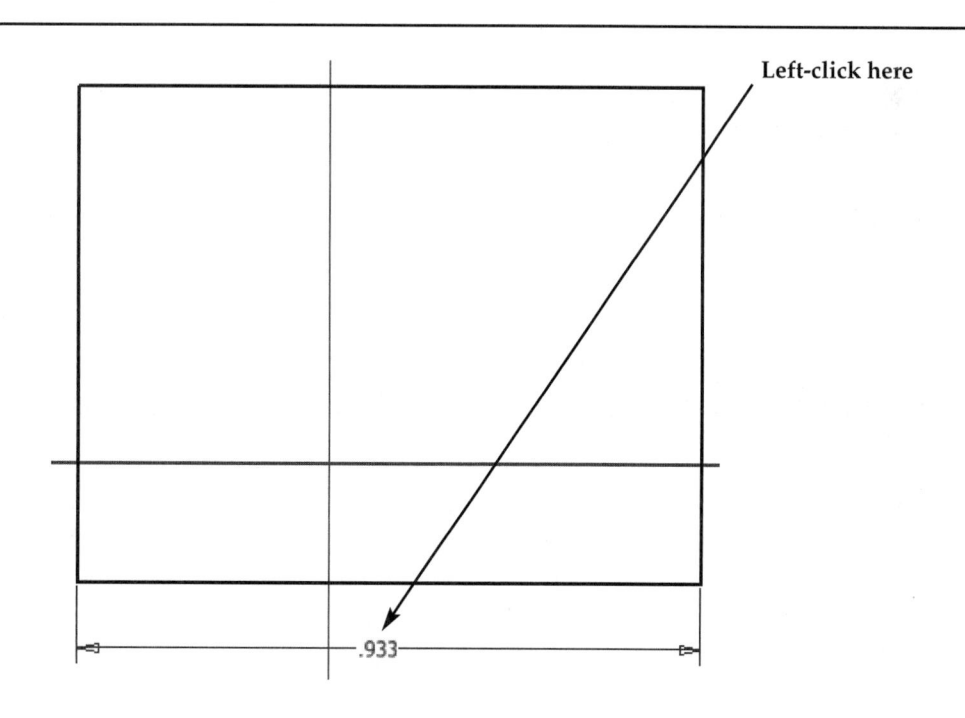

Left-click here

.933

24.   Move the cursor to where the dimension will be placed and left-click once. While the dimension is still red, left-click once. The Edit Dimension dialog box will appear as shown in Figure 1-18.

25.   To edit the dimension, enter **2.00** in the Edit Dimension dialog box (while the current dimension is highlighted) and press **Enter** on the keyboard.

FIGURE 1-18

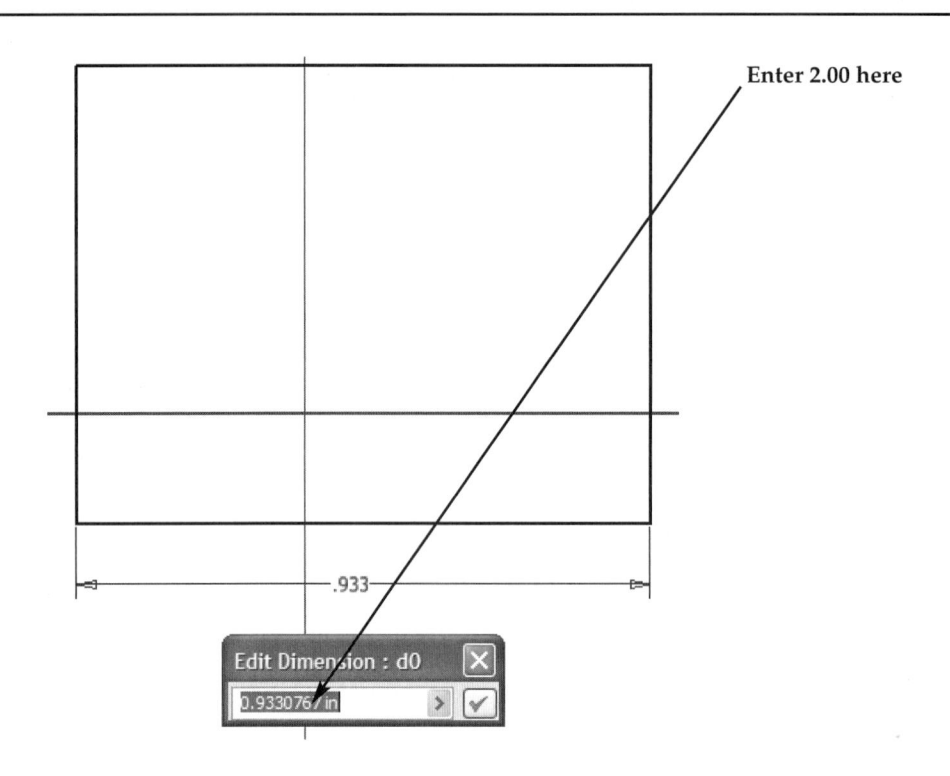

26.   The dimension of the line will become 2 inches as shown in Figure 1-19.

FIGURE 1-19

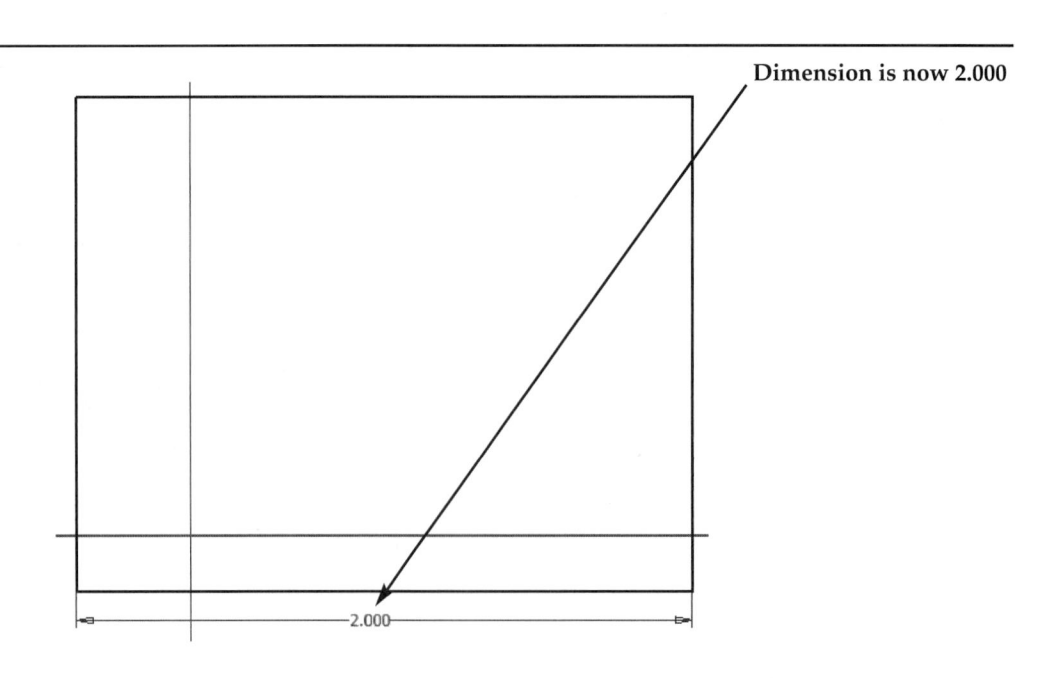

27.   Move the cursor to the upper middle portion of the screen and left-click on **Dimension** as shown in Figure 1-20.

FIGURE 1-20

Left-click here

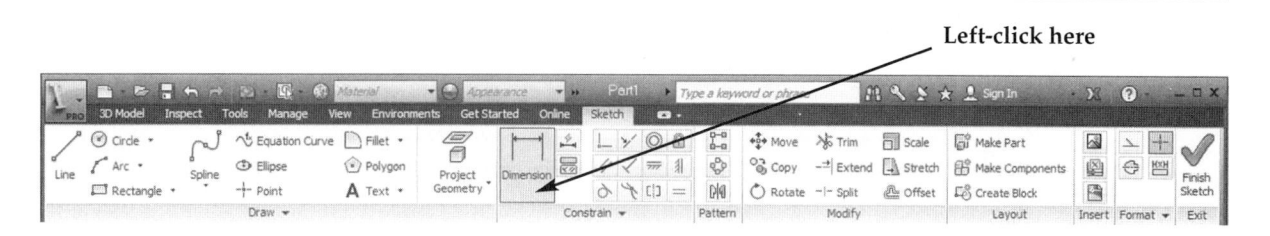

28.   After selecting **Dimension,** move the cursor over the right side vertical line. The line will turn red as shown in Figure 1-21. Left-click once on the line.

FIGURE 1-21

Turned red

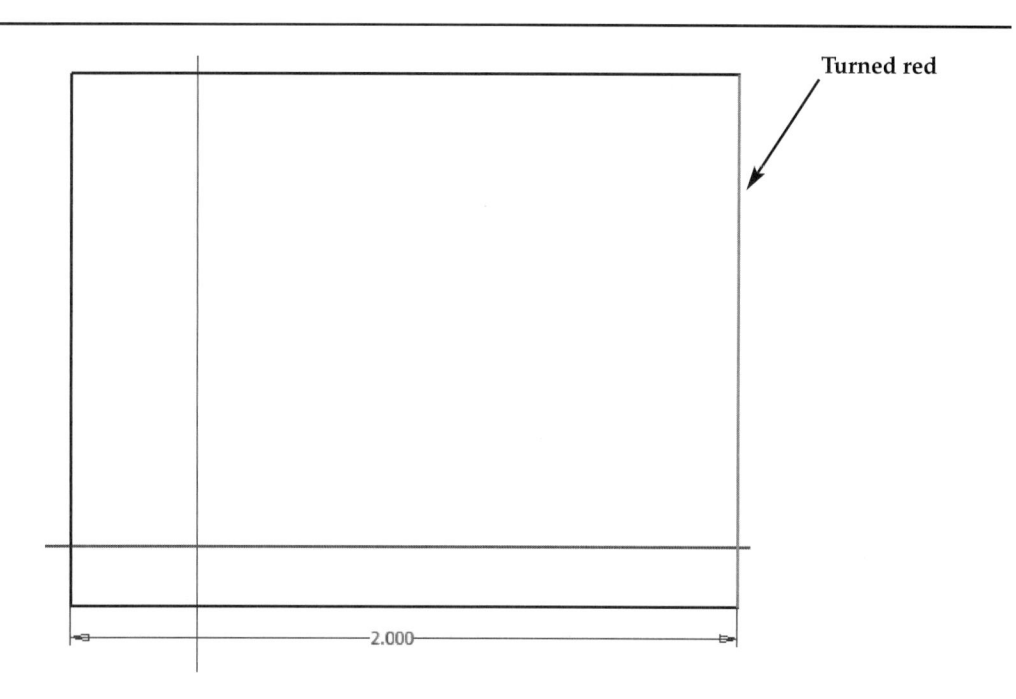

—2.000—

29.   The dimension is attached to the cursor. Move the cursor up and down to verify it is attached. Move the cursor to the right of the line where the dimension will be placed and left-click once. While the dimension is still red, left-click the mouse once. The Edit Dimension dialog box will appear as shown in Figure 1-22.

FIGURE 1-22

Enter .25 here

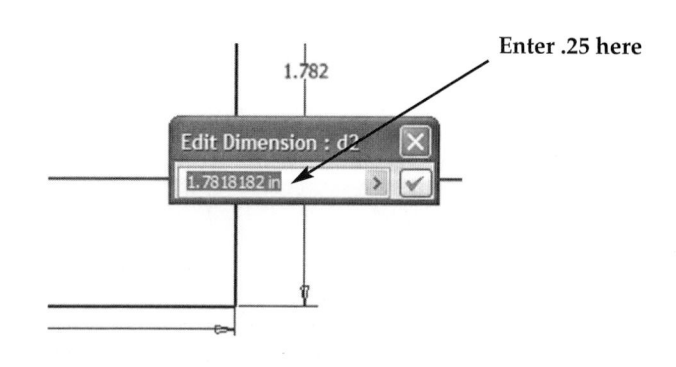

1.782

Edit Dimension : d2   ☒

1.7818182 in   ✓

30.   To edit the dimension, enter **.25** in the Edit Dimension dialog box (while the current dimension is highlighted) and press **Enter** on the keyboard.

31.   The screen should look similar to Figure 1-23.

FIGURE 1-23

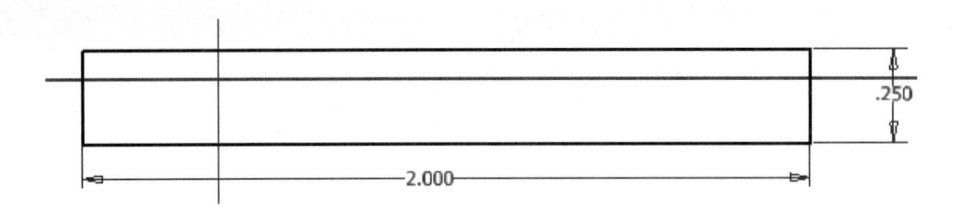

32.   Complete the remainder of the sketch as shown in Figure 1-24.

FIGURE 1-24

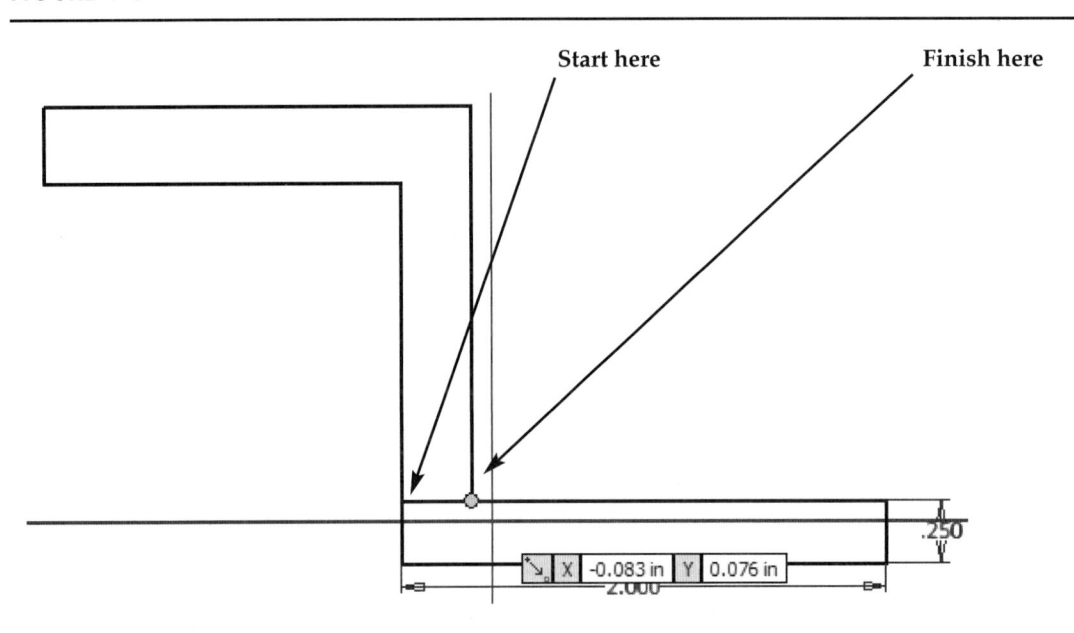

33.   Move the cursor to the upper middle portion of the screen and left-click on **Trim** as shown in Figure 1-25.

FIGURE 1-25

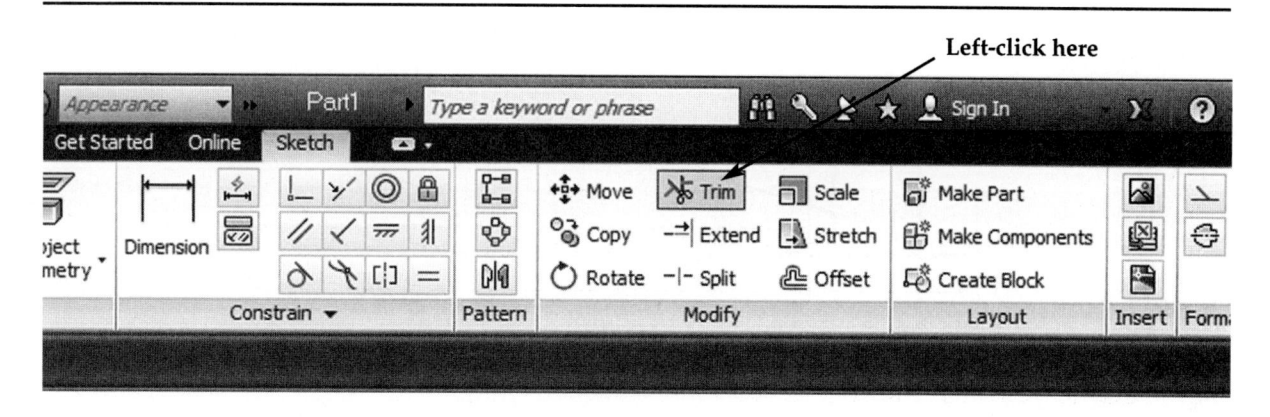

34.   Move the cursor over the portion of the line that is shown in Figure 1-26. The line will become dashed. Inventor is guessing that this line will be trimmed.

FIGURE 1-26

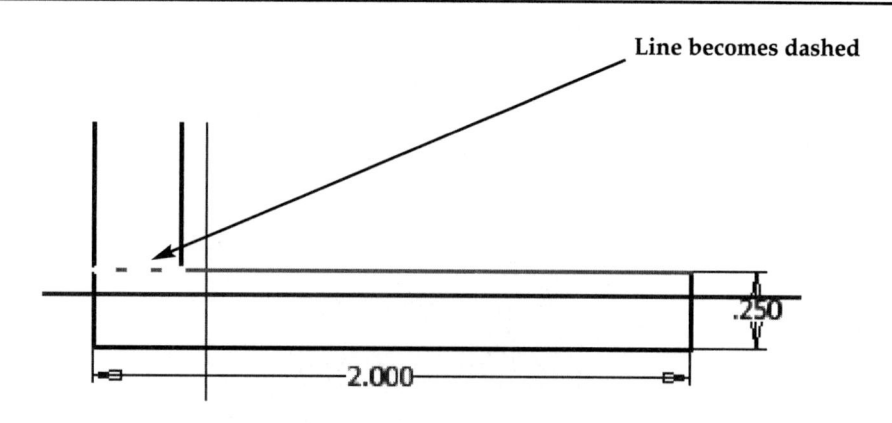

35.   While the line is dashed, left-click on the dashed portion. The line will be trimmed as shown in Figure 1-27.

FIGURE 1-27

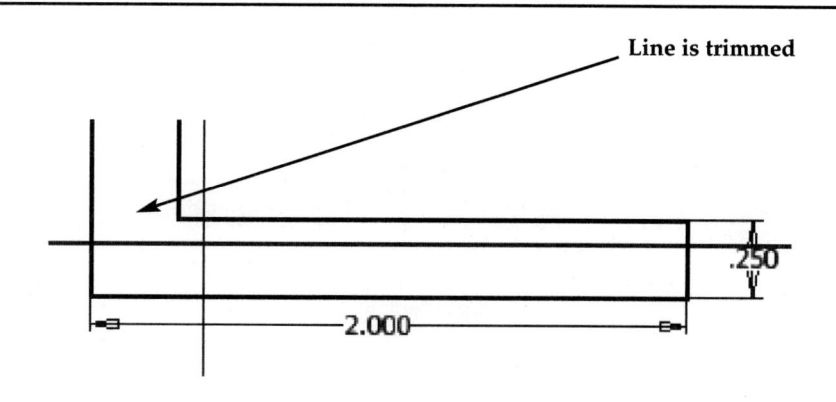

36. Move the cursor over the line in the lower left corner of the drawing as shown in Figure 1-28. The line will turn red. This particular line will have to be deleted so that the line above can be extended the full length.

FIGURE 1-28

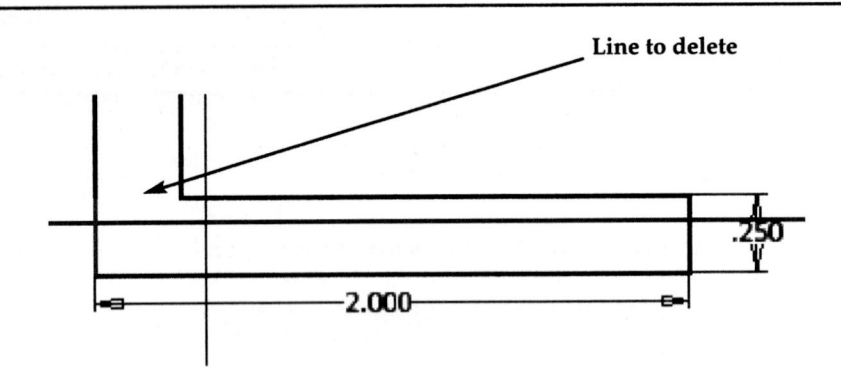

37. After the line turns red, right-click the mouse. A pop-up menu will appear. Left-click on **Delete** as shown in Figure 1-29.

FIGURE 1-29

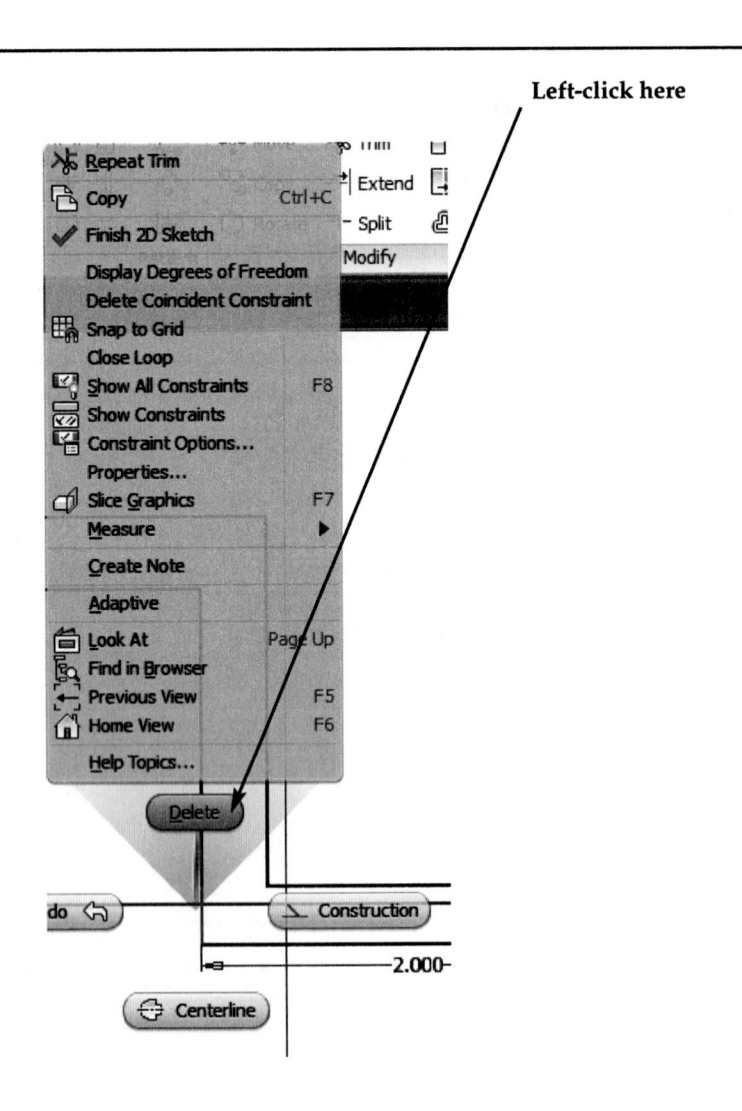

38.   The line will be deleted as shown in Figure 1-30.

FIGURE 1-30

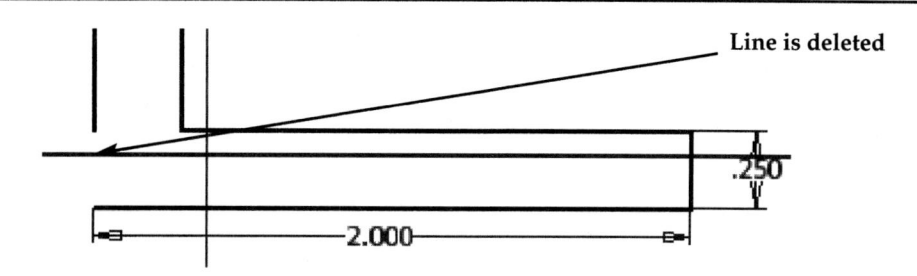

39.   Move the cursor to the upper middle portion of the screen and left-click on **Extend** as shown in Figure 1-31.

FIGURE 1-31

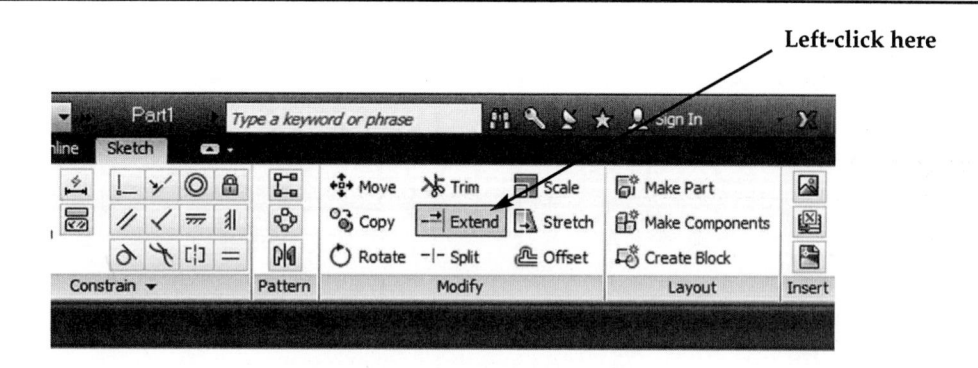

40.   Move the cursor to the line above the recently deleted line. This is the line that will be extended. After the cursor is over the line it will turn red and extend a line downward, as shown in Figure 1-32.

FIGURE 1-32

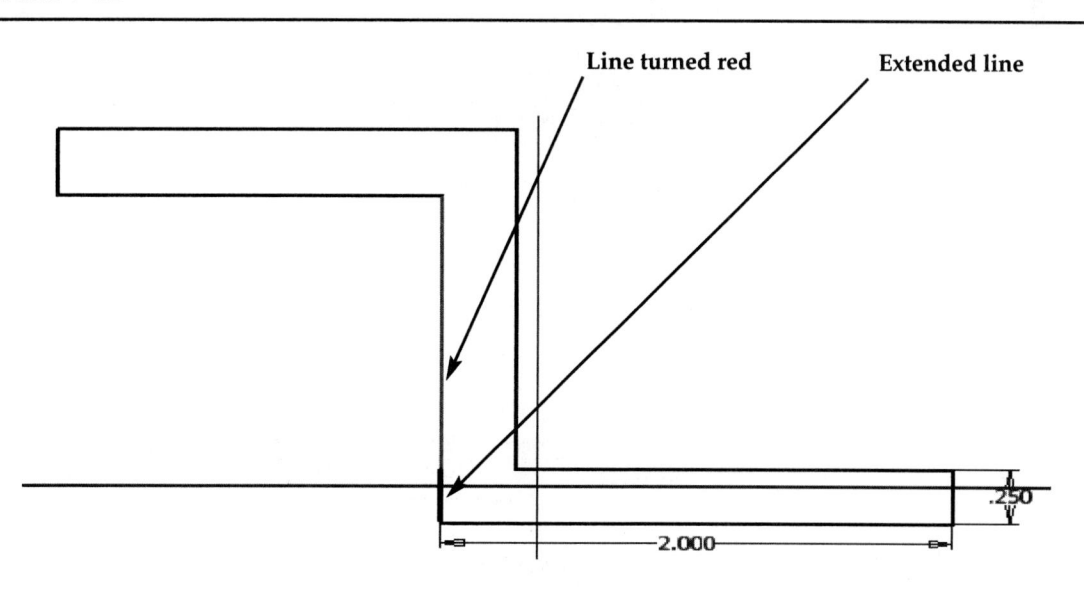

41. After the extended line appears, left-click the mouse. The line will extend, creating one continuous line as shown in Figure 1-33. This is crucial in creating a sketch that can be extruded into a solid model.

FIGURE 1-33

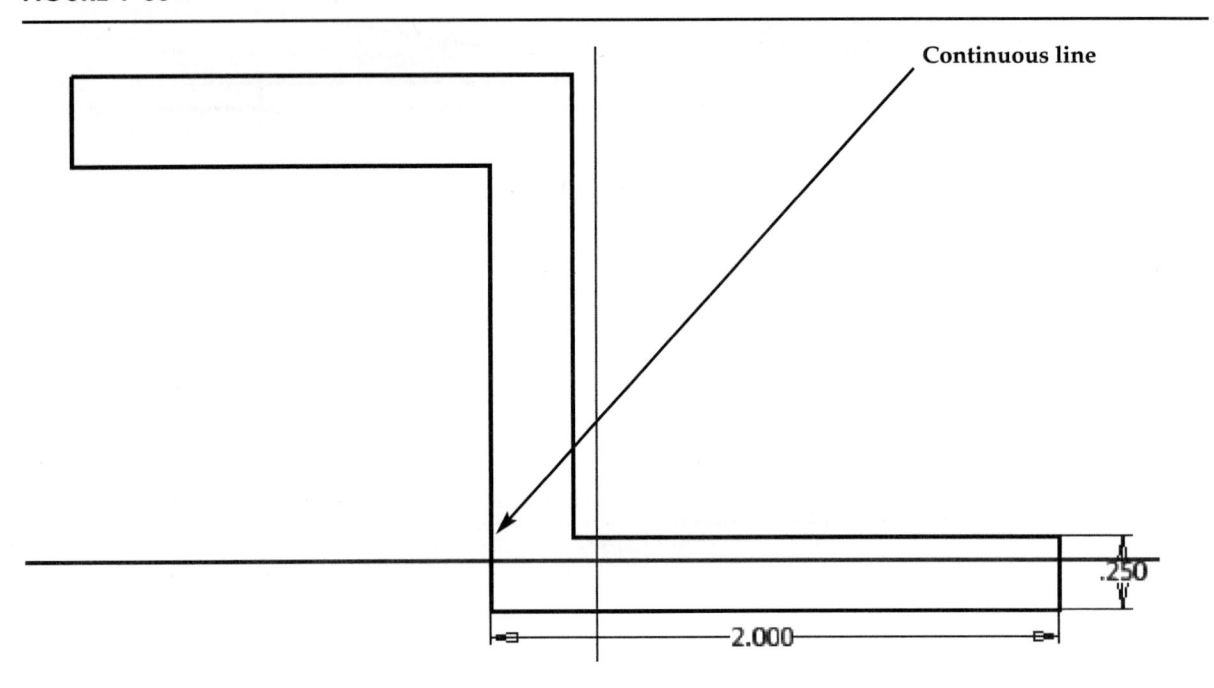

42. Move the cursor to the upper middle portion of the screen and left-click on **Dimension** as shown in Figure 1-34.

FIGURE 1-34

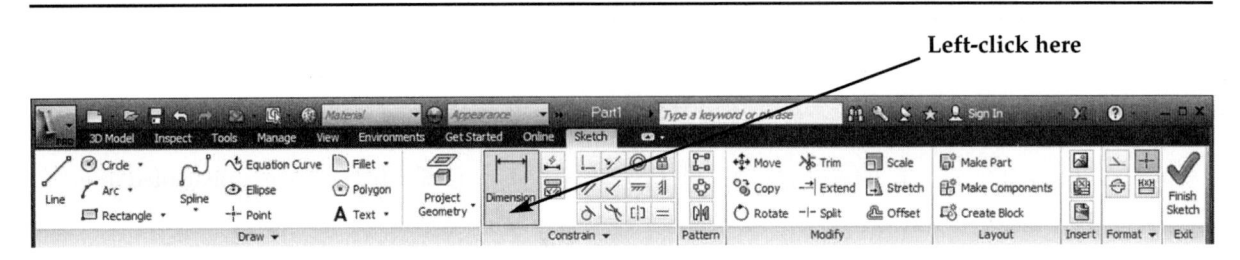

43. After selecting **Dimension**, move the cursor over the left vertical line. The line will turn red as shown in Figure 1-35. Left-click once on the line.

FIGURE 1-35

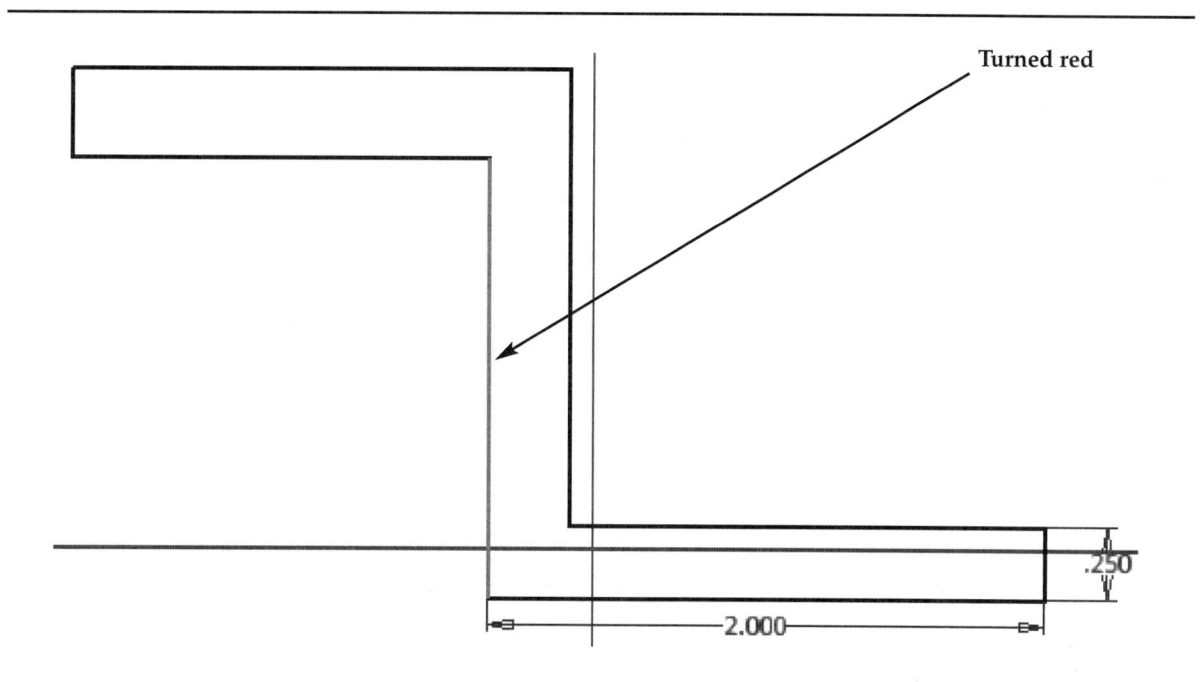

44. Even though a dimension will be attached to the cursor simply ignore it and move the cursor to the right vertical line and left-click after it turns red as shown in Figure 1-36.

FIGURE 1-36

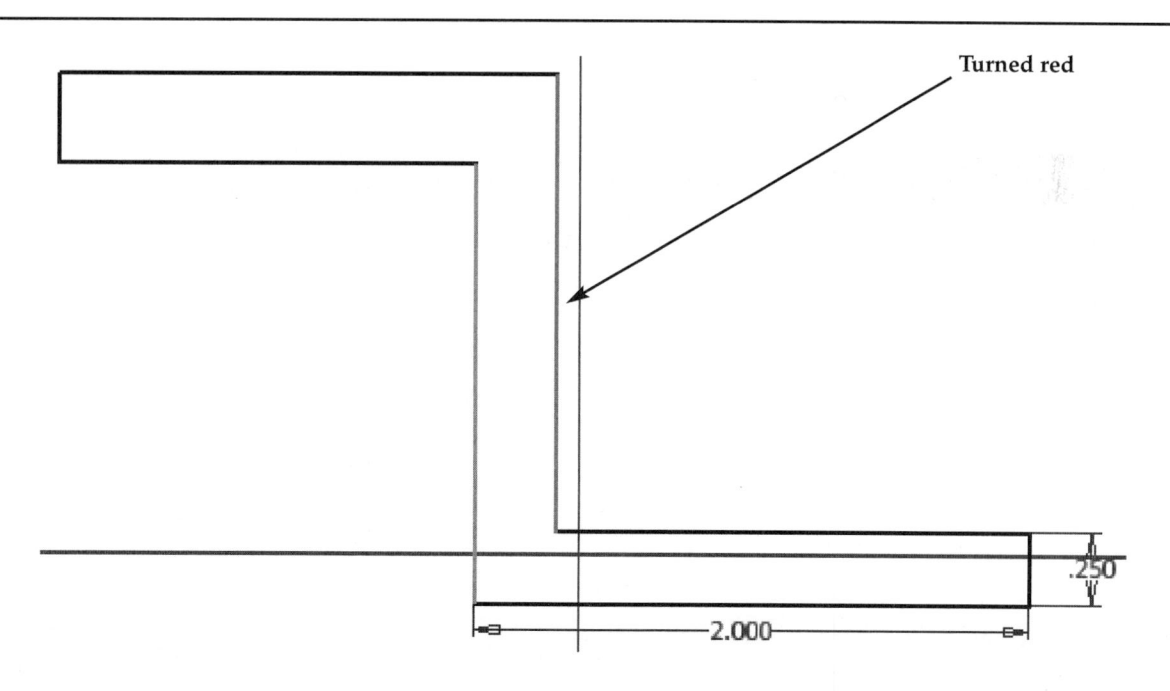

45. Enter **.25** in the Edit Dimension dialog box (while the current dimension is highlighted) and press **Enter** on the keyboard.

46.   Complete the remainder of the sketch as shown in Figure 1-37.

FIGURE 1-37

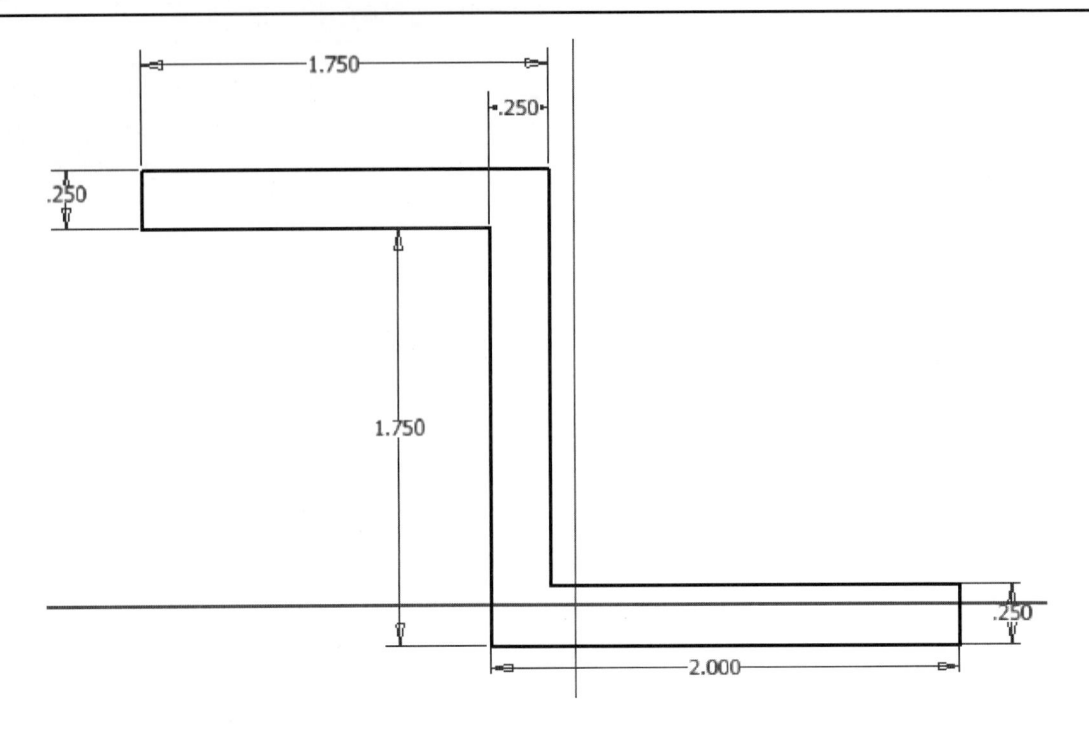

47.   After the sketch is complete, it is time to extrude the sketch into a solid. Right click anywhere around the sketch. A pop-up menu will appear. Left-click on **OK** as shown in Figure 1-38. Pressing the **Esc** key will also "get out" of the Dimension command.

FIGURE 1-38

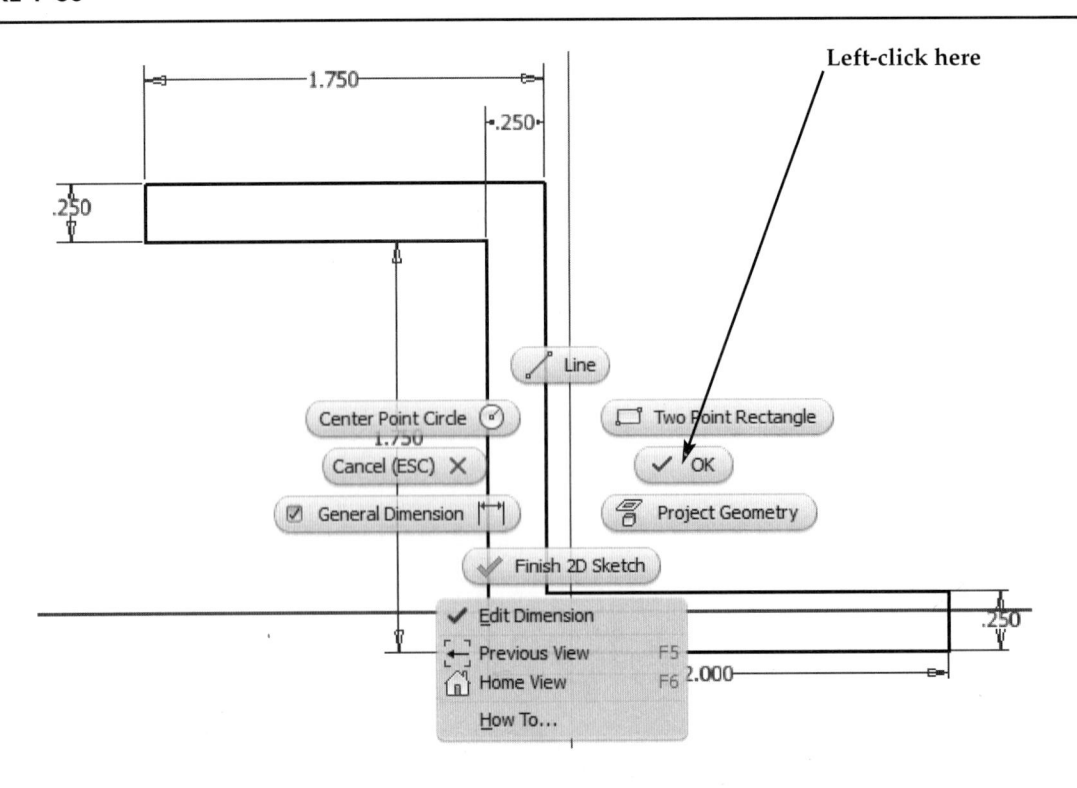

48. After you have verified that no commands are active, right-click anywhere on the sketch. A pop-up menu will appear. Left-click on **Finish 2D Sketch**. Left-click **Home View** as shown in Figure 1-39.

FIGURE 1-39

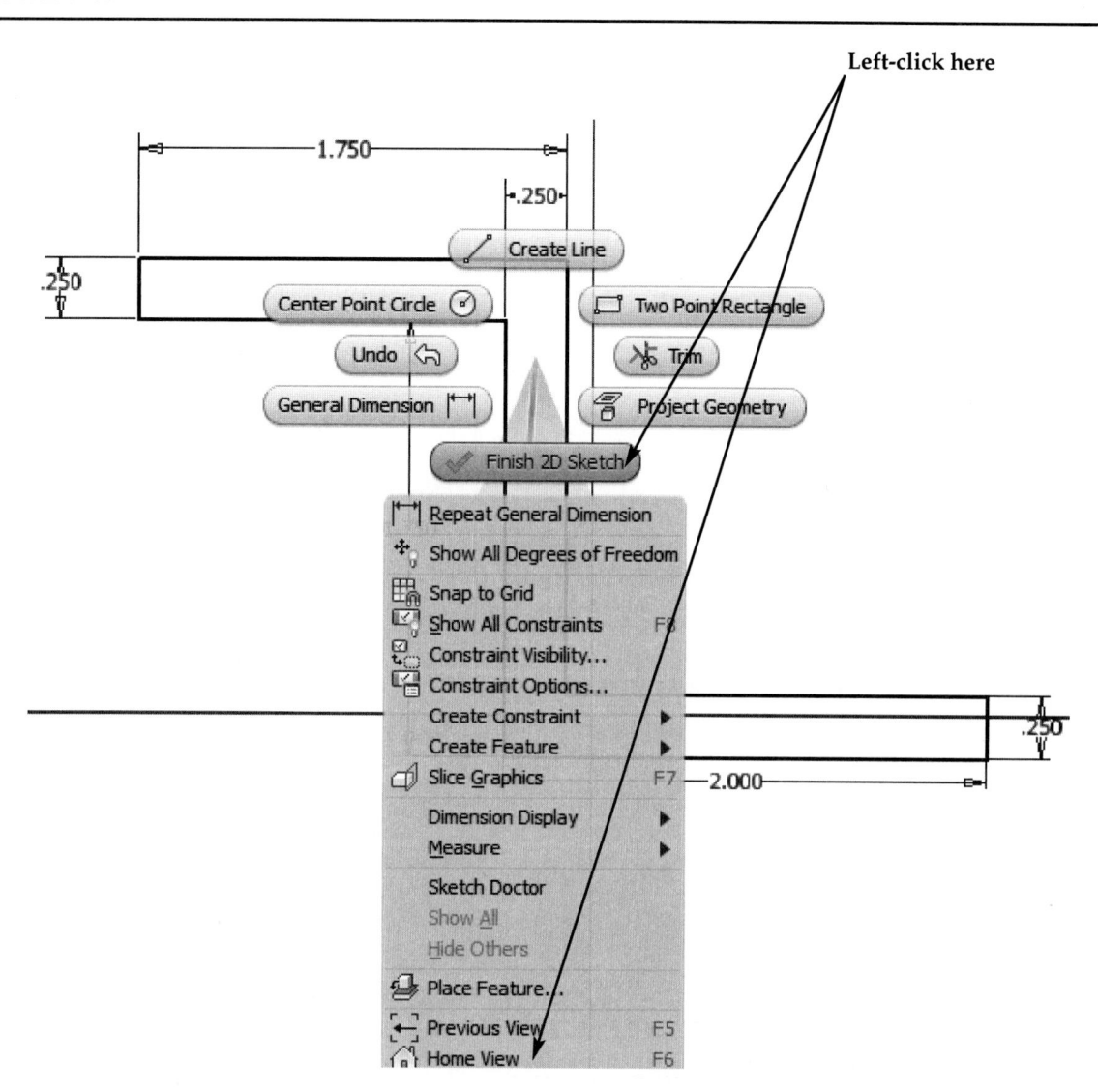

49. Inventor is now out of the Sketch Panel and into the Part Features Panel. Notice that the commands at the top of the screen are now different. To work in the Part Features Panel, a sketch must be present and have no opens (non-connected lines). If there are any opens in the sketch, an error message will appear. The view shown below is an Isometric/Home View. Your screen should look similar to Figure 1-40.

FIGURE 1-40

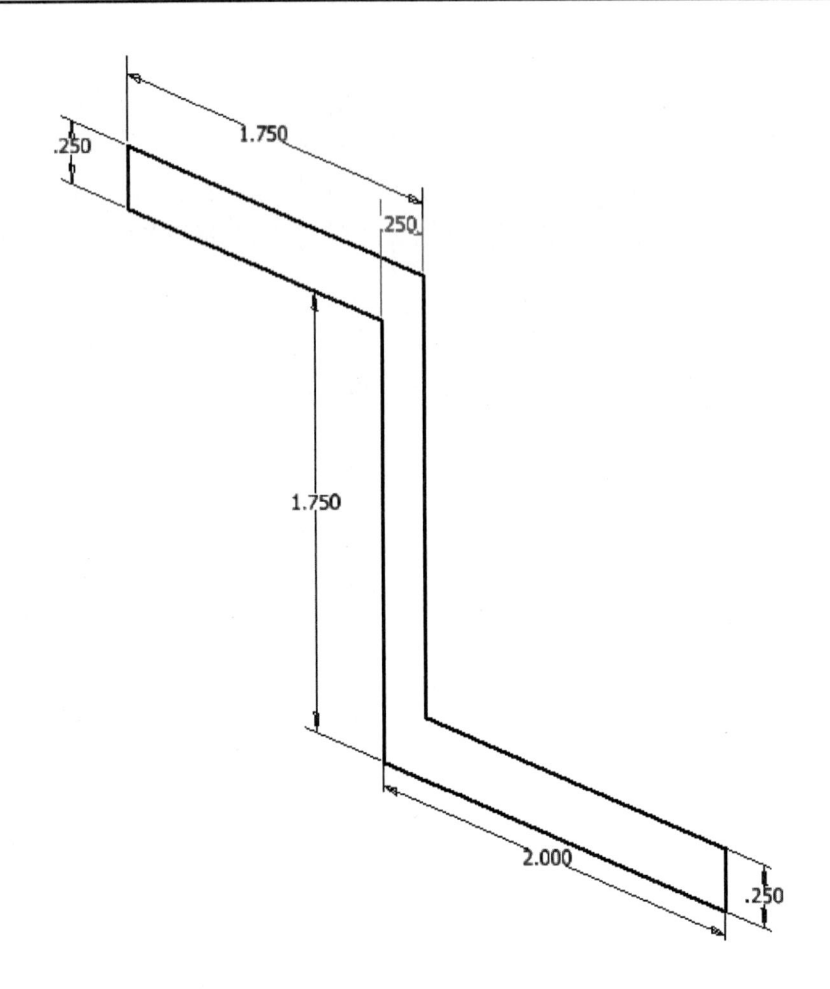

EXTRUDE A SKETCH IN THE PART FEATURES PANEL USING THE EXTRUDE COMMAND

50. Move the cursor to the upper left portion of the screen and left-click on **Extrude**. The Extrude dialog box will appear in collapsed form. Left-click on the drop-down arrow in the center of the collapsed dialog box. This will cause the dialog box to expand. Inventor also provides a preview of the extrusion. If Inventor gave you an error message, there are opens (non-connected lines) somewhere on the sketch. Check each intersection for opens by using the **Extend** and **Trim** commands. Your screen should look similar to Figure 1-41.

FIGURE 1-41

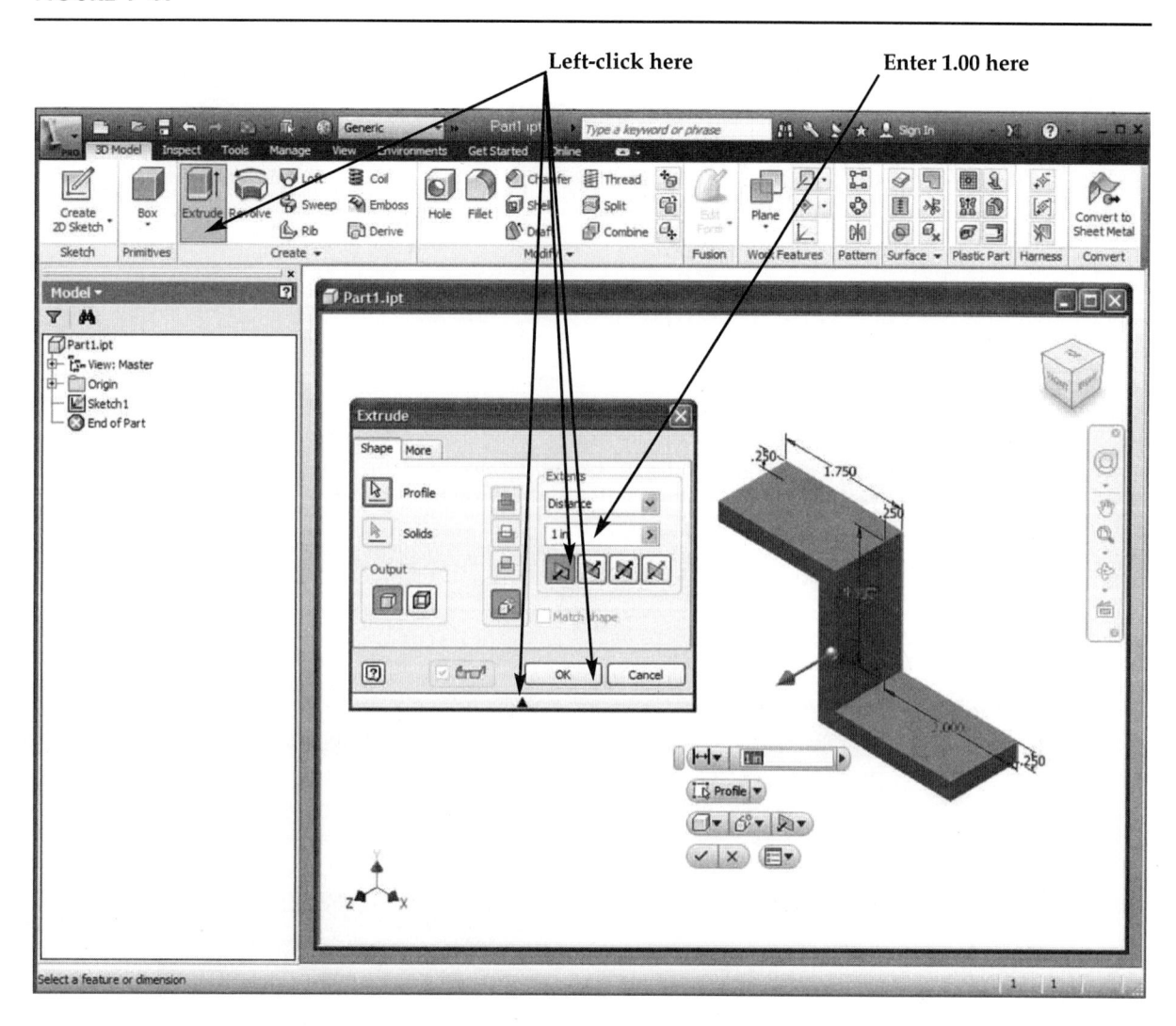

51. Enter **1.00** as shown and left-click on **OK**. Inventor will create a solid from the sketch as shown in Figure 1-41.

CREATE A FILLET IN THE PART FEATURES PANEL USING THE FILLET COMMAND

52. Your screen should look similar to Figure 1-42.

FIGURE 1-42

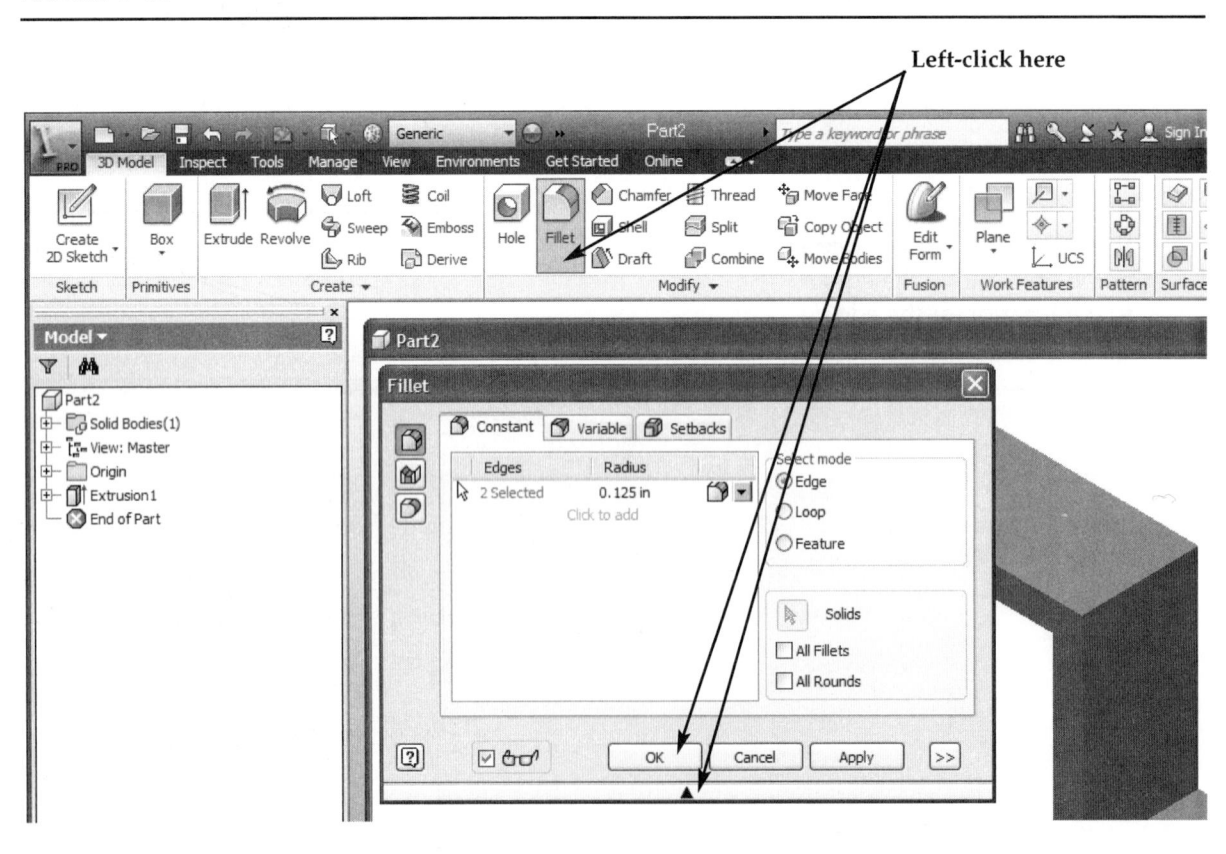

53. Move the cursor to the upper middle portion of the screen and left-click on **Fillet**. The Fillet dialog box will appear in collapsed form. Left-click on the drop-down arrow in the center of the collapsed dialog box. This will cause the dialog box to expand as shown in Figure 1-43.

FIGURE 1-43

54.   Move the cursor to the lower left edge of the part. After the edge turns red, left-click once as shown in Figure 1-44.

FIGURE 1-44

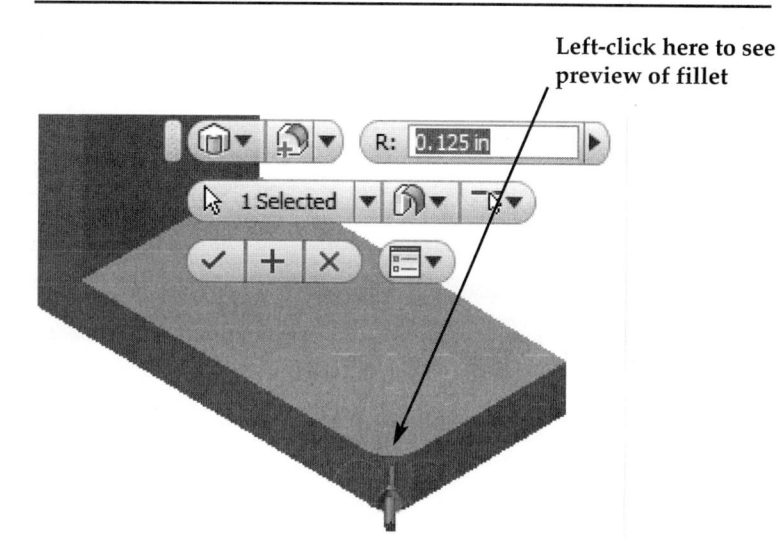

Left-click here to see preview of fillet

55.   Notice the blue mesh illustrating a preview of the fillet. Left-click on the opposite edge as shown in Figure 1-45.

FIGURE 1-45

Left-click here to see preview of fillets

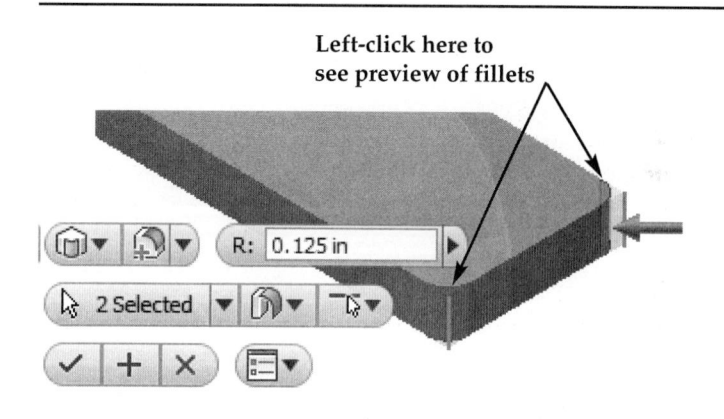

56.   Left-click on the two upper remaining edges. Even though the far upper edge is not visible, move the cursor to the location of the edge and Inventor will find it as shown in Figure 1-46.

FIGURE 1-46

Preview of fillets

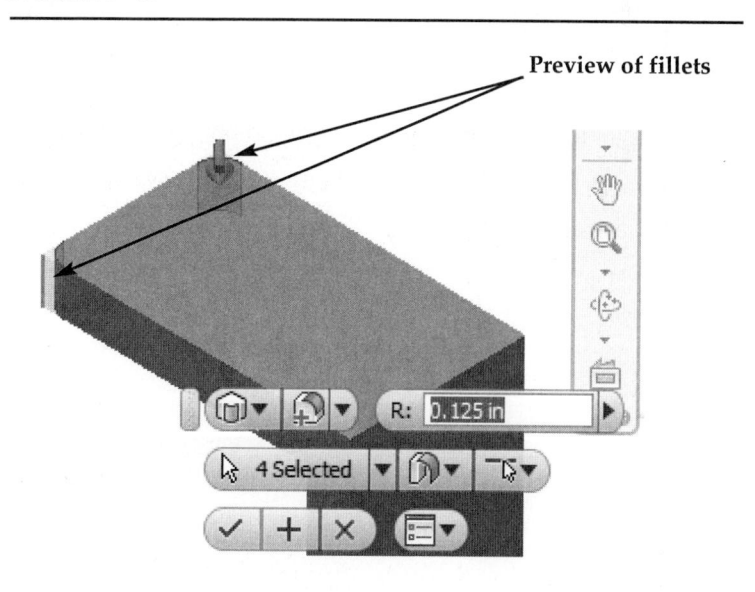

57.   Move the cursor to the dimension located in the dimension box. Enter **.5** and press **Enter** on the keyboard as shown in Figure 1-47.

FIGURE 1-47

Enter .5 here, press Enter   Left-click here

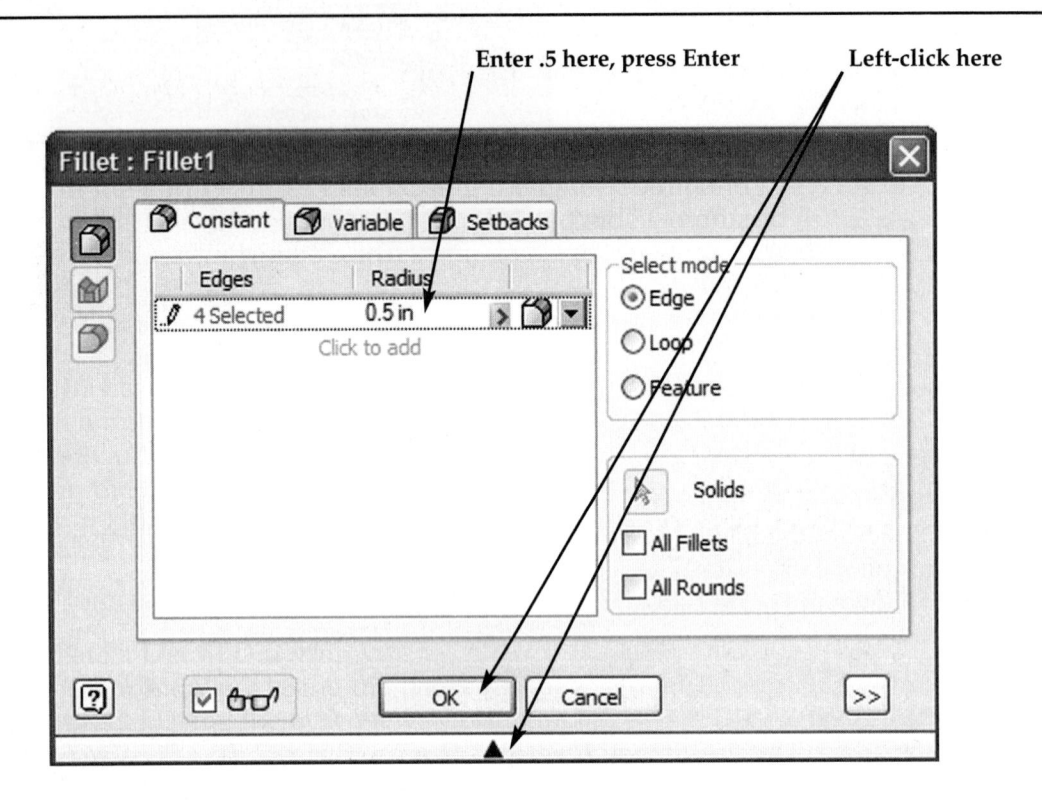

58.   Your screen should look similar to Figure 1-48.

FIGURE 1-48

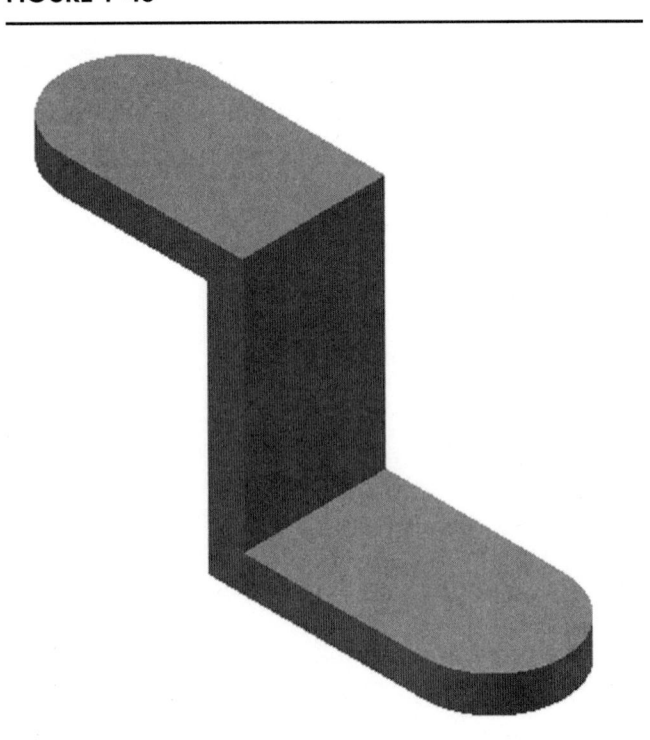

59. The next task will include cutting a hole in each of the ends. To accomplish this, a sketch will need to be constructed on this surface. Move the cursor to the surface that will have the new sketch as shown in Figure 1-49. Notice the edges of the surface become outlined in red.

FIGURE 1-49

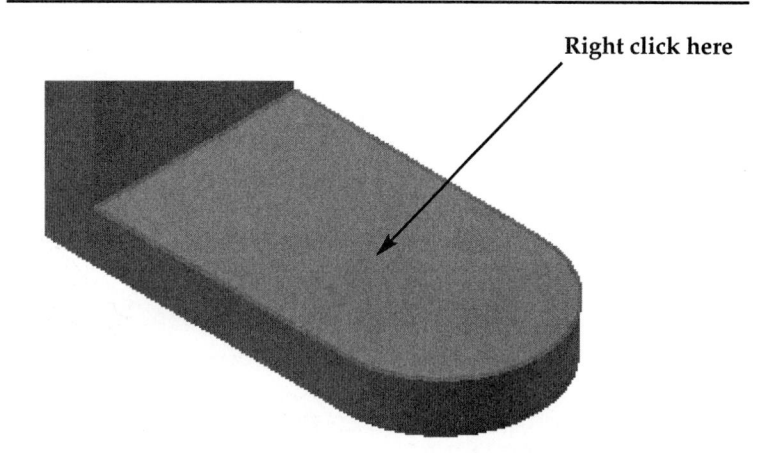

Right click here

60. After the edges of the surface turn red, right-click on the surface. The surface will change color. A pop-up menu will also appear. Left-click on **New Sketch** as shown in Figure 1-50.

FIGURE 1-50

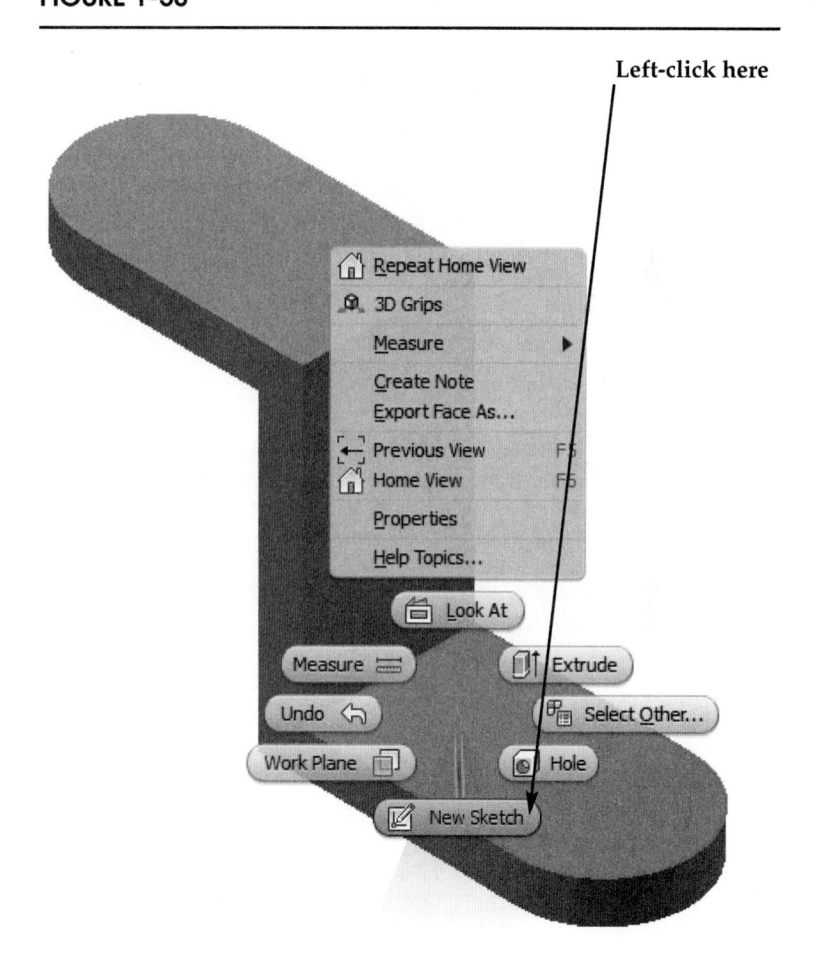

Left-click here

61. Inventor will create a "sketch" on that particular surface. Notice the menu at the top of the screen has changed back to the options available in the Sketch Panel. Inventor has now returned to the Sketch Panel.

62. Your screen should look similar to Figure 1-51.

FIGURE 1-51

63. Move the cursor to the upper left portion of the screen and left-click on **Circle** as shown in Figure 1-52.

FIGURE 1-52

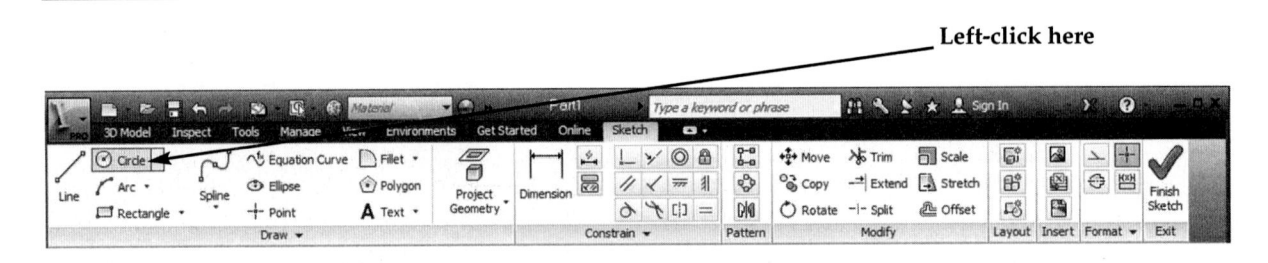

64. A green dot will appear at the center of the Fillet radius as shown in Figure 1-53.

FIGURE 1-53

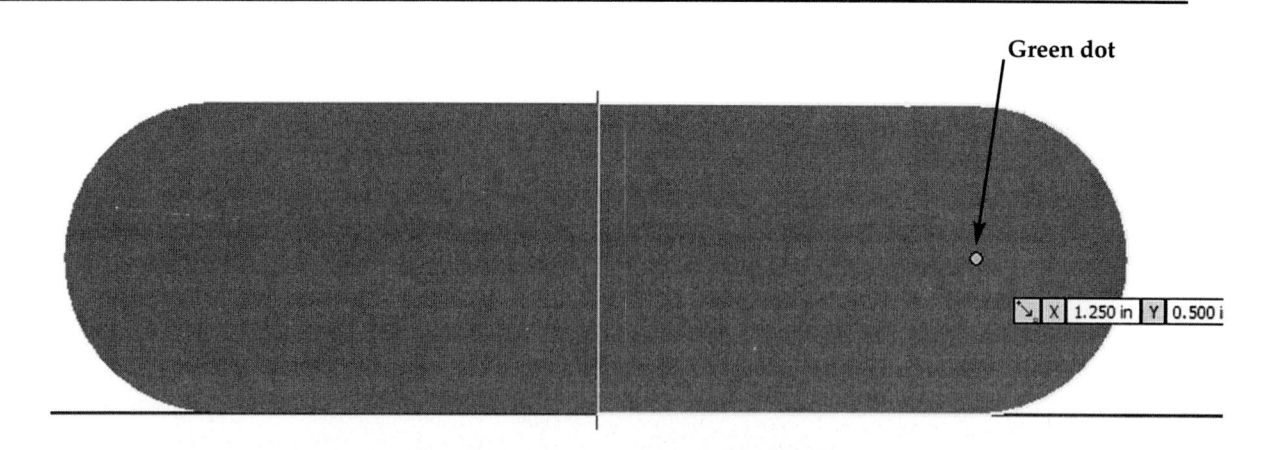

Green dot

X 1.250 in Y 0.500 i

CREATE A HOLE IN THE PART FEATURES PANEL USING THE EXTRUDE COMMAND

65. After the green dot appears, left-click once. This will be the center of a circle, which will later become a thru hole. Move the cursor out to the side. The hole will become larger. Move the cursor out far enough to create a hole size similar to Figure 1-54.

FIGURE 1-54

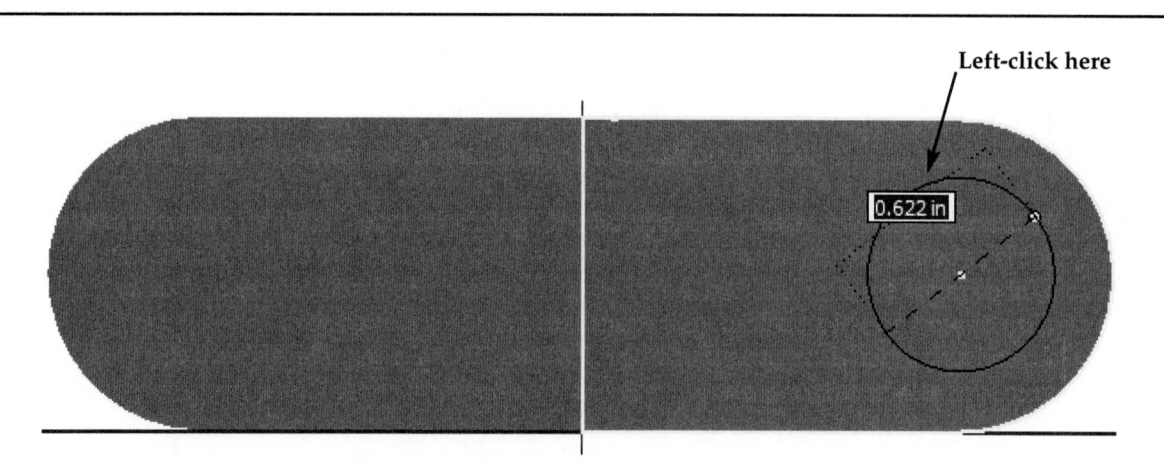

Left-click here

0.622 in

66. After the hole size looks similar to Figure 1-54, left-click once.

67. Using the **Dimension** command, enter **.50** in the Edit Dimension dialog box and press **Enter** on the keyboard. The diameter of the hole will become .50 inches. Press the **Esc** key to ensure that no commands are active.

68.   Right click anywhere on the drawing. A pop-up menu will appear. Left-click on **Finish 2D Sketch** or left-click on **Finish Sketch/Exit** at the upper right portion of the screen. Left-click on **Home View** as shown in Figure 1-55.

FIGURE 1-55

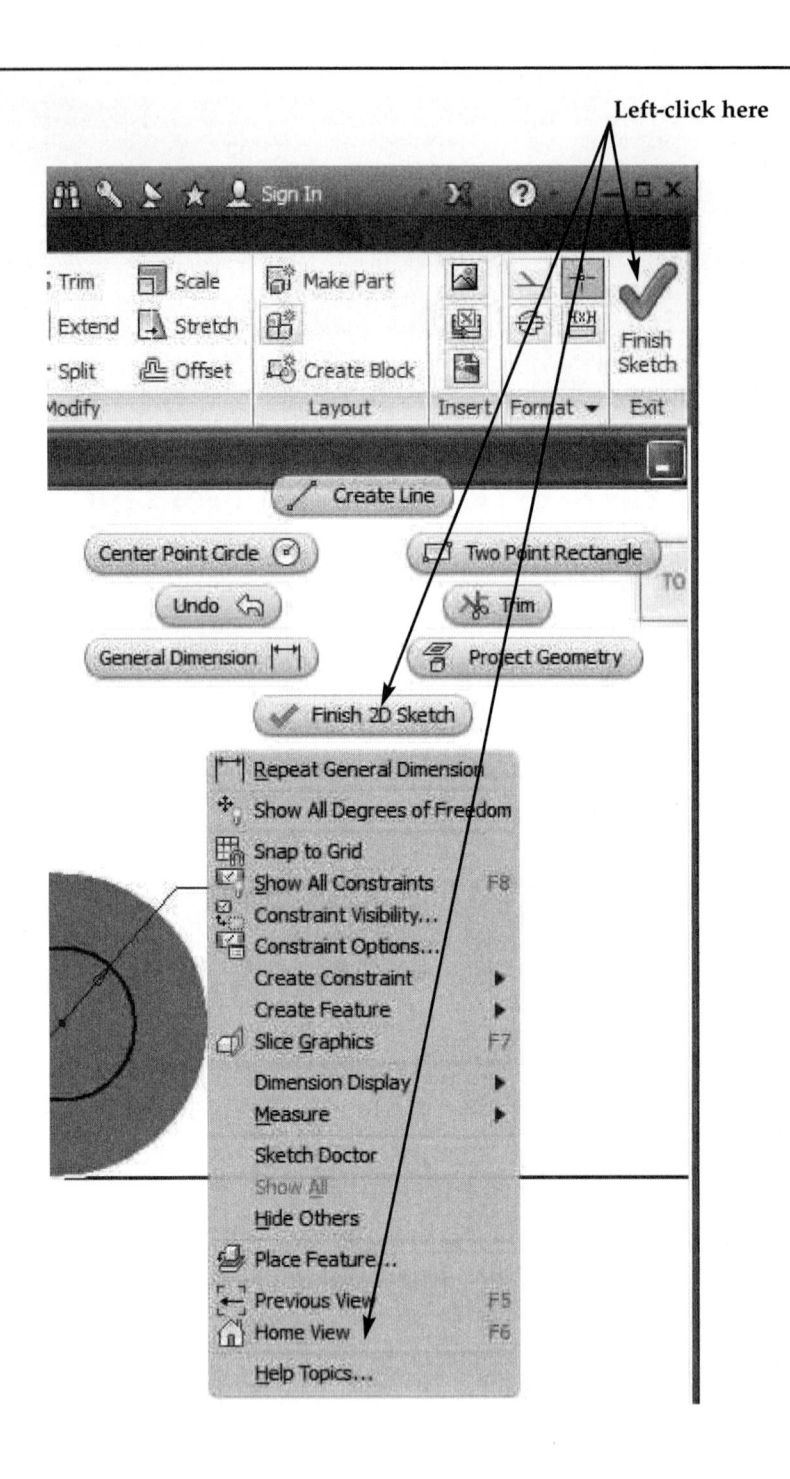

69. Inventor is now out of the Sketch Panel and into the Part Features Panel. Your screen should look similar to Figure 1-56.

FIGURE 1-56

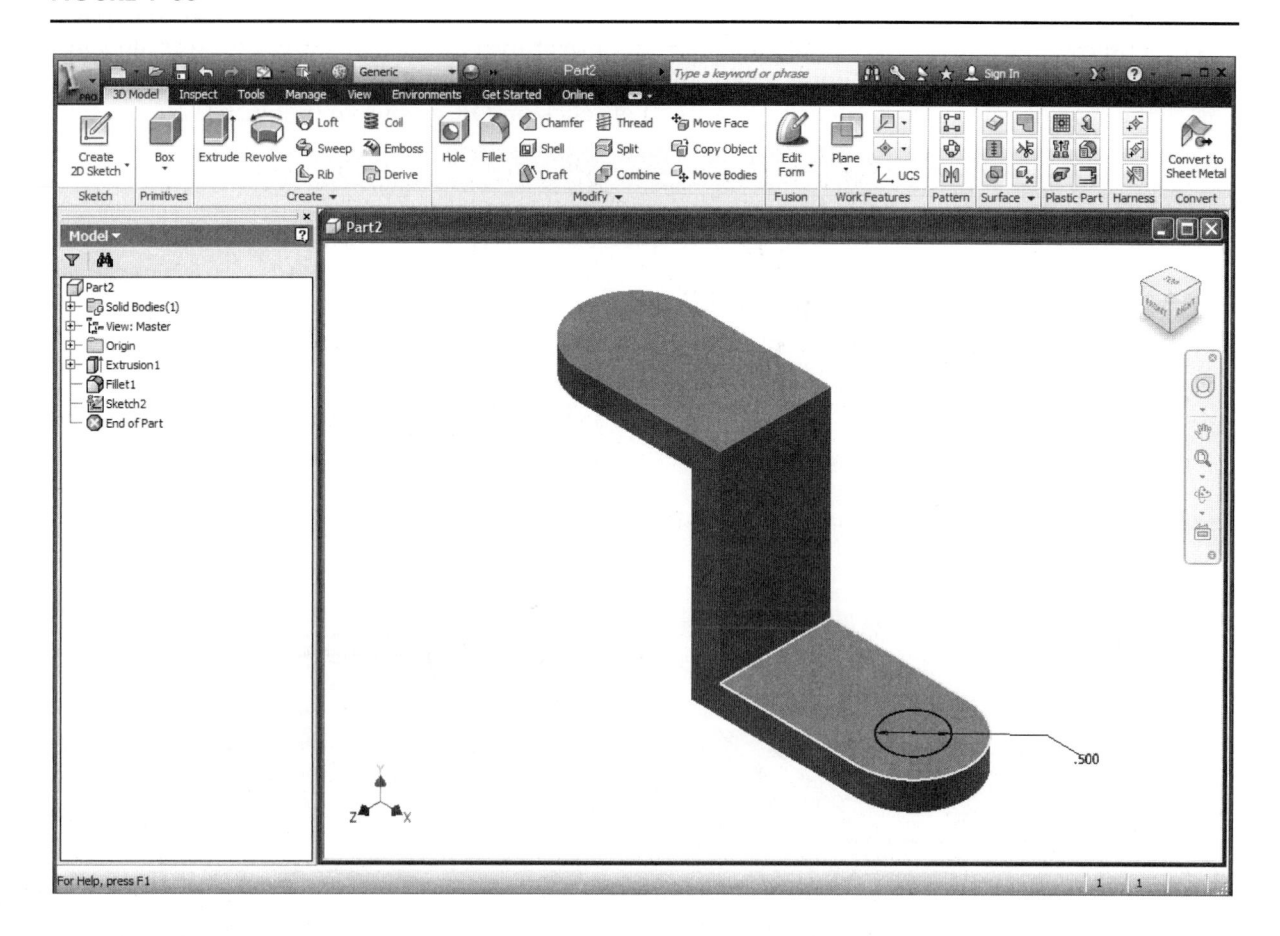

CREATE A COUNTER BORE IN THE PART
FEATURES PANEL USING THE HOLE COMMAND

70.   Move the cursor to the upper left portion of the screen and left-click on **Extrude**. You may have to left-click on the drop-down arrow located in the lower center of the dialog box to expand it as shown in Figure 1-57. Now move the cursor over the circle in the drawing. After the circle turns red, left-click once. Enter **.25** for the depth. Left-click on the directional arrow to define the type of extrusion. Left-click on "Cut." Left-click on **OK** as shown in Figure 1-57.

FIGURE 1-57

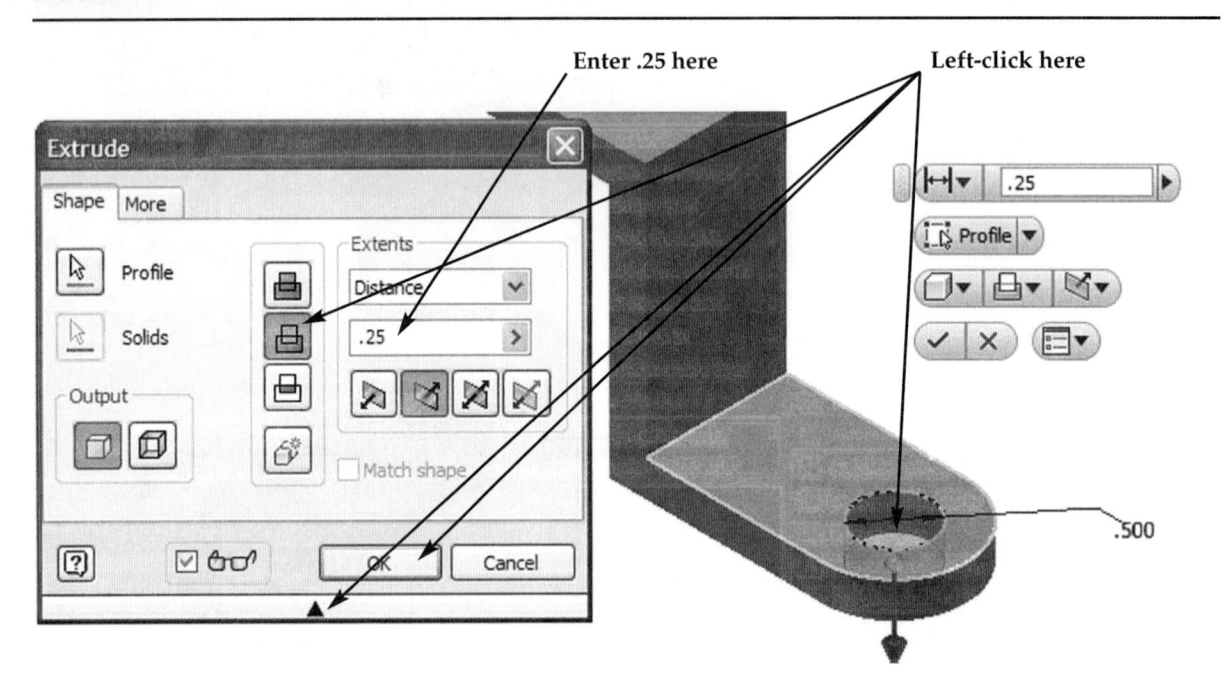

71.   Your screen should look similar to Figure 1-58.

FIGURE 1-58

72   Another method of creating a hole is to use the **Point, Center Point** command.

73.   To use the **Point, Center Point** command, Inventor will need to be in the Sketch Panel. Right click on the surface as shown in Figure 1-59.

FIGURE 1-59

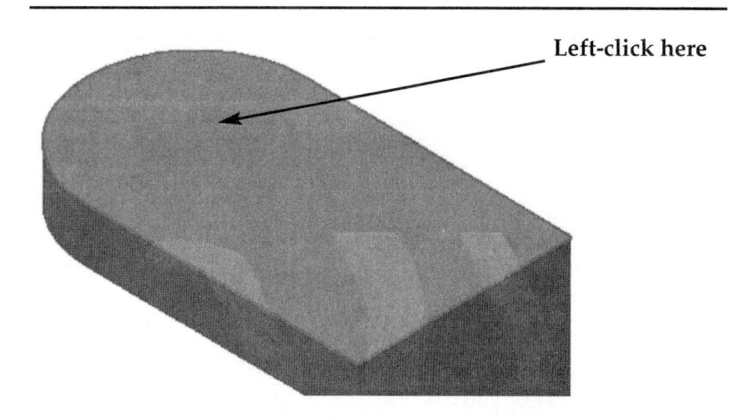

Left-click here

74.   The surface will change color. A pop-up menu will also appear. Left-click on **New Sketch** as shown in Figure 1-60.

FIGURE 1-60

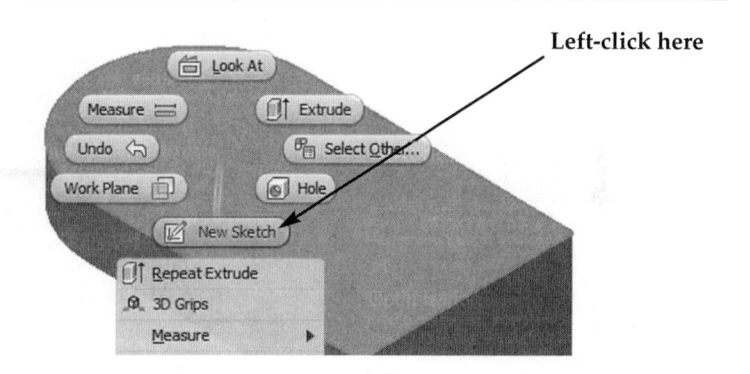

Left-click here

75.   Inventor will return to the Sketch Panel as shown in Figure 1-61.

FIGURE 1-61

76.   Move the cursor to the middle left portion of the screen and left-click on **Point** as shown in Figure 1-62.

FIGURE 1-62

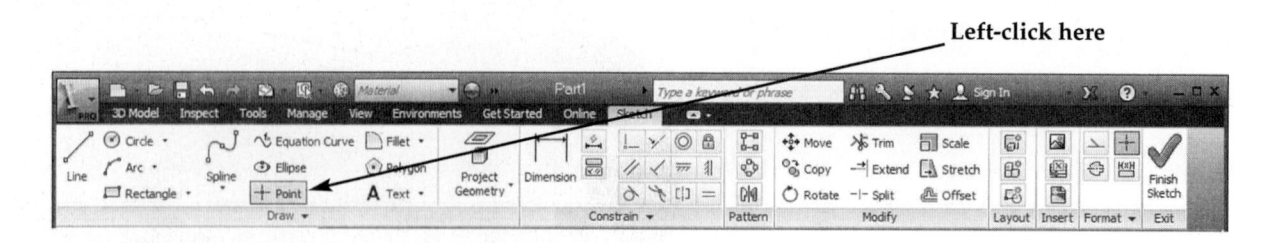

Left-click here

77.   A green dot will appear at the center of the Fillet radius. Left-click on the green dot as shown in Figure 1-63.

FIGURE 1-63

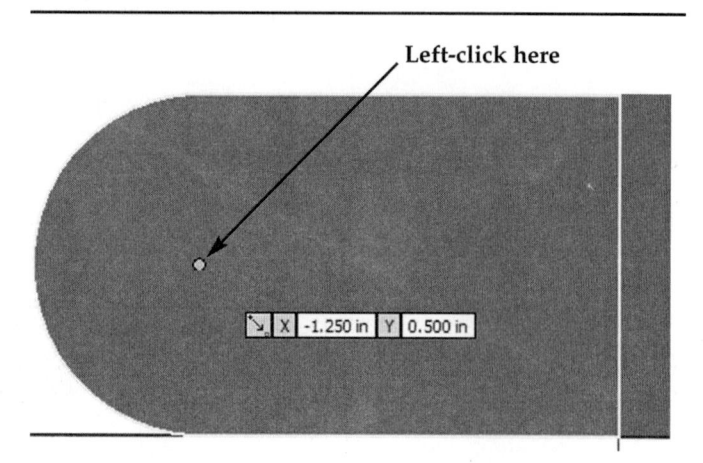

Left-click here

X -1.250 in   Y 0.500 in

78.   After left-clicking on the center point, Inventor will place a small center marker on the center of the Fillet radius as shown in Figure 1-64. Press the **Esc** key once.

FIGURE 1-64

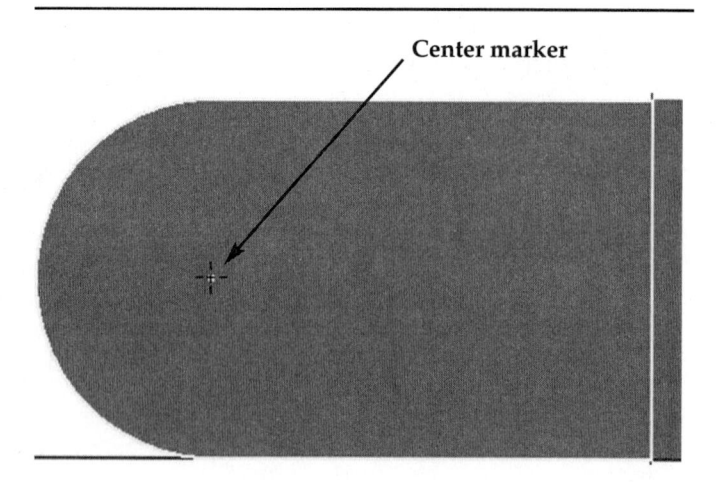

Center marker

79. Right click anywhere on the drawing. A pop-up menu will appear. Left-click on **Finish 2D Sketch** as shown in Figure 1-65.

FIGURE 1-65

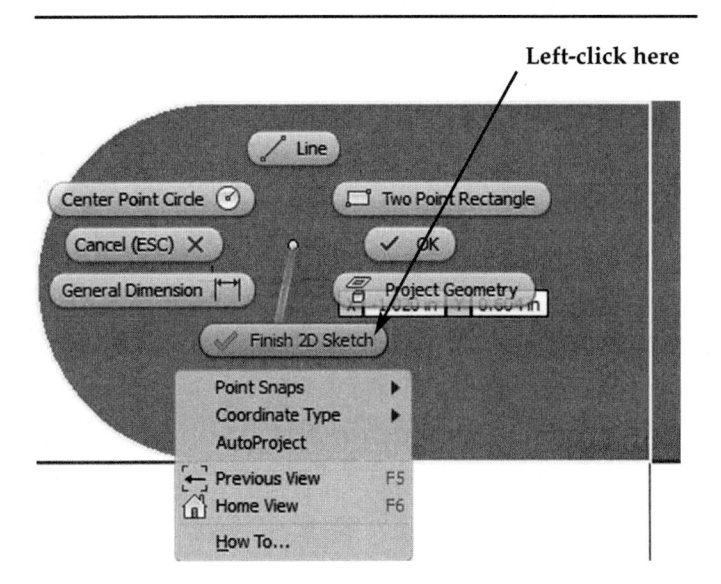

80. Inventor is now out of the Sketch Panel and into the Part Features Panel. Your screen should look similar to Figure 1-66.

FIGURE 1-66

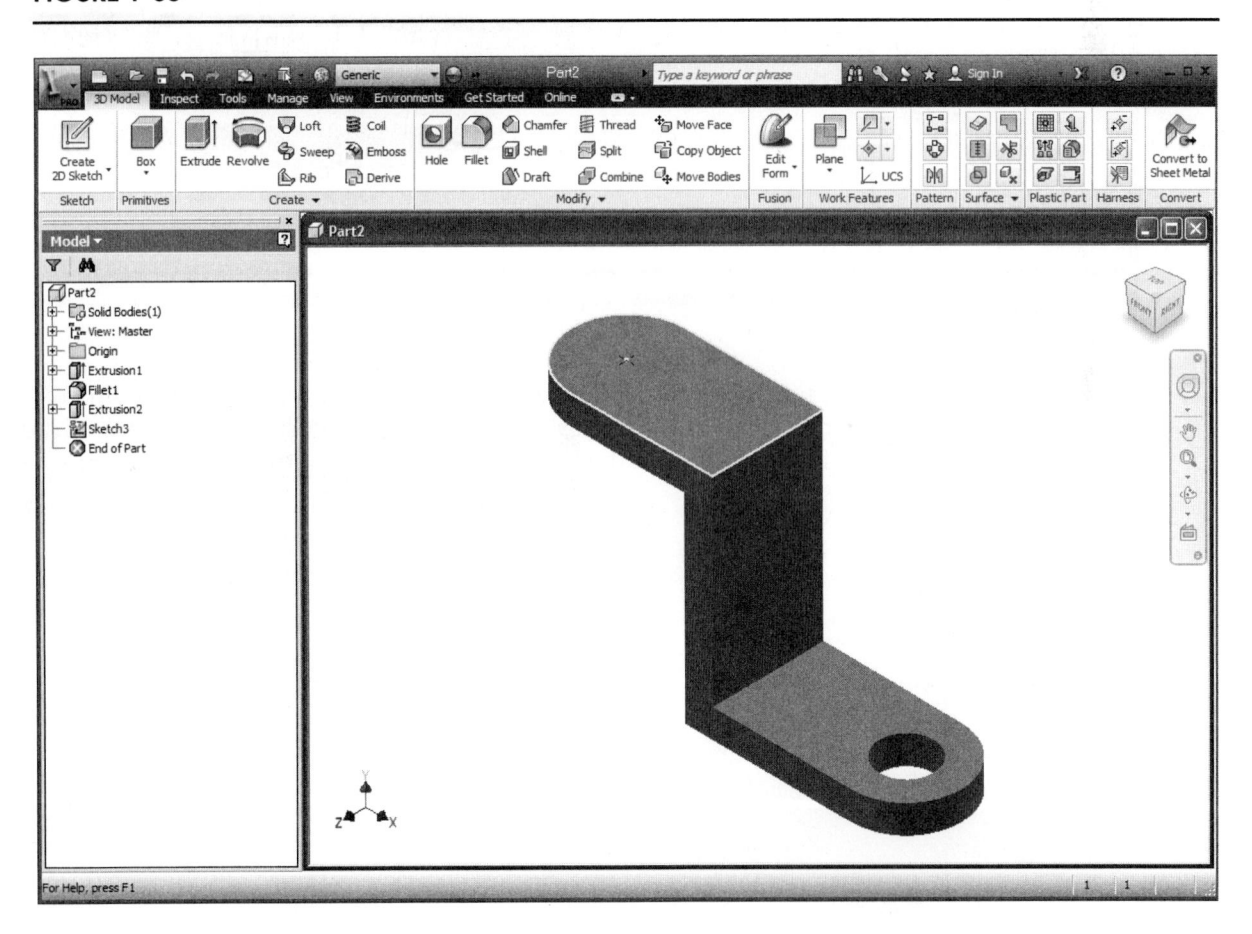

81.   Move the cursor to the upper middle portion of the screen and left-click on **Hole**. The Hole dialog box will appear as shown in Figure 1-67.

FIGURE 1-67

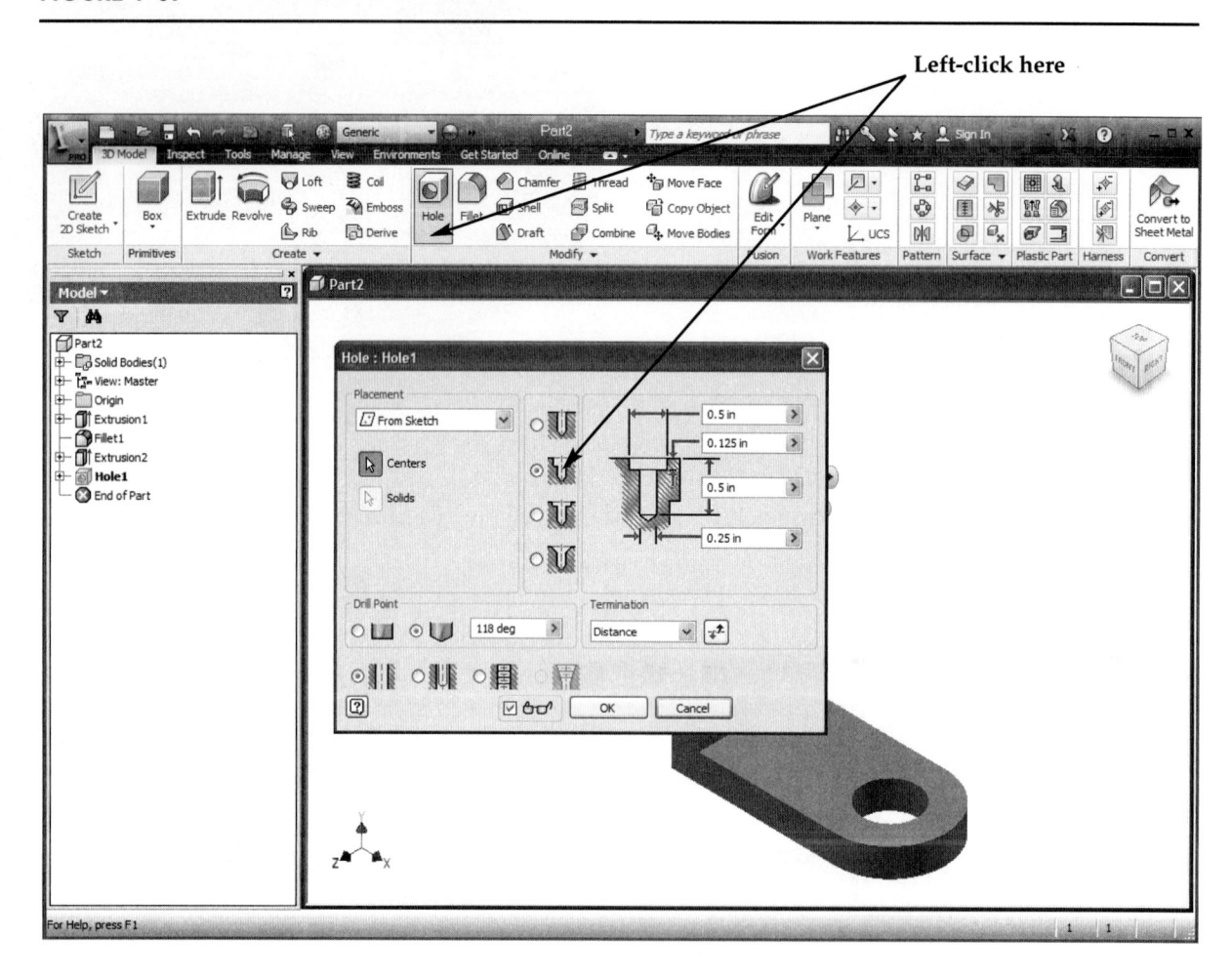

82.   Place a dot as shown in Figure 1-67. This is the "Counter Bore" icon. A preview and dimensions of the hole type are provided on the right side of the Holes dialog box. Select one of the other hole types and watch the preview of the hole in the right side of the Hole dialog box change.

83.   To edit the dimensions of the counter bore hole, use the cursor to highlight the desired dimension as shown in Figure 1-68.

FIGURE 1-68

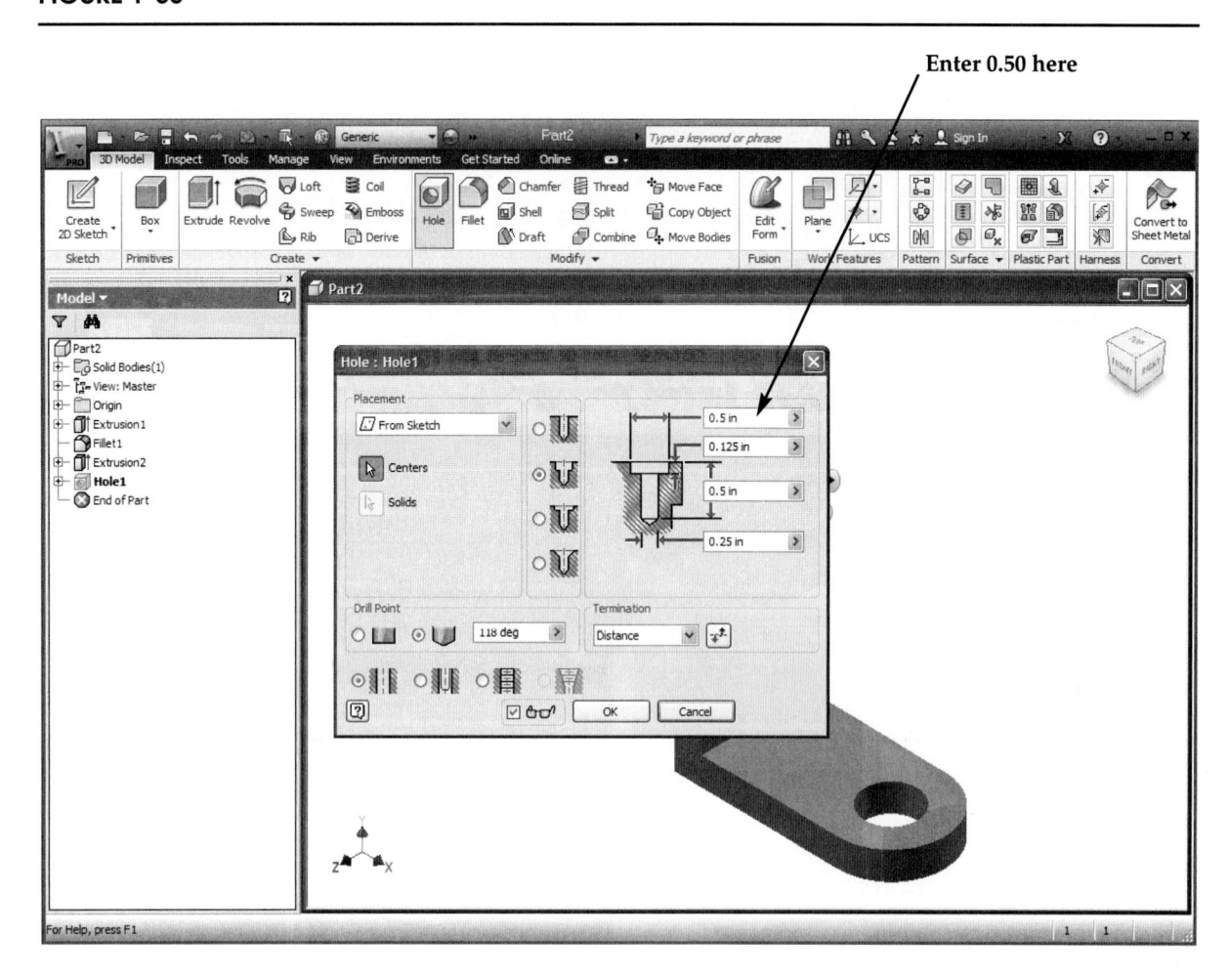

84.   After highlighting the dimension, enter in **.50** (if .50 is not already entered in) for the counter bore diameter.

85.   Enter in **.125** for the counter bore depth, **.50** for the overall depth, **.25** for the hole diameter. Left-click on **Distance** under Termination and left-click on **OK** as shown in Figure 1-69.

FIGURE 1-69

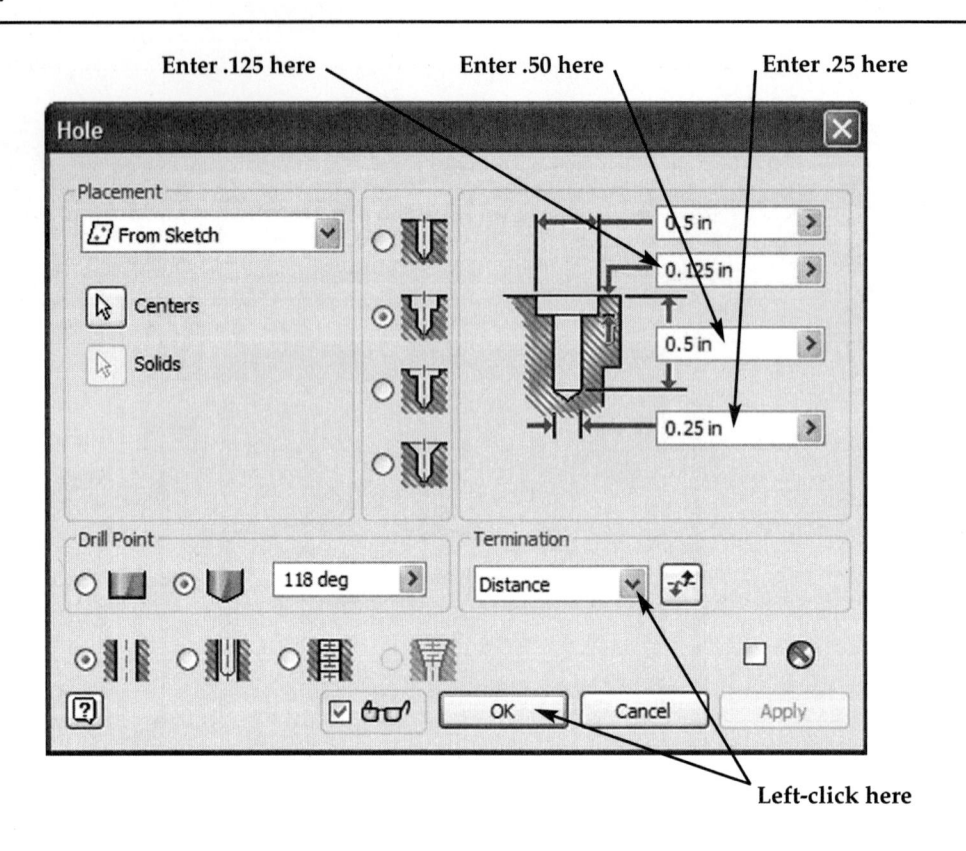

86.   Your screen should look similar to Figure 1-70.

FIGURE 1-70

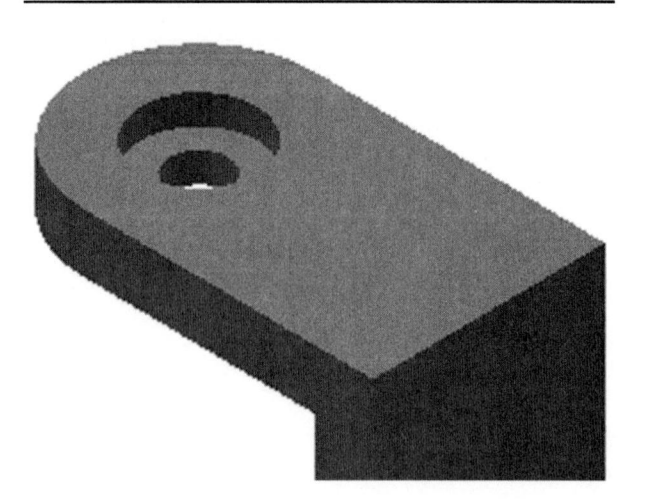

87. To ensure that the hole is correct, move the cursor to the upper left portion of the screen and left-click on the **View** tab. Left-click on the "Free Orbit/Orbit/Rotate" icon (previous version of Inventor display, "Orbit or Rotate") as shown in Figure 1-71.

FIGURE 1-71

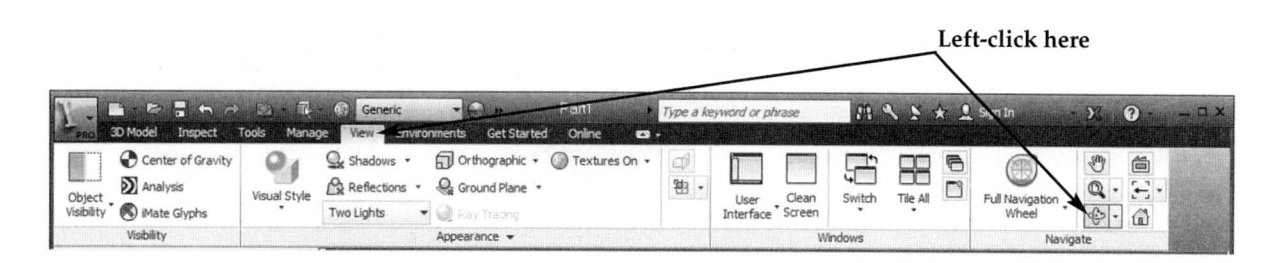

88. The Free Orbit/Rotate command will become active. Left-click anywhere inside the white circle, hold the left mouse button down, and drag the cursor upward. The part will rotate upward as shown in Figure 1-72.

FIGURE 1-72

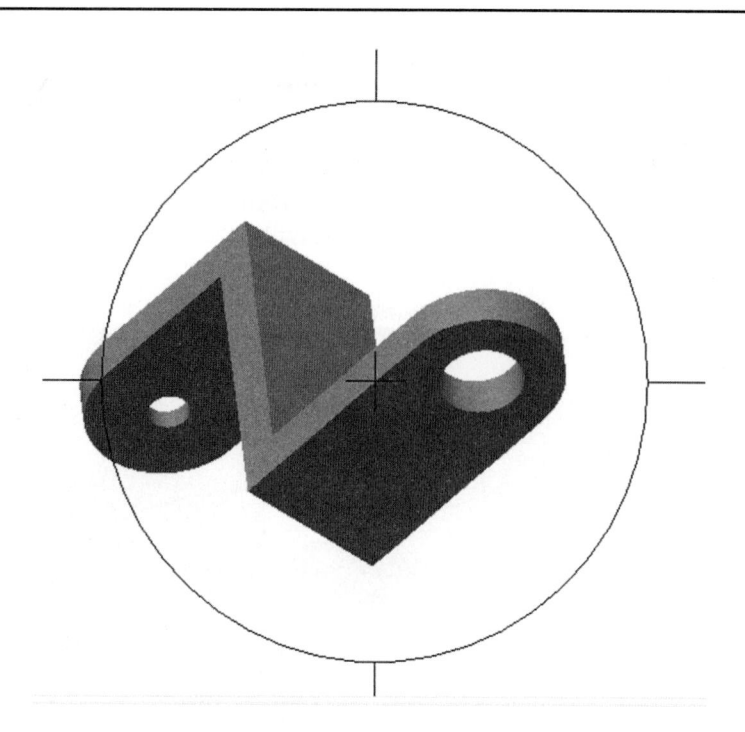

89.   Holding the left mouse button down keeps the part attached to the cursor. To view the part in Isometric, right-click anywhere on the screen and left-click on **Home View** from the pop-up menu as shown in Figure 1-73.

FIGURE 1-73

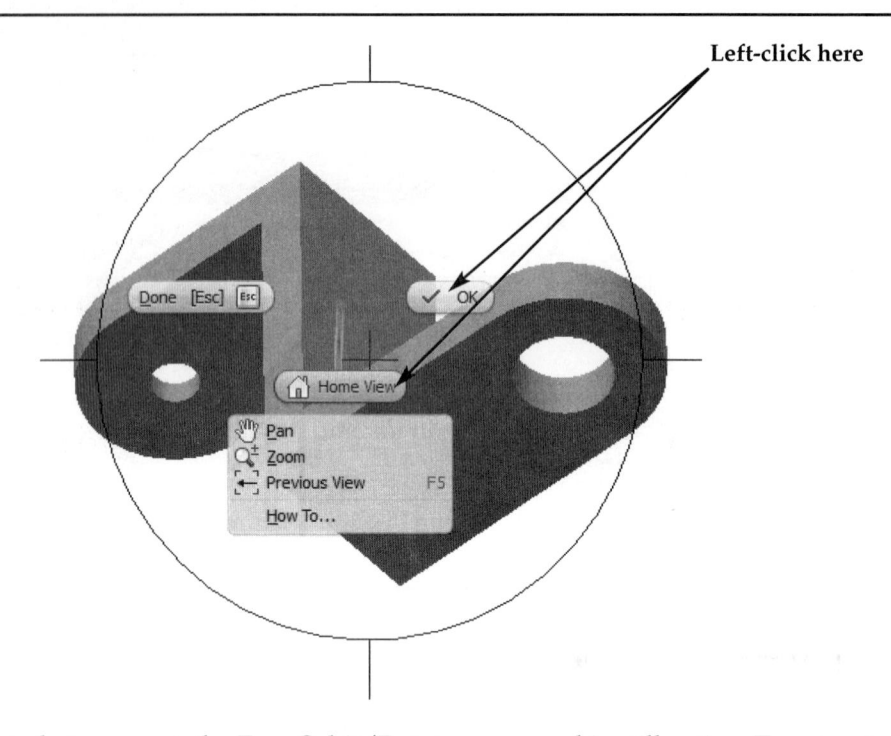

90.   As long as the white circle is present, the Free Orbit/Rotate command is still active. To get out of the Orbit/Rotate command, either use the keyboard and press **Esc** once or twice, or right-click anywhere on the screen. A pop-up menu will appear. Left-click on **OK** from the pop-up menu shown in Figure 1-74.

FIGURE 1-74

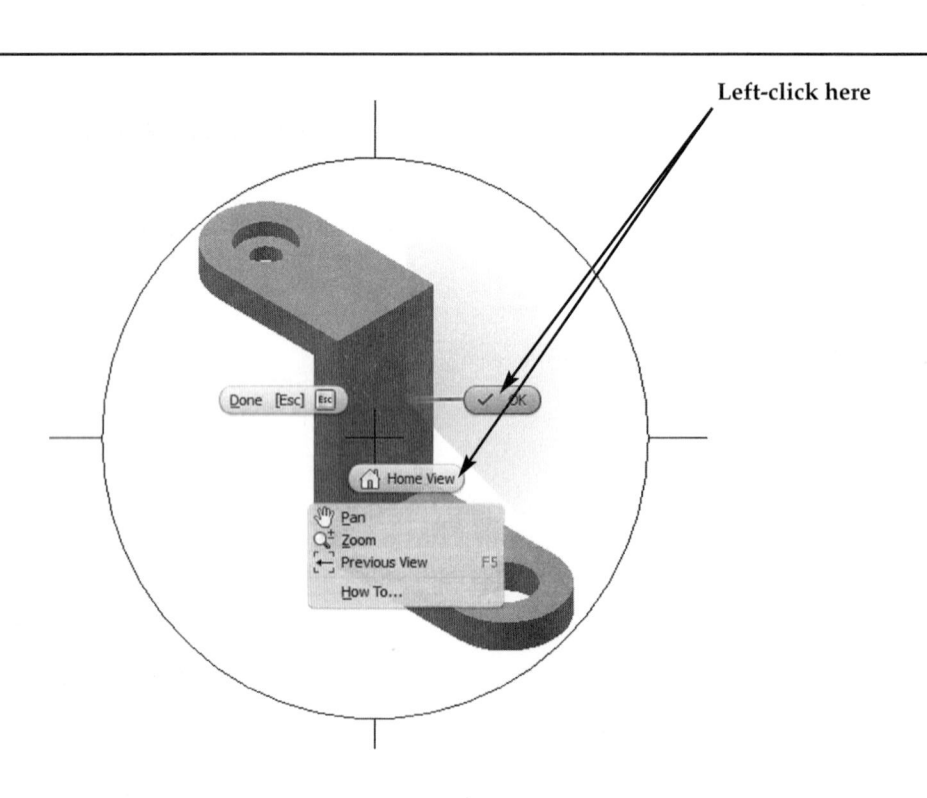

91. Other commands for viewing are at the upper right portion of the screen. Left-click on the **View** tab to access each command. You can also use the icons located at the far right. Each icon has a drop-down arrow to locate more viewing options. Left-click on the drop-down arrow under each command to become more familiar with what options are available as shown in Figure 1-75. Probably the most convenient tool for zooming in and out is the mouse wheel. Simply roll the wheel forward to move the view away. To bring the view closer, simply roll the wheel towards you.

FIGURE 1-75

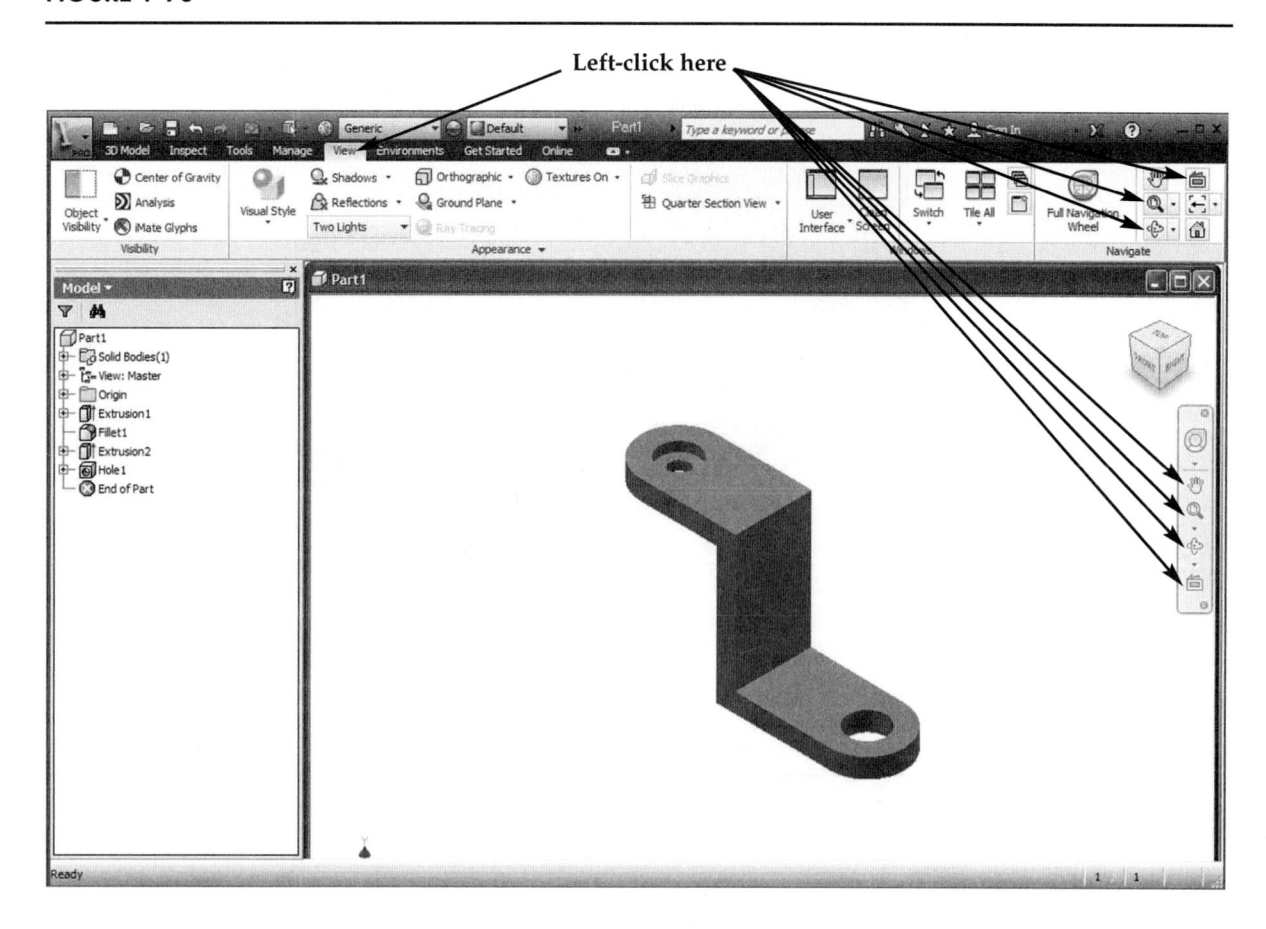

92. To save the model, move the cursor to the upper left portion of the screen and left-click on the "**I**" at the far upper left. A drop-down menu will appear as shown in Figure 1-76. Save the file where it can be retrieved later.

FIGURE 1-76

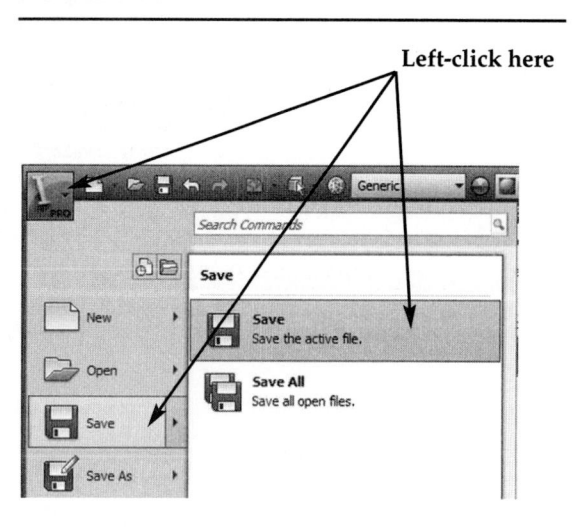

CHAPTER PROBLEMS

Use the Extrude and Extrude-Cut commands to complete the following.

PROBLEM 1-1

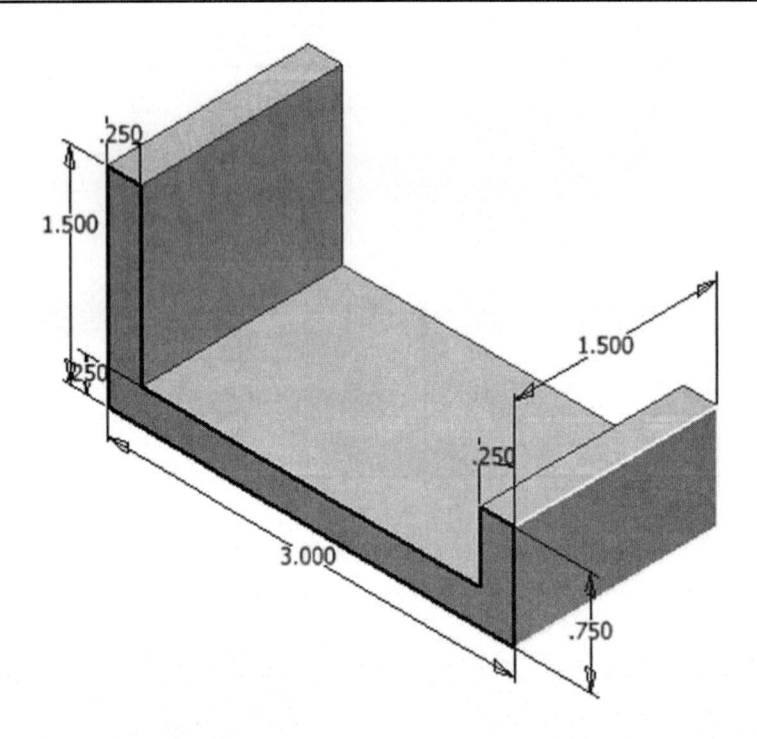

PROBLEM 1-2

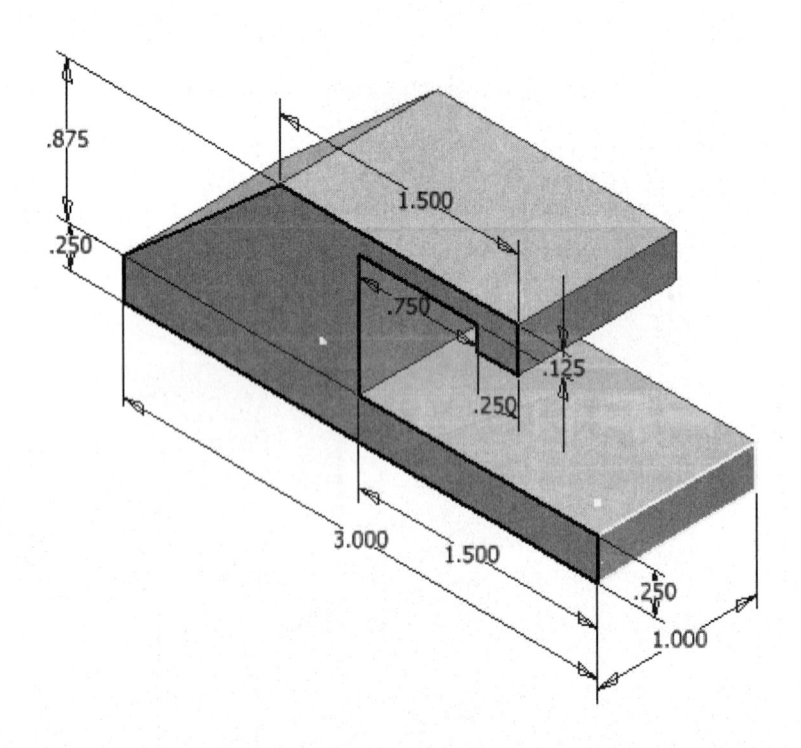

PROBLEM 1-3

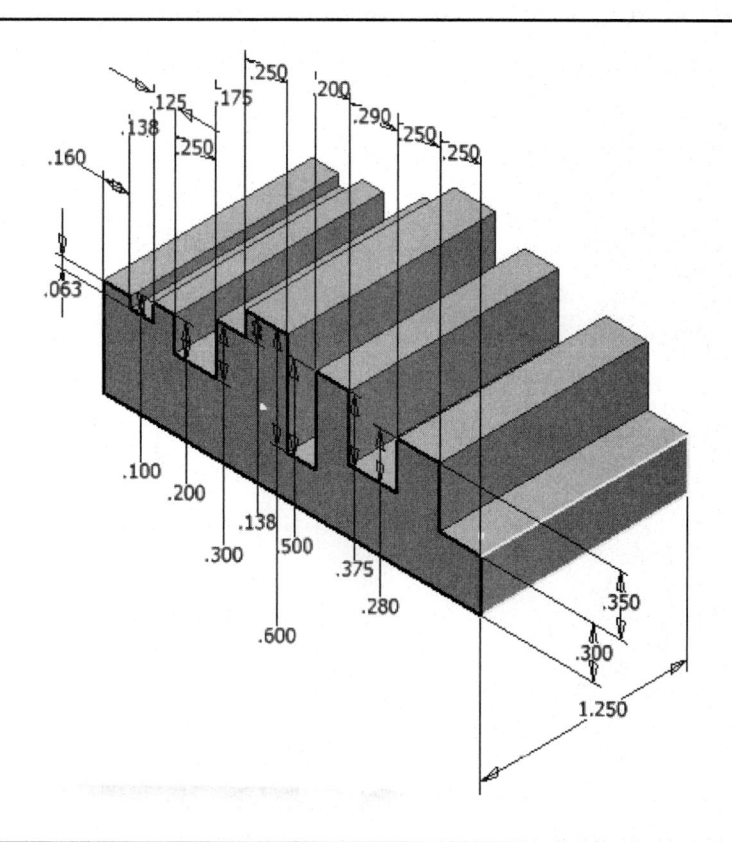

PROBLEM 1-4

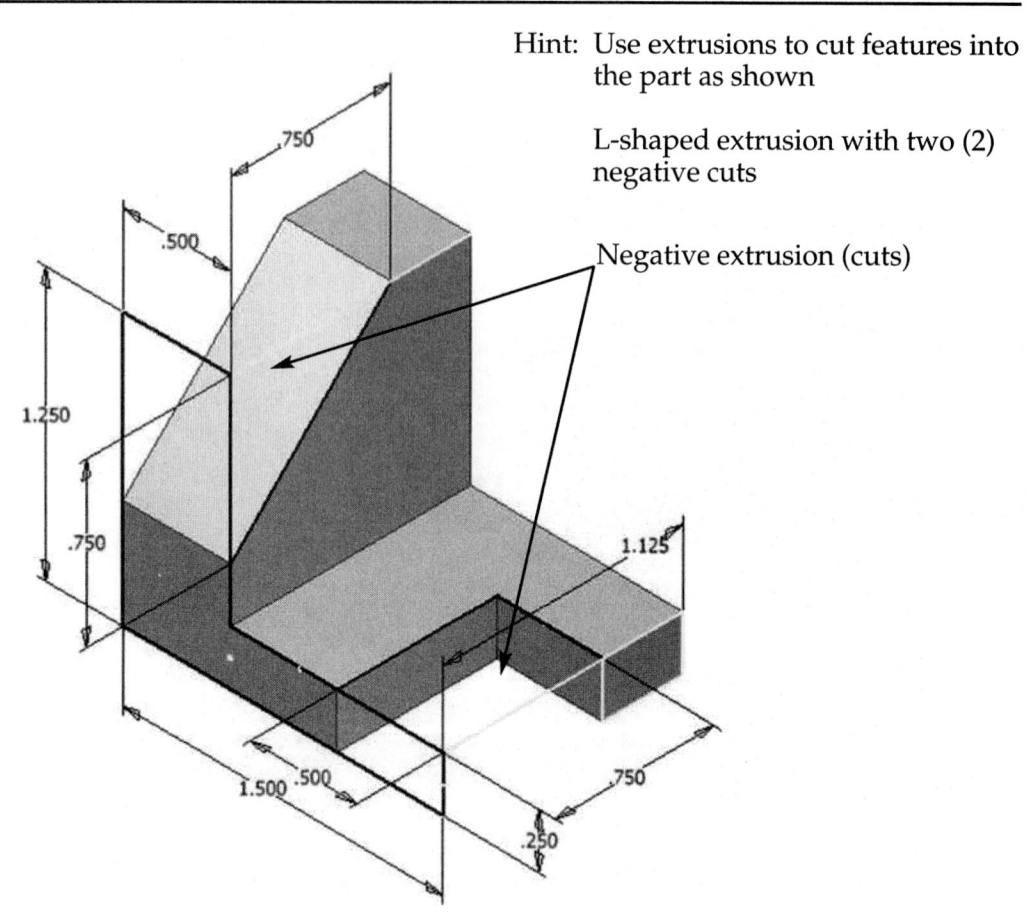

Hint: Use extrusions to cut features into the part as shown

L-shaped extrusion with two (2) negative cuts

Negative extrusion (cuts)

PROBLEM 1-5

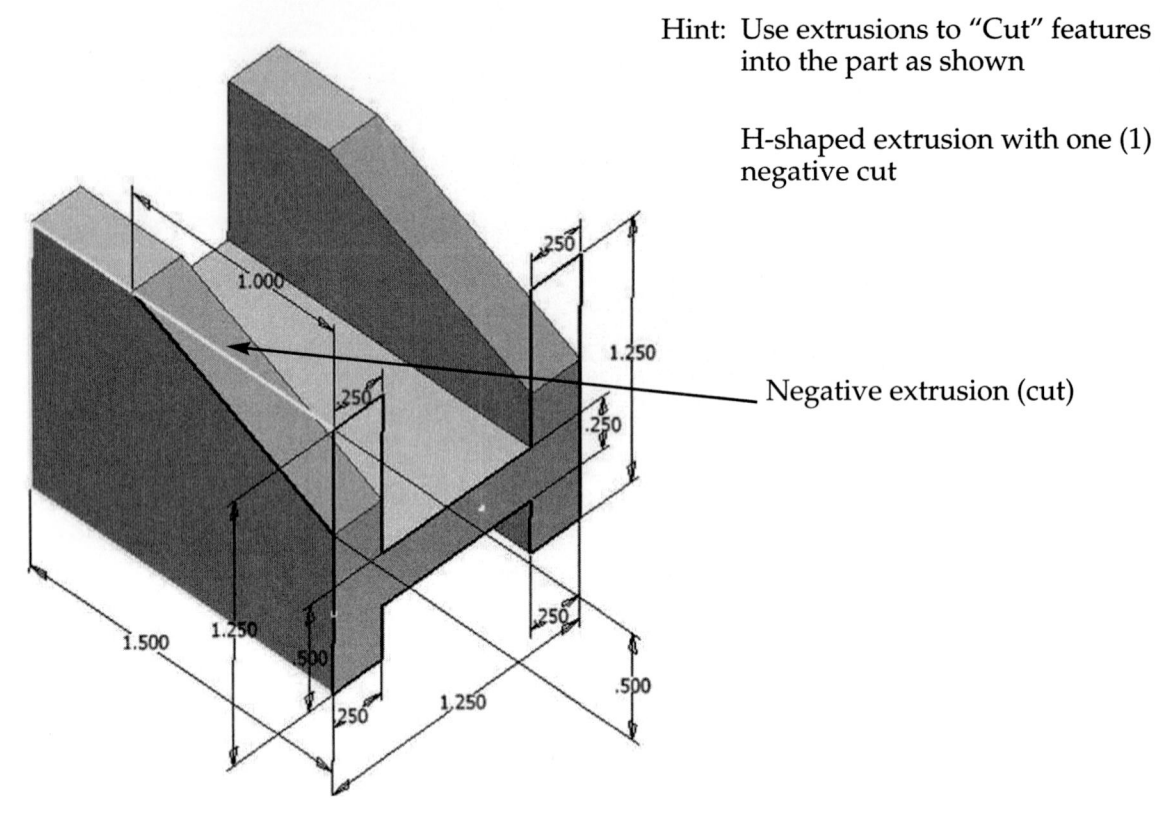

Hint: Use extrusions to "Cut" features
into the part as shown

H-shaped extrusion with one (1)
negative cut

Negative extrusion (cut)

PROBLEM 1-6

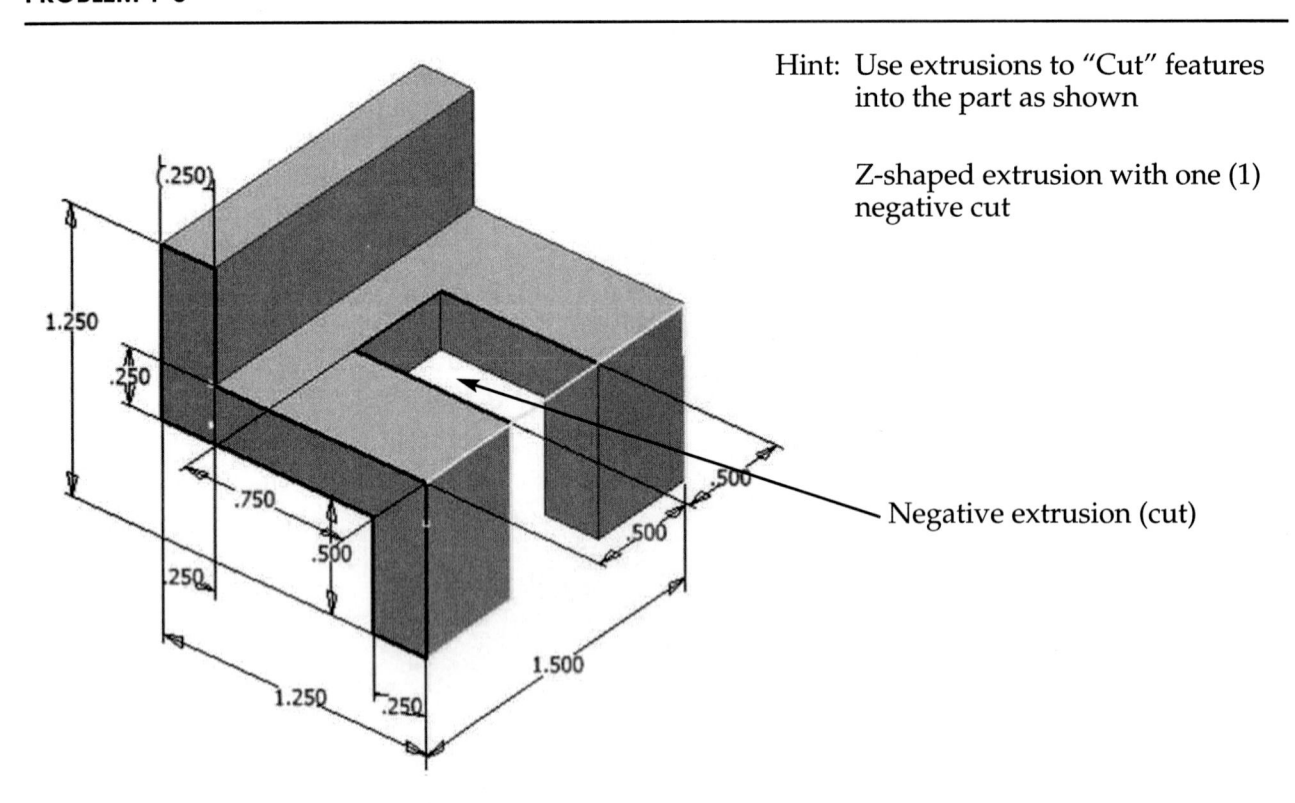

Hint: Use extrusions to "Cut" features
into the part as shown

Z-shaped extrusion with one (1)
negative cut

Negative extrusion (cut)

PROBLEM 1-7

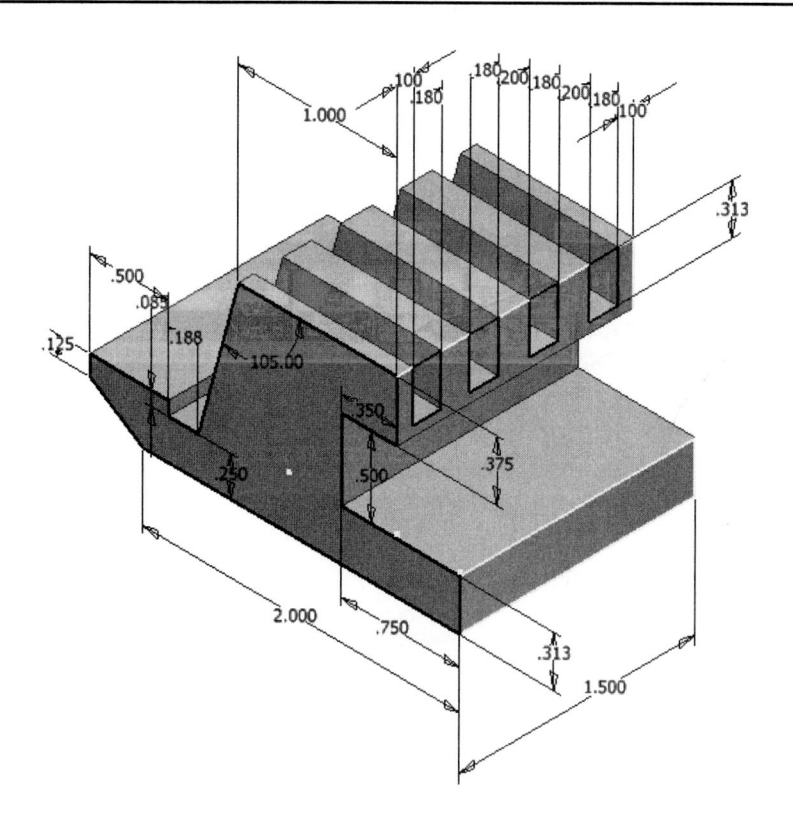

PROBLEM 1-8

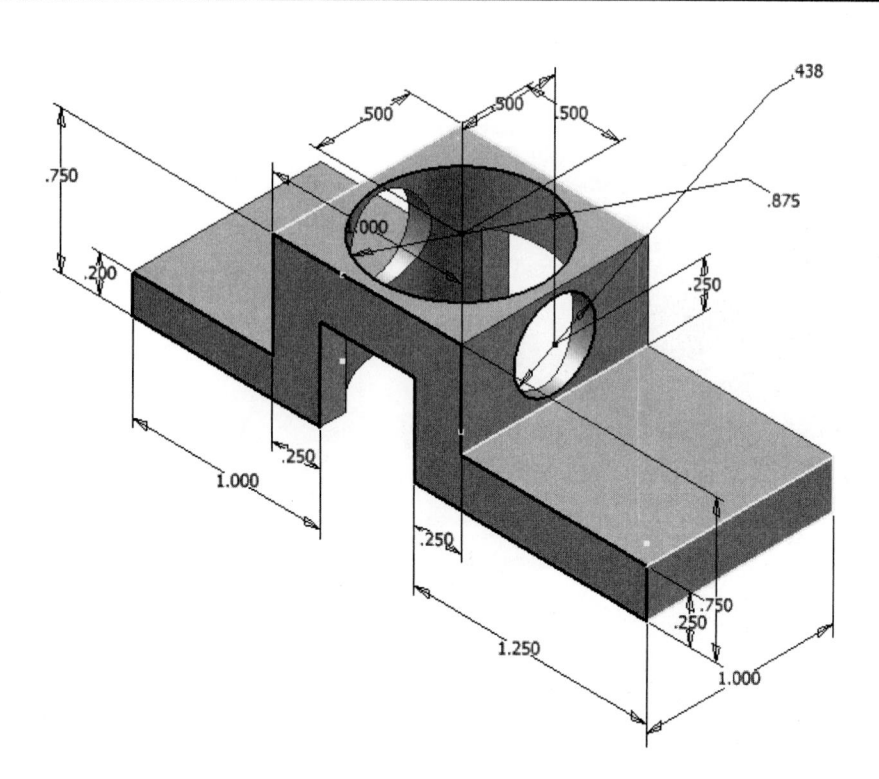

PROBLEM 1-9

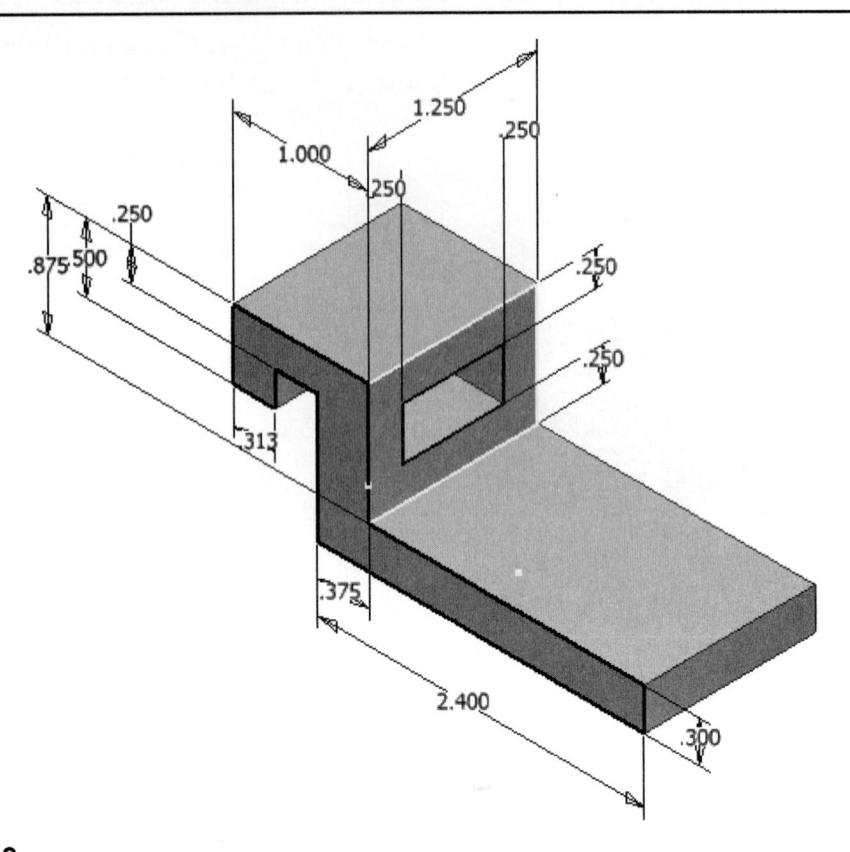

PROBLEM 1-10

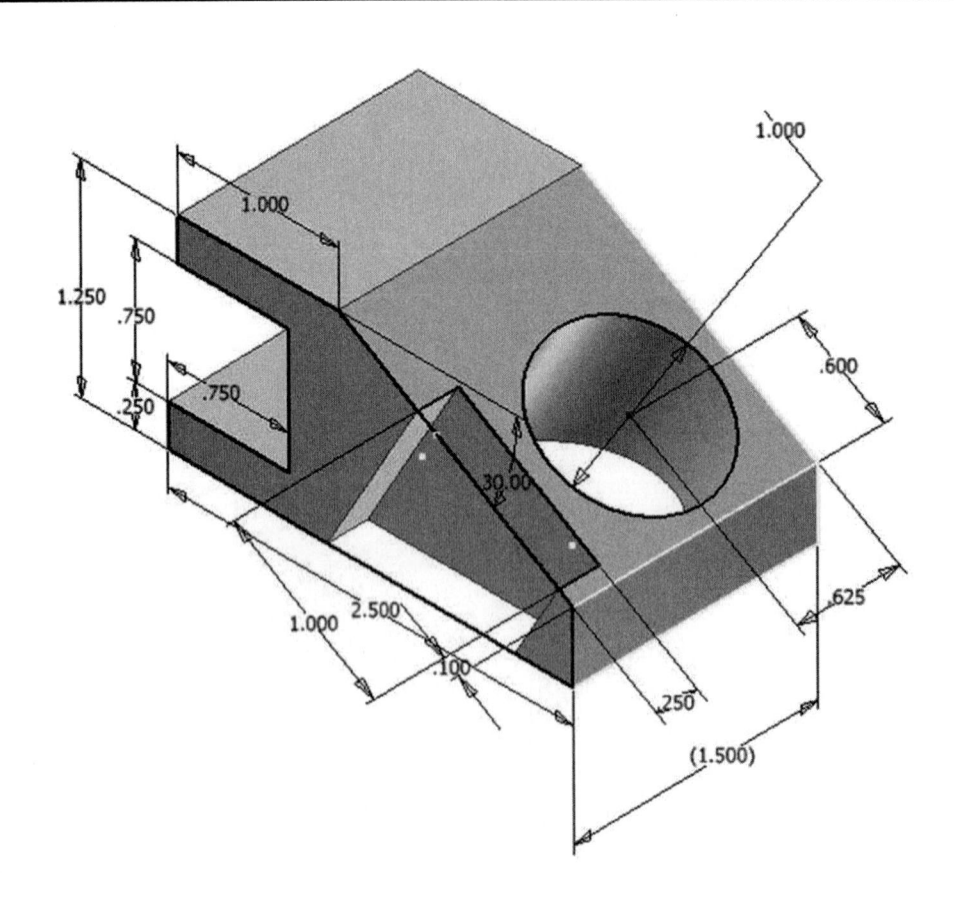

PROBLEM 1-11

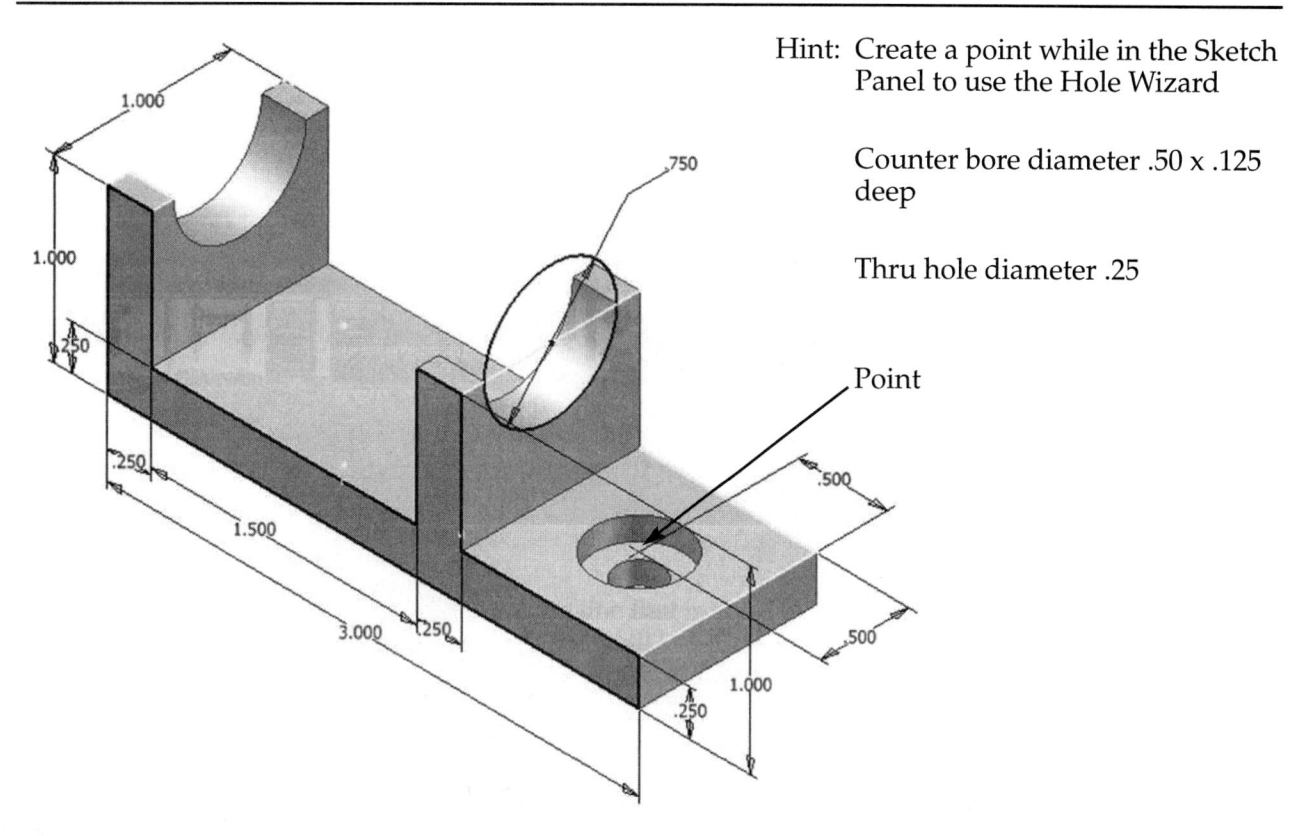

Hint: Create a point while in the Sketch Panel to use the Hole Wizard

Counter bore diameter .50 x .125 deep

Thru hole diameter .25

Point

PROBLEM 1-12

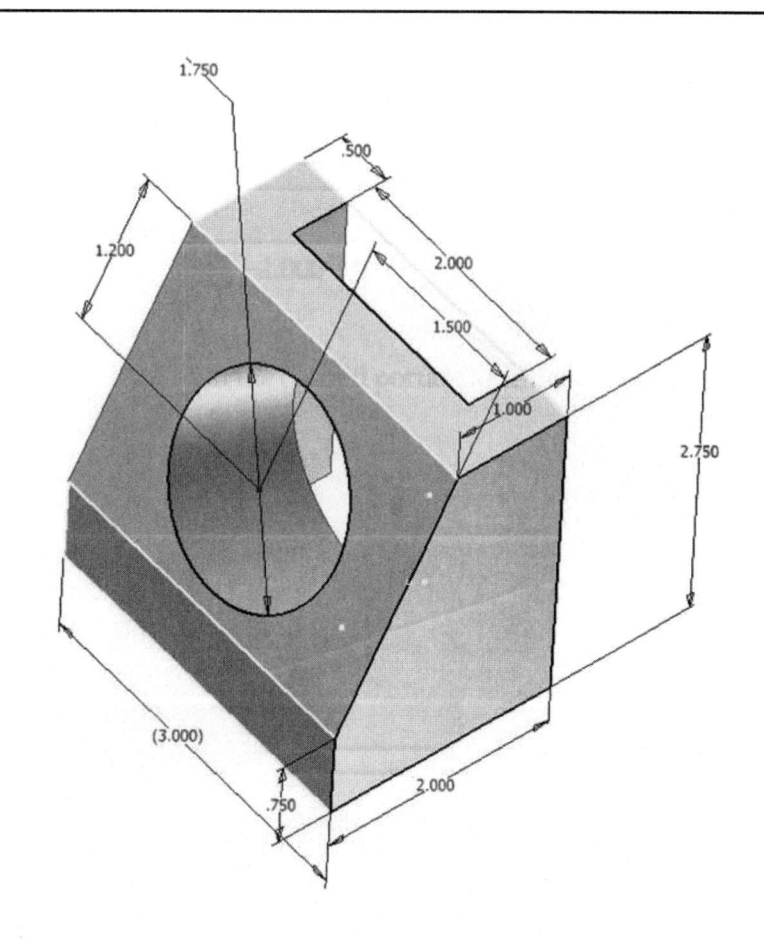

PROBLEM 1-13

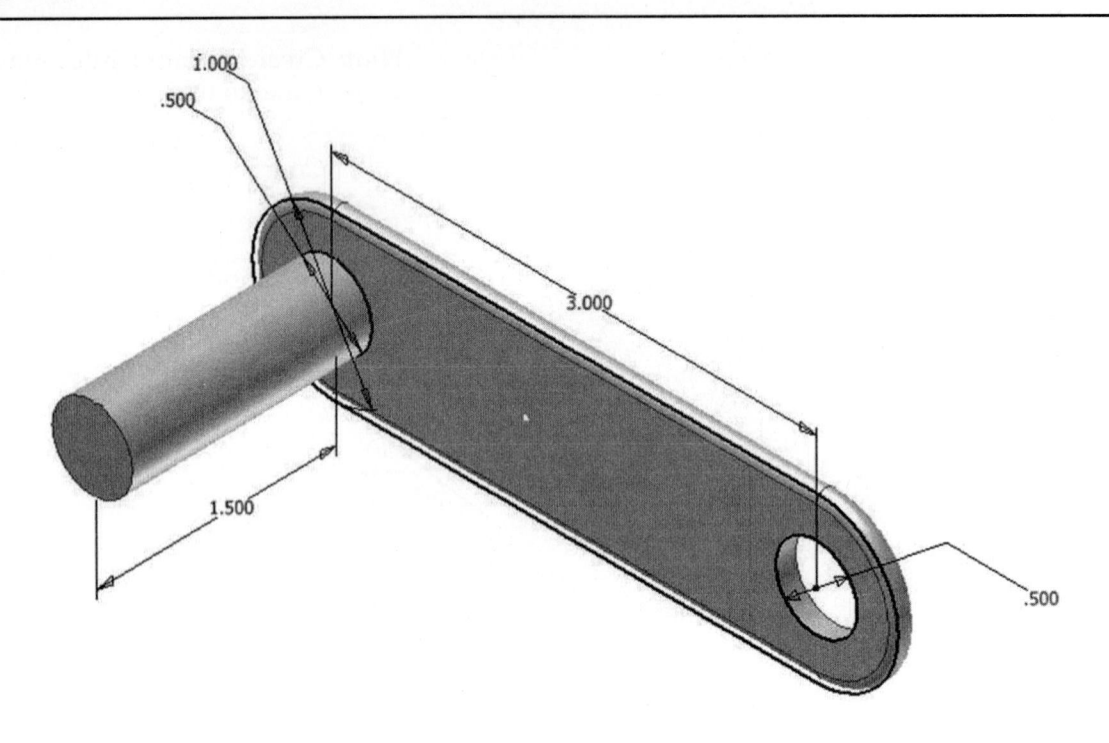

PROBLEM 1-14

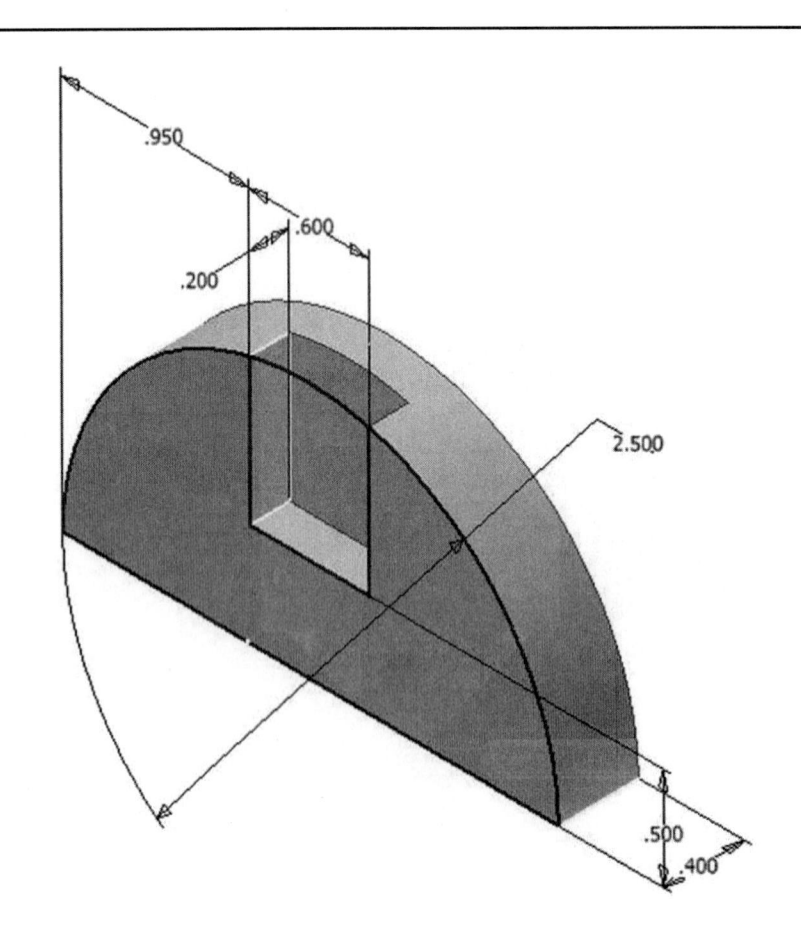

PROBLEM 1-15

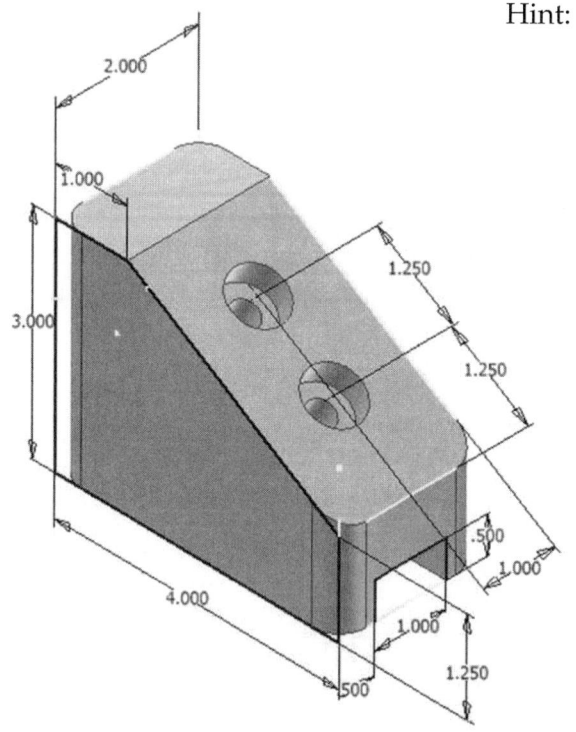

Hint: (2) Counter bored holes

.75 diameter x .375 deep,
thru hole diameter = .375

Fillet radius .375

PROBLEM 1-16

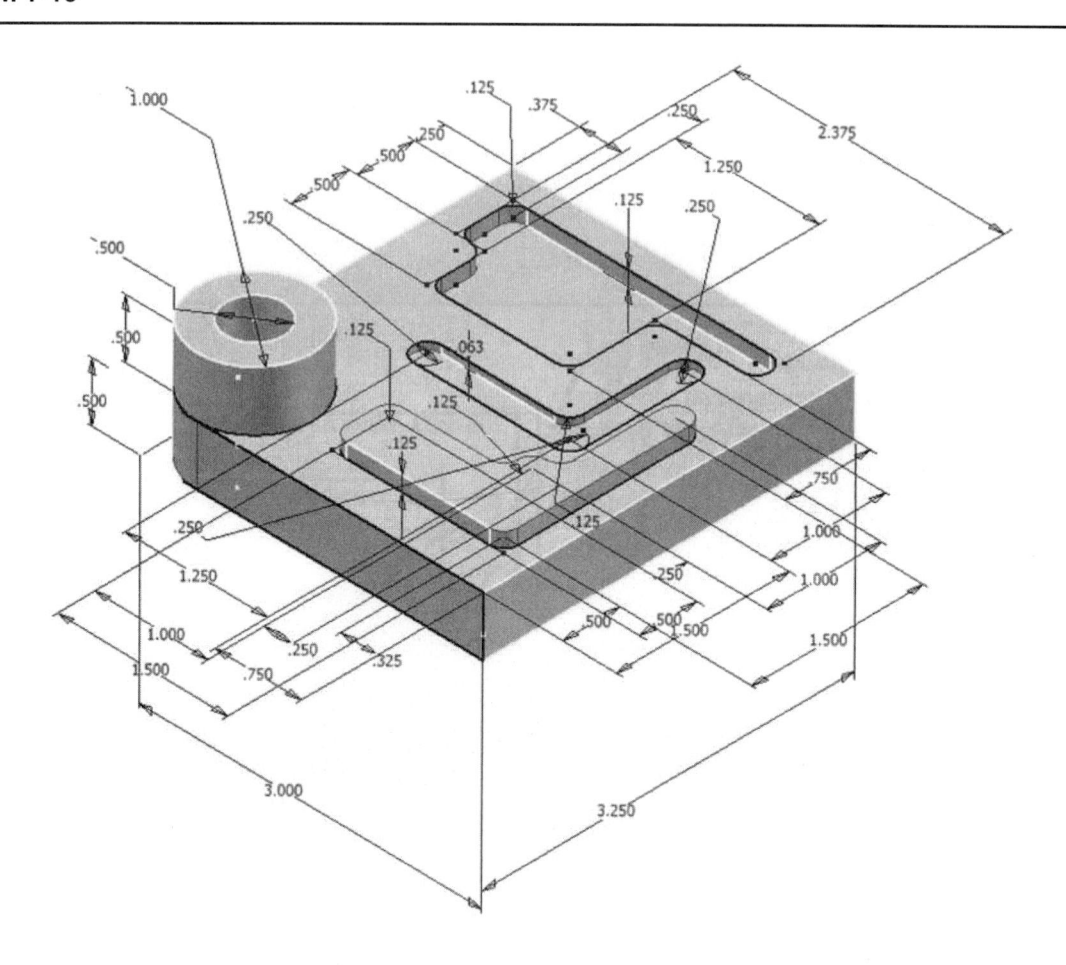

2

LEARNING MORE BASICS

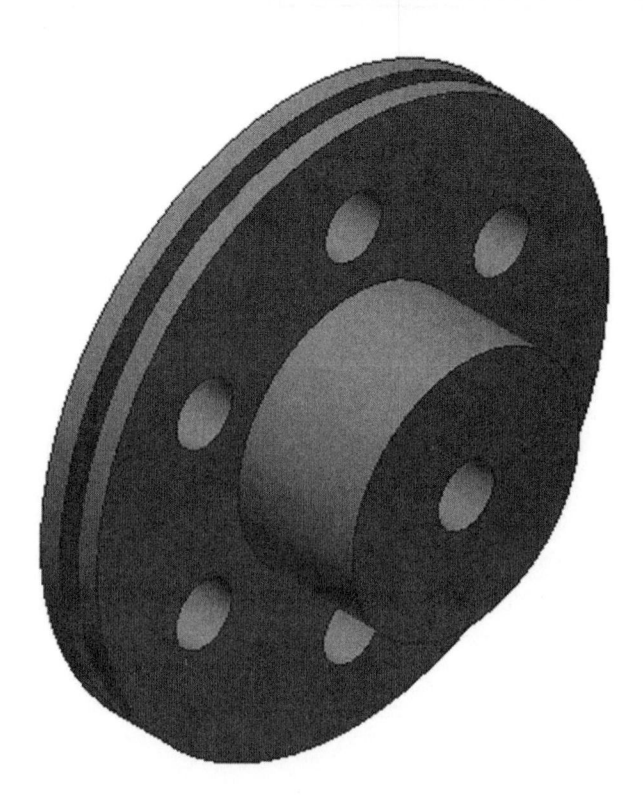

Chapter 2 includes instruction on how to design the part shown above.

OBJECTIVES

1. Create a simple sketch using the Sketch Panel

2. Dimension a sketch using the Dimension command

3. Revolve a sketch in the Model/Part Features Panel using the Revolve command

4. Create a groove using the Revolve Cut command

5. Create a hole in the Model/Part Features Panel using the Extrude command

6. Create a series of holes in the Model/Part Features Panel using the Circular Pattern and command

1.   Start Autodesk Inventor 2013 by referring to Chapter 1.

2.   After Autodesk Inventor 2013 is running, begin a new sketch.

3.   Complete the sketch (including dimensions) as shown. Include the line above the sketch as shown in Figure 2-1. This line will be used to revolve the sketch around.

FIGURE 2-1

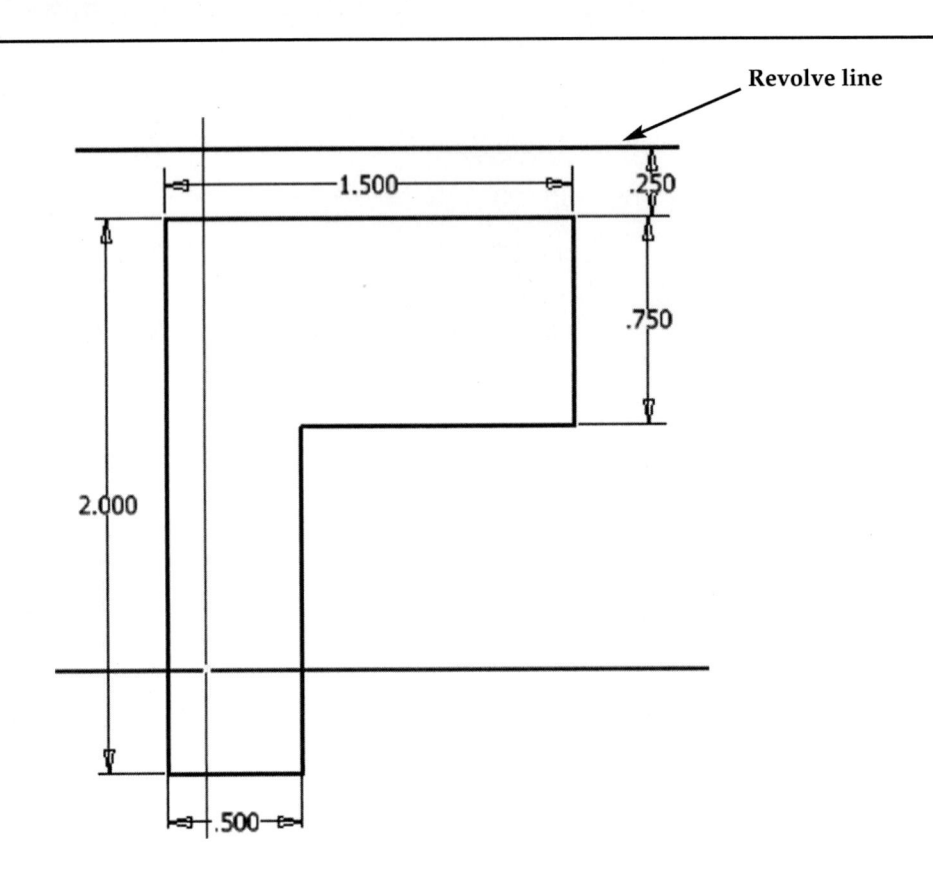

4. Right-click around the sketch. A pop-up menu will appear. Left-click on **Home View**, then **Finish 2D Sketch** as shown in Figure 2-2.

FIGURE 2-2

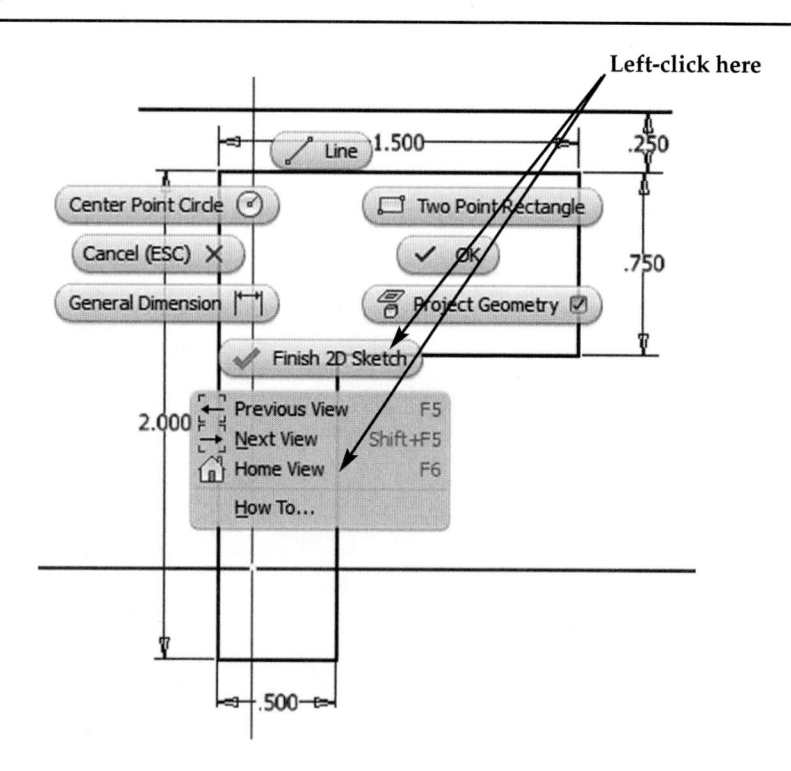

5. The view will become Isometric as shown in Figure 2-3.

FIGURE 2-3

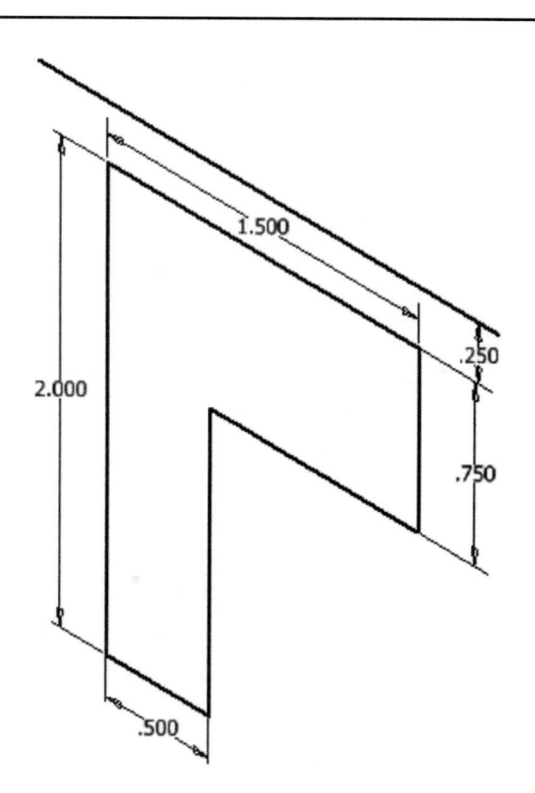

REVOLVE A SKETCH IN THE PART FEATURES PANEL
USING THE REVOLVE COMMAND

6.   Move the cursor to the upper left portion of the screen and left-click on **Revolve**. Left-click on the drop-down arrow in the center of the collapsed dialog box. This will cause the dialog box to expand. If Inventor gave you an error message, there are opens (non-connected lines) somewhere on the sketch or the view is not Isometric. Check each intersection for opens by using the **Extend** and **Trim** commands and make sure the view is Isometric. Your screen should look similar to Figure 2-4.

FIGURE 2-4

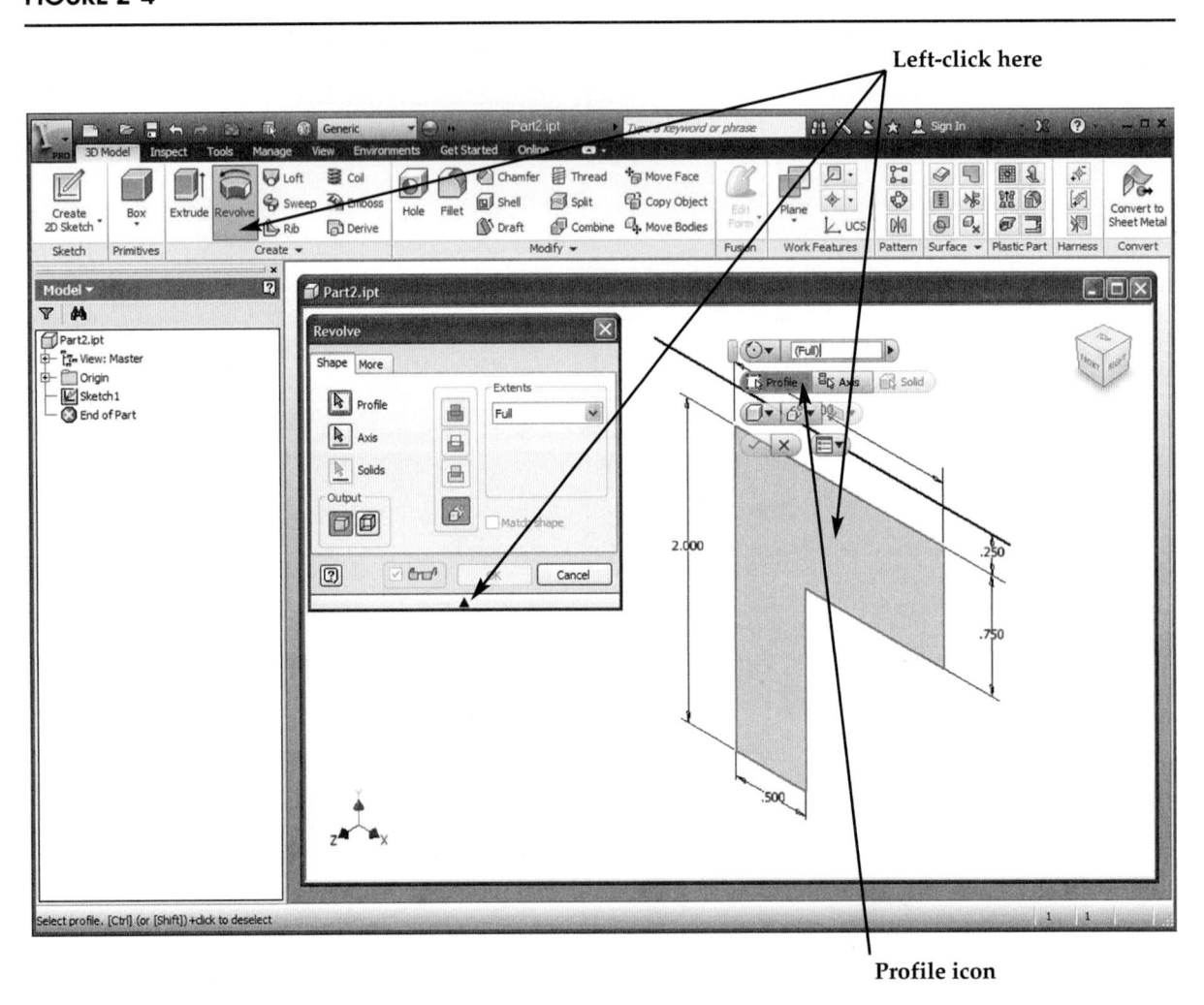

7.   Notice that the Profile icon has already been selected. Because there is only one profile present, Inventor assumes that particular profile will be selected. If the drawing contained more than one profile, you would have to first select the **Profile** icon in the Revolve dialog box, then use the cursor to select the desired profile.

8. Left-click on the **Axis** icon. Move the cursor over the axis, causing it to turn red and left-click once as shown in Figure 2-5.

FIGURE 2-5

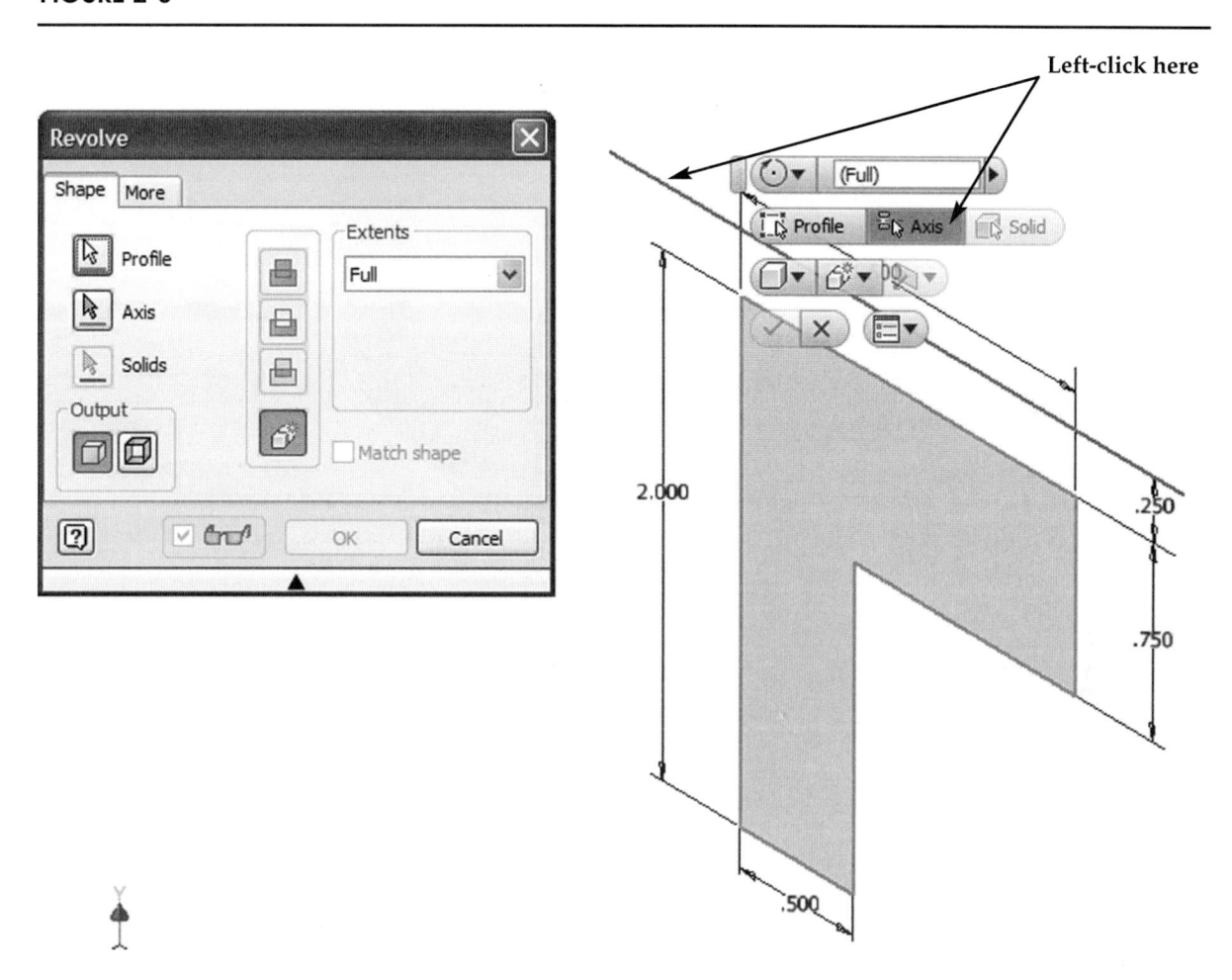

9.   A preview of the revolve will appear as shown in Figure 2-6.

FIGURE 2-6

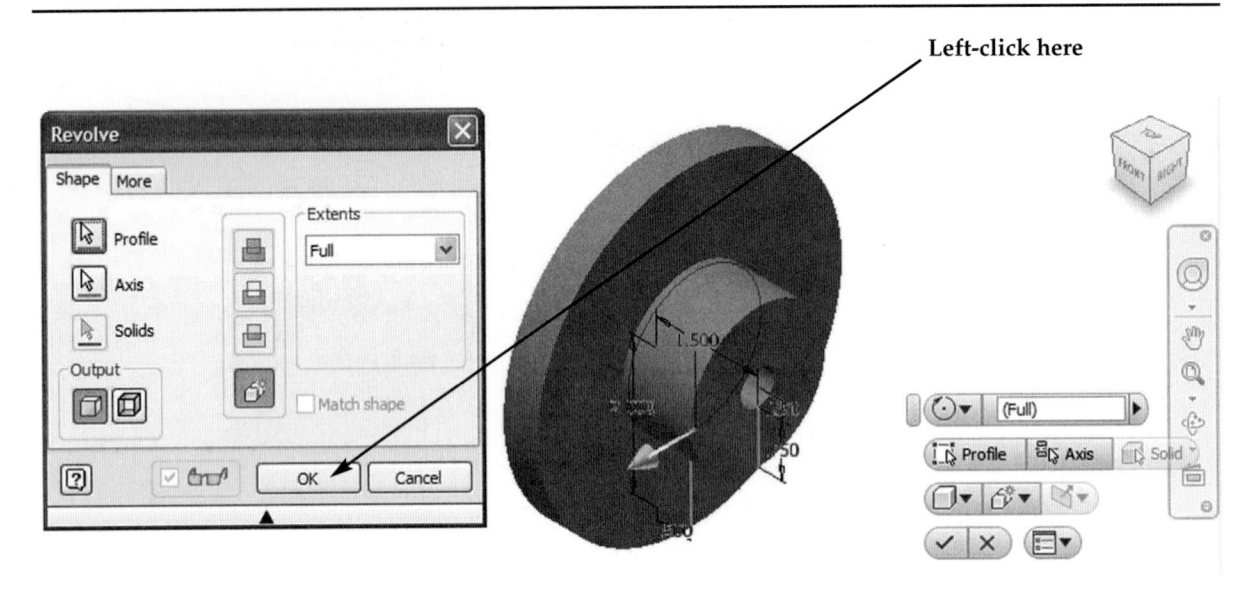

10.   Left-click on **OK**.

11.   Your screen should look similar to Figure 2-7. You may have to zoom out to view the entire part.

FIGURE 2-7

12. Move the cursor over **XY Plane** located in the upper left portion of the screen and right-click once. A pop-up menu will appear. Left-click on **New Sketch** as shown in Figure 2-8.

FIGURE 2-8

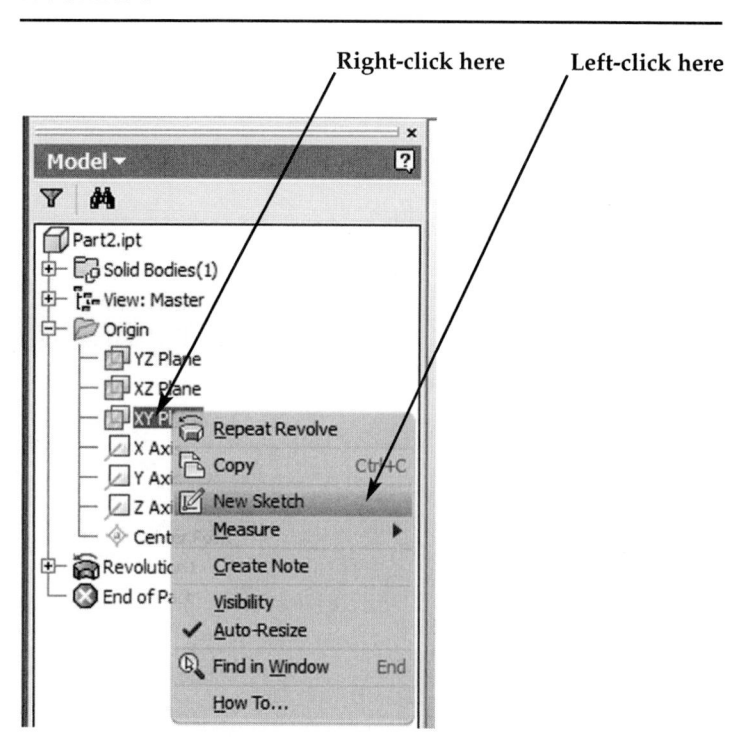

13. Inventor will create a work plane in the center of the part. You may have to select **Home View** to see the plane as shown in Figure 2-9.

FIGURE 2-9

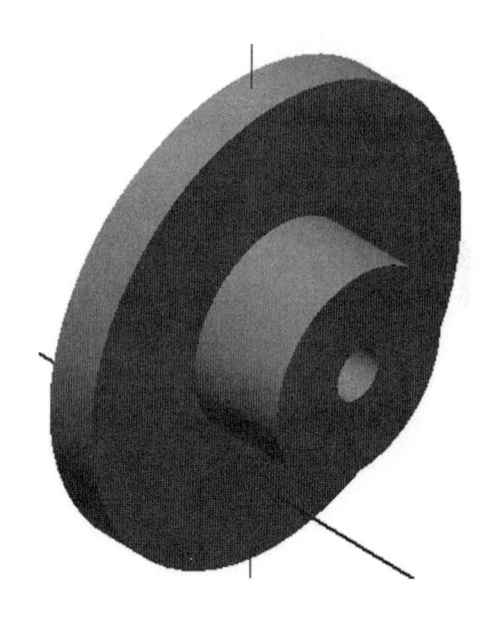

14.   Move the cursor to the upper middle portion of the screen and left-click on the **View** tab. Left-click on the arrow under **Visual Style**. A drop-down menu will appear. Left-click on **Wireframe** as shown in Figure 2-10.

FIGURE 2-10

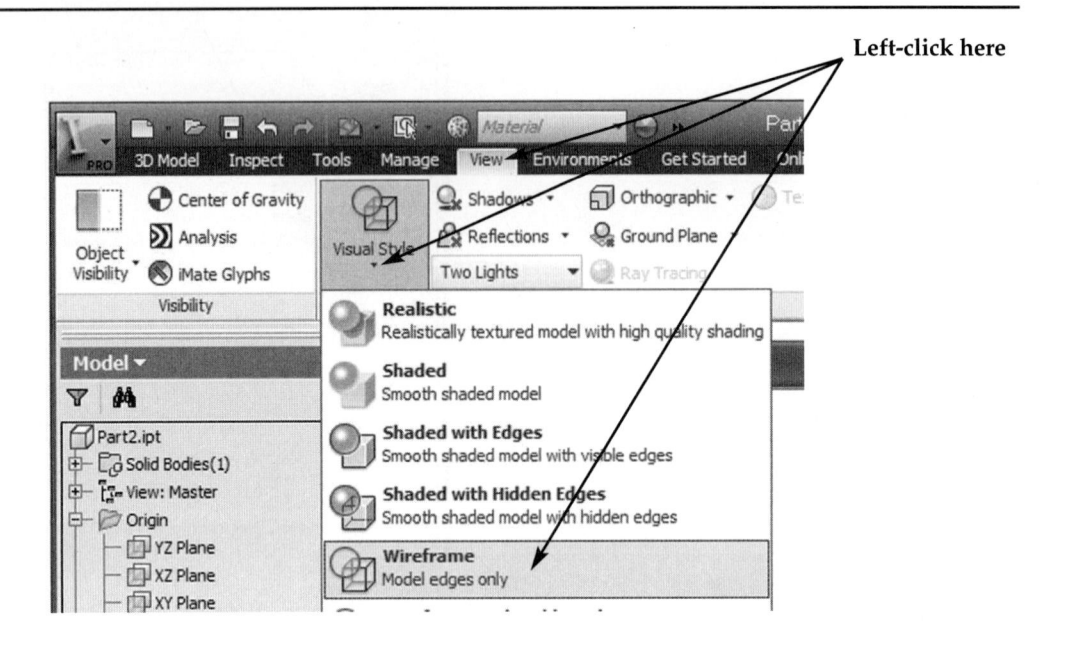

15.   Inventor will change the display of the model to wireframe as shown in Figure 2-11.

FIGURE 2-11

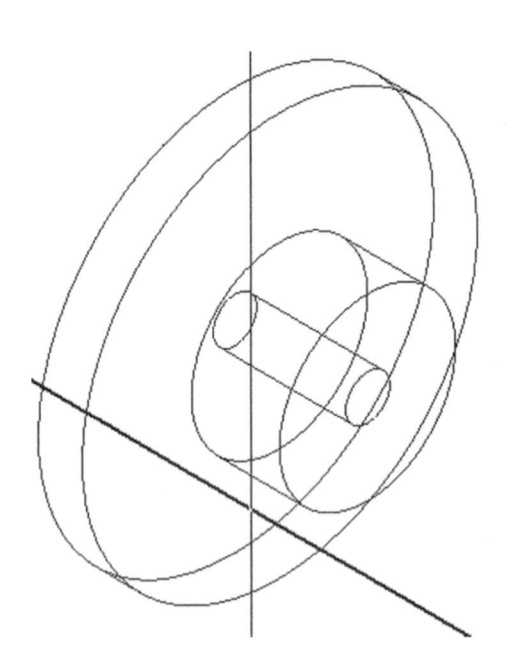

16. Move the cursor to the middle right portion of the screen and left-click on the "Face View/Look At" icon as shown in Figure 2-12.

FIGURE 2-12

Left-click here

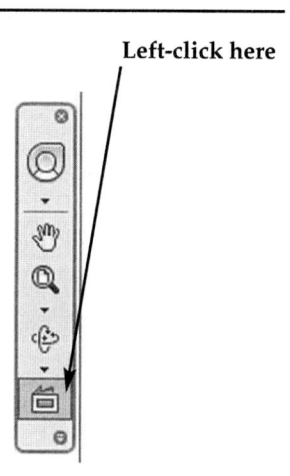

17. Move the cursor to the upper left portion of the screen and left-click on the **XY Plane** as shown in Figure 2-13.

FIGURE 2-13

Left-click here

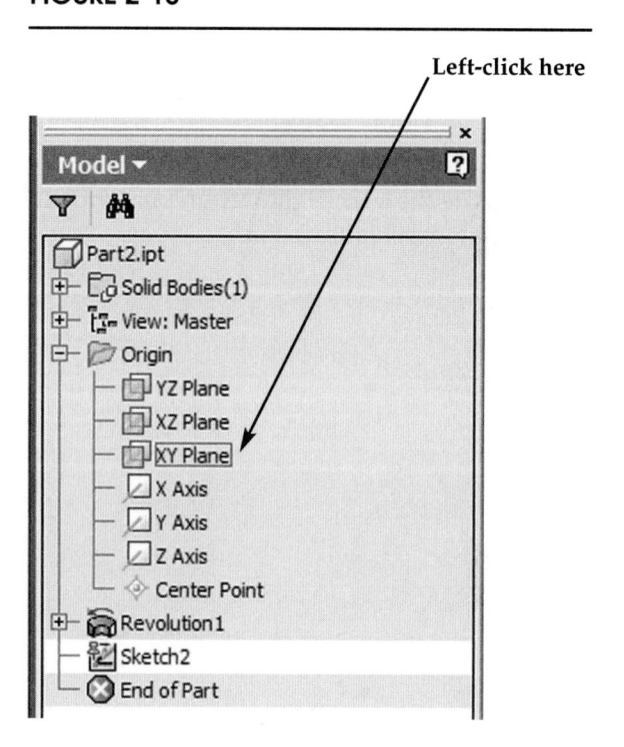

18.   Inventor will rotate the model around, providing a perpendicular view of the XY Plane as shown in Figure 2-14.

FIGURE 2-14

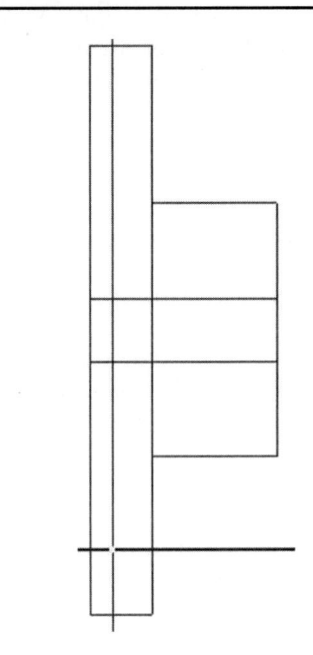

19.   Use the "Free Orbit/Rotate" command to rotate the part slightly from perpendicular as shown in Figure 2-15.

FIGURE 2-15

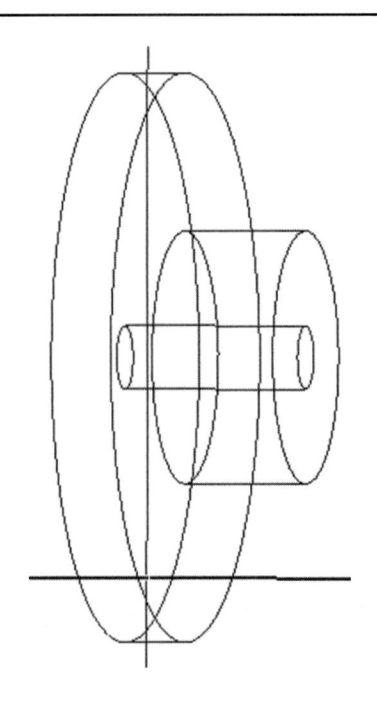

USE THE REVOLVE CUT COMMAND TO CREATE A GROOVE

20.   Move the cursor to the upper middle portion of the screen and left-click on **Project Geometry**. You may need to left-click on the **Sketch** tab if the Project Geometry icon is not visible as shown in Figure 2-16.

FIGURE 2-16

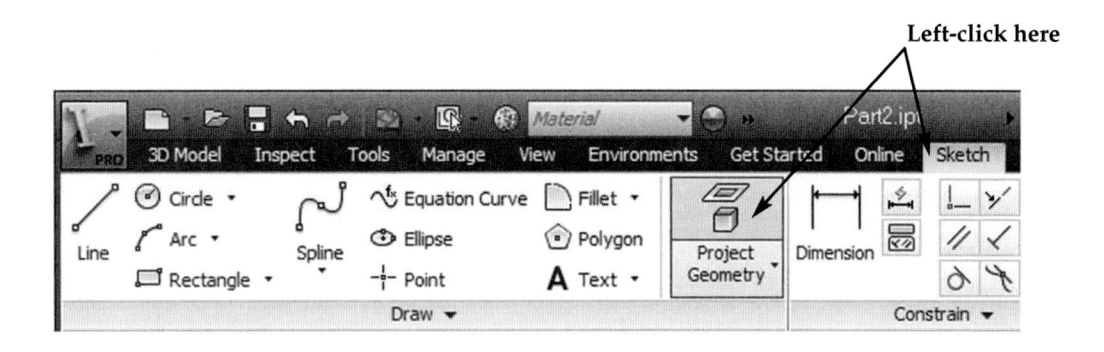

21.   Move the cursor over the left side line. When it becomes highlighted (turns red), left-click on it once as shown in Figure 2-17.

FIGURE 2-17

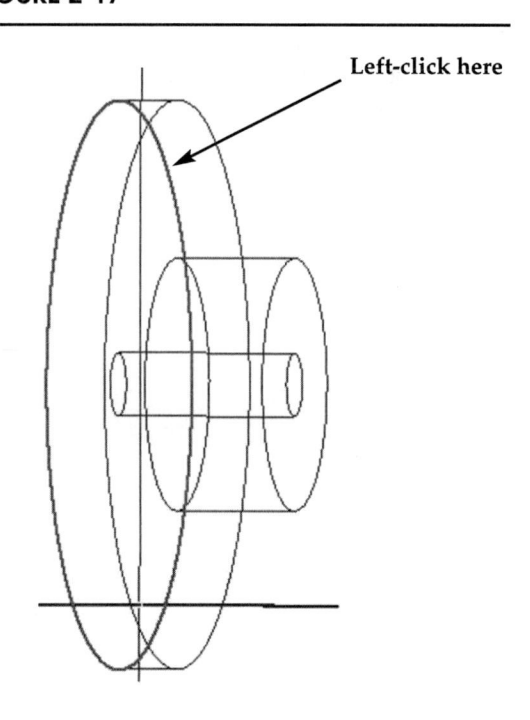

22.   Inventor will project this line onto the new
       sketch as shown in Figure 2-18.

FIGURE 2-18

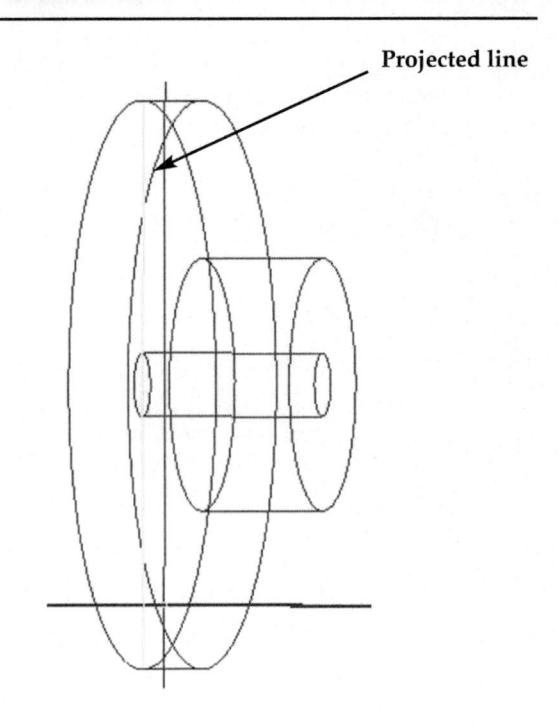

Projected line

23.   Move the cursor to the upper middle portion of the screen and left-click on **Project Geometry**
       as shown in Figure 2-19.

FIGURE 2-19

Left-click here

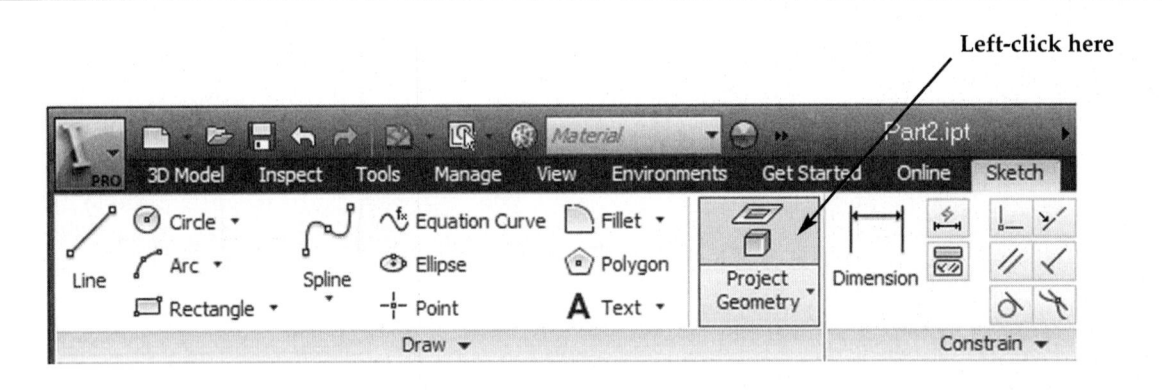

24. Move the cursor of the right side line. When it becomes highlighted (turns red), left-click on it once as shown in Figure 2-20.

FIGURE 2-20

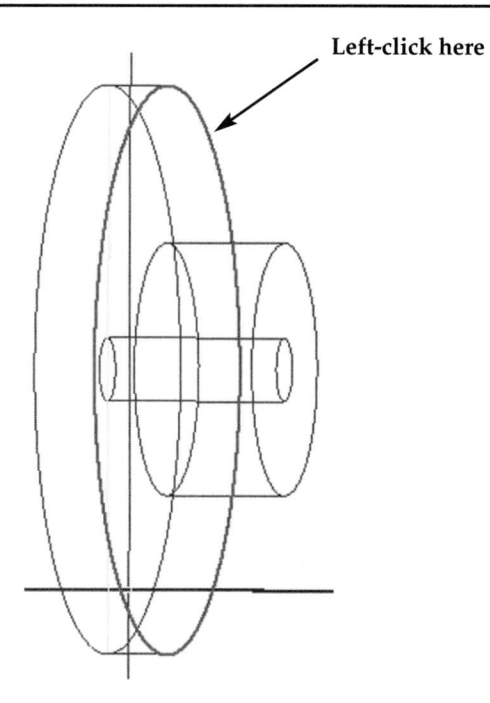

Left-click here

25. Inventor will project this line onto the new sketch as shown in Figure 2-21.

FIGURE 2-21

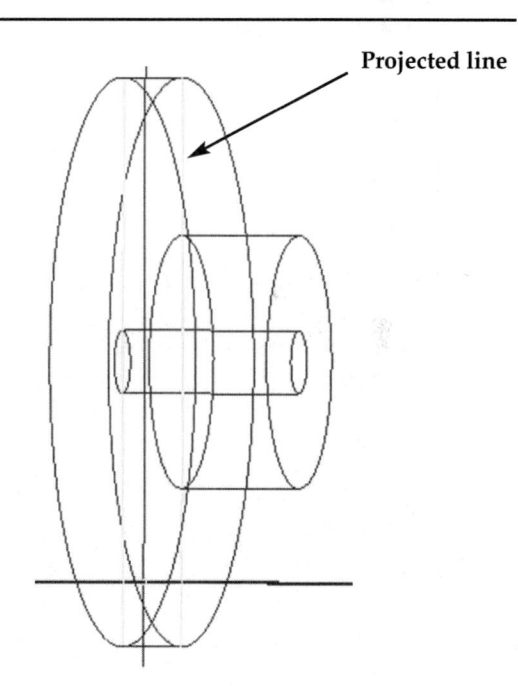

Projected line

26. Move the cursor to the upper portion of the screen and left-click on the drop-down arrow under **Project Geometry**. Left-click on **Project Cut Edges** as shown in Figure 2-22.

FIGURE 2-22

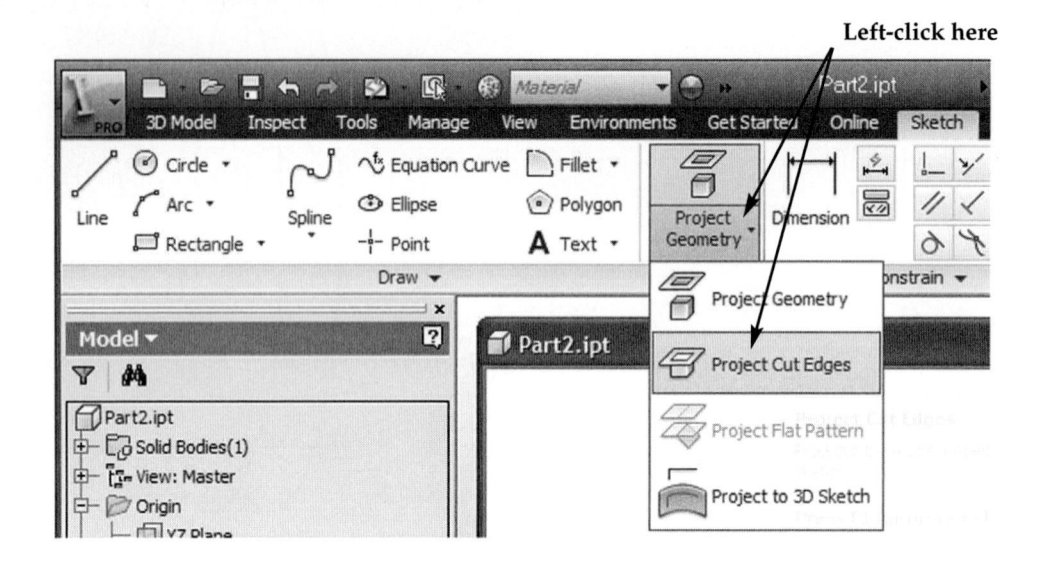

27. Inventor will project all cut edges onto the new sketch as shown in Figure 2-23. Due to the image being in grayscale for production purposes, all lines are not visible in Figure 2-23.

FIGURE 2-23

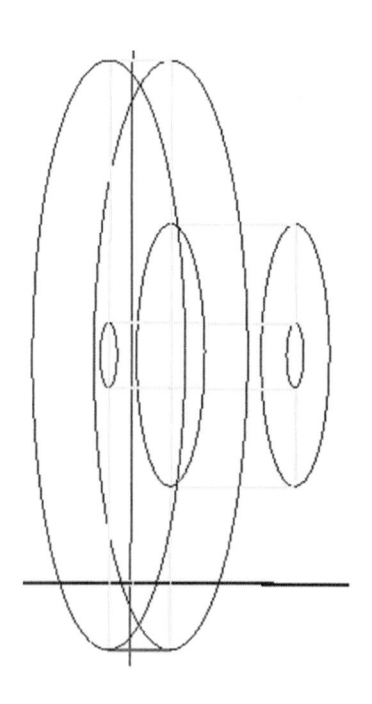

28. Move the cursor to the upper right portion of the screen and left-click on **Front** as shown in Figure 2-24.

FIGURE 2-24

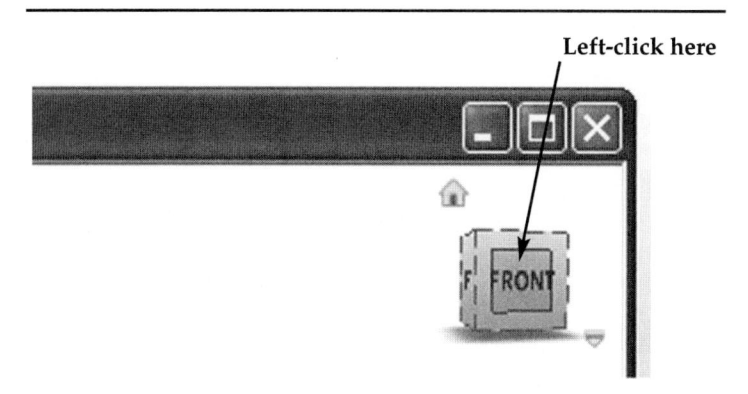

Left-click here

29. Inventor will provide a perpendicular view of the lines that were just projected onto the sketch as shown in Figure 2-25.

FIGURE 2-25

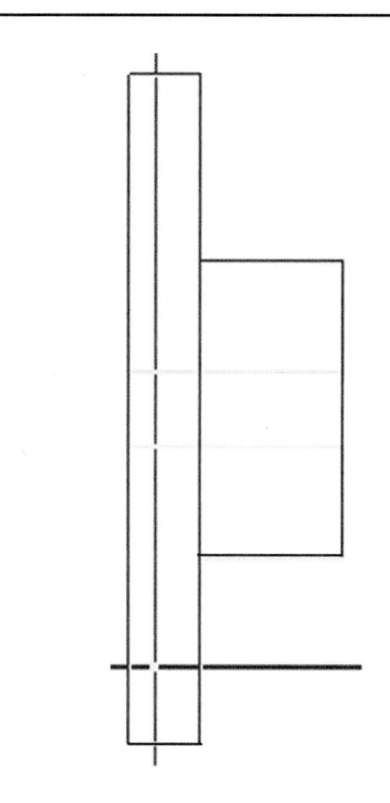

30.   Complete the sketch as shown (draw a triangle) at the bottom of the part. Once the triangle is complete, delete all of the projected lines (including ones not shown) as shown in Figure 2-26.

FIGURE 2-26

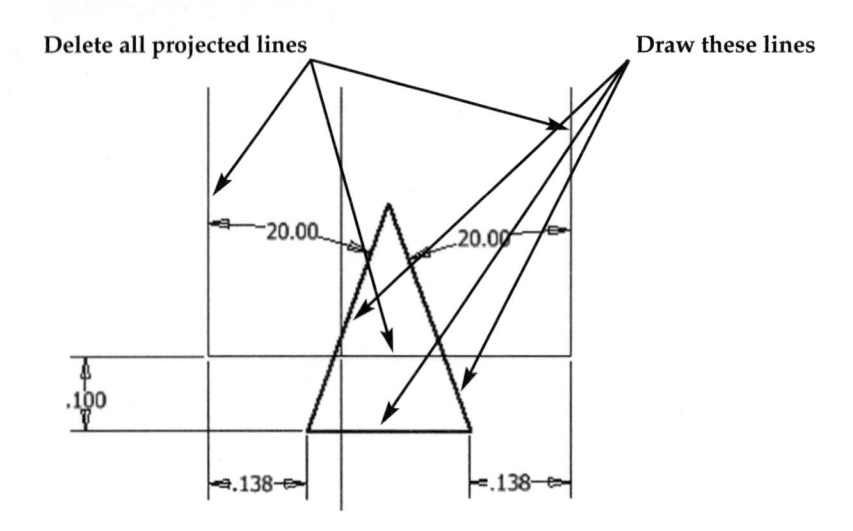

31.   Draw an axis line/revolve line in the center of the part. Finish deleting all projected lines. Once deleted, projected lines may still appear in yellow. This is not a problem as shown in Figure 2-27.

FIGURE 2-27

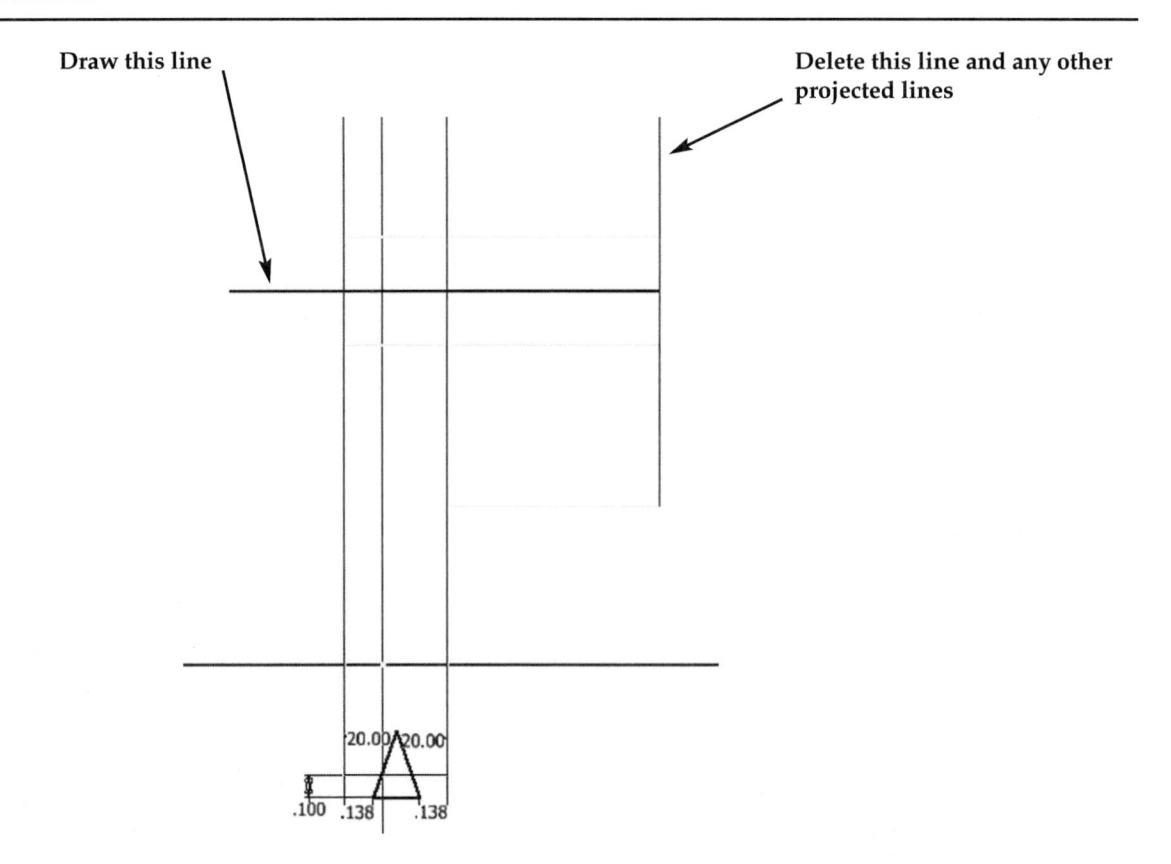

32.   When deleting projected lines, if the option to delete a projected line does not appear in pop-up menu, then the projected line is already deleted, even though it still appears in yellow as shown in Figure 2-28.

FIGURE 2-28

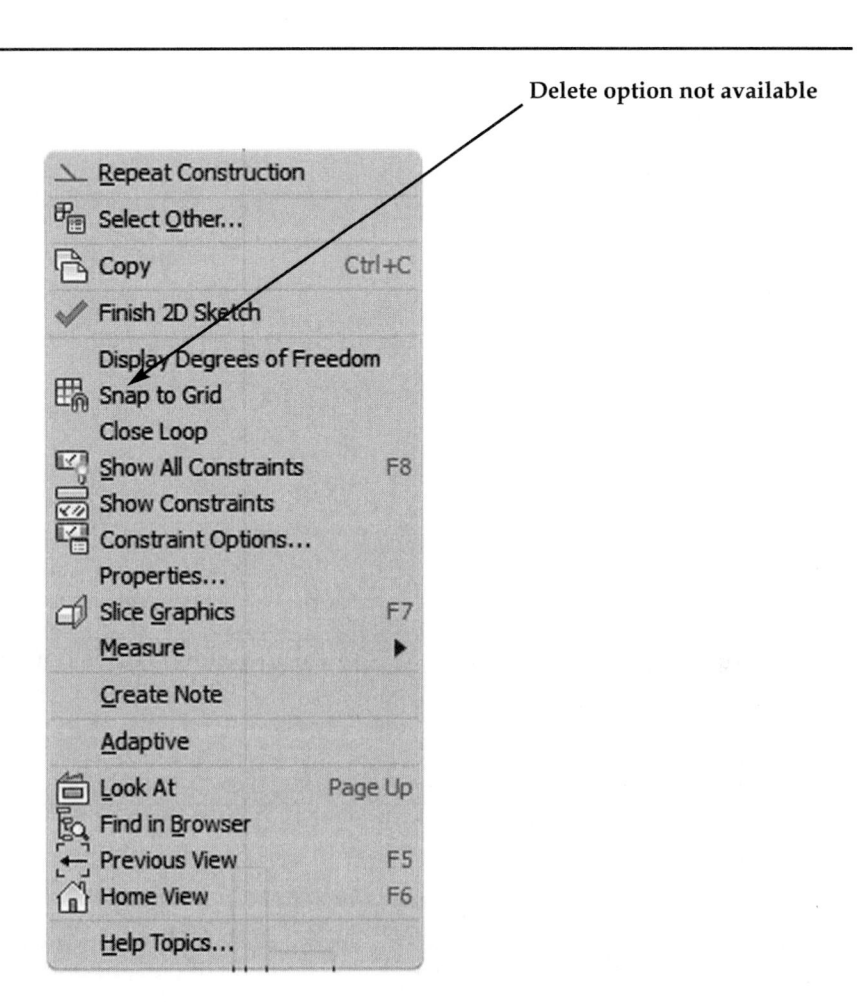

33. Exit out of the Sketch Panel using the **Finish 2D Sketch** command. Use the "Orbit/Rotate" command to rotate the part off to the side. Your screen should look similar to Figure 2-29.

FIGURE 2-29

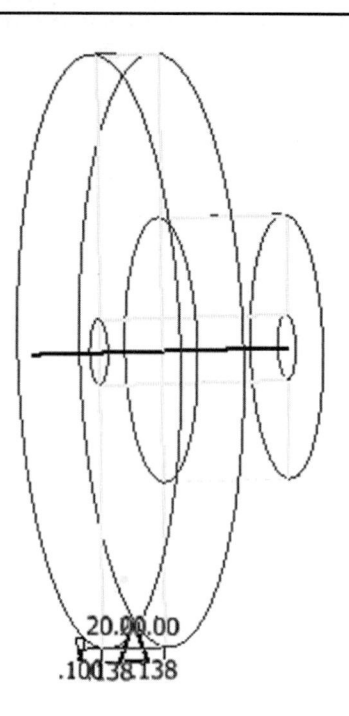

34. Move the cursor to the upper left portion of the screen and left-click on the **3D Model** tab. Left-click on **Revolve** as shown in Figure 2-30.

FIGURE 2-30

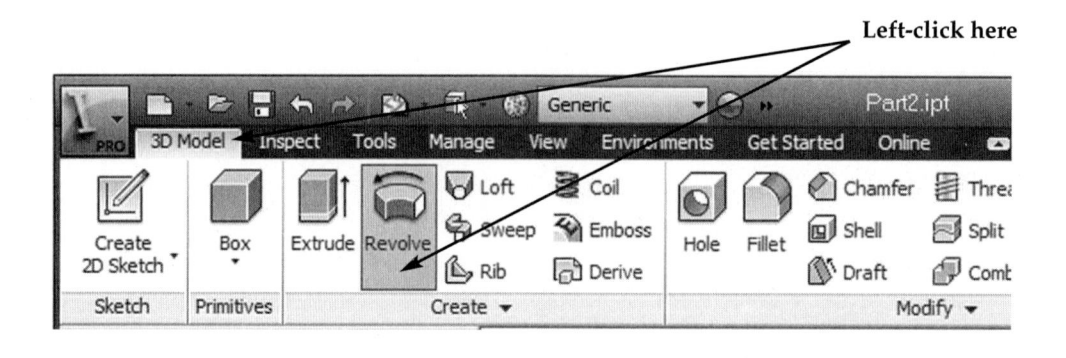

35.   The Revolve dialog box will appear. Left-click on the **Profile** icon and then on the small triangle created while in the Sketch Panel as shown in Figure 2-31.

FIGURE 2-31

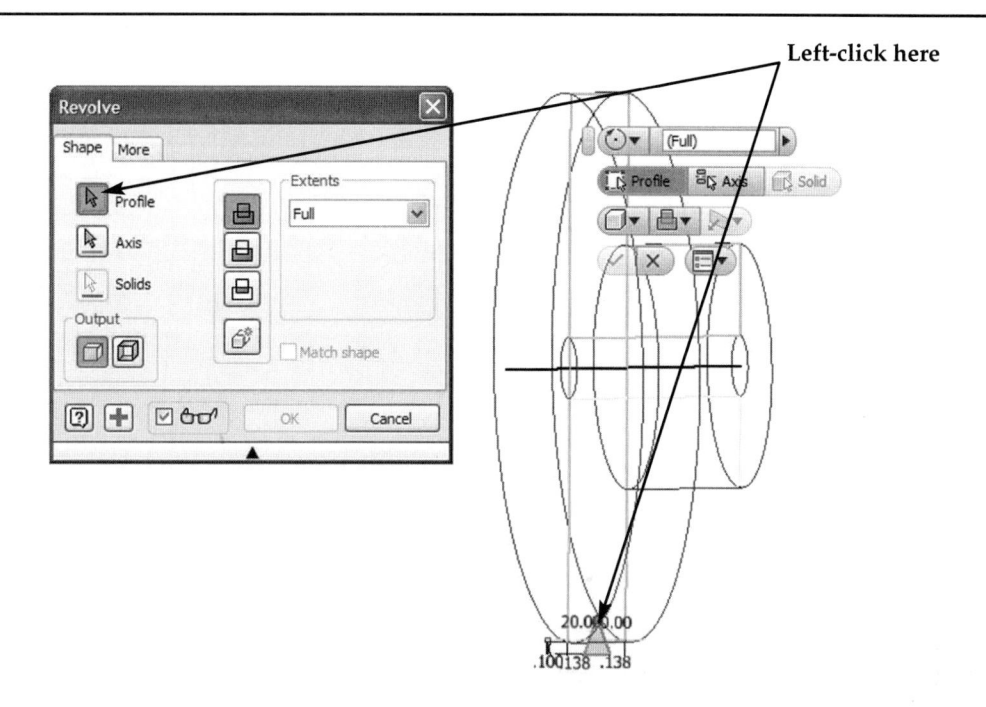

36.   Left-click on the **Axis** icon. Left-click on the axis you created. Left-click on the "Cut" icon. Left-click on **OK** as shown in Figure 2-32.

FIGURE 2-32

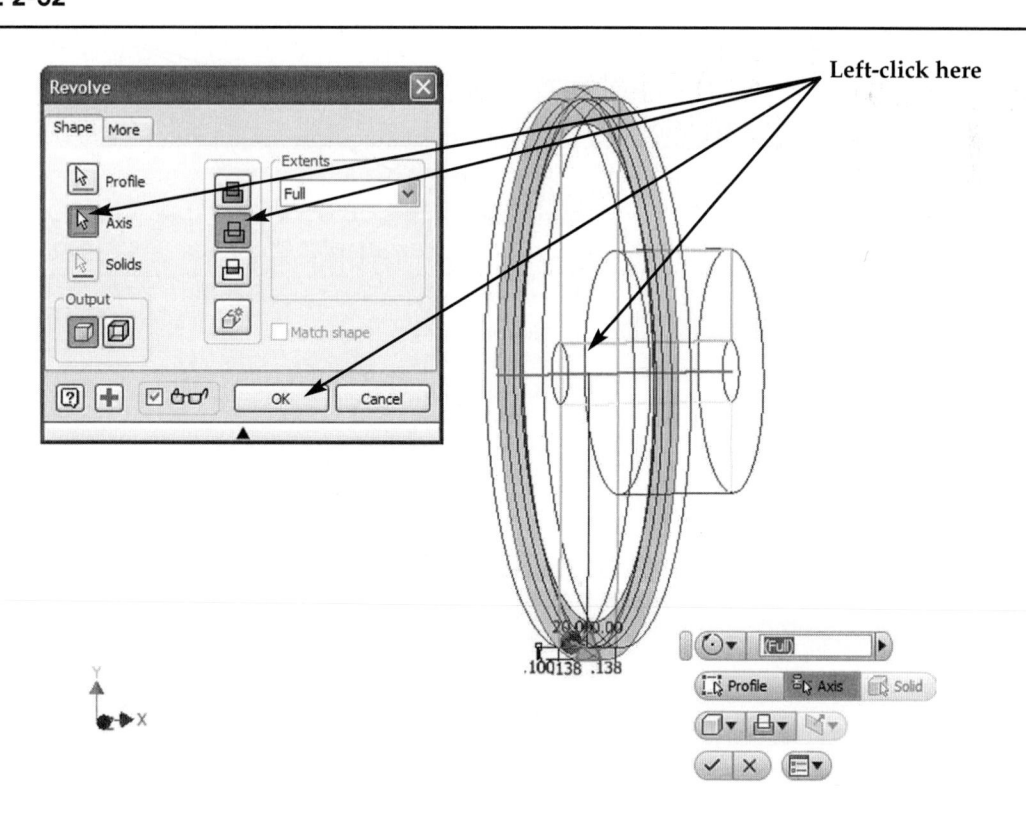

37.   Your screen should look similar to Figure 2-33.

FIGURE 2-33

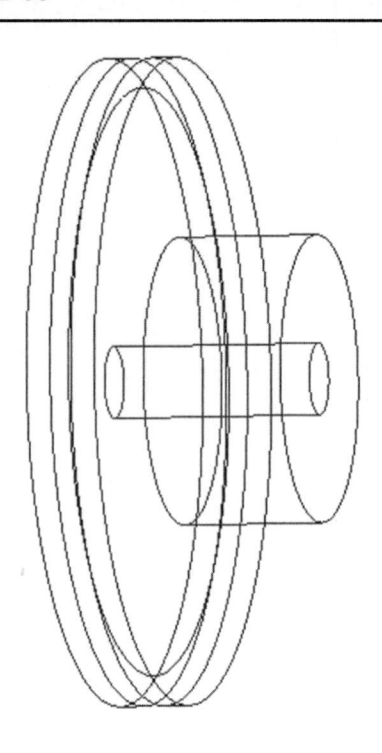

38.   Move the cursor to the upper middle portion of the screen and left-click on the **View** tab. Left-click on the arrow under **Visual Style**. A drop-down menu will appear. Left-click on **Shaded with Edges** as shown in Figure 2-34.

FIGURE 2-34

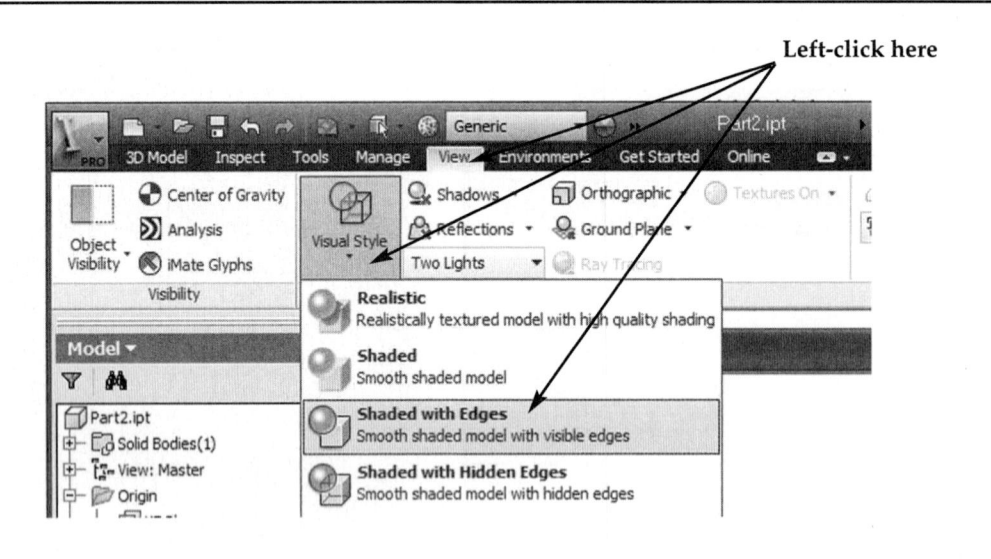

CREATE A HOLE IN THE PART FEATURES PANEL USING THE EXTRUDE COMMAND

39.   Your screen should look similar to Figure 2-35.

FIGURE 2-35

40.   Rotate the part around using the **Home View** command. Move the cursor to the surface of the part, causing the edge to turn red. After the edge becomes red, right-click on the surface as shown in Figure 2-36.

FIGURE 2-36

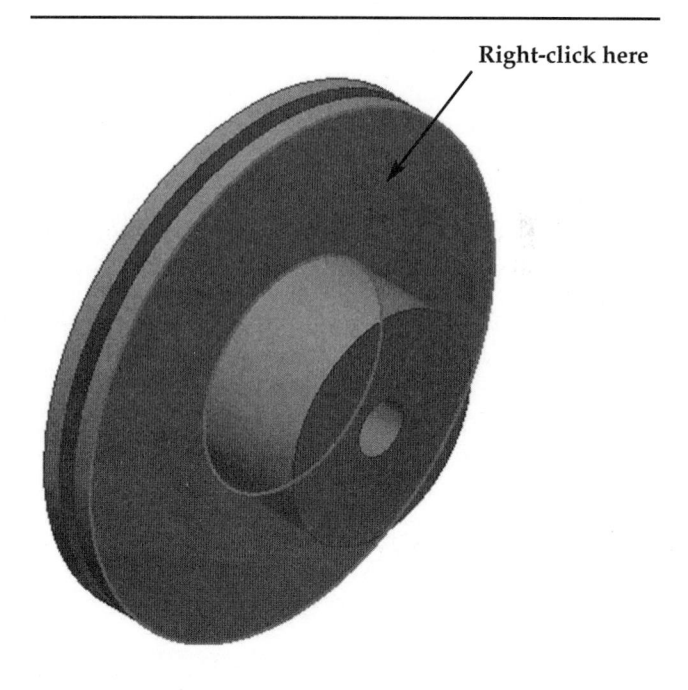

Right-click here

41.   The surface will turn blue and a pop-up menu will appear. Left-click on **New Sketch** as shown in Figure 2-37.

FIGURE 2-37

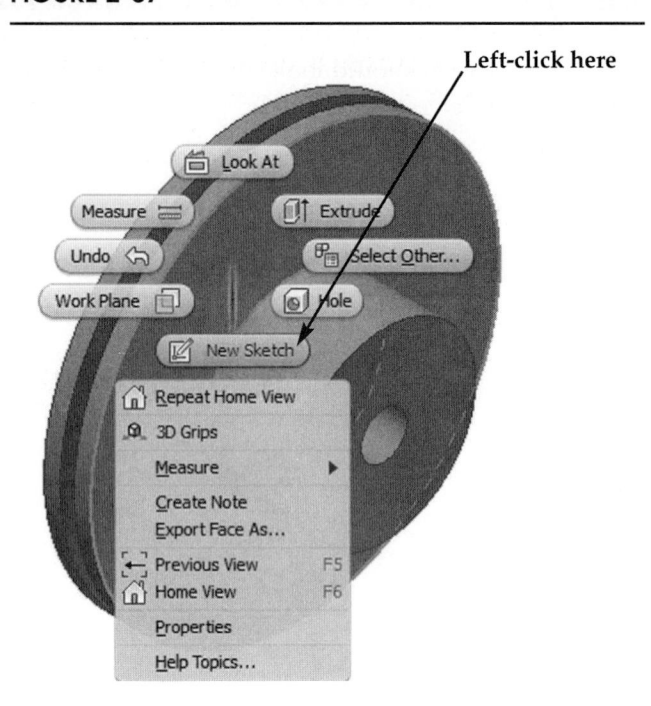

42.   Inventor will begin a new sketch on the selected surface. Inventor will turn the view to the Front. You will need to use the **Home View** command to provide an Isometric View. Your screen should look similar to Figure 2-38.

FIGURE 2-38

43.   To gain a better look at the selected surface, move the cursor to the top center portion of the screen and left-click on the **View** tab. Left-click on the "Face View/Look At" icon. You can also left-click on the "right" portion of the View Cube as shown in Figure 2-39. The View Cube can also be used to rotate the part around (by holding the left mouse button down), similar to using the Orbit/Rotate icon.

FIGURE 2-39

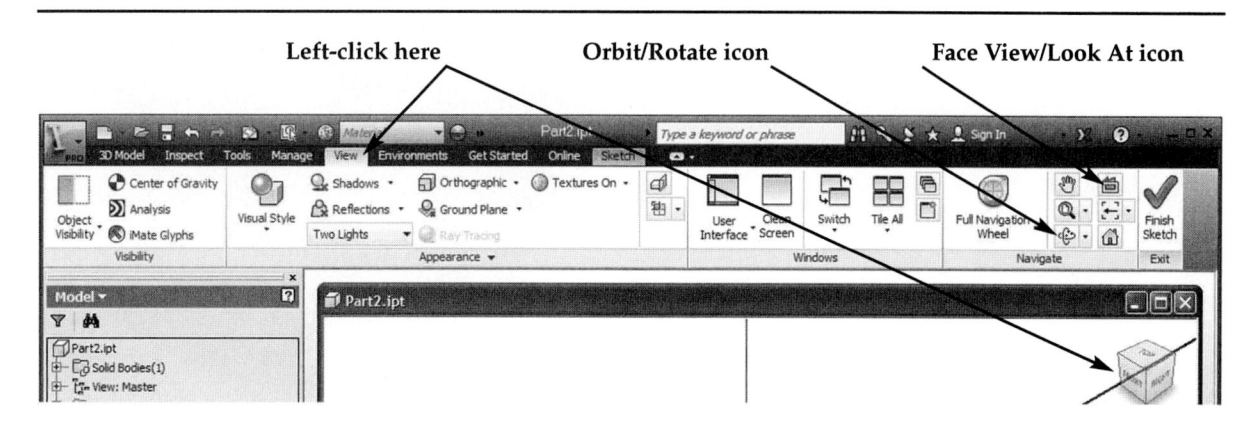

44.   Left-click on the surface where the new sketch will be constructed as shown in Figure 2-40.

FIGURE 2-40

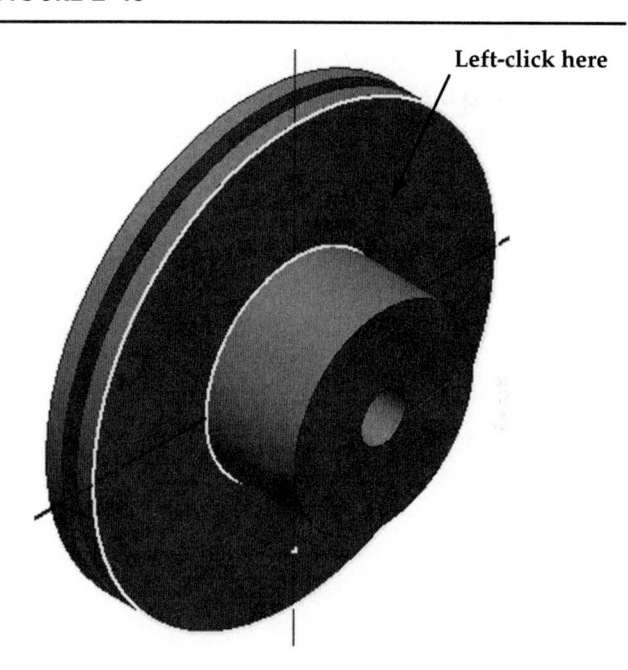

45.   Inventor will rotate the part to provide a perpendicular view of the selected surface as shown in Figure 2-41.

FIGURE 2-41

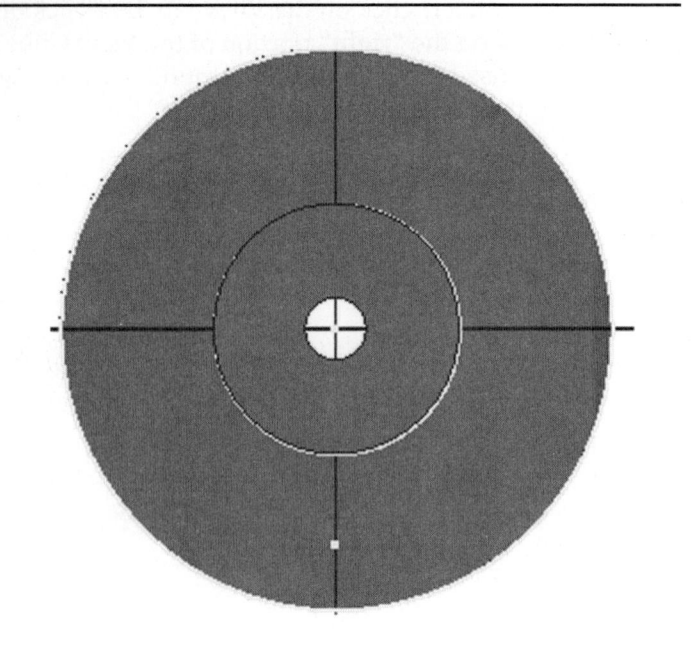

46.   Move the cursor to the upper middle portion of the screen and left-click on the **Sketch** tab. Left-click on **Line** as shown in Figure 2-42.

FIGURE 2-42

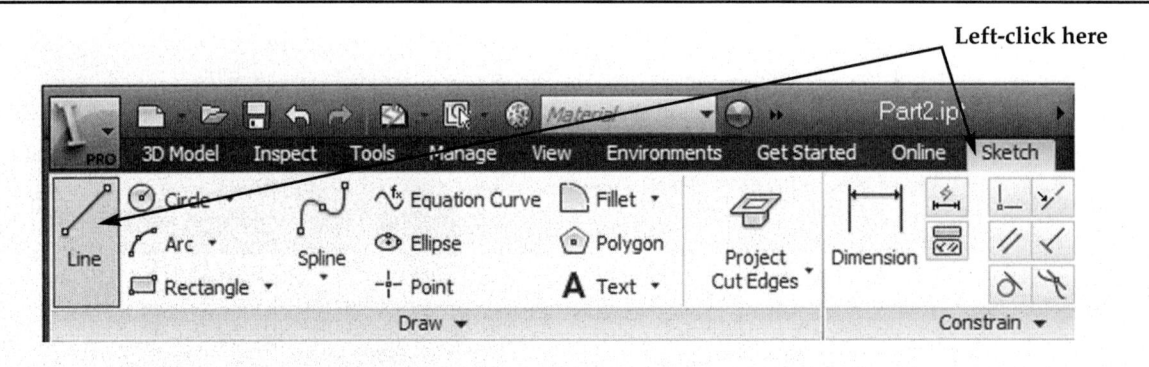

47. Left-click on the center of the hole. Ensure that a green dot appears as shown in Figure 2-43.

FIGURE 2-43

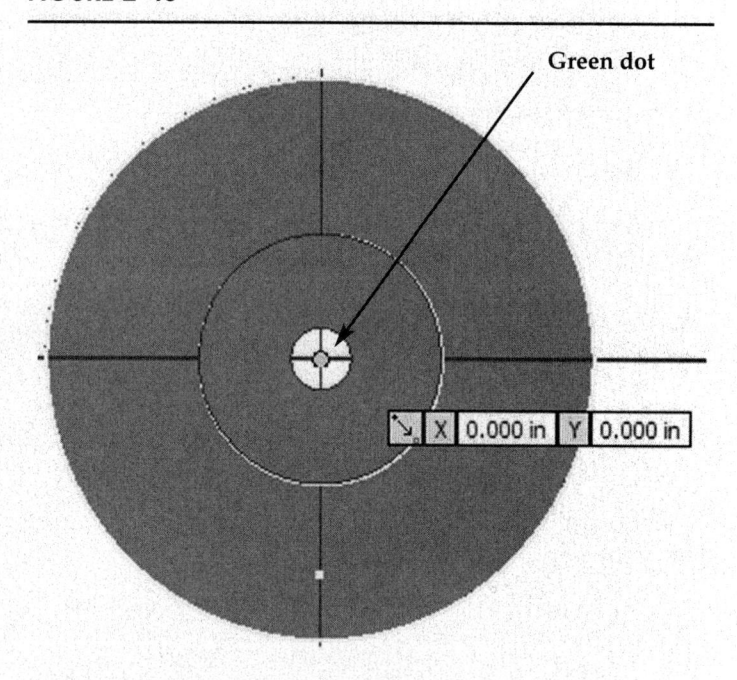

48. Move the cursor straight up and left-click as shown in Figure 2-44.

FIGURE 2-44

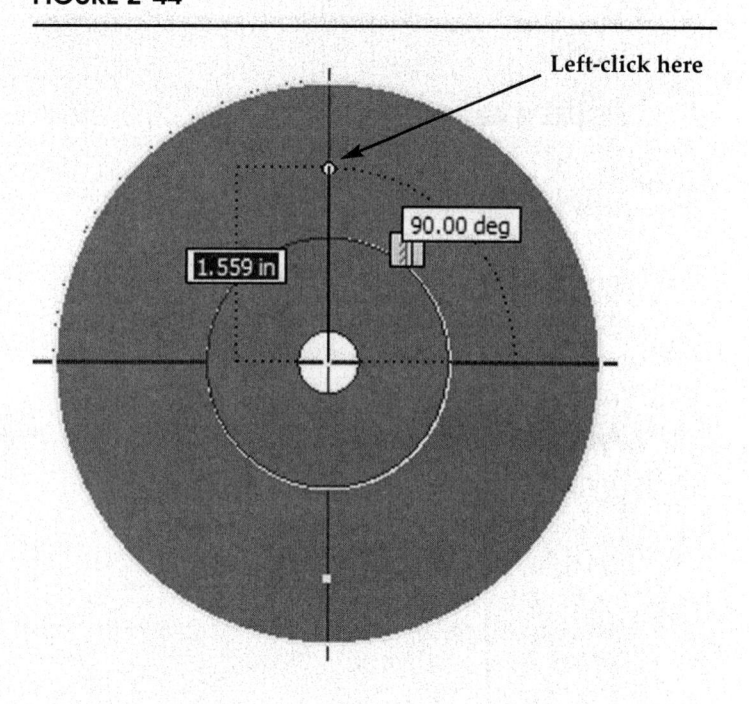

49.   Right-click. A pop-up menu will appear. Left-click on **OK** as shown in Figure 2-45.

FIGURE 2-45

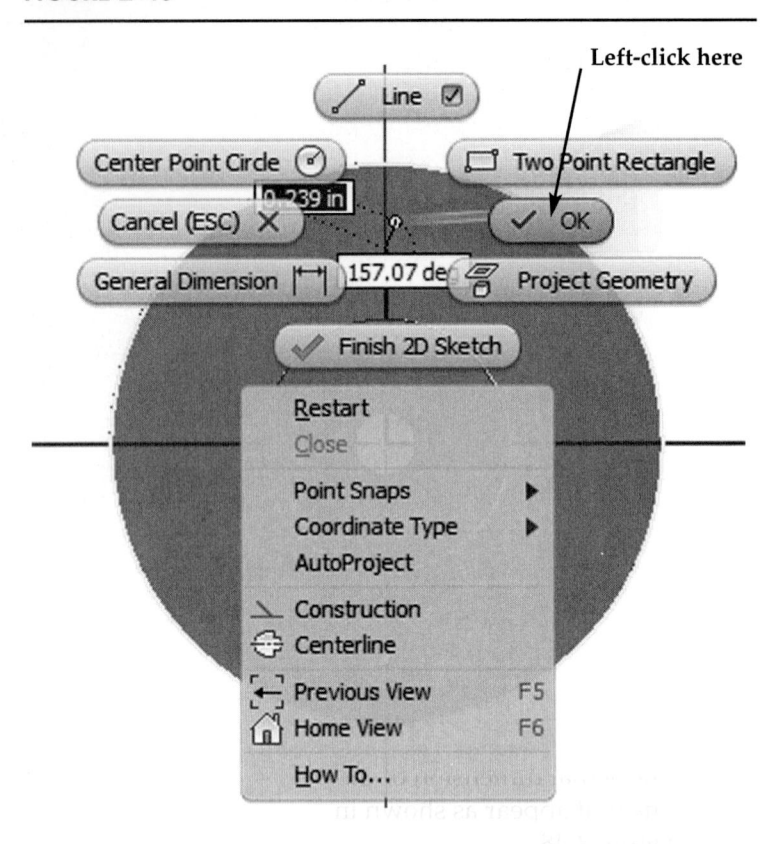

50.   Move the cursor to the upper middle portion of the screen and left-click on **Dimension** as shown in Figure 2-46.

FIGURE 2-46

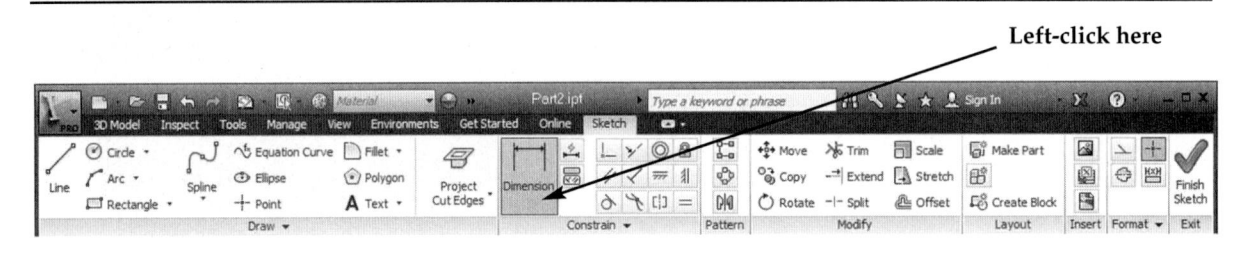

51. After selecting **Dimension**, move the cursor to the line that was just drawn. The line will turn red as shown in Figure 2-47. Select the line by left-clicking anywhere on the line <u>or</u> on each of the end points. To use the end points of the line, move the cursor over one of the end points. A small red square will appear. Left-click once and move the cursor to the other end point. After the red square appears, left-click once. The dimension will be attached to the cursor.

FIGURE 2-47

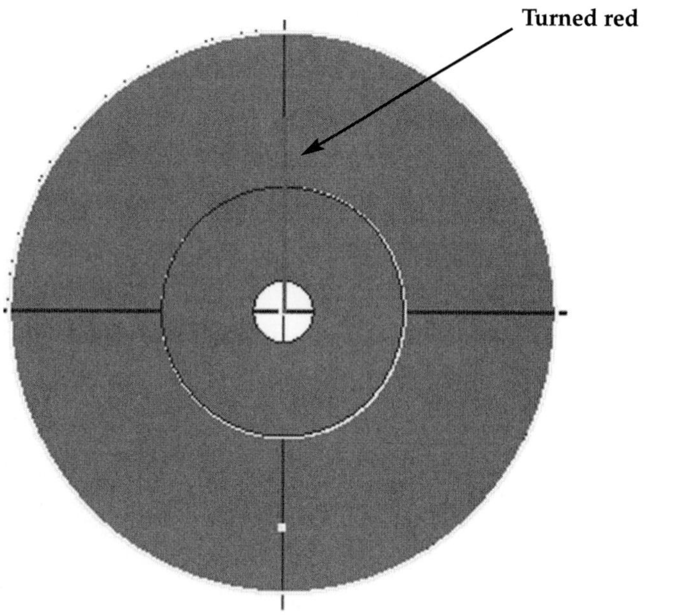

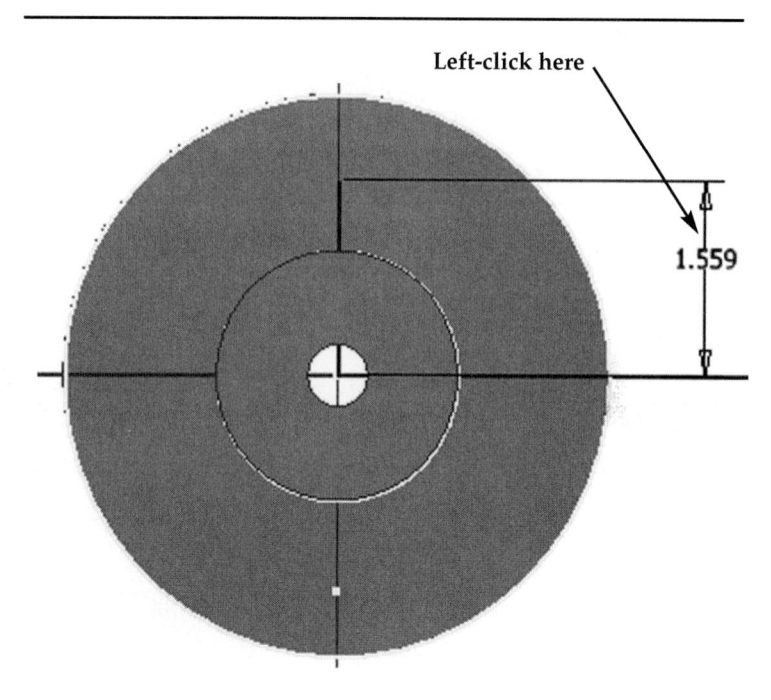

52. Move the cursor to the side. The actual dimension of the line will appear as shown in Figure 2-48.

FIGURE 2-48

53. Move the cursor to where the dimension will be placed and left-click once. While the dimension is still in red, left-click once. The Edit Dimension dialog box will appear as shown in Figure 2-49.

FIGURE 2-49

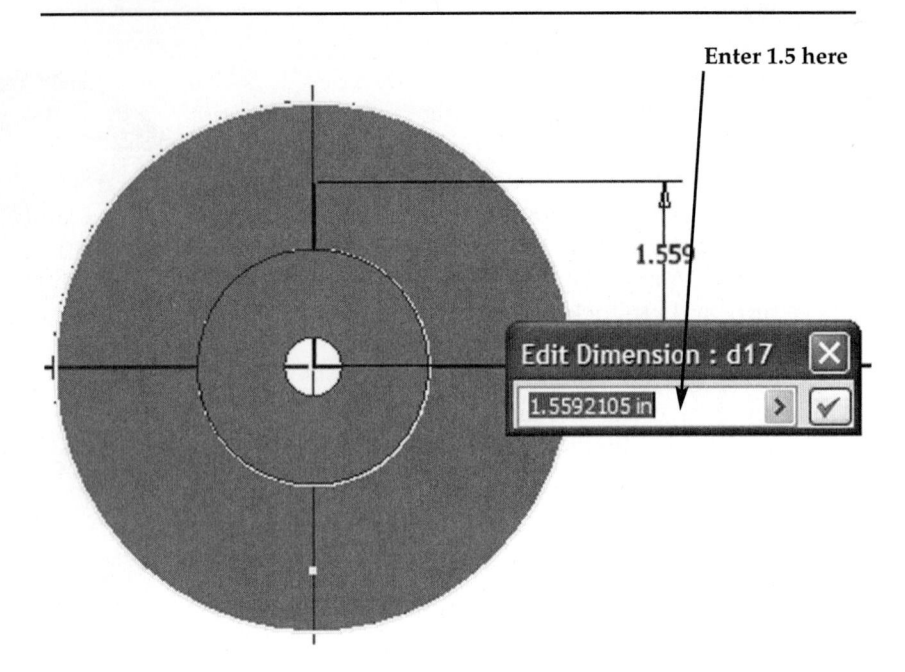

54. To edit the dimension, enter **1.5** in the Edit Dimension dialog box (while the current dimension is highlighted) and press **Enter** on the keyboard.

55. The dimension of the line will become 1.5 inches as shown in Figure 2-50. Use the "Zoom" icons to zoom out if necessary.

FIGURE 2-50

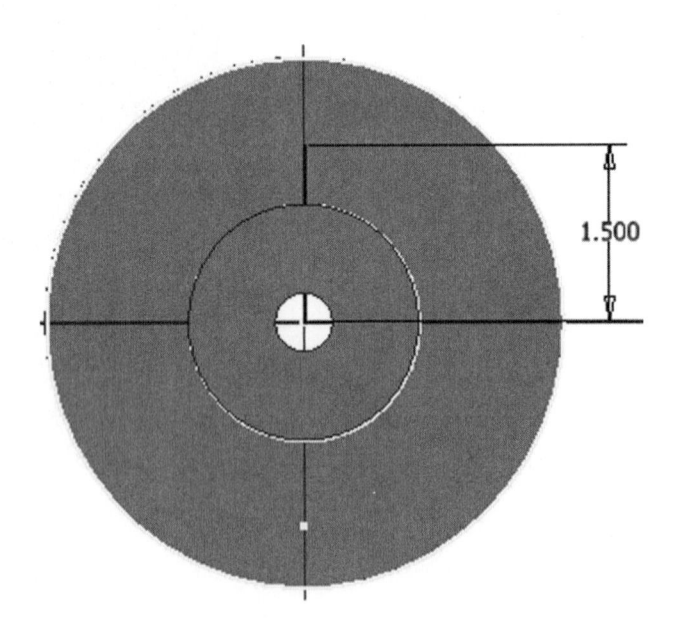

56. Move the cursor to the upper left portion of the screen and left-click on **Circle** as shown in Figure 2-51.

FIGURE 2-51

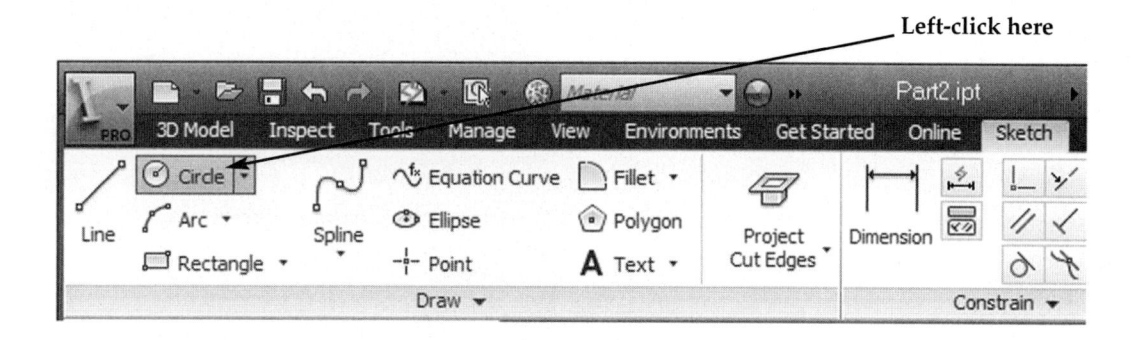

57. Left-click on the end point of the line as shown in Figure 2-52.

FIGURE 2-52

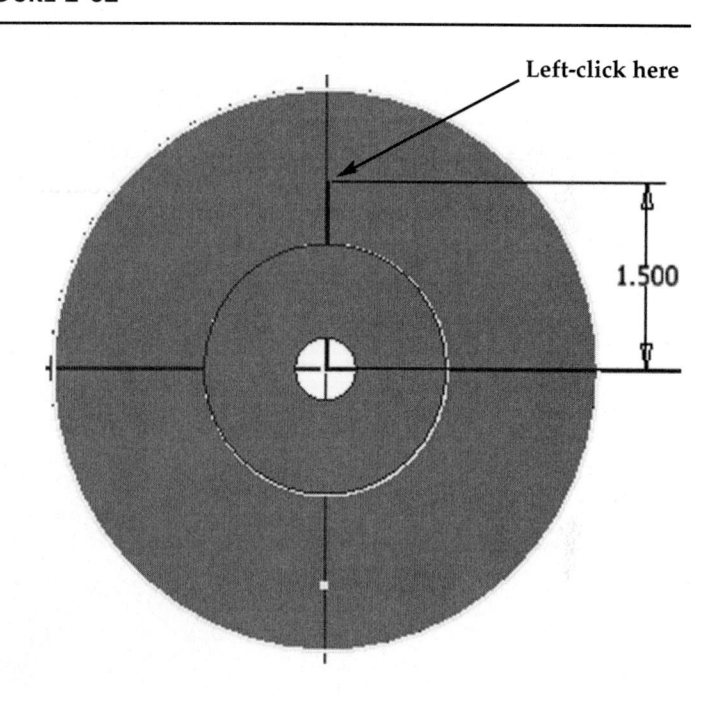

58. Move the cursor out to the right to create a circle as shown in Figure 2-53.

FIGURE 2-53

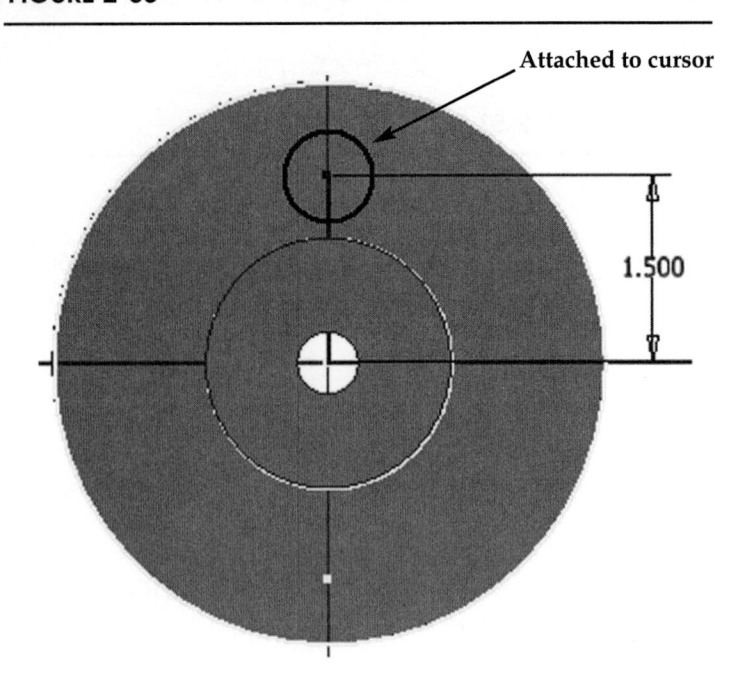

59. Left-click as shown in Figure 2-54. Press the **Esc** key once.

FIGURE 2-54

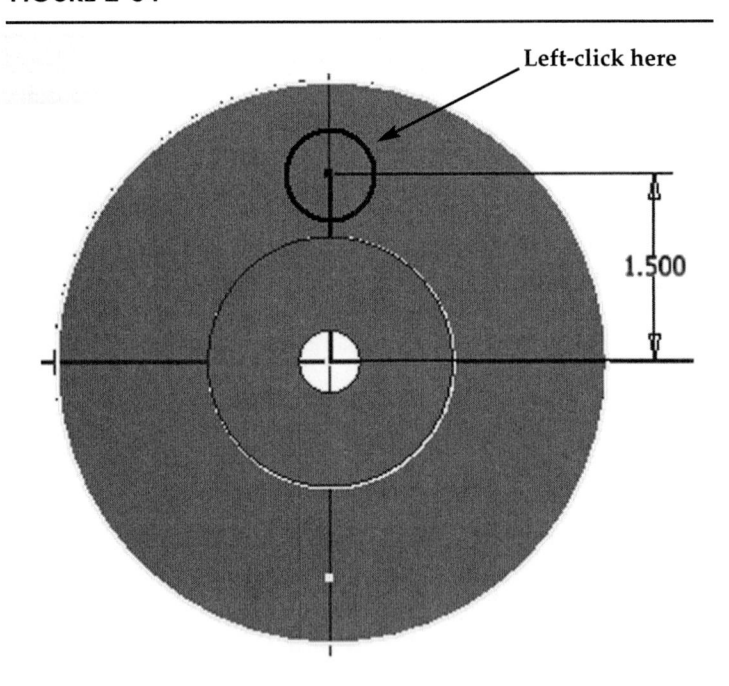

60. Move the cursor to the upper middle portion of the screen and left-click on **Dimension** as shown in Figure 2-55.

FIGURE 2-55

Left-click here

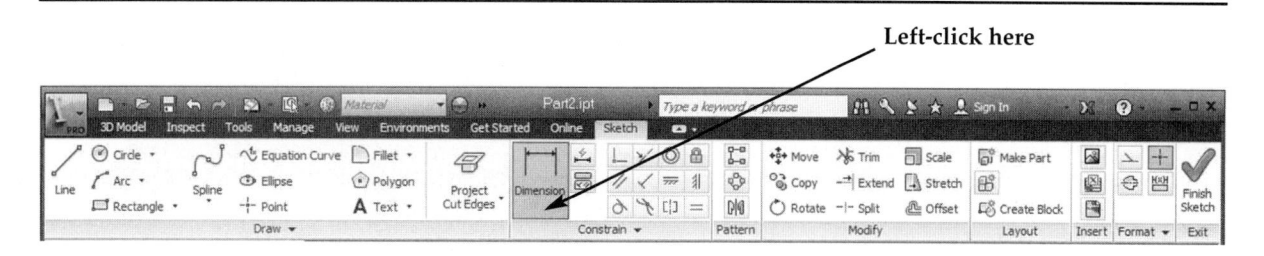

61. After selecting **Dimension**, move the cursor to the edge of the circle that was just drawn. The circle will turn red. Select the circle by left-clicking anywhere on the circle (not the center) as shown in Figure 2-56. The dimension will be attached to the cursor.

FIGURE 2-56

Left-click here

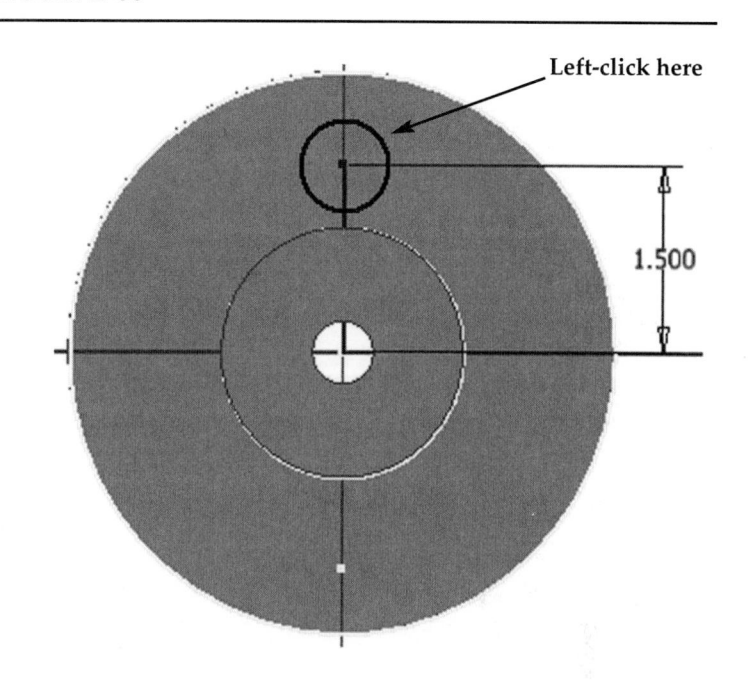

1.500

62. Move the cursor to the side. The actual dimension of the line will appear as shown in Figure 2-57.

FIGURE 2-57

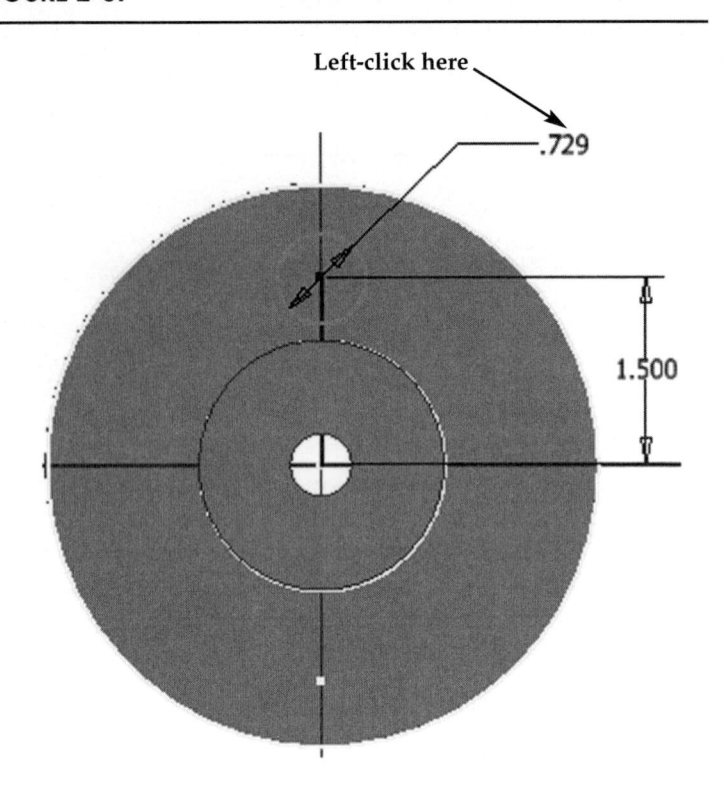

63. Move the cursor to where the dimension will be placed and left-click once. While the dimension is still in red, left-click once. The Edit Dimension dialog box will appear as shown in Figure 2-58.

FIGURE 2-58

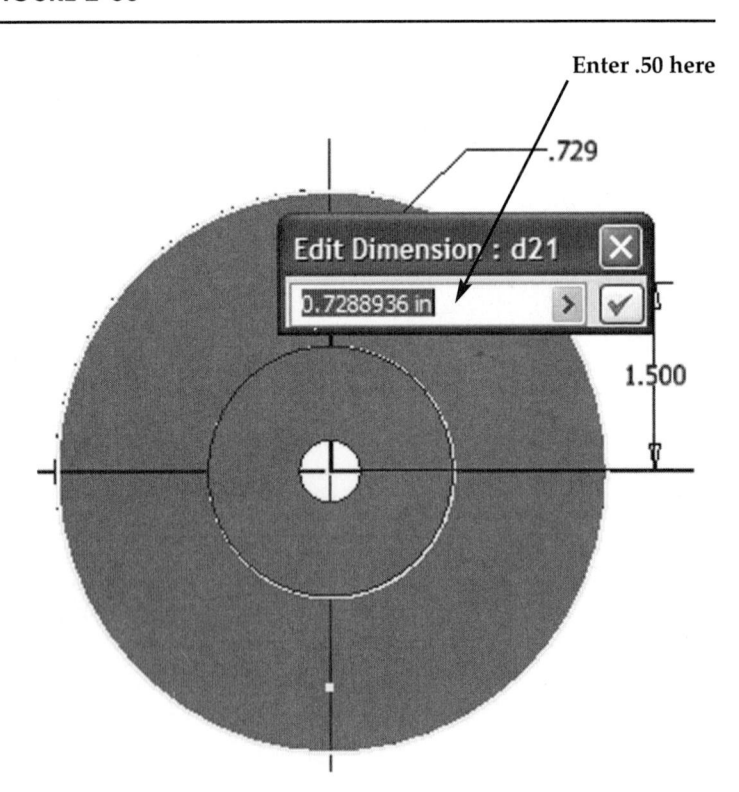

64. To edit the dimension, enter **.50** in the Edit Dimension dialog box (while the current dimension is highlighted) and press **Enter** on the keyboard. Press the **Esc** key once or twice.

65. The dimension of the line will become .50 inches as shown in Figure 2-59. Use the "Zoom" icons to zoom out if necessary.

FIGURE 2-59

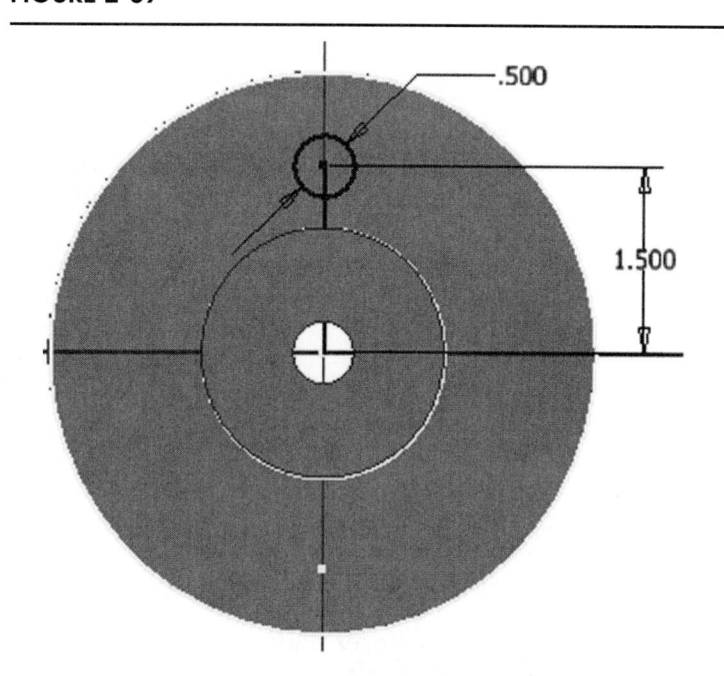

66. Move the cursor to the line that was used to locate the center of the circle. The line will turn red as shown in Figure 2-60.

FIGURE 2-60

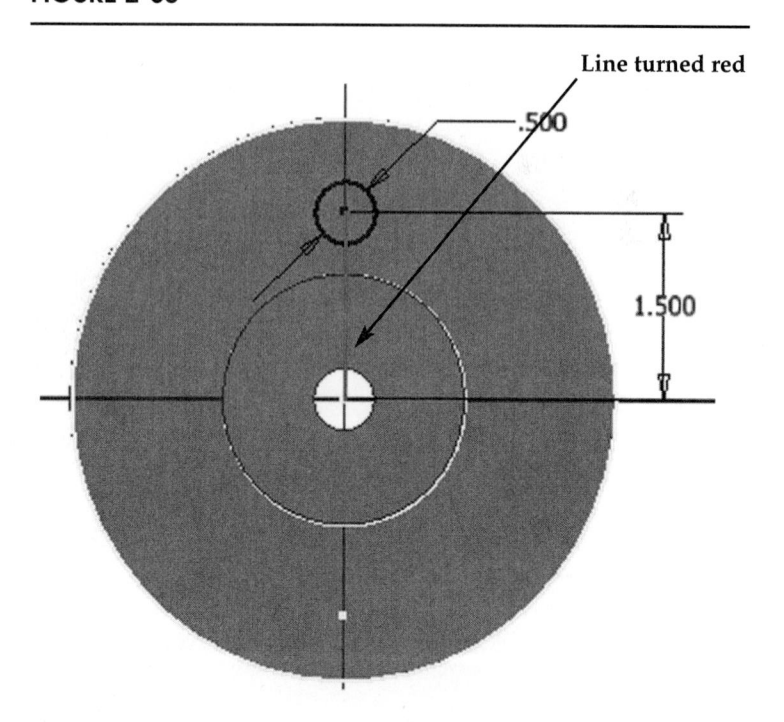

67. Right-click on the line after it turns red. A pop-up menu will appear. Left-click on **Delete** as shown in Figure 2-61.

FIGURE 2-61

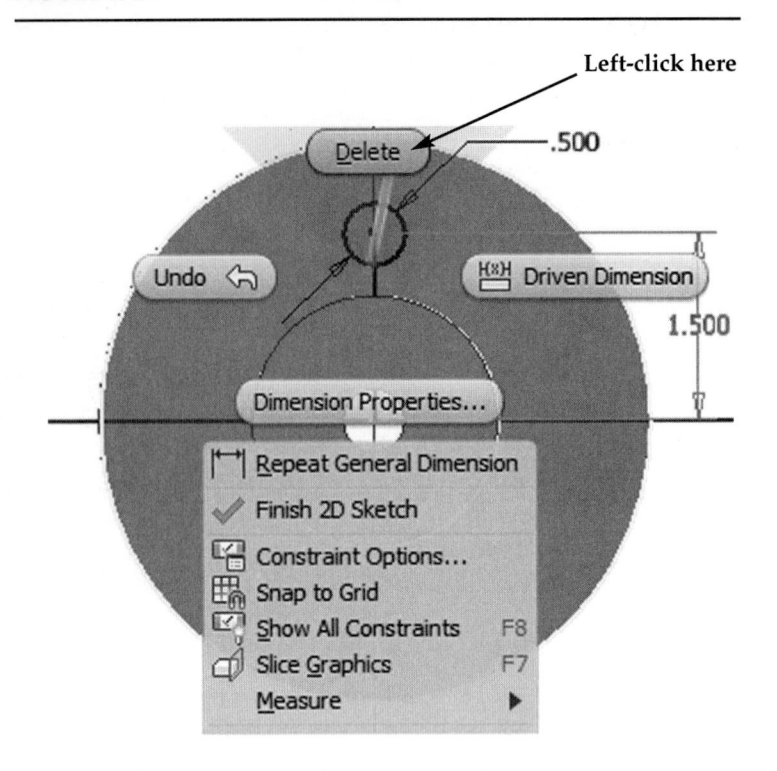

68. Press **Esc** once or twice, or right-click around the drawing. A pop-up menu will appear. Left click on **OK** or **Cancel (Esc)** as shown in Figure 2-62.

FIGURE 2-62

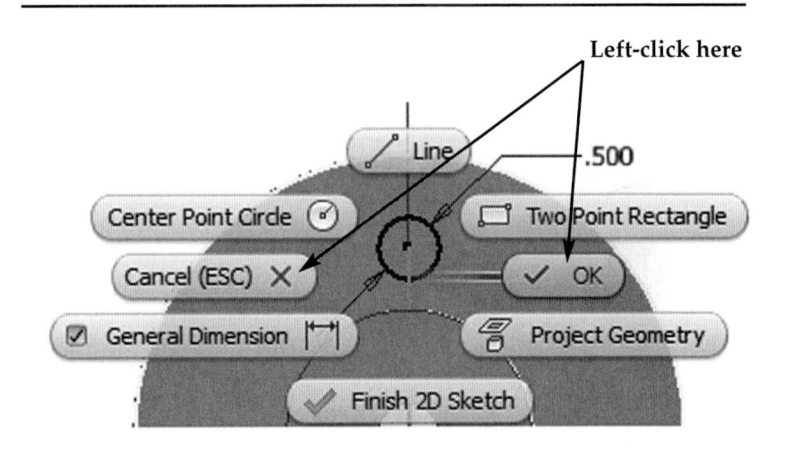

69. After you have verified that no commands are active, right-click anywhere on the sketch. A pop-up menu will appear. Left-click on **Finish 2D Sketch** as shown in Figure 2-63.

FIGURE 2-63

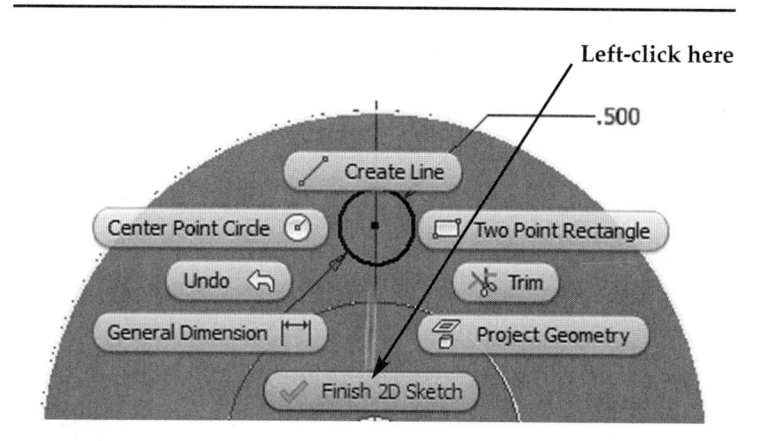

70. Inventor is now out of the Sketch Panel and into the Part Features Panel. Notice that the commands at the top of the screen are now different. Your screen should look similar to Figure 2-64.

FIGURE 2-64

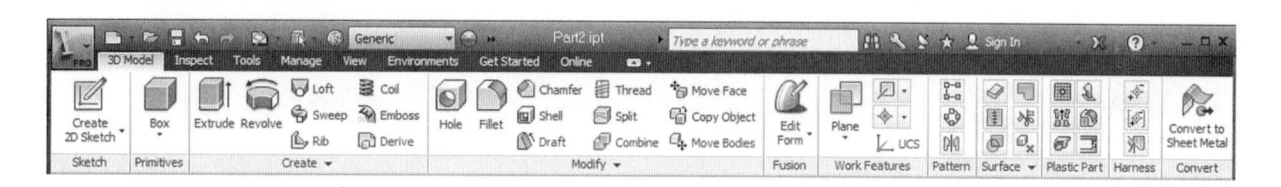

71.   Right-click around the part. A
       pop-up menu will appear. Left-
       click on **Home View** as shown in
       Figure 2-65.

FIGURE 2-65

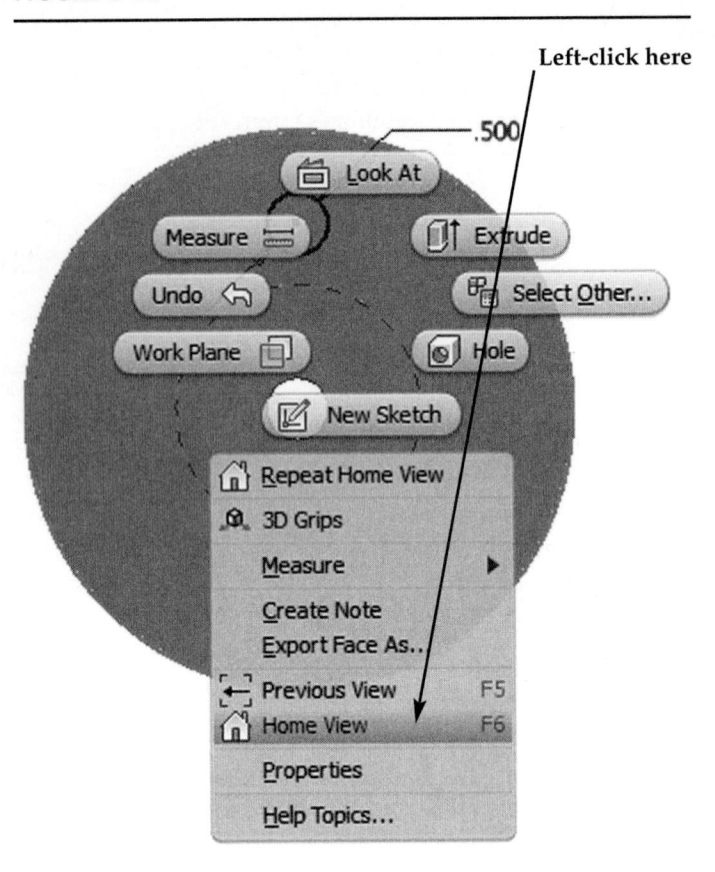

72.   The view will become Isometric as
       shown in Figure 2-66.

FIGURE 2-66

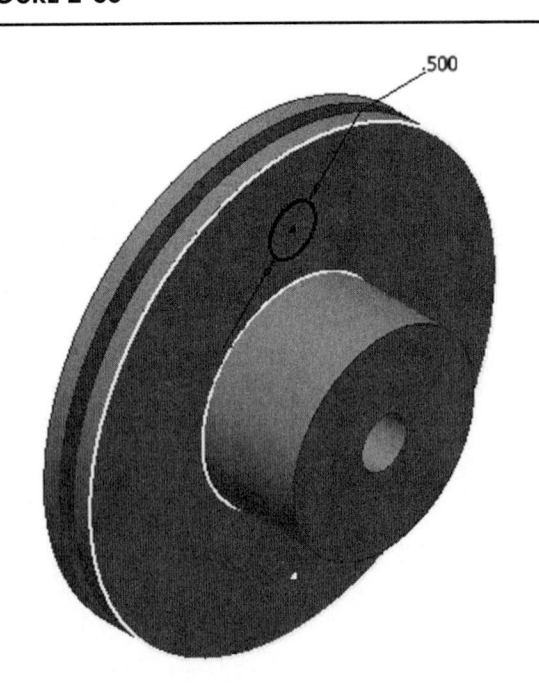

73. Move the cursor to the upper left portion of the screen and left-click on **Extrude**. The Extrude dialog box will appear as shown in Figure 2-67.

FIGURE 2-67

Left-click here

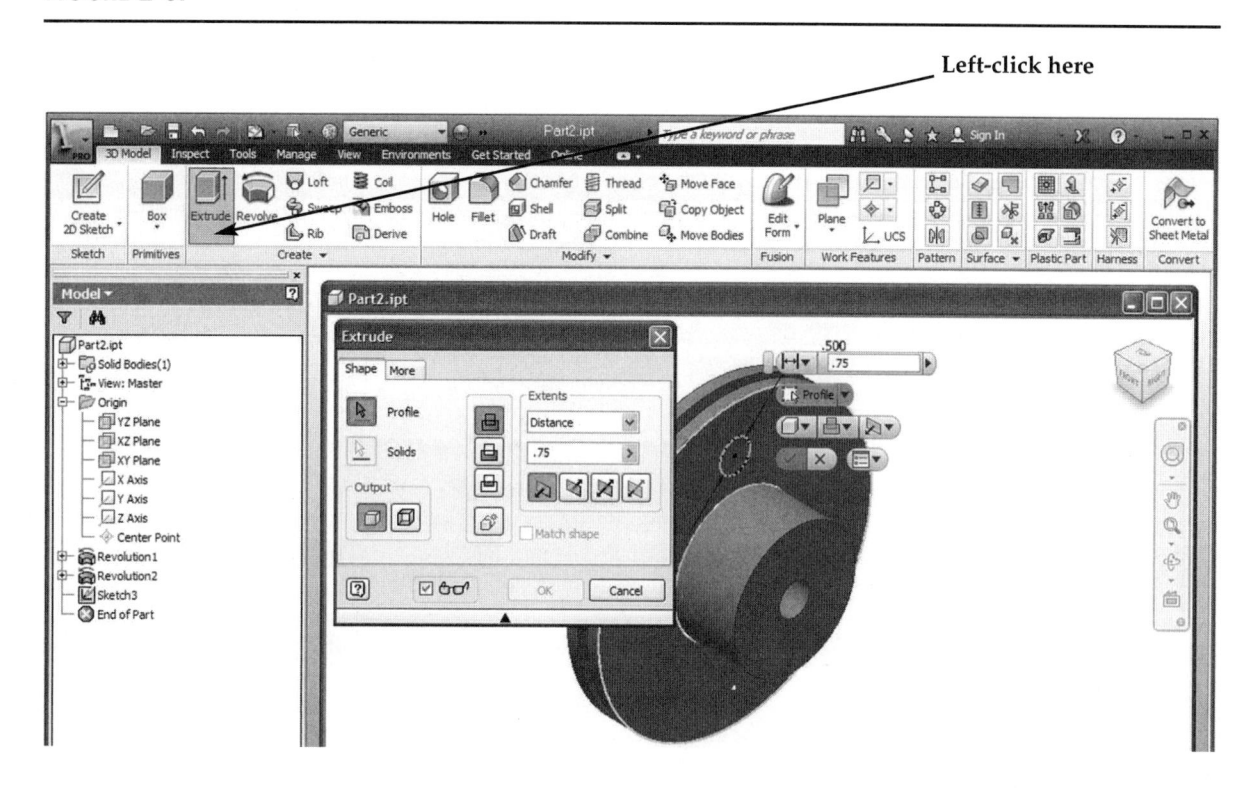

74. Move the cursor to the inside of the circle, causing it to turn red as shown in Figure 2-68.

FIGURE 2-68

Move cursor here

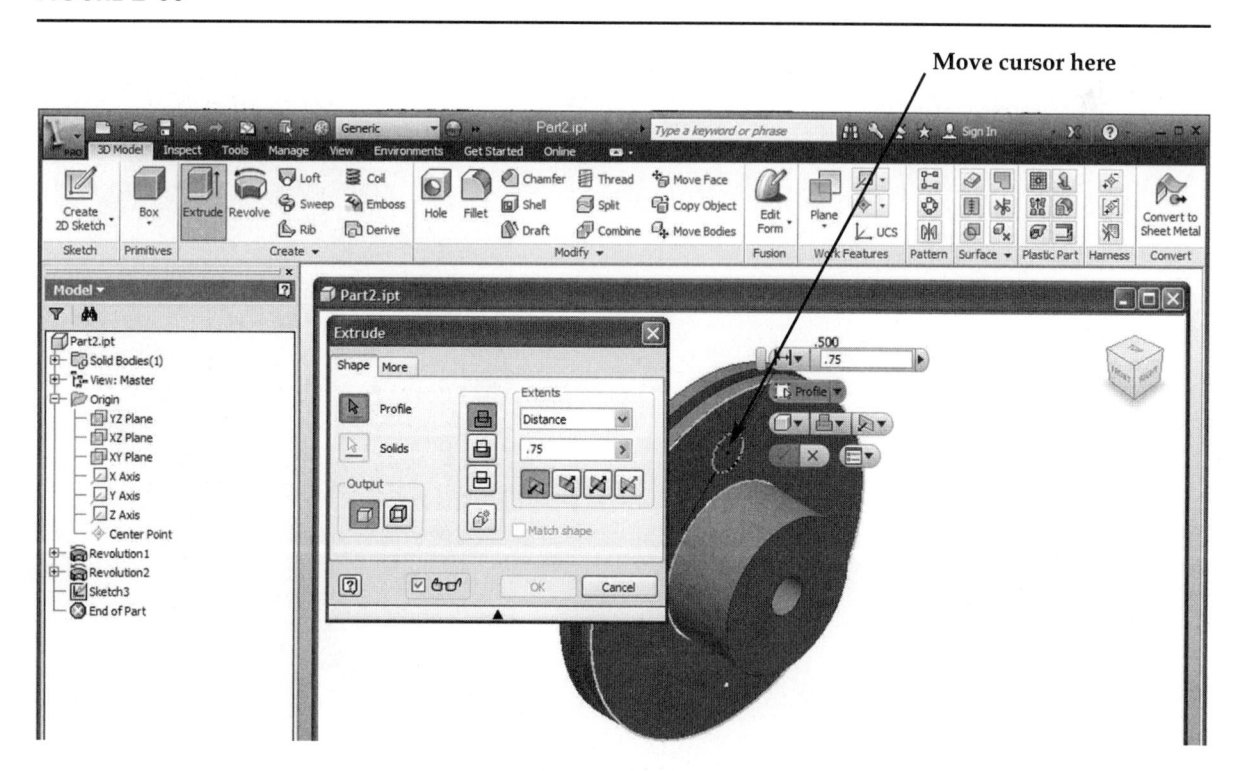

75. After the hole turns red, left-click once. Select the "Cut" icon in the Extrude dialog box. Select the "Direction" icon to ensure the extrusion occurs in the right direction and left-click on **OK** as shown in Figure 2-69.

FIGURE 2-69

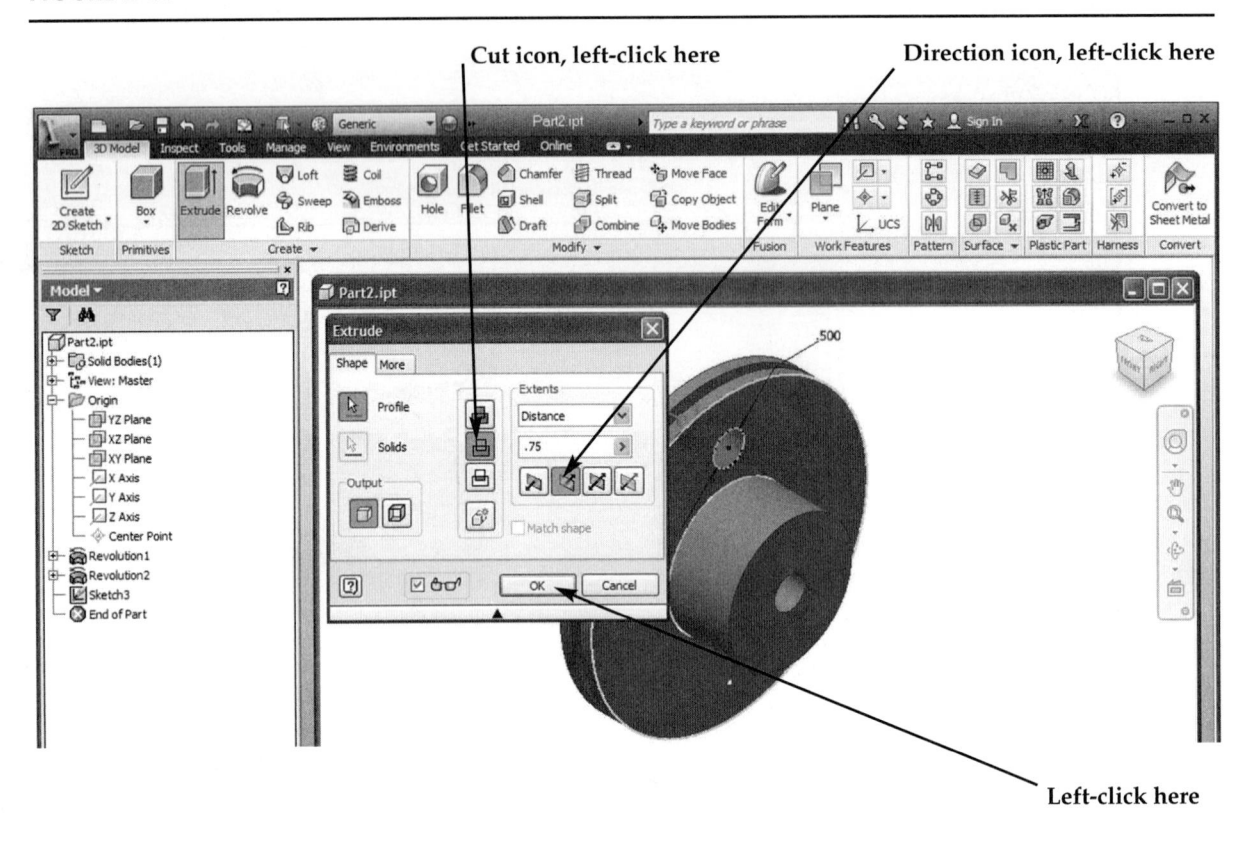

76. Your screen should look similar to Figure 2-70.

FIGURE 2-70

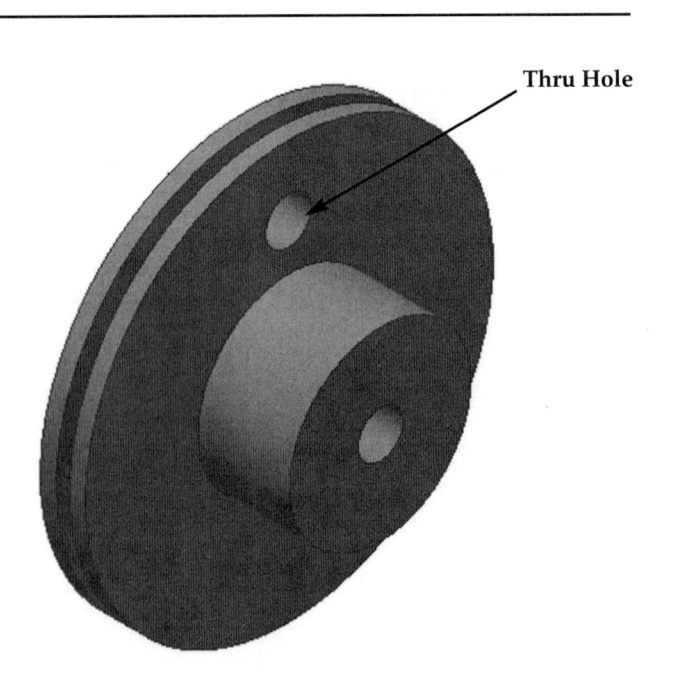

CREATE A SERIES OF HOLES USING THE CIRCULAR PATTERN COMMAND

77.   Move the cursor to the upper right portion of the screen and left-click on the "Circular" icon. Earlier versions of Inventor have the Circular Pattern icon located at the middle left portion of the screen with text on the icon. If this is the case, you may have to scroll down to see the command. The Circular Pattern dialog box will appear as shown in Figure 2-71.

FIGURE 2-71

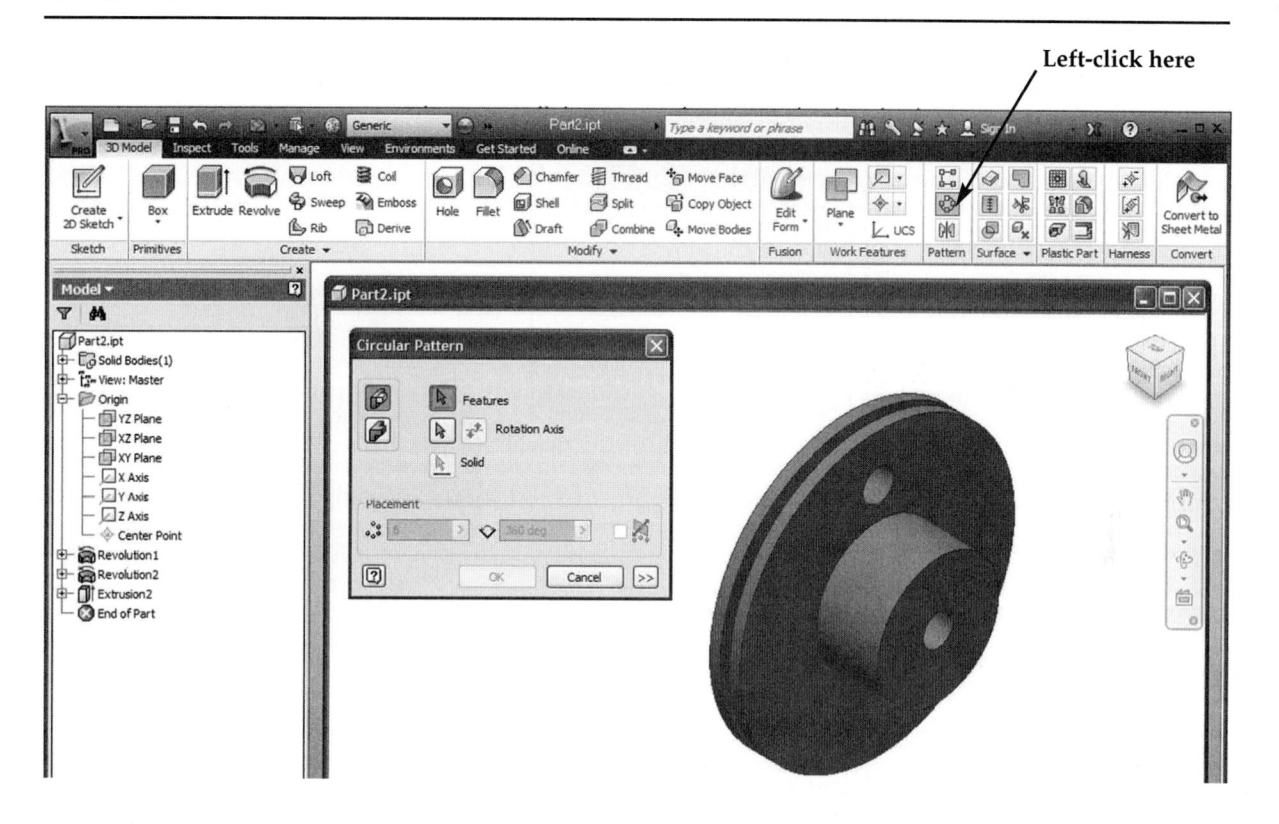

78.   Move the cursor to the center of the hole, causing red dashed lines to appear and left-click once. The part must be displayed in Home/Isometric View for Inventor to find the hole as shown in Figure 2-72.

FIGURE 2-72

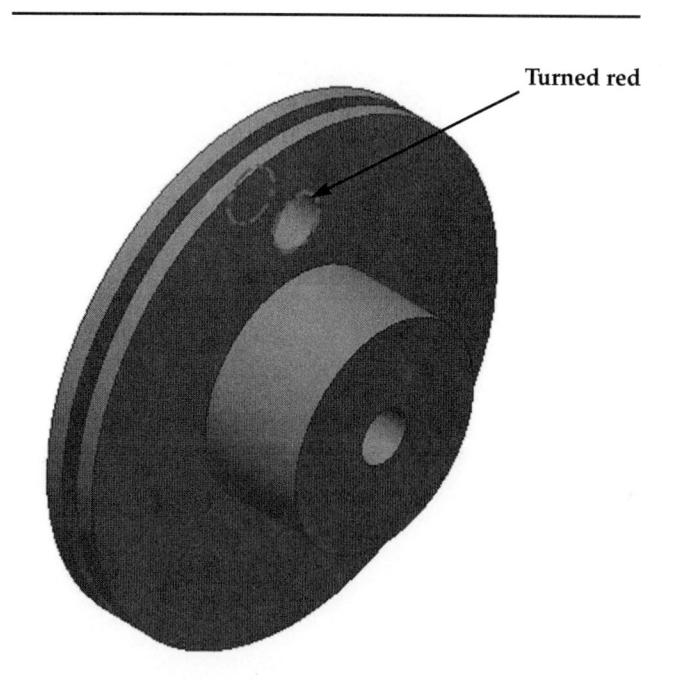

79.   Left-click on **Rotation Axis** in the dialog box as shown in Figure 2-73.

FIGURE 2-73

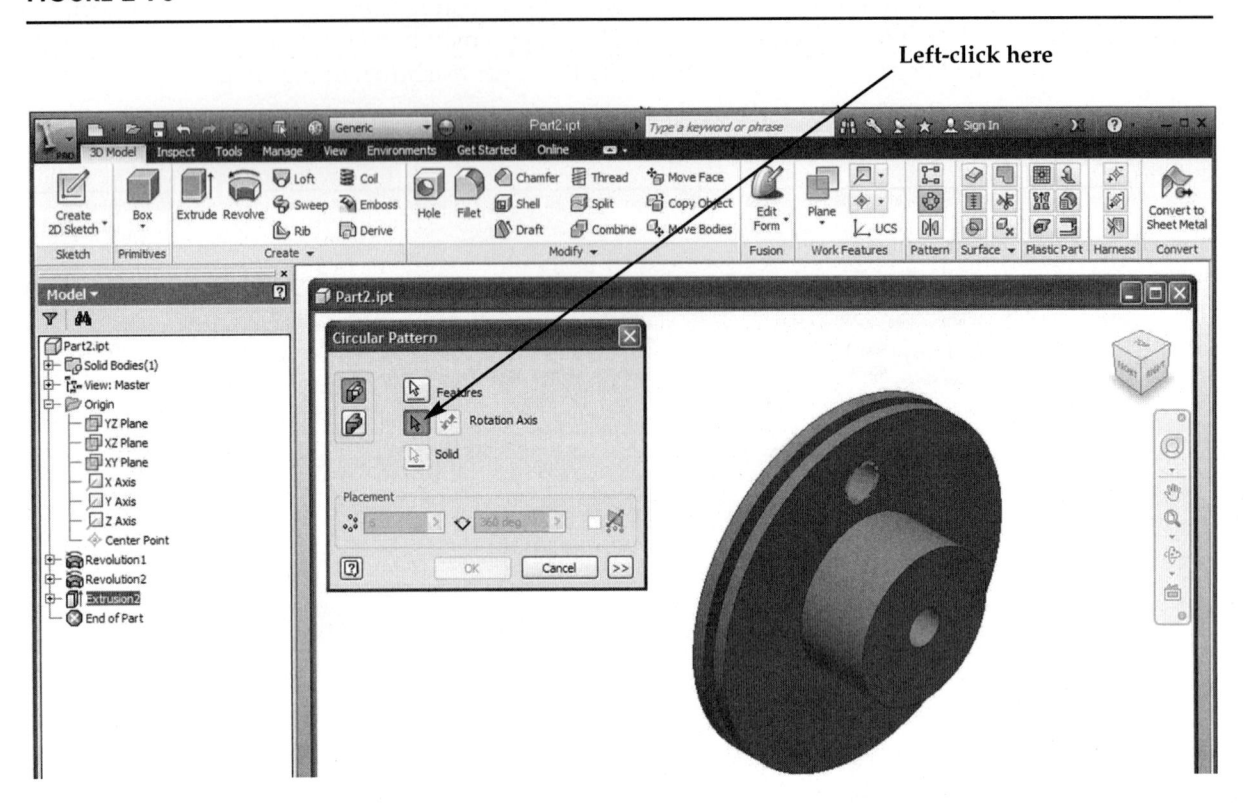

80.   Move the cursor to the edge of the part. The edges will turn red as shown in Figure 2-74.

FIGURE 2-74

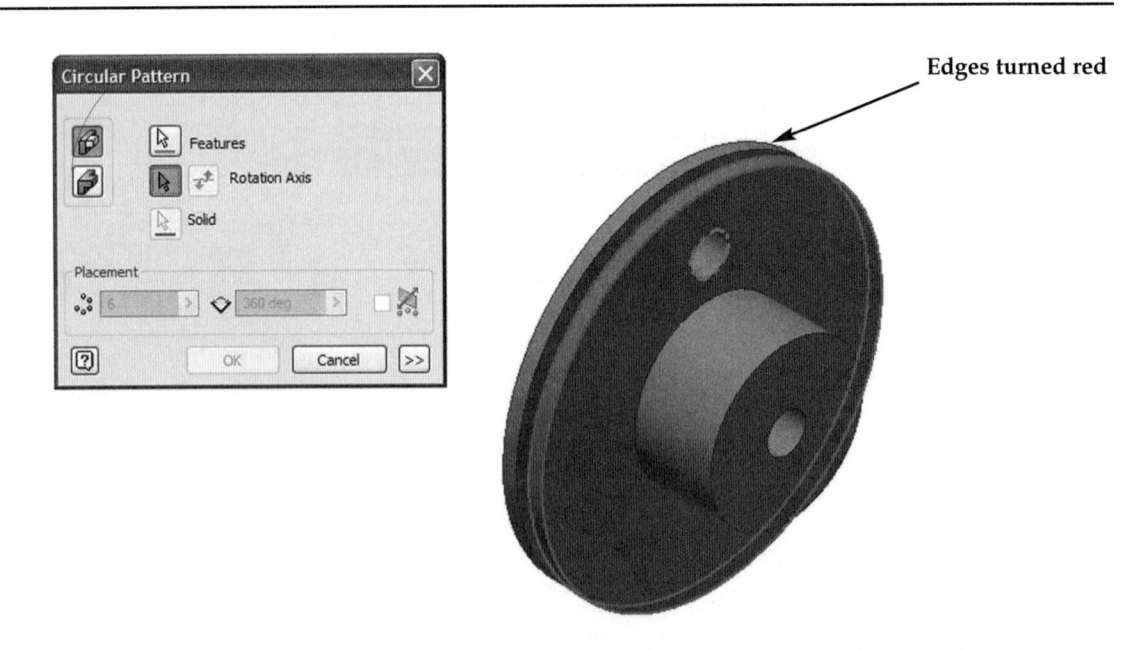

81.   After the edges turn red, left-click once. Inventor will provide a preview of the hole pattern as shown in Figure 2-75.

FIGURE 2-75

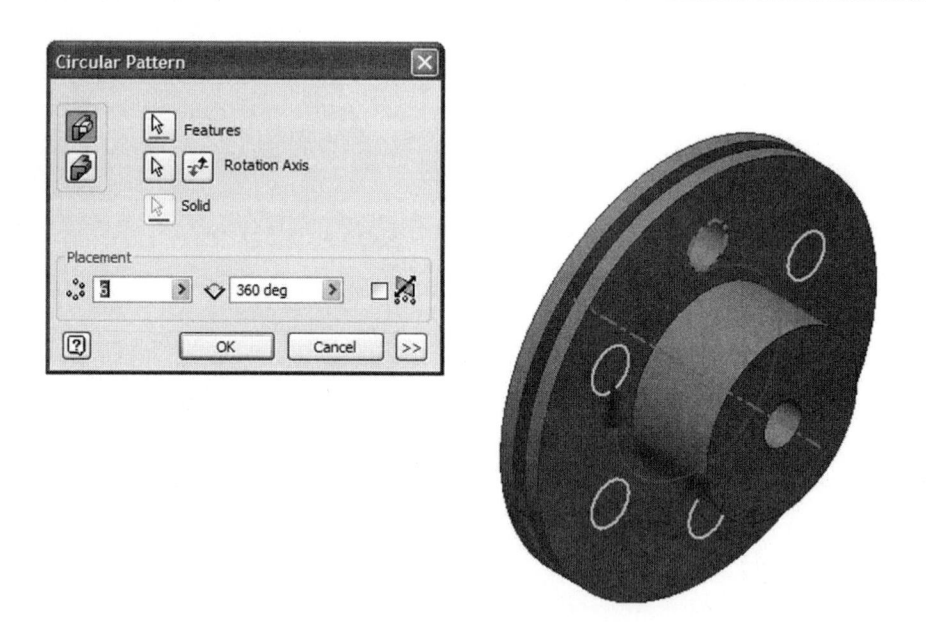

82.   There are options in the Circular Pattern dialog box that are used for dictating the number of holes to be produced and the number of degrees between the holes. Verify that **6** is displayed for the number of holes. Verify that **360 deg** is displayed for the number of degrees. Left-click on **OK** as shown in Figure 2-76.

FIGURE 2-76

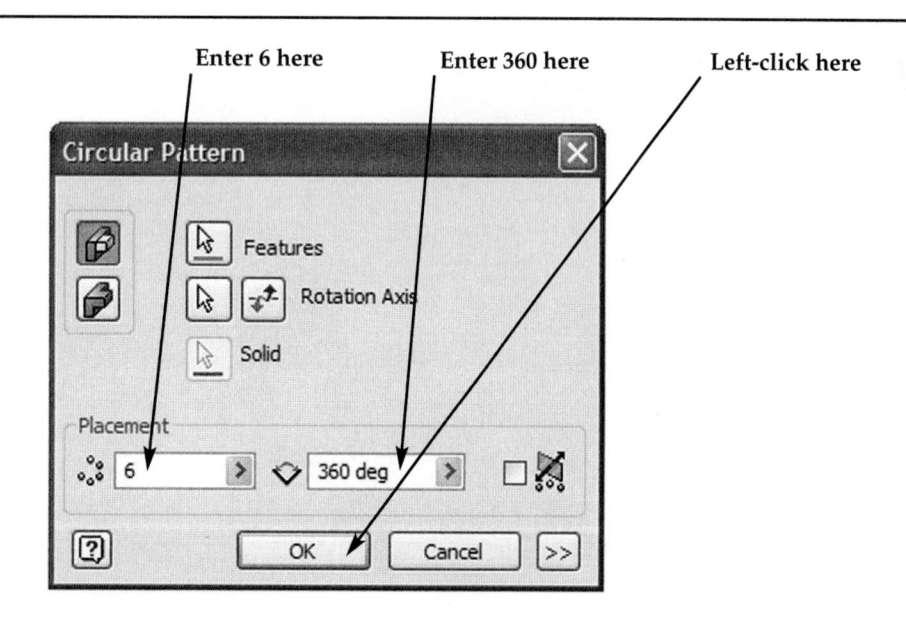

83.   Your screen should look similar to Figure 2-77.

FIGURE 2-77

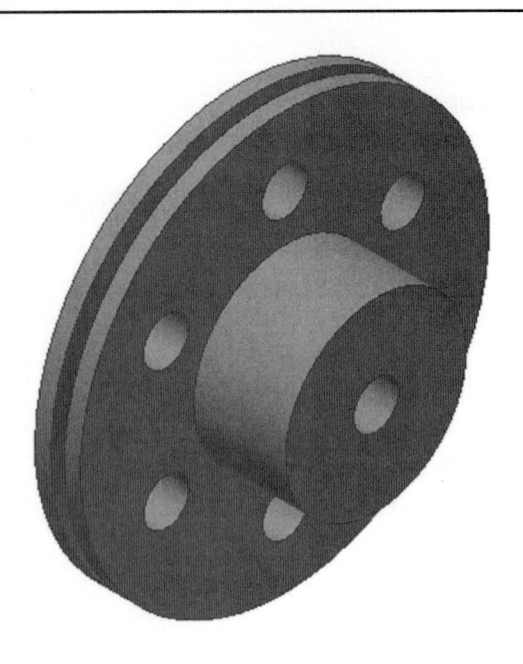

![black bar]

CHAPTER PROBLEMS

Use the Revolve and Revolve Cut commands to complete the following.

PROBLEM 2-1

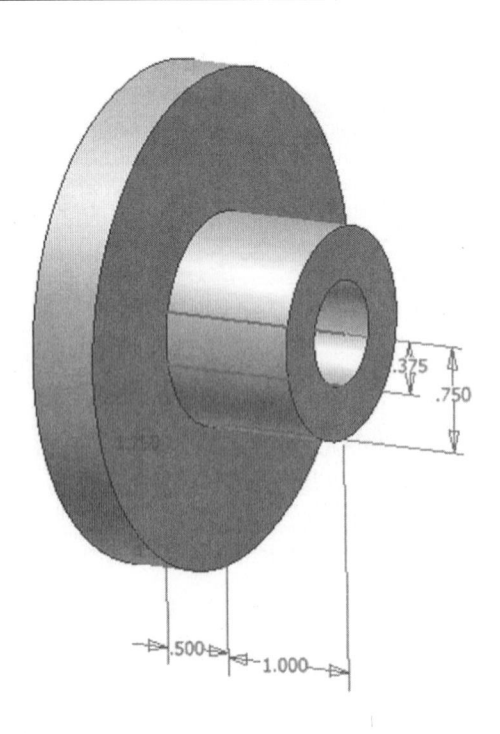

PROBLEM 2-2

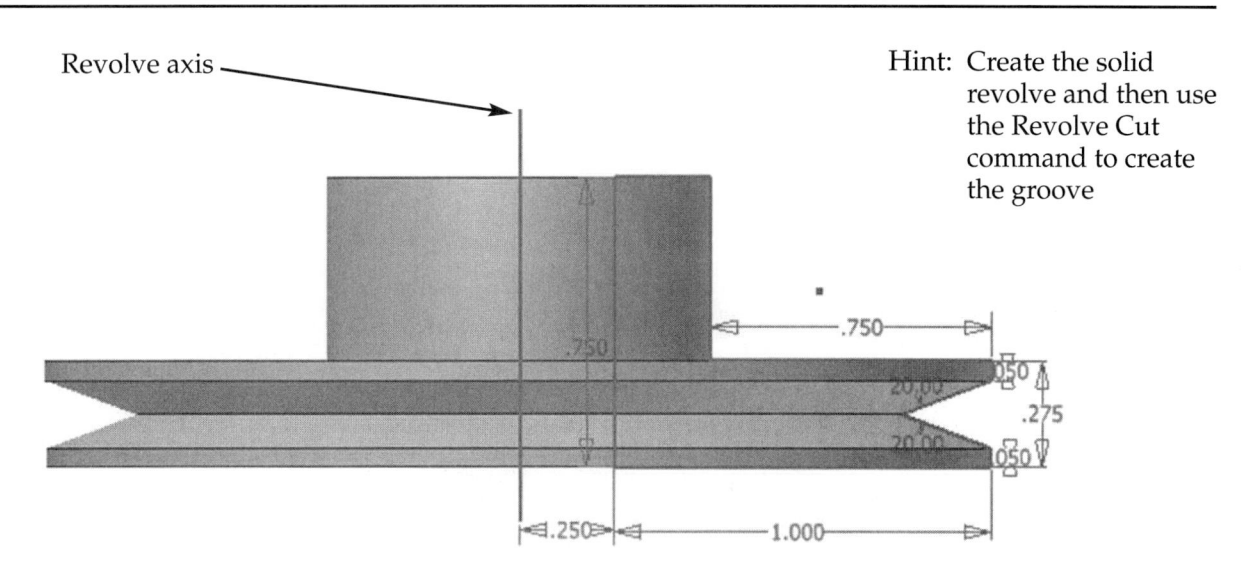

Revolve axis

Hint: Create the solid revolve and then use the Revolve Cut command to create the groove

.750

20.00

.750

.275

20.00

.050

.050

.250

1.000

PROBLEM 2-3

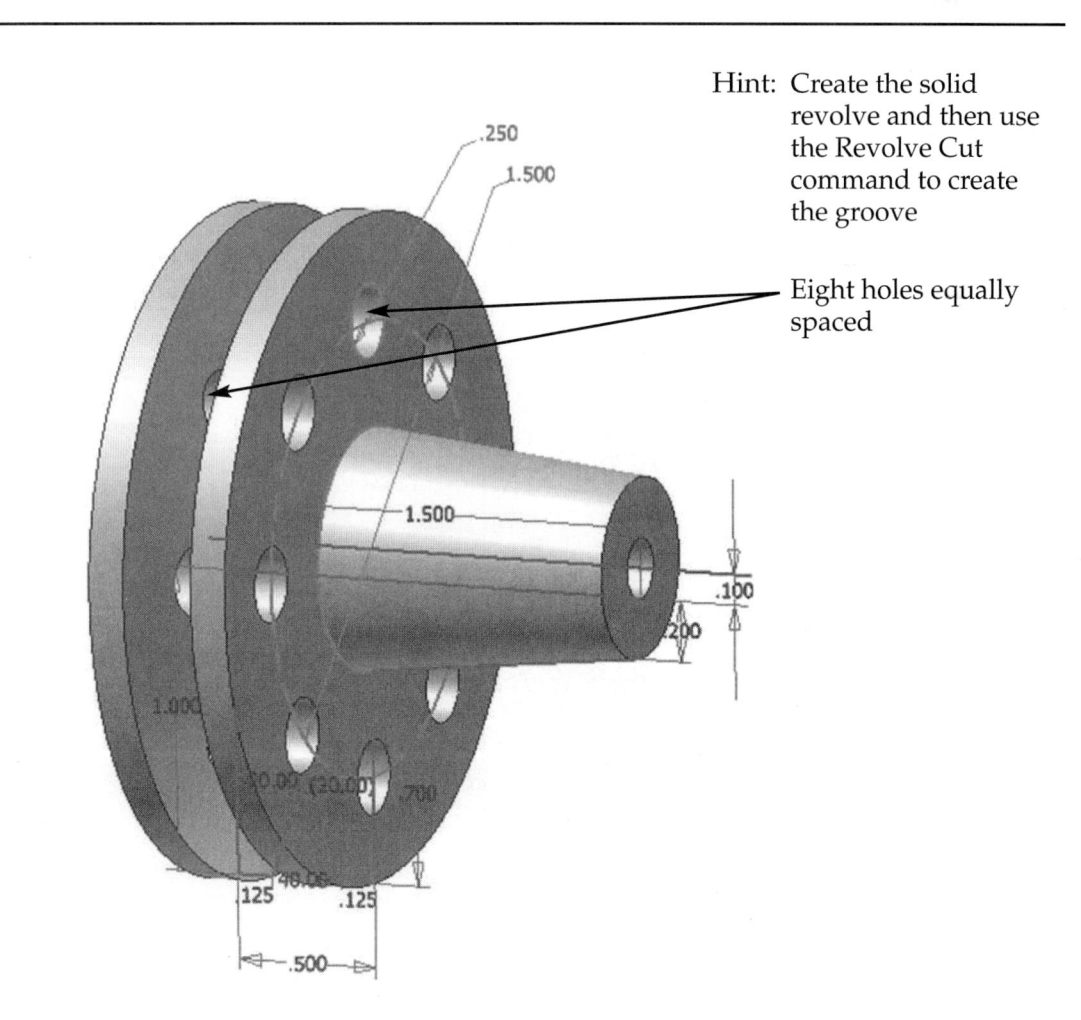

Hint: Create the solid revolve and then use the Revolve Cut command to create the groove

Eight holes equally spaced

.250

1.500

1.500

.100

.200

1.000

20.00 (20.00)   .700

.125   40.00   .125

.500

PROBLEM 2-4

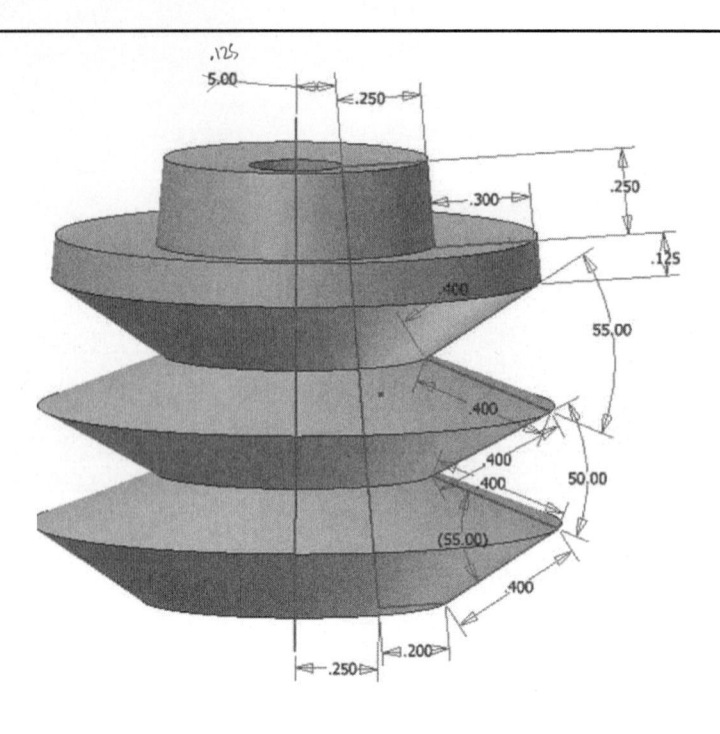

PROBLEM 2-5

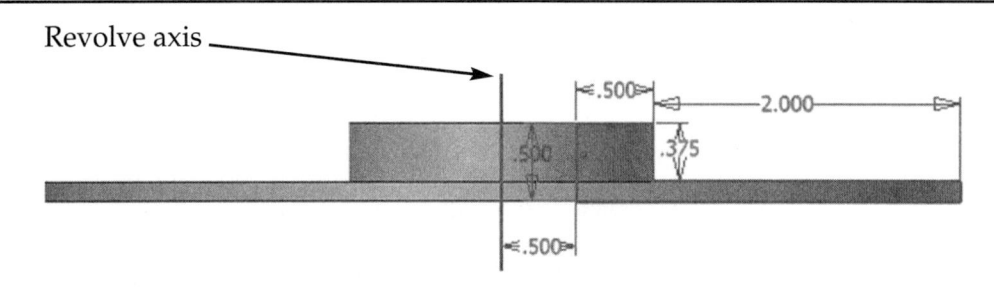

PROBLEM 2-6

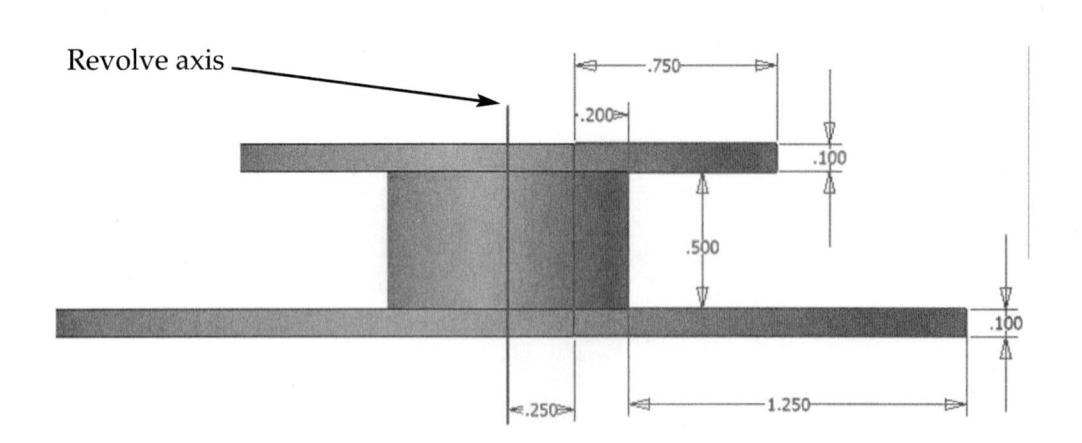

PROBLEM 2-7

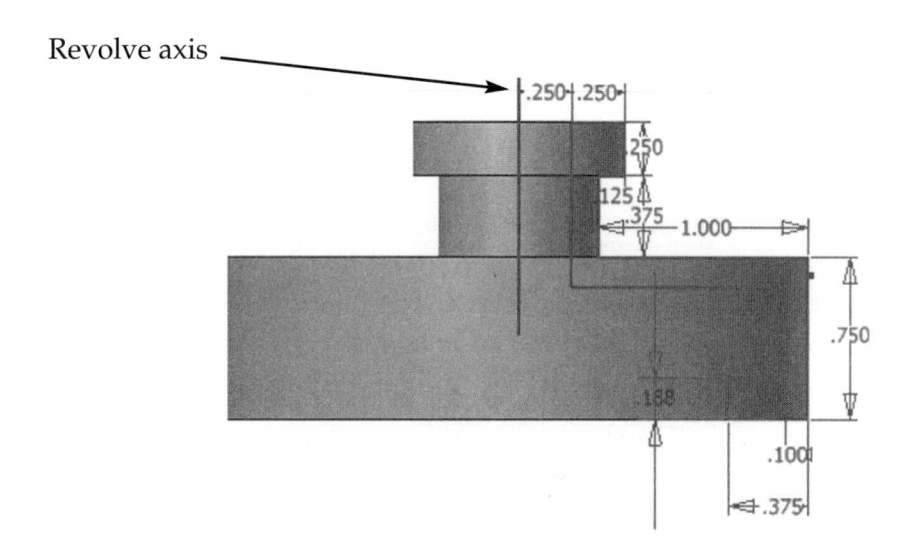

Revolve axis

PROBLEM 2-8

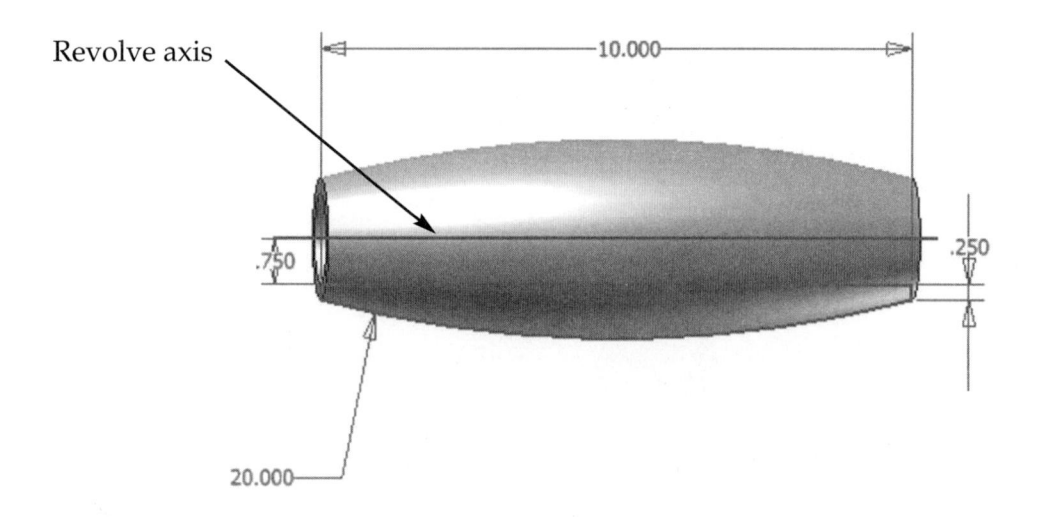

Revolve axis

PROBLEM 2-9

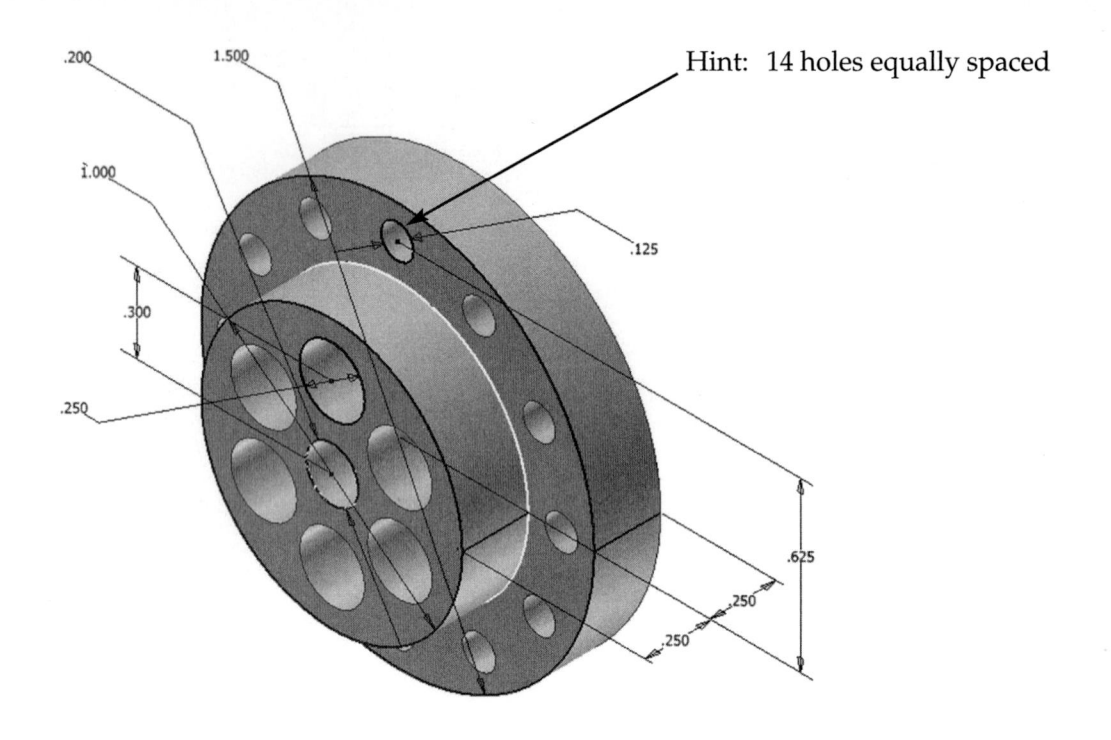

Hint: 14 holes equally spaced

PROBLEM 2-10

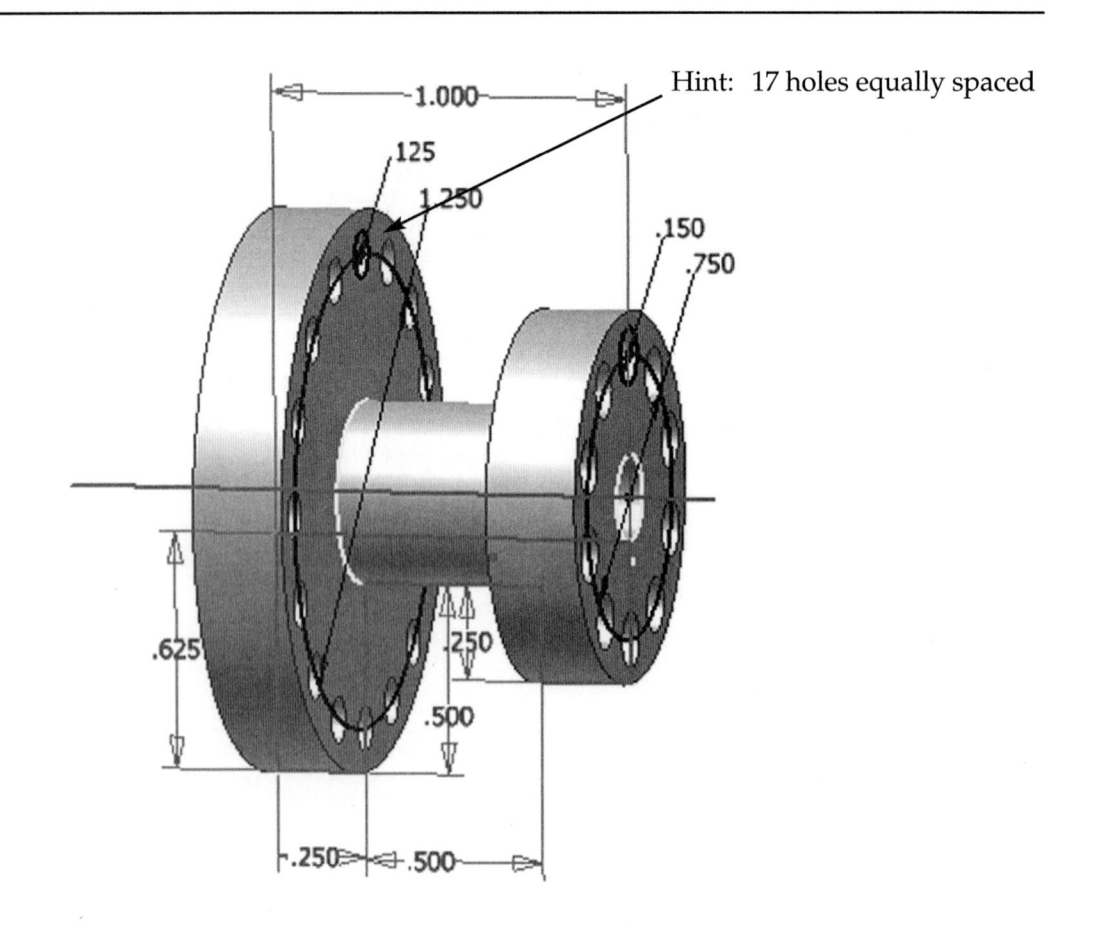

Hint: 17 holes equally spaced

PROBLEM 2-11

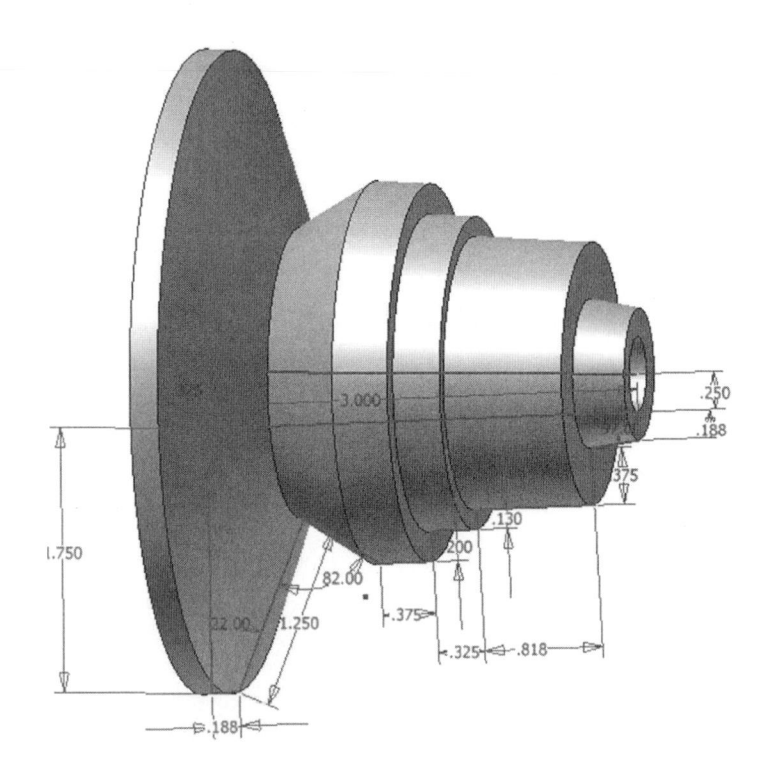

PROBLEM 2-12

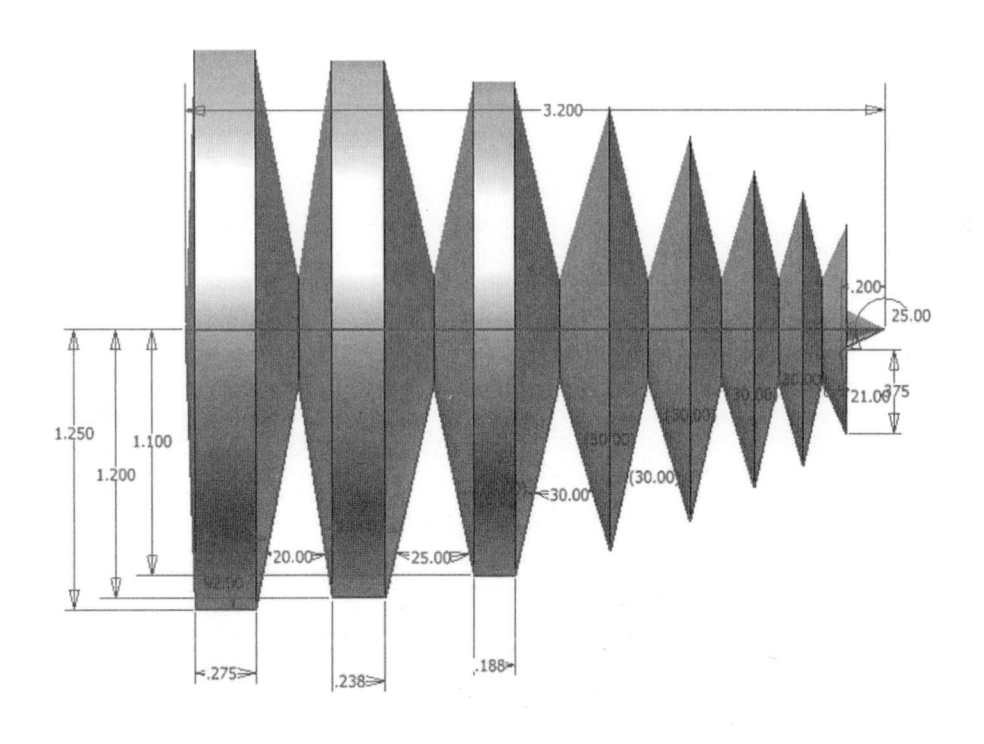

CHAPTER 3

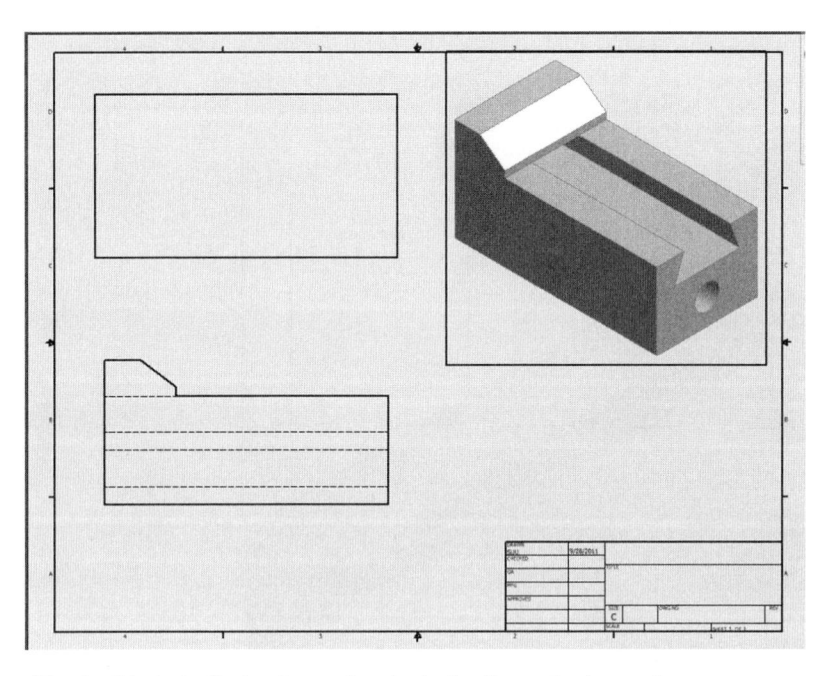

Chapter 3 includes instruction on how to design the parts shown above.

LEARNING TO CREATE A DETAIL DRAWING

OBJECTIVES

1. Create a simple sketch using the Sketch Panel

2. Extrude a sketch into a solid using the Model/Part Features Panel

3. Create an Orthographic View using the Place Views/Drawing Views Panel

4. Edit the appearance of a Solid Model using the Edit Views command

1.   Start Autodesk Inventor 2013 by referring to Chapter 1.

2.   After Autodesk Inventor 2013 is running, begin a new sketch.

3.   Complete the drawing shown in Figure 3-1.

FIGURE 3-1

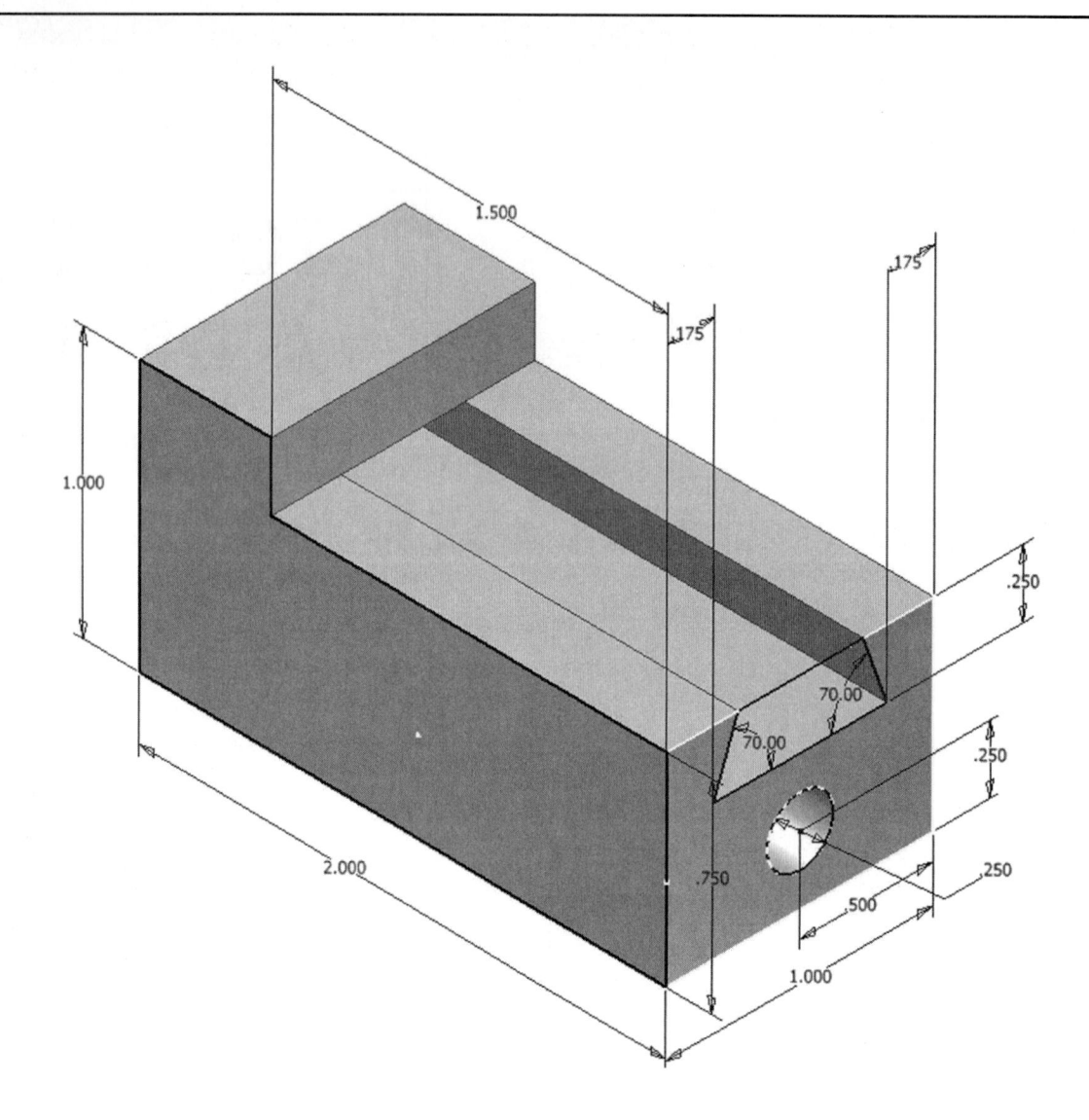

4. Move the cursor to the upper middle portion of the screen and left-click on Chamfer. The **Chamfer** dialog box will appear. Left-click on the drop-down arrow to fully expand the dialog box as shown in Figure 3-2.

FIGURE 3-2

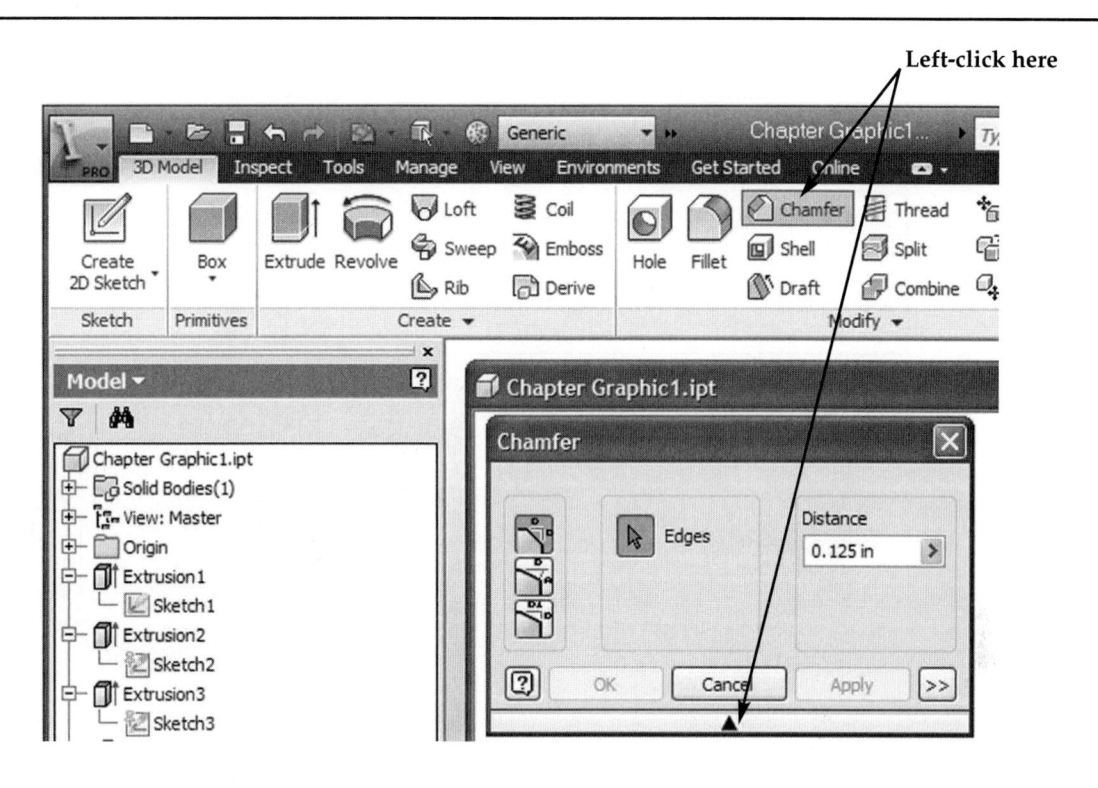

5. After selecting **Chamfer**, left-click on the "Two Distance Chamfer" icon. Left-click on the **Edge** icon as shown in Figure 3-3.

FIGURE 3-3

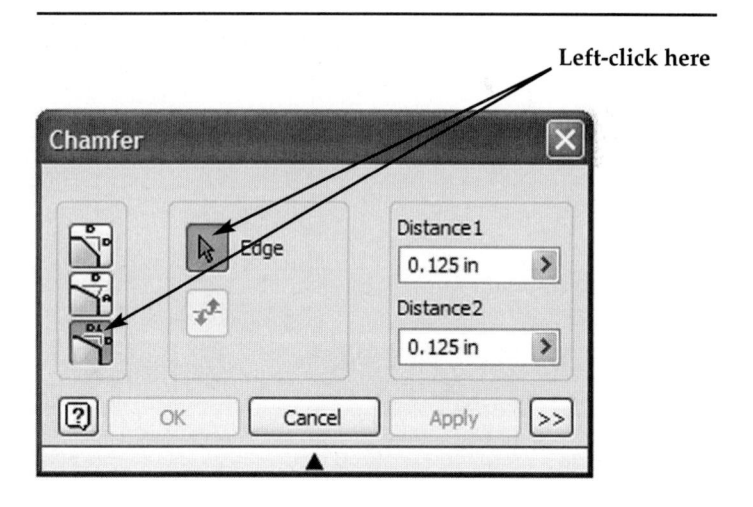

6.   Move the cursor to the front upper corner. A red line will appear as shown in Figure 3-4.

FIGURE 3-4

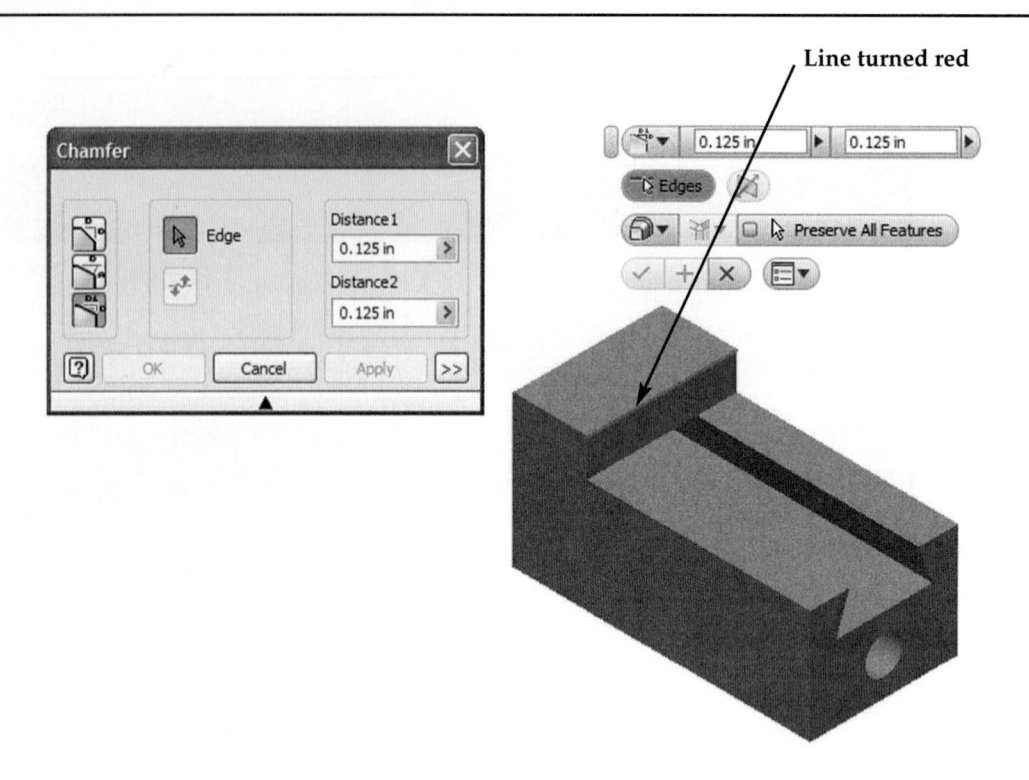

7.   Inventor will provide a preview of the anticipated chamfer as shown in Figure 3-5.

FIGURE 3-5

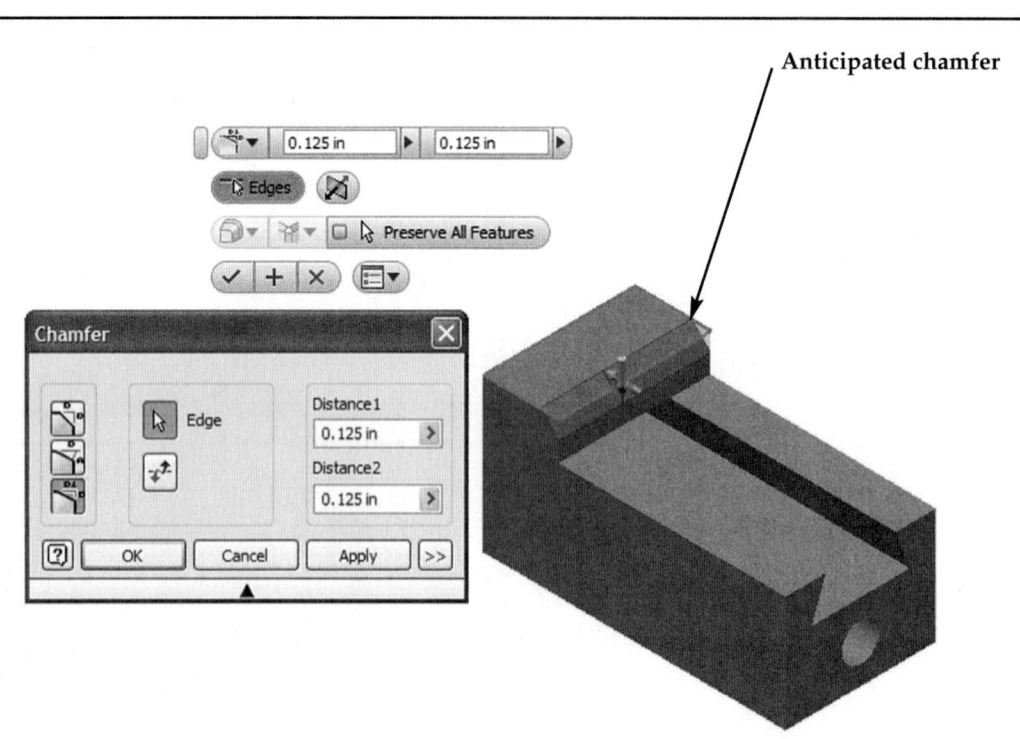

8. Move the cursor to Distance 1 in the dialog box and highlight the text. Enter **.25** in the dialog box. Inventor will provide a preview of the chamfer as shown in Figure 3-6.

FIGURE 3-6

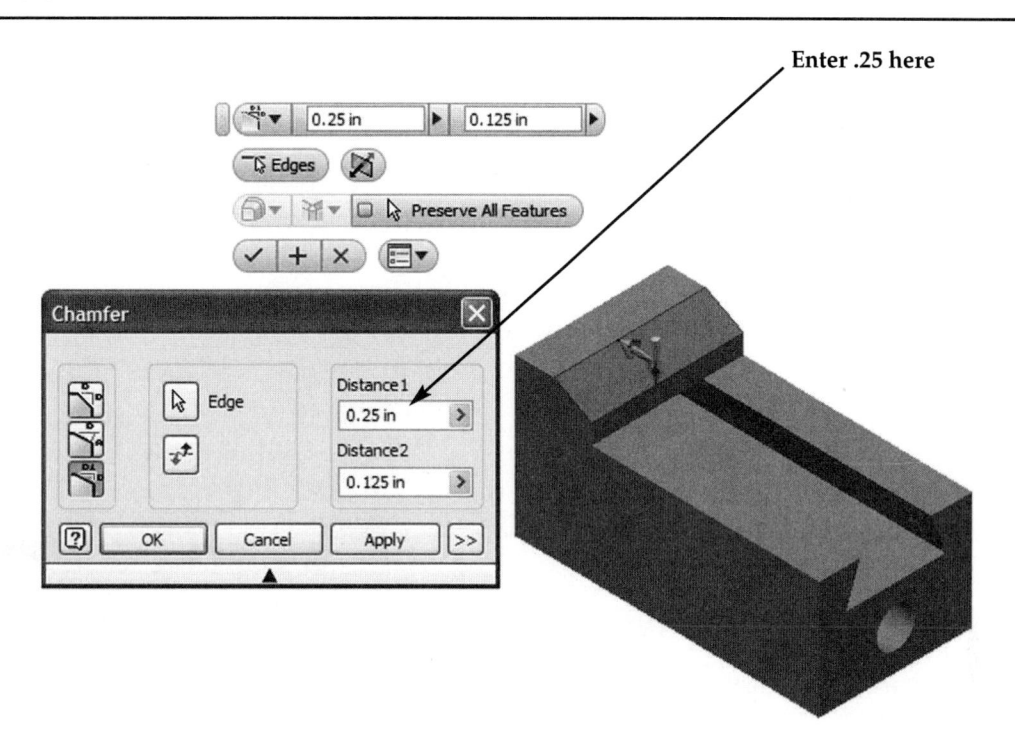

9. Move the cursor to Distance 2 in the dialog box and highlight the text. Enter **.1875** in the dialog box. Inventor will provide a preview of the chamfer as shown in Figure 3-7.

FIGURE 3-7

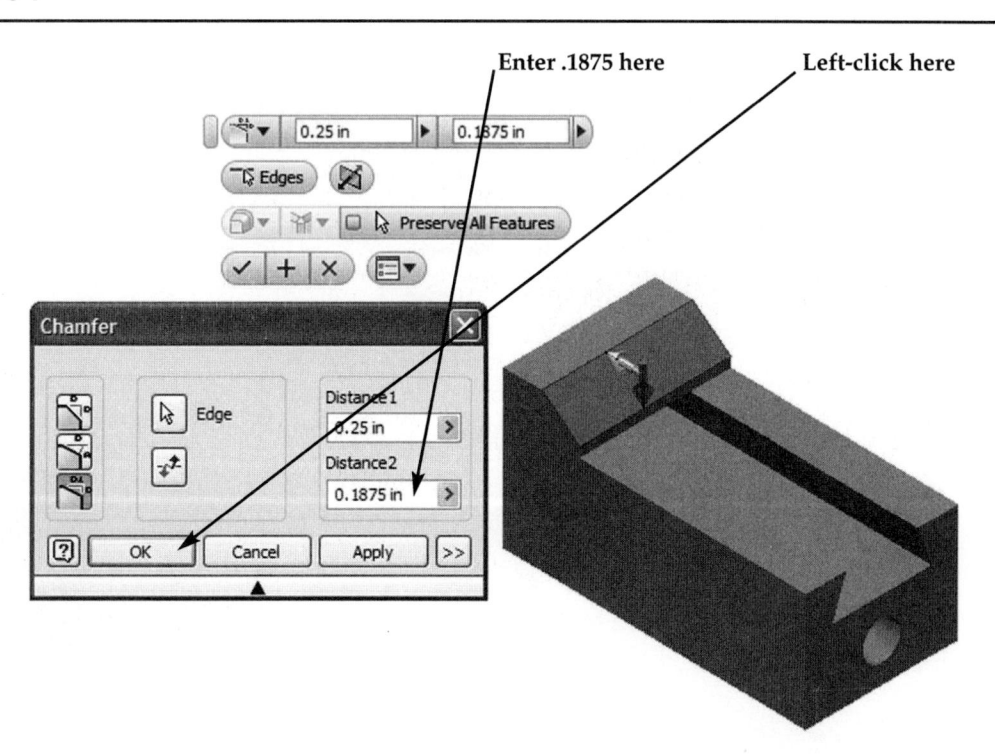

10. Left-click on **OK**. Your screen should look similar to Figure 3-8.

FIGURE 3-8

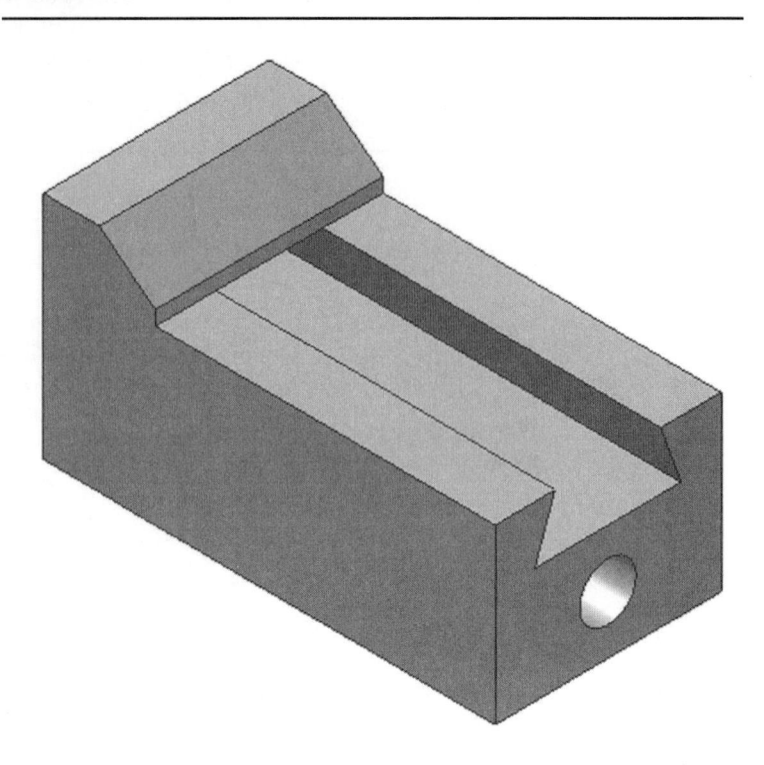

11. Save the part file for easy retrieval to be used in the following section. Do not close the part file.

12. After the part file has been saved, move the cursor to the upper left portion of the screen and left-click on the "New" icon as shown in Figure 3-9.

FIGURE 3-9

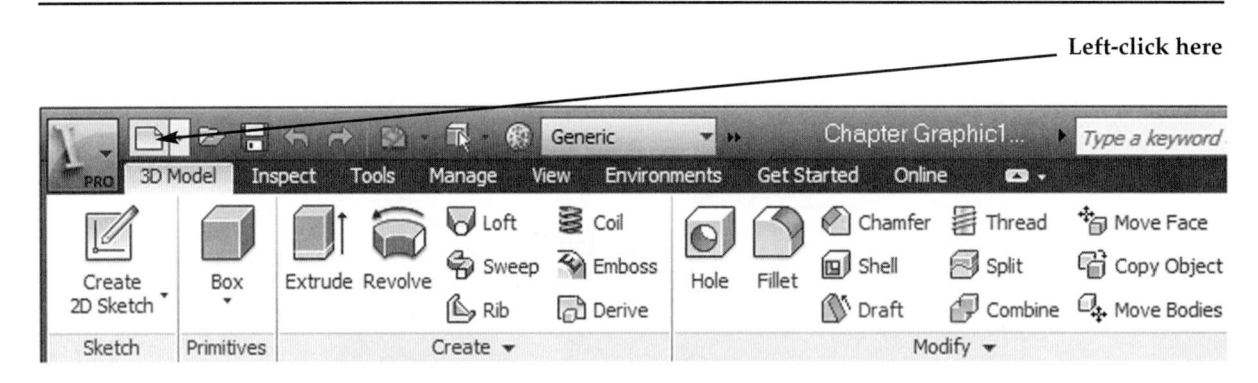

CREATE AN ORTHOGRAPHIC VIEW USING THE DRAWING VIEWS PANEL

13.   The Create New File dialog box will appear. Left-click on the **English** folder. Left-click on the **ANSI (in).idw.** Left-click on **Create** as shown in Figure 3-10.

FIGURE 3-10

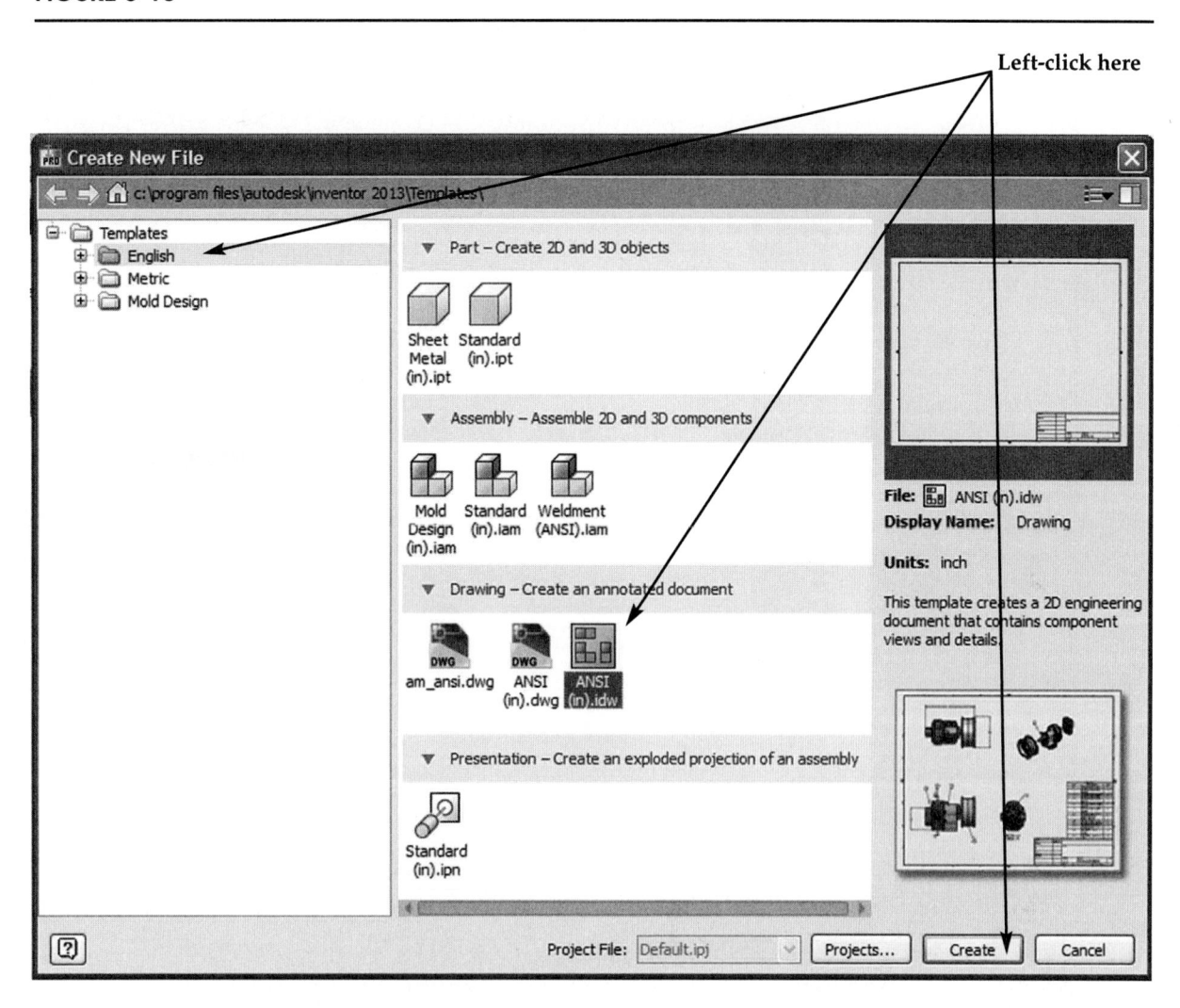

14.   Your screen should look similar to Figure 3-11.

FIGURE 3-11

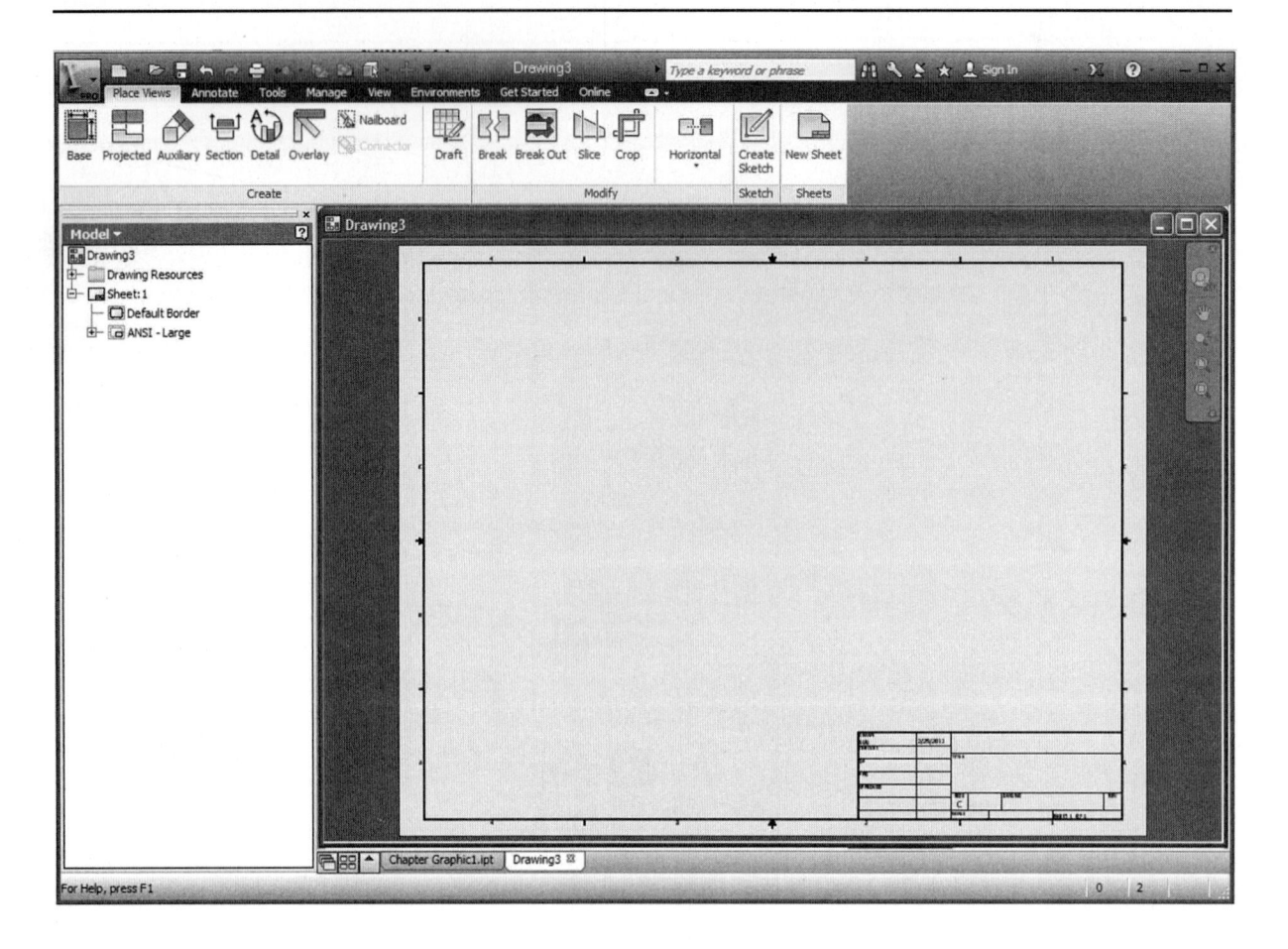

15.   Inventor is now in the Place Views Panel. Notice the commands at the top of the screen are now different. The width of the screen has been reduced to add instructional clarification.

16. Move the cursor to the upper left portion of the screen and left-click on **Base** as shown in Figure 3-12.

FIGURE 3-12

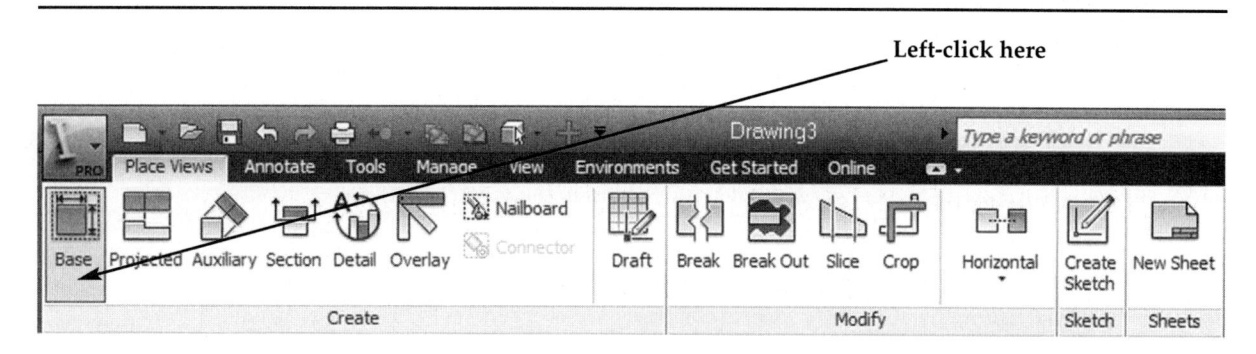

17. The drawing of the wedge block should appear attached to the cursor. Move the cursor around to verify it is attached. If the part does not appear attached to the cursor, use the "Explore" icon to locate the part file as shown in Figure 3-13.

FIGURE 3-13

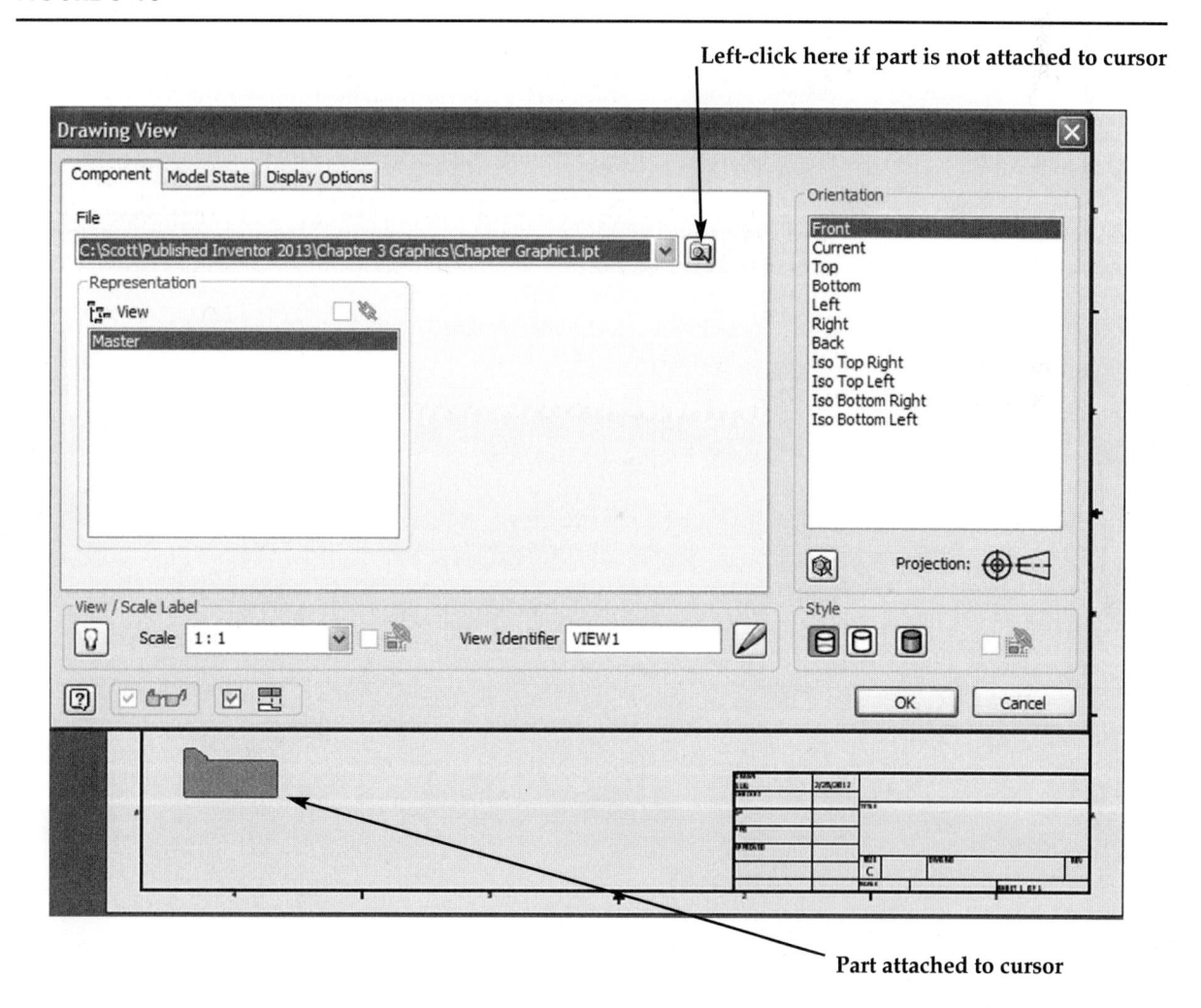

18.   Different views can be selected for the front, top, and side views. Select the desired view from the **Orientation** selection box as shown in Figure 3-14. To understand how the orientation selection works, left-click on **Top** or **Left** to have the top view or left view as the front (base) view.

FIGURE 3-14

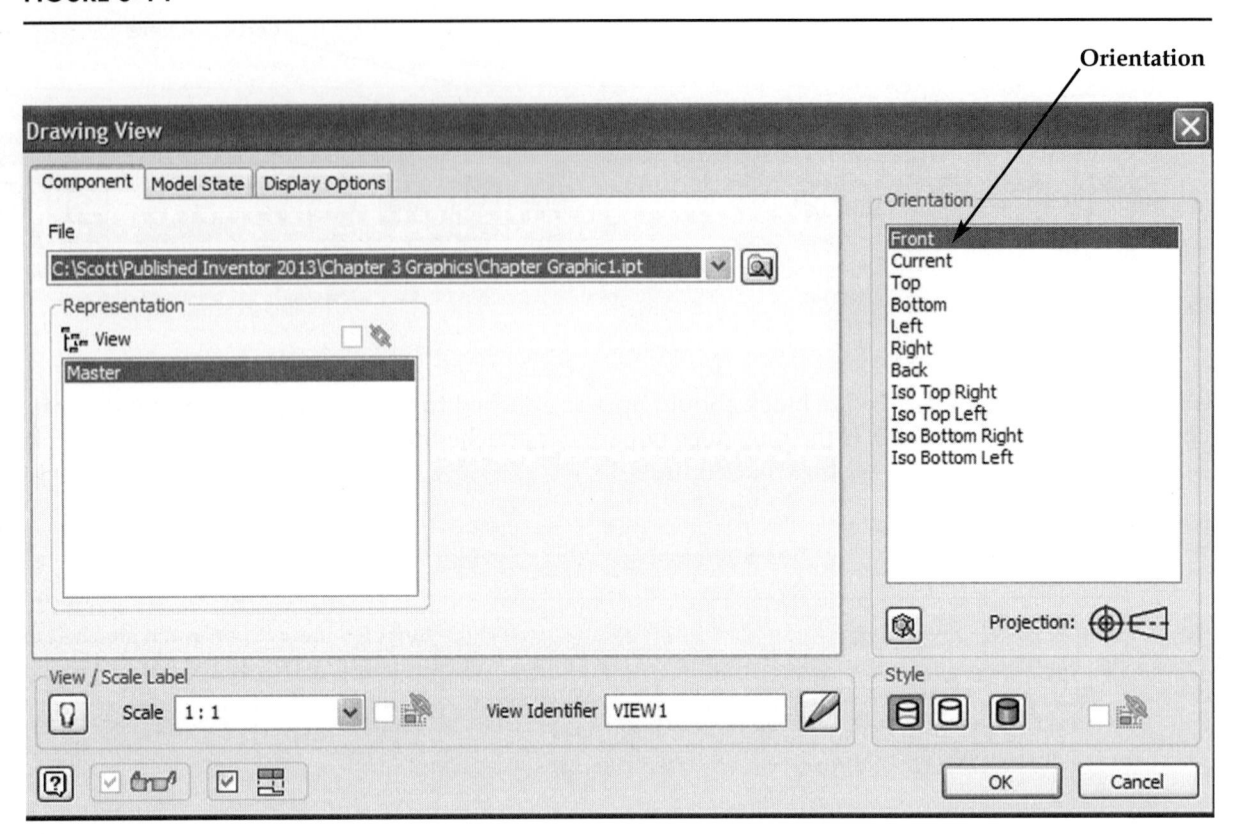

19. Select the **Front** view for the base view. Left-click on the **Scale** drop-down box and set the drawing scale to **4:1**. The size of the wedge block will become larger as shown in Figure 3-15.

FIGURE 3-15

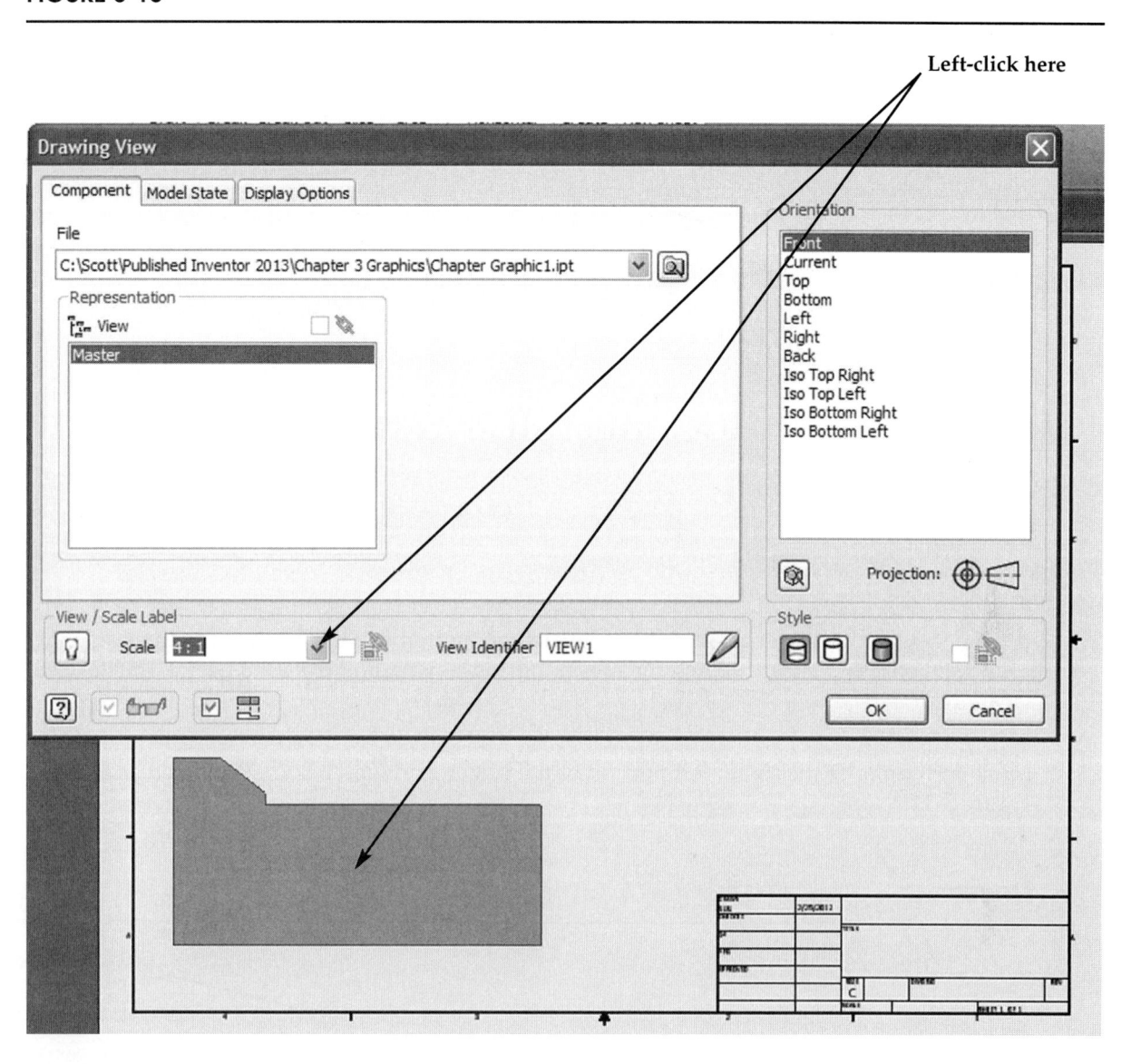

20. Place the part just above the title block that is in the lower right corner of the screen and left-click once. This will place the part as shown in Figure 3-16.

FIGURE 3-16

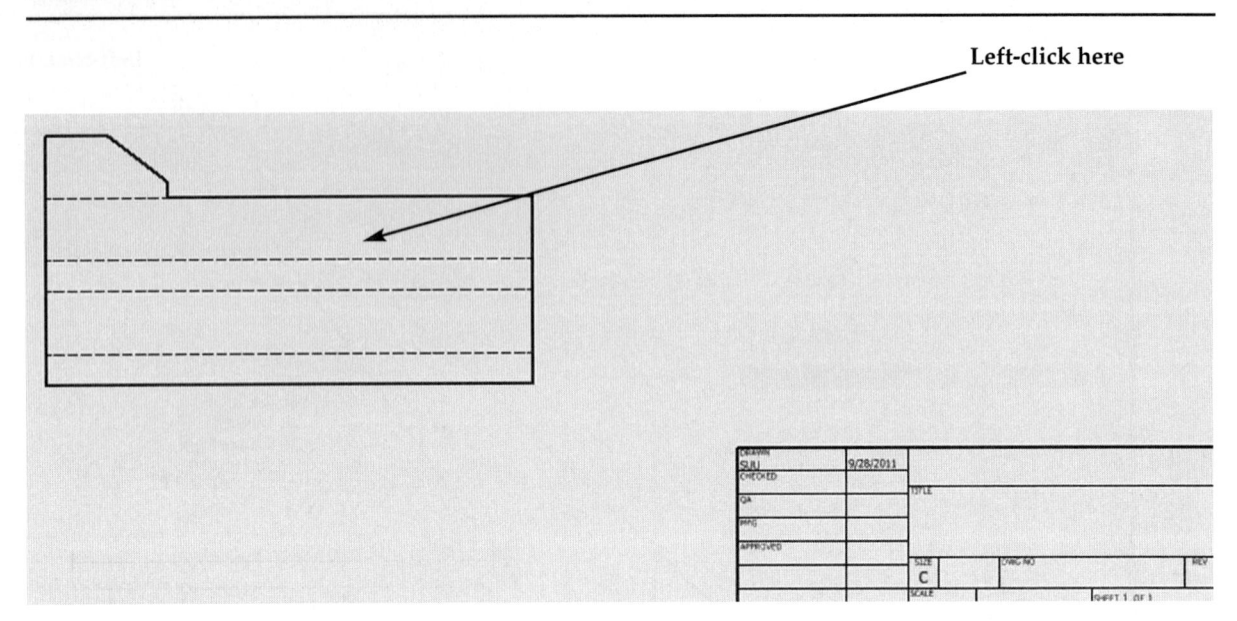

21. If the part was inadvertently placed too low or too high, move the cursor over the dots that surround the part, left-click (holding the mouse button down), and drag the part to the desired location.

22. Move the cursor to the upper left portion of the screen and left-click on **Projected** as shown in Figure 3-17.

FIGURE 3-17

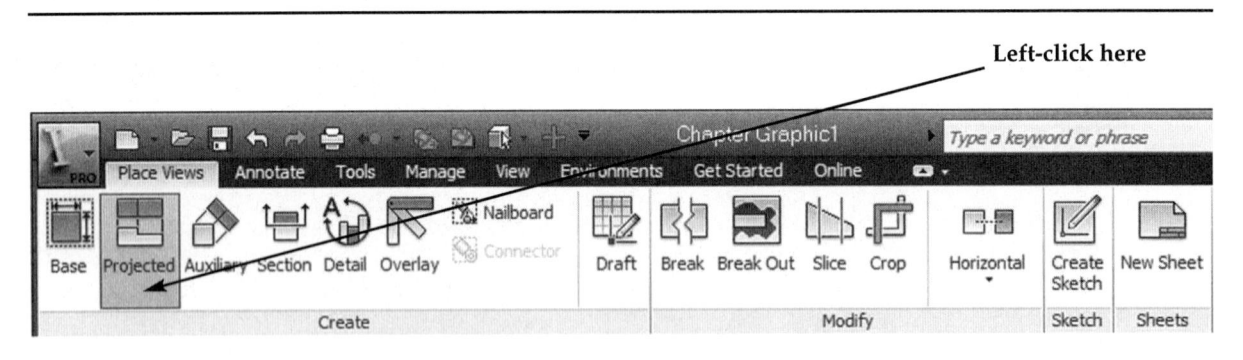

23. The part will be attached to the cursor. Move the cursor upward and left-click as shown in Figure 3-18.

FIGURE 3-18

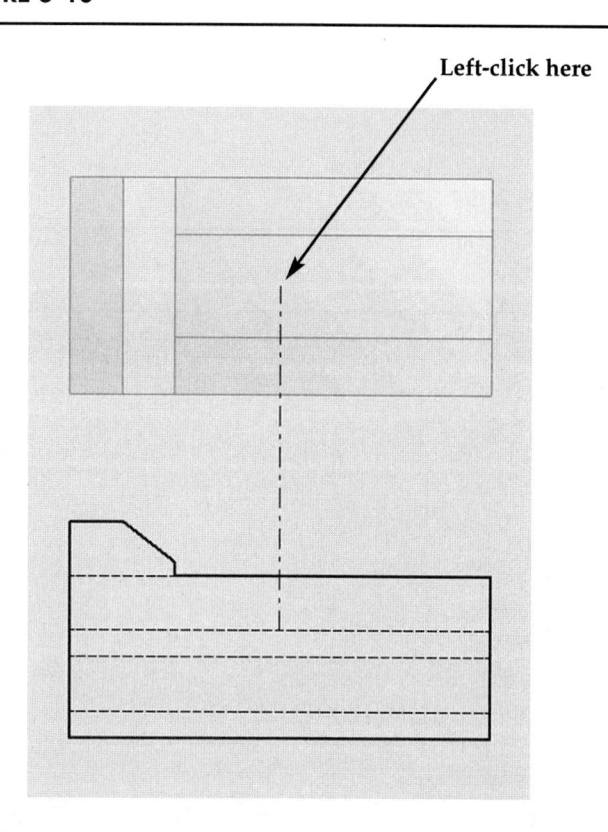

Left-click here

24. Place the part the desired distance from the front (base) view and left-click once. Notice the black lines around the top view as shown in Figure 3-19. This indicates that the view has been placed.

FIGURE 3-19

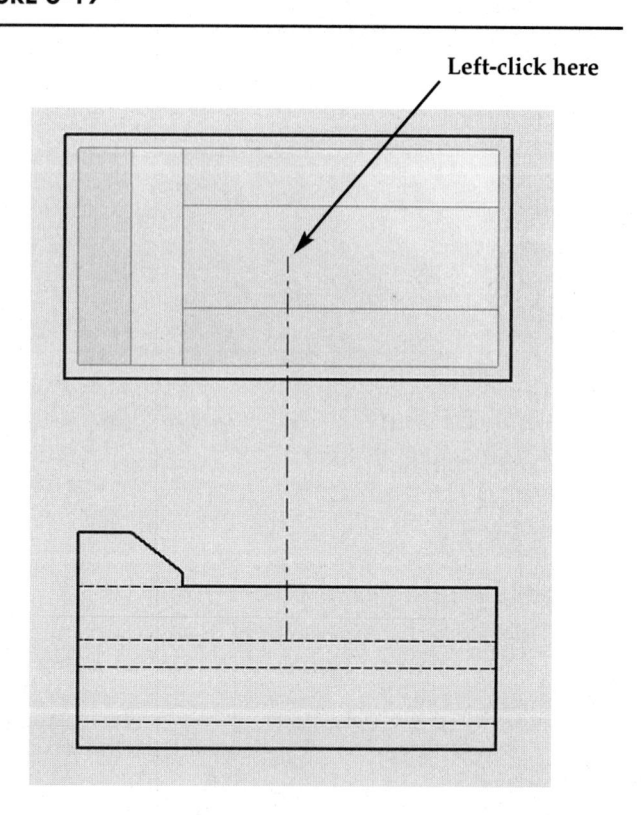

Left-click here

CREATE A SOLID MODEL USING THE EDIT VIEWS COMMAND

25.   Move the cursor over to the upper right corner of the page and left-click once as shown in Figure 3-20.

FIGURE 3-20

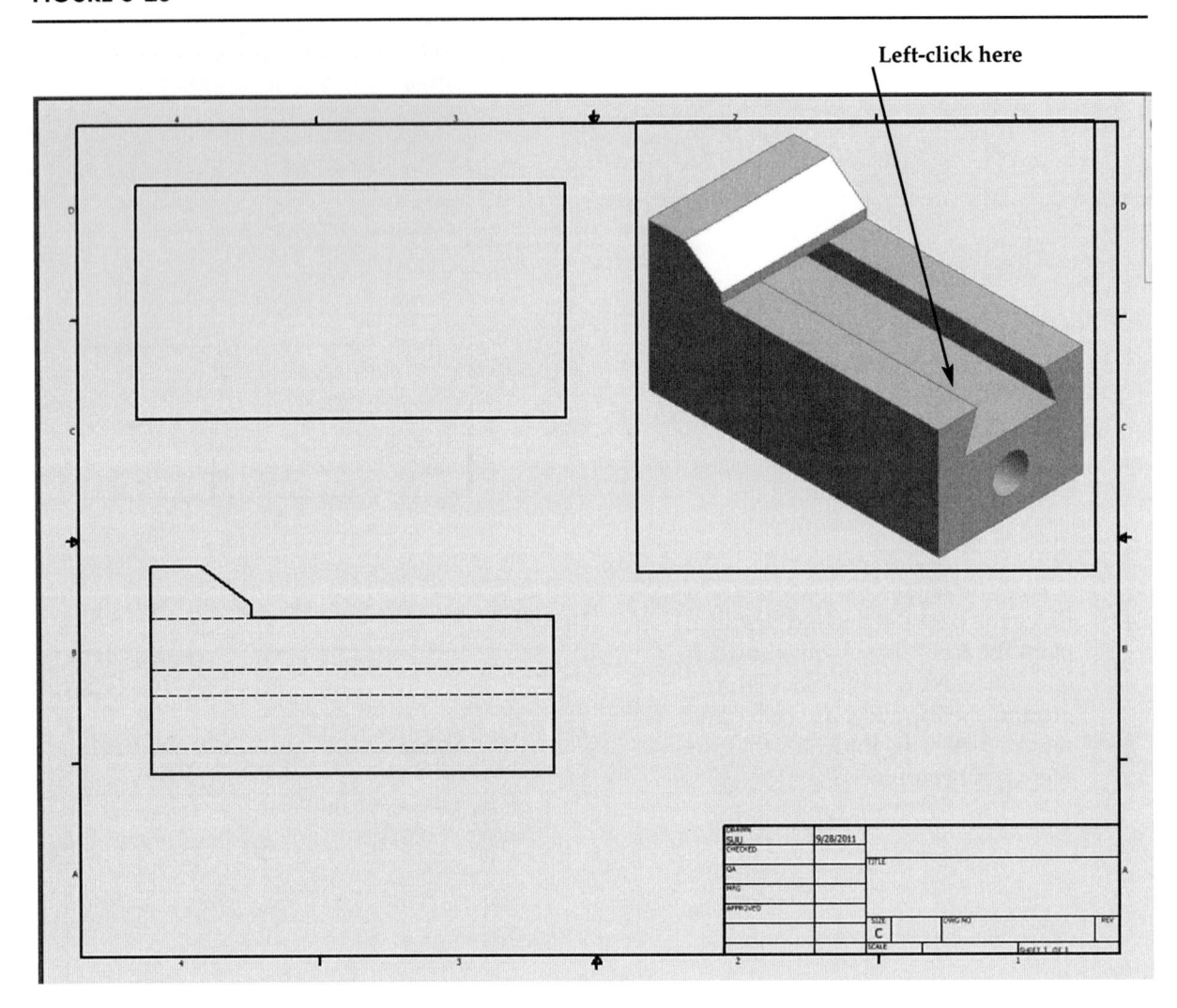

Left-click here

26. Move the cursor down to where the side view will be located and left-click once as shown in Figure 3-21.

FIGURE 3-21

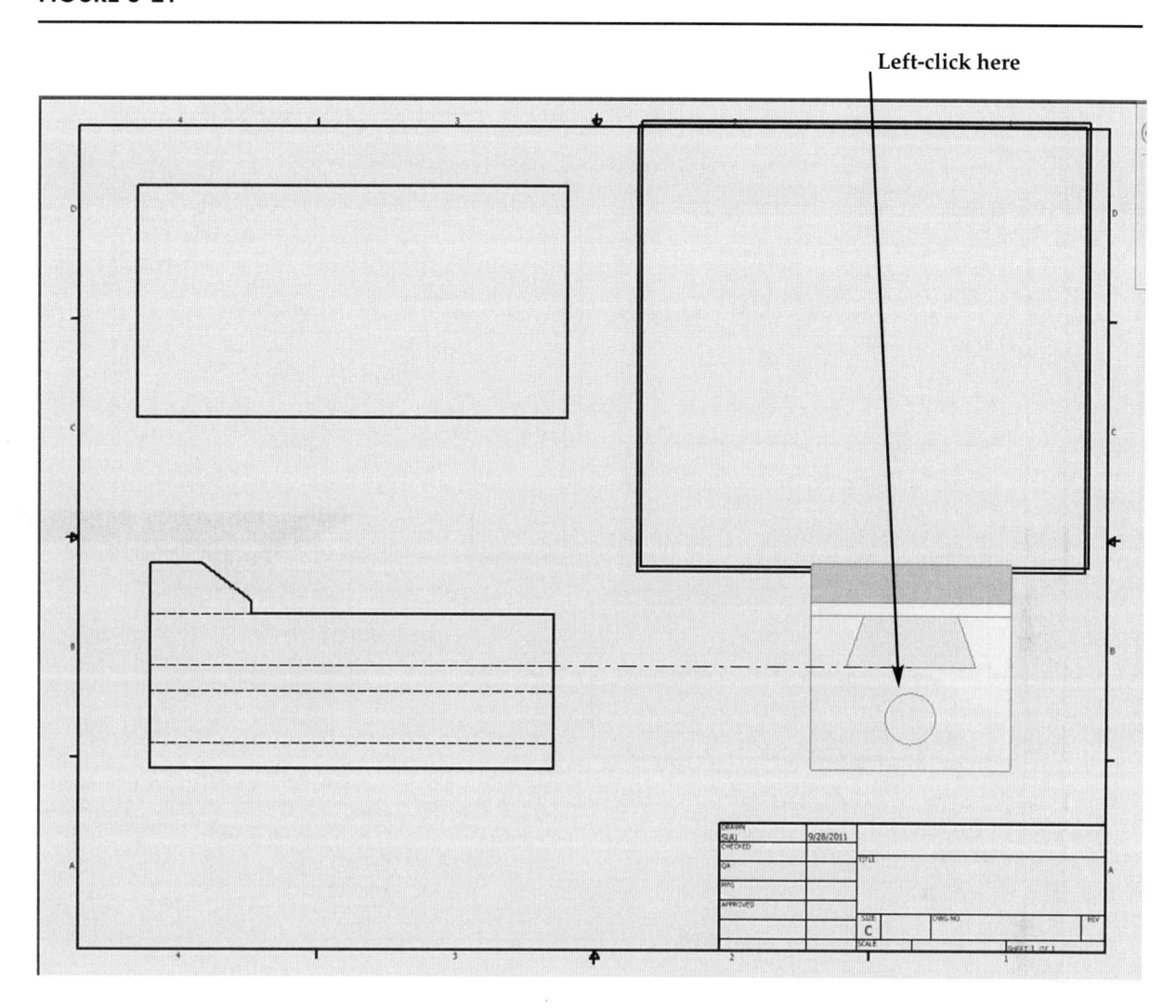

27. Right click on the last view created (side view). A pop-up menu will appear. Left-click on **Create** as shown in Figure 3-22.

FIGURE 3-22

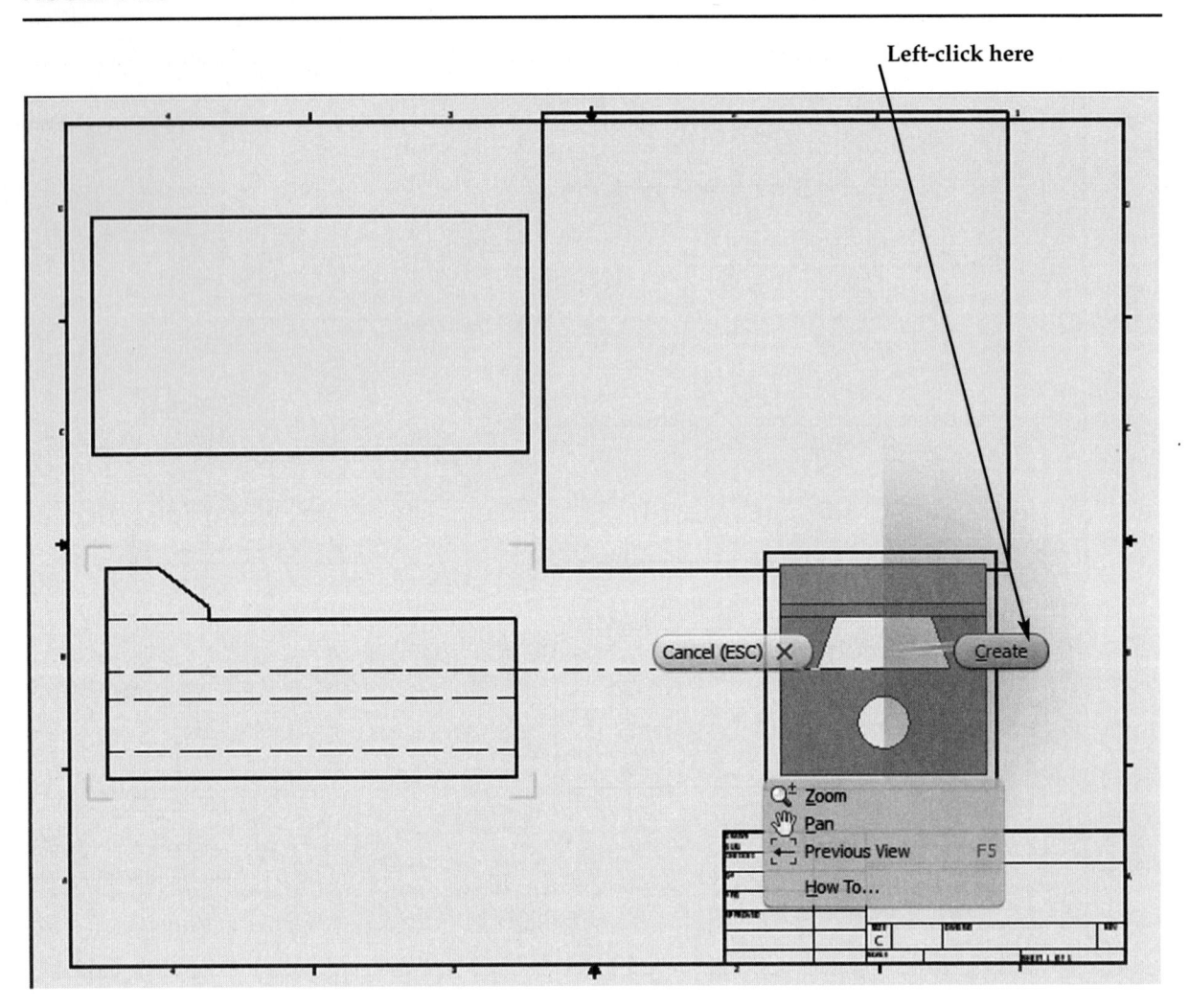

28.   Your screen should look similar to Figure 3-23.

FIGURE 3-23

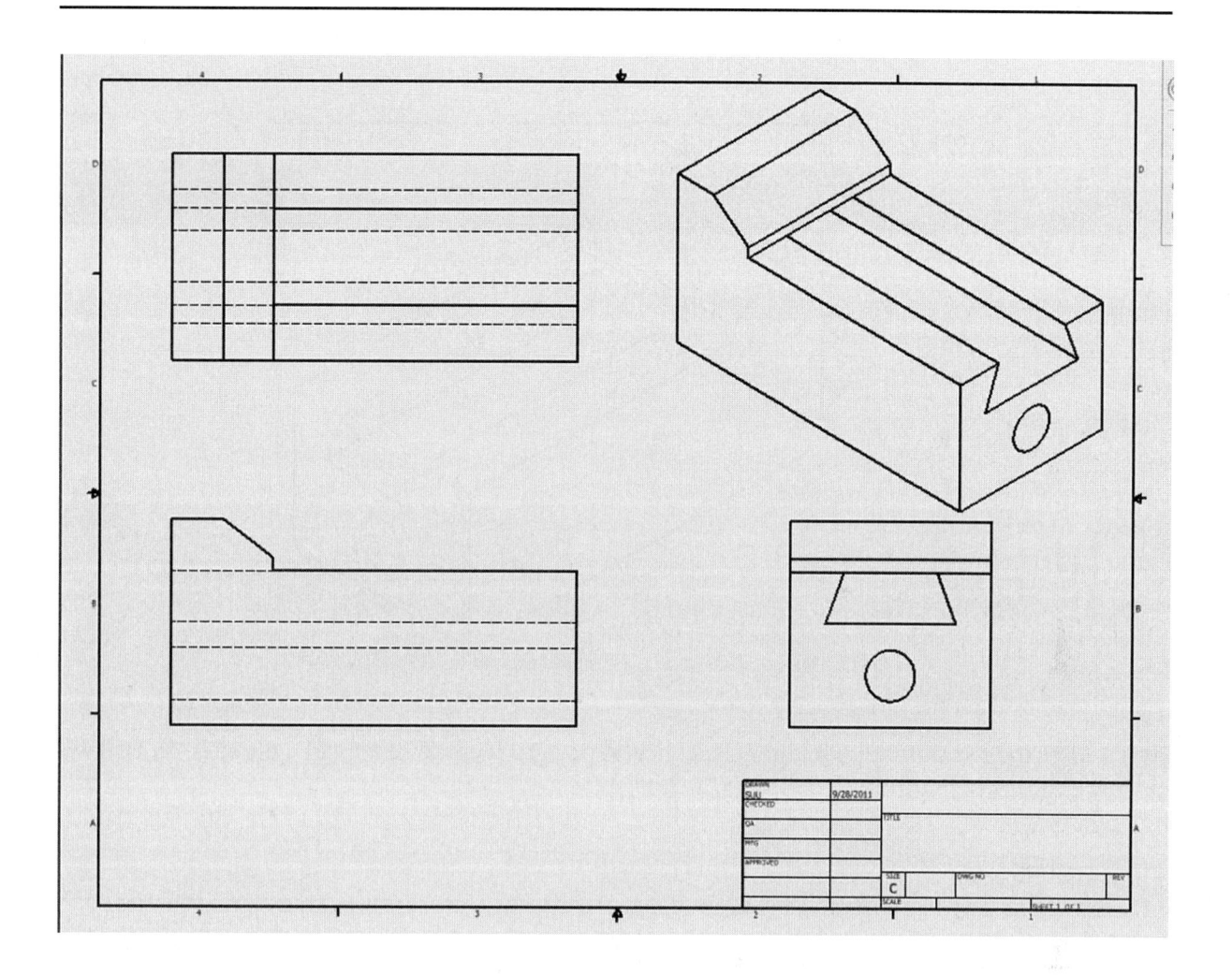

29.   Move the cursor over the Isometric View in the upper right corner of the drawing. Red dots will appear as shown in Figure 3-24.

FIGURE 3-24

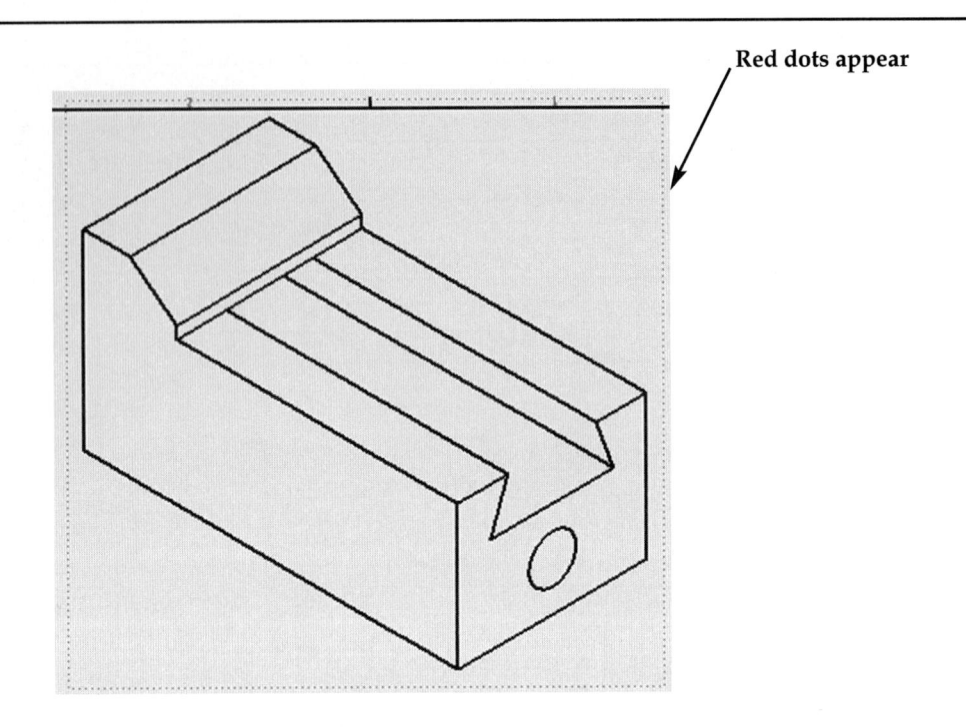

Red dots appear

30.   After the red dots appear, right click once. A pop-up menu will appear. Left-click on **Edit View** as shown in Figure 3-25.

FIGURE 3-25

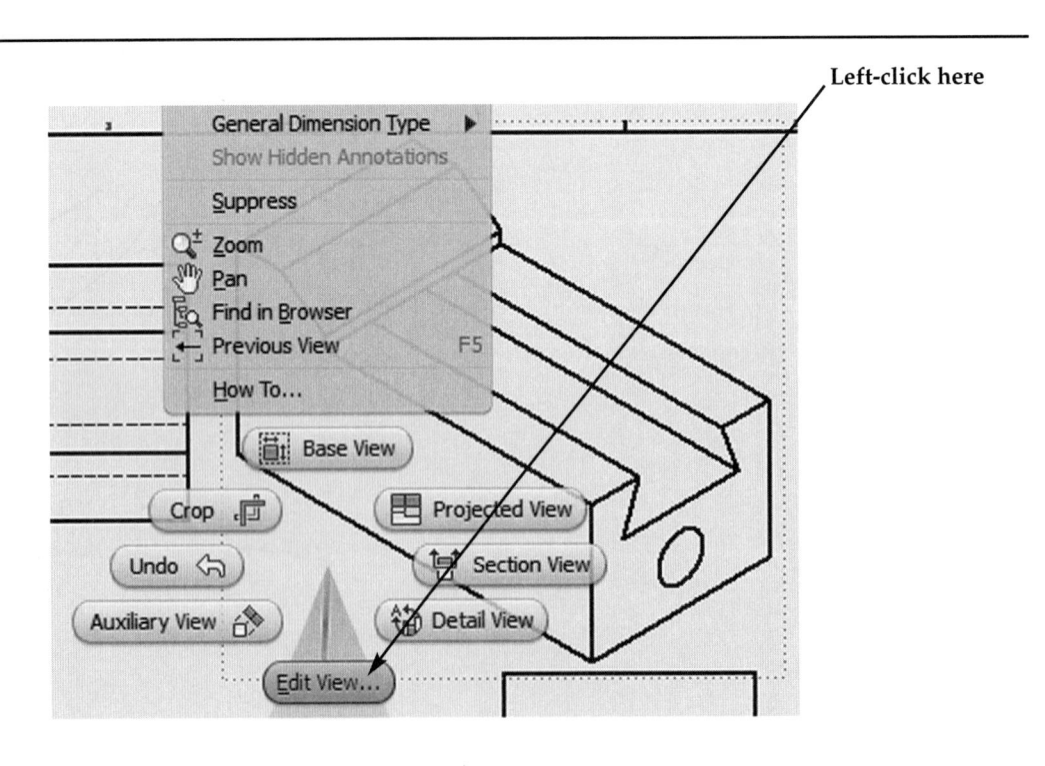

Left-click here

31. The Drawing View dialog box will appear. Left-click on the blue barrel to the right under Style. Left-click on **OK** as shown in Figure 3-26.

FIGURE 3-26

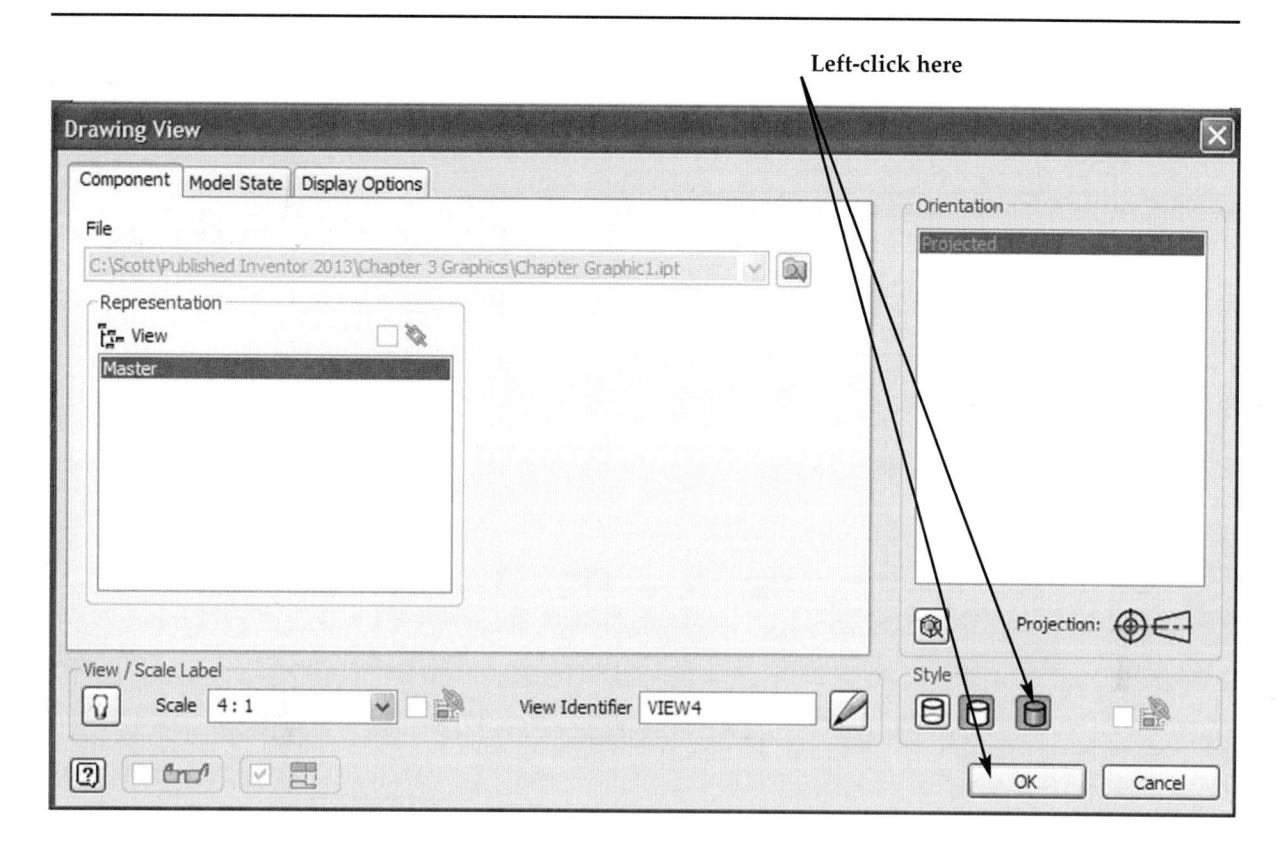

32. The Isometric View will become a miniature solid model as shown in Figure 3-27.

FIGURE 3-27

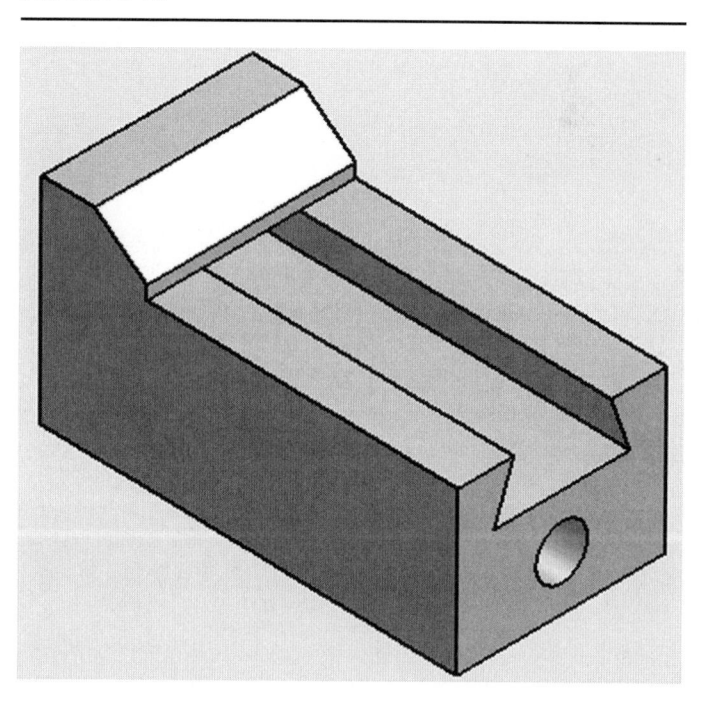

33.   Your screen should look similar to Figure 3-28.

FIGURE 3-28

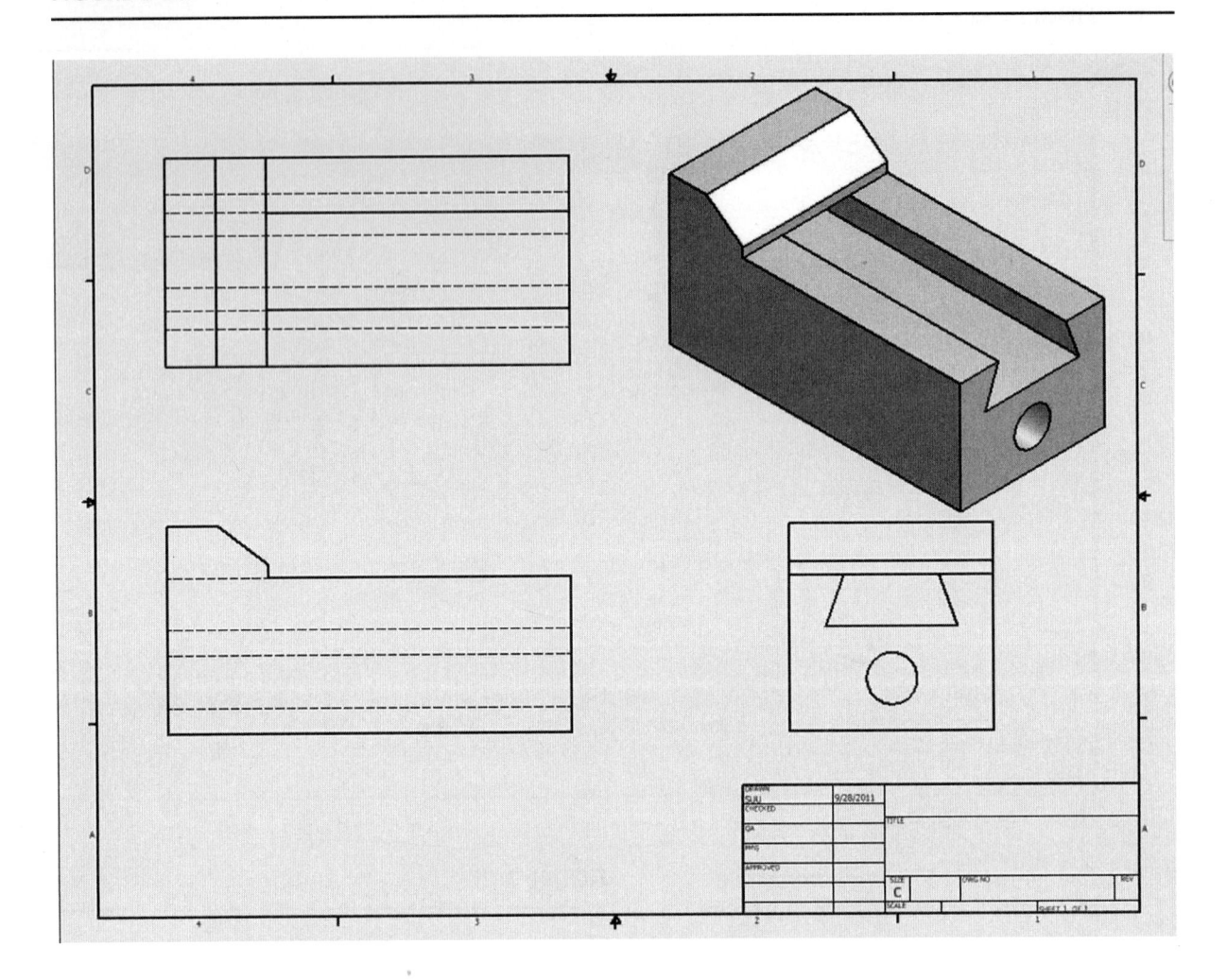

CHAPTER PROBLEMS

Create 3 view/multi-view drawings of the following parts.

PROBLEM 3-1

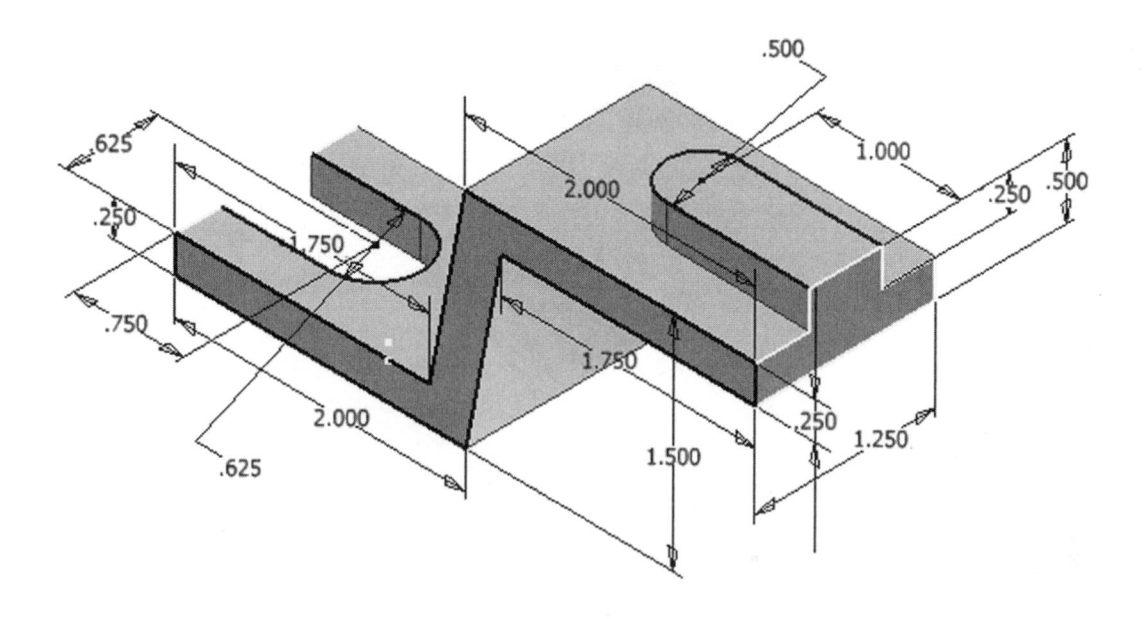

PROBLEM 3-2

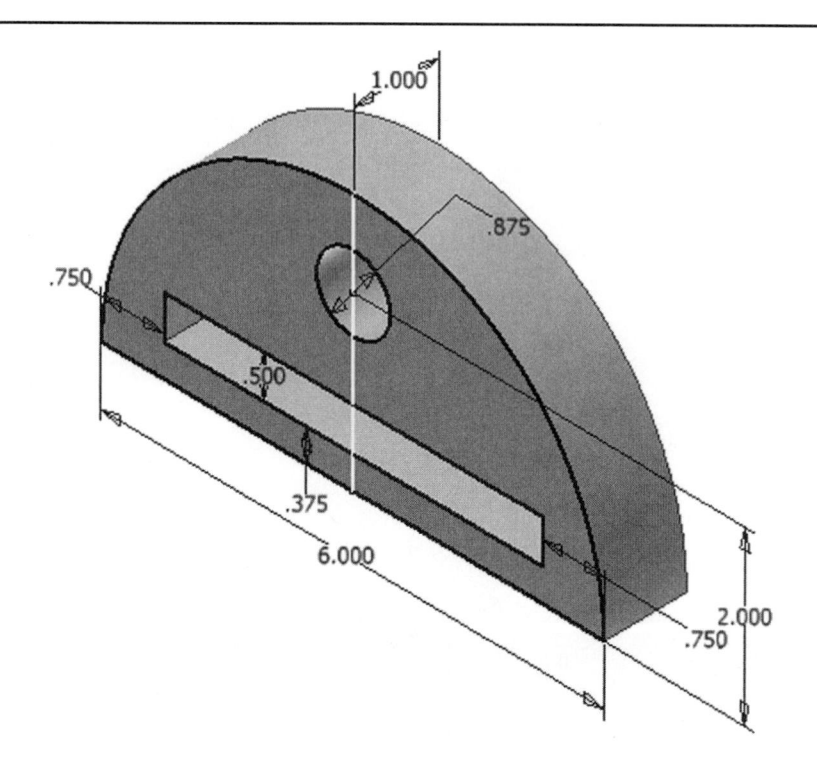

PROBLEM 3-3

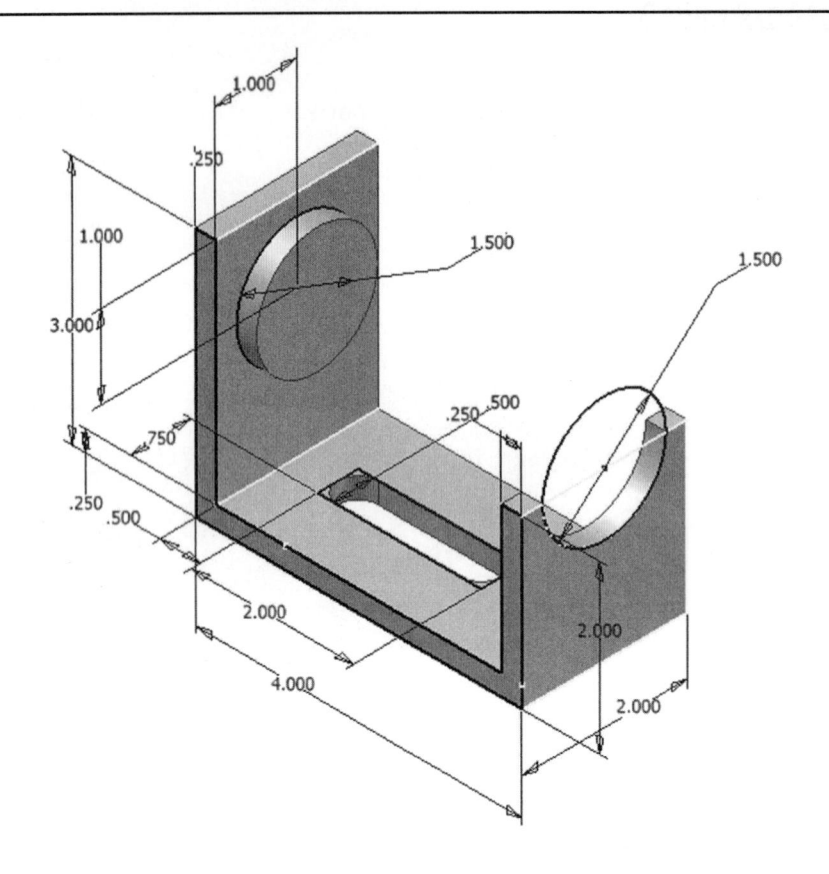

PROBLEM 3-4

Hint: Use the 2 directional
Extrude-Cut to create
the 1.625 diameter hole

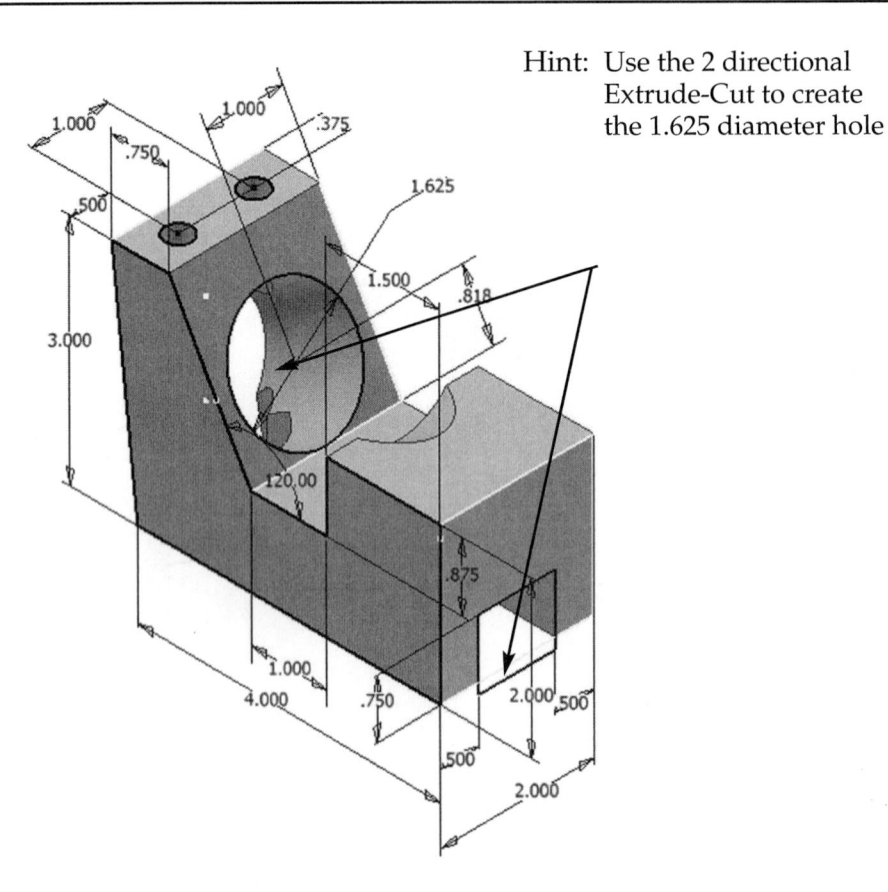

PROBLEM 3-5

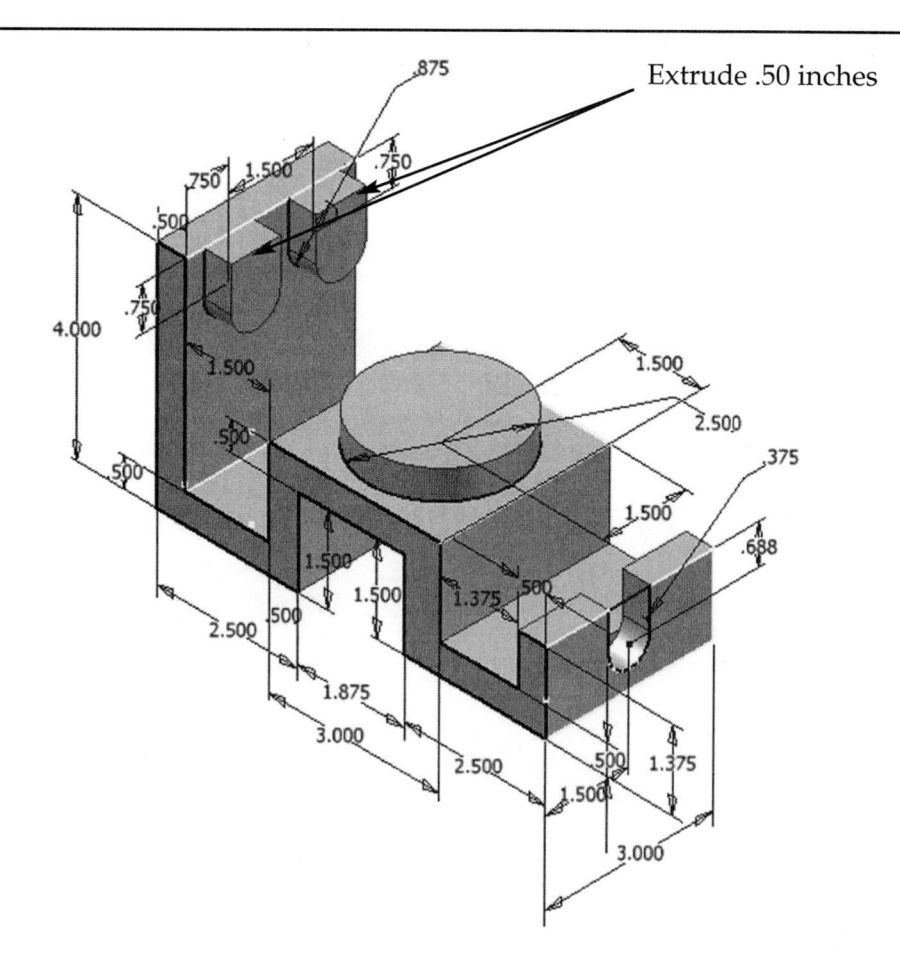

Extrude .50 inches

PROBLEM 3-6

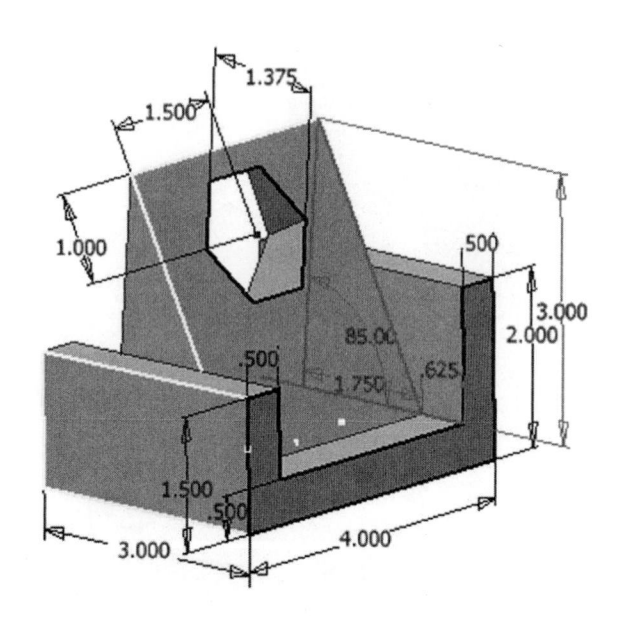

PROBLEM 3-7

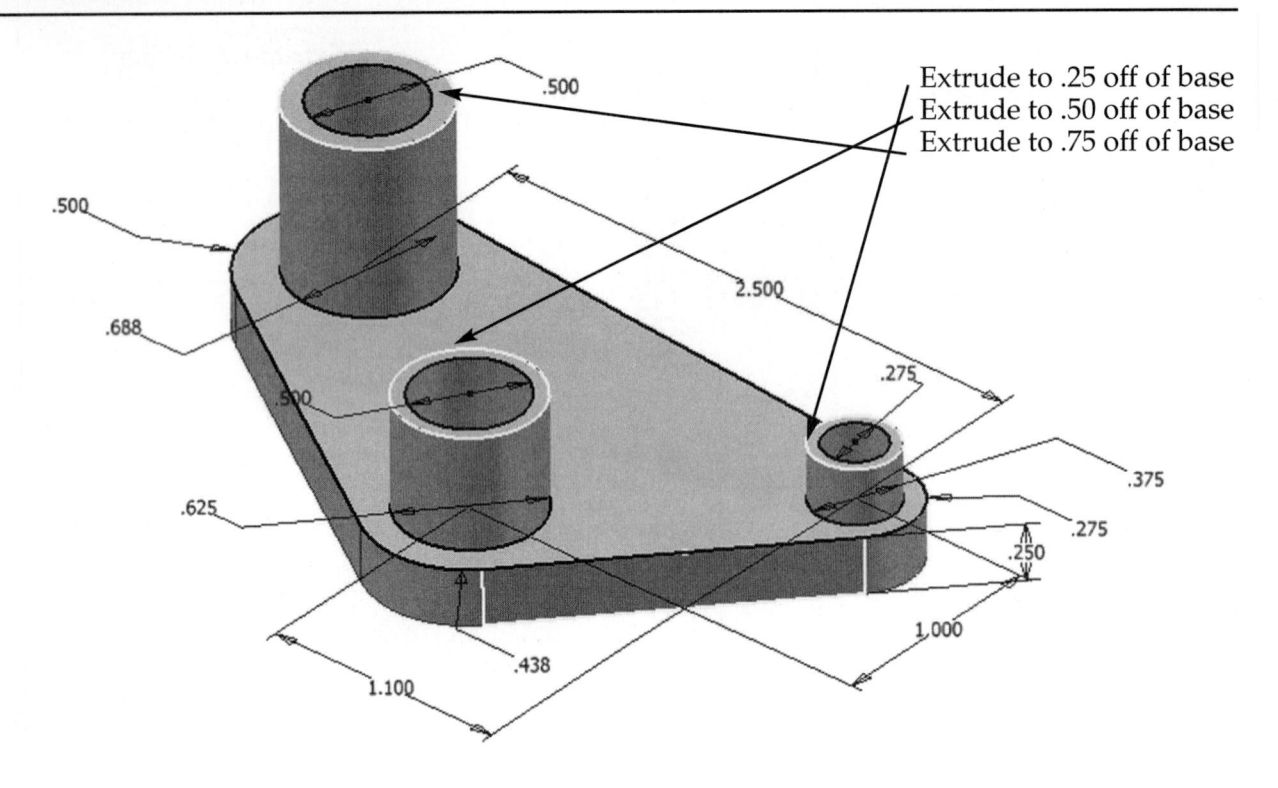

Extrude to .25 off of base
Extrude to .50 off of base
Extrude to .75 off of base

PROBLEM 3-8

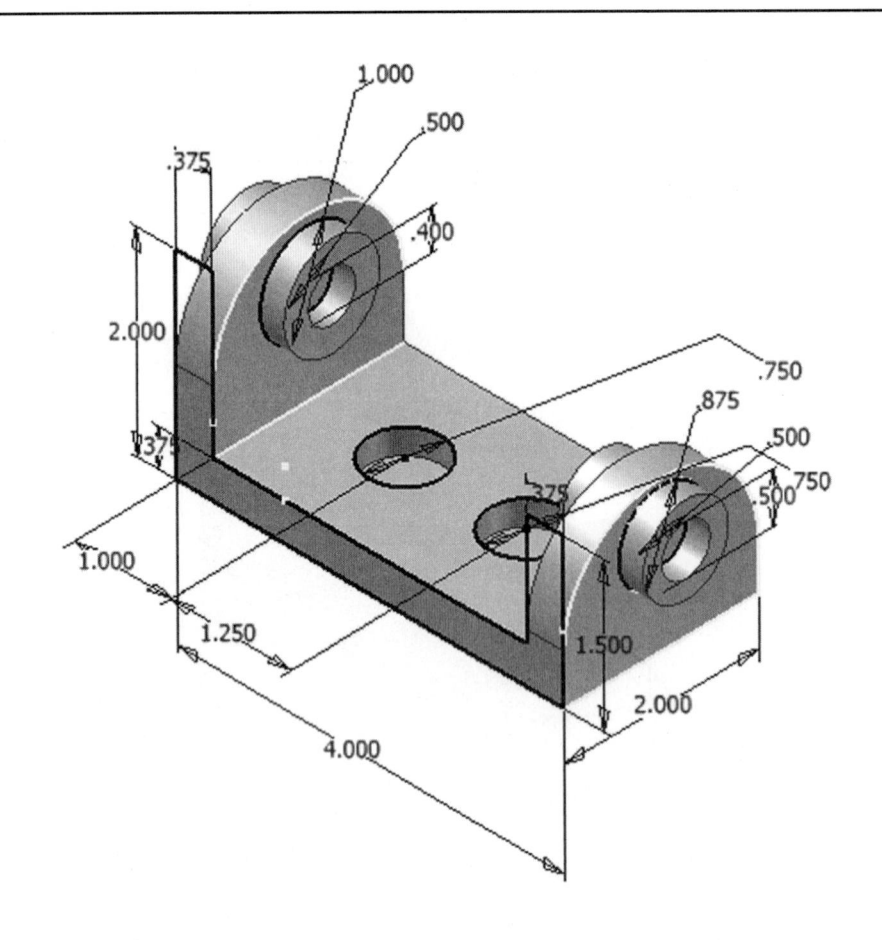

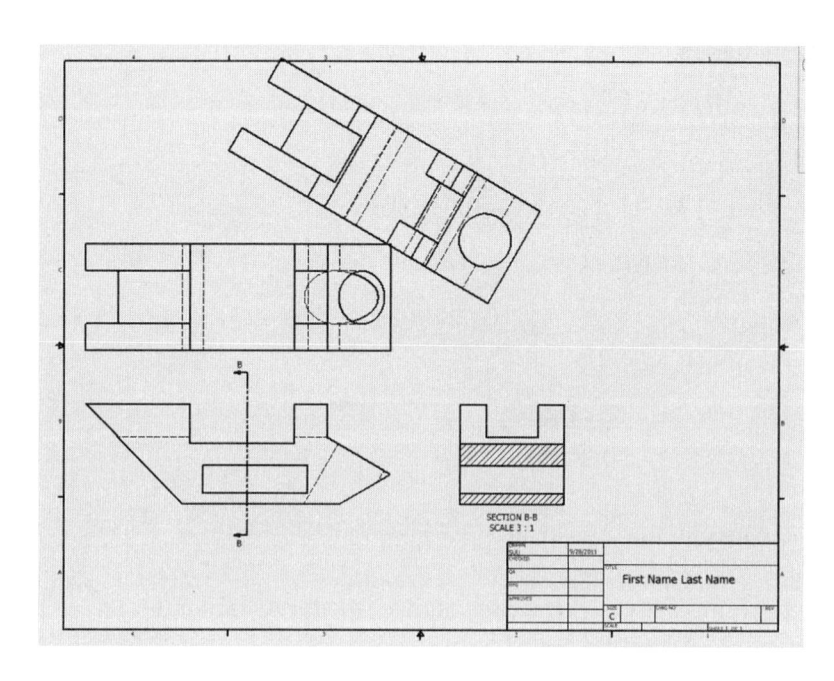

Chapter 4 includes instruction on how to create the drawings shown above.

ADVANCED DETAILED DRAWING PROCEDURES

OBJECTIVES

1. Create an Auxiliary View using the Place Views/Drawing Views Panel

2. Create a Section View using the Place Views /Drawing Views Panel

3. Dimension Views using the Annotation/ Drawing Annotation Panel

4. Create text using the Annotation/Drawing Annotation Panel

1.   Start Autodesk Inventor 2013 by referring to Chapter 1.

2.   After Autodesk Inventor 2013 is running, complete the following part as shown in Figure 4-1.

FIGURE 4-1

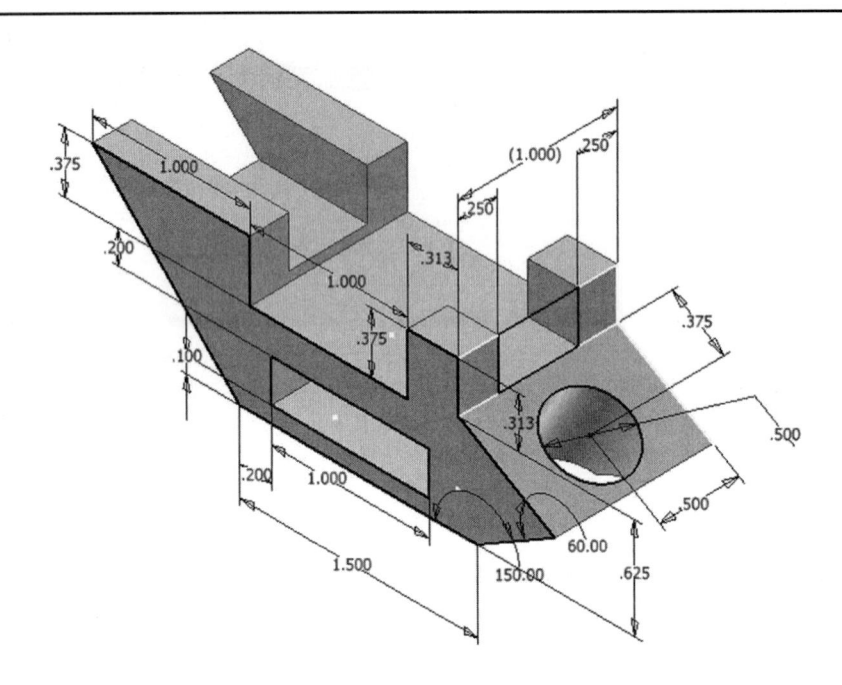

3.   Once the part is complete, project the part into a 3 view drawing as discussed in Chapter 3 using a 3:1 scale as shown in Figure 4-2.

FIGURE 4-2

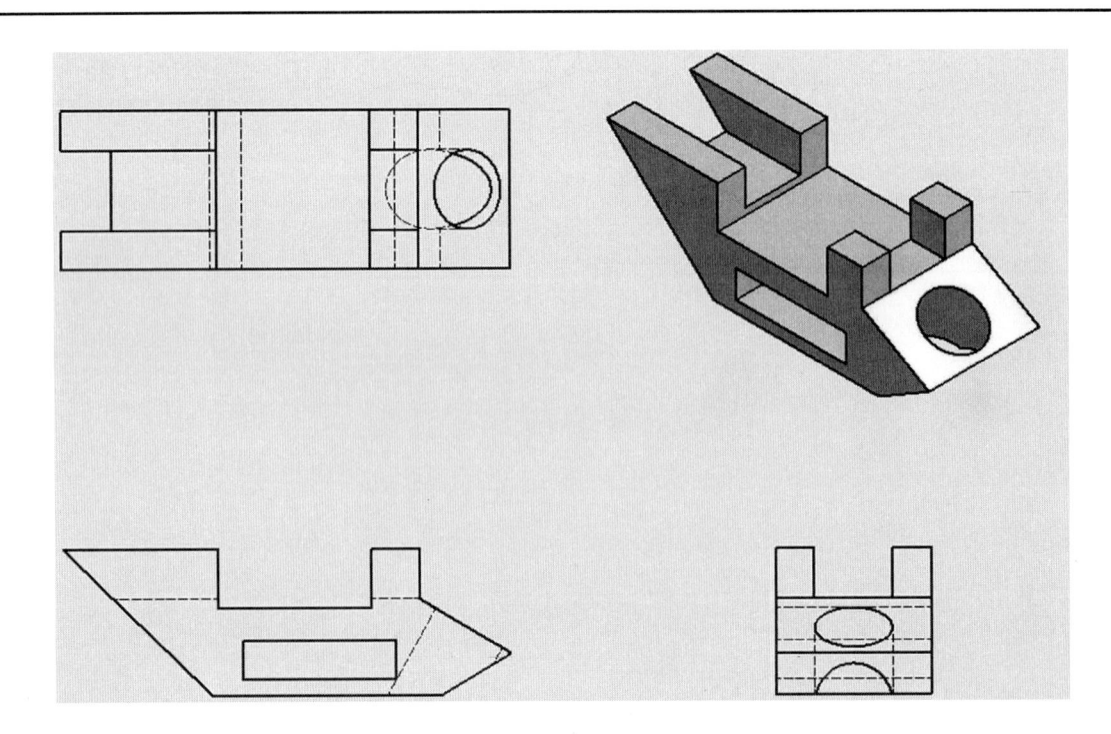

4. Start by moving the views closer together to provide additional room on the drawing. Move the cursor over the top view, causing dots to appear around the view. After the dots appear, left-click on the dots (holding the left mouse button down) and drag the view down closer to the front (base) view as shown in Figure 4-3.

FIGURE 4-3

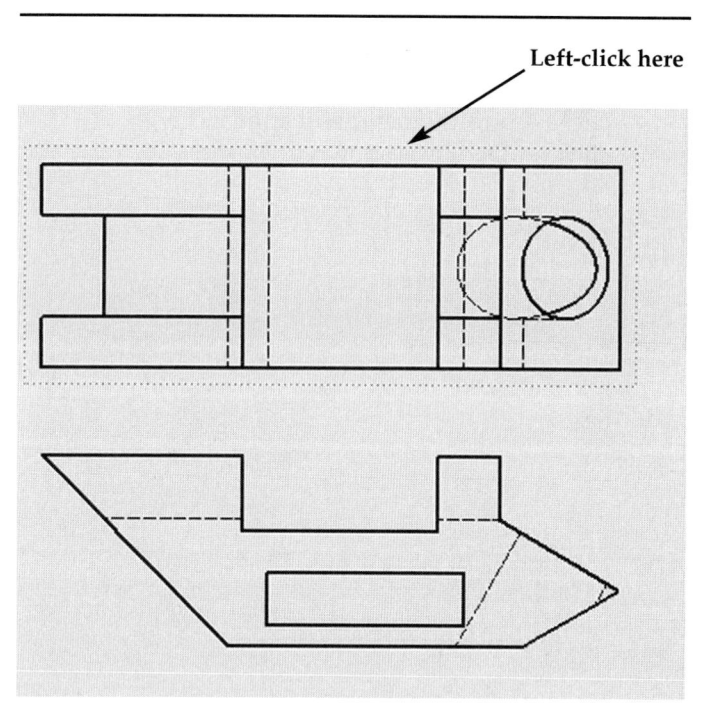

5. Move the side view closer to the front (base) view. Start by moving the cursor over the side view, causing dots to appear around the view. After the dots appear, left-click on the dots (hold the left mouse button down) and drag the view closer to the front (base) view as shown in Figure 4-4.

FIGURE 4-4

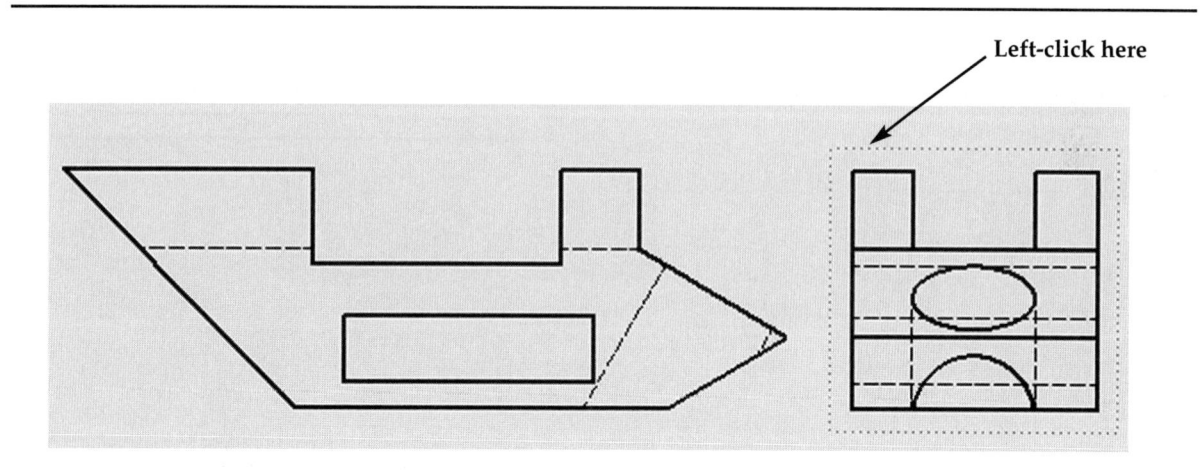

6.   You will need to delete the Isometric View that was created in Chapter 3. Move the cursor near the Isometric View, causing red dots to appear. Right-click. A pop-up menu will appear. Left-click on **Delete** as shown in Figure 4-5.

FIGURE 4-5

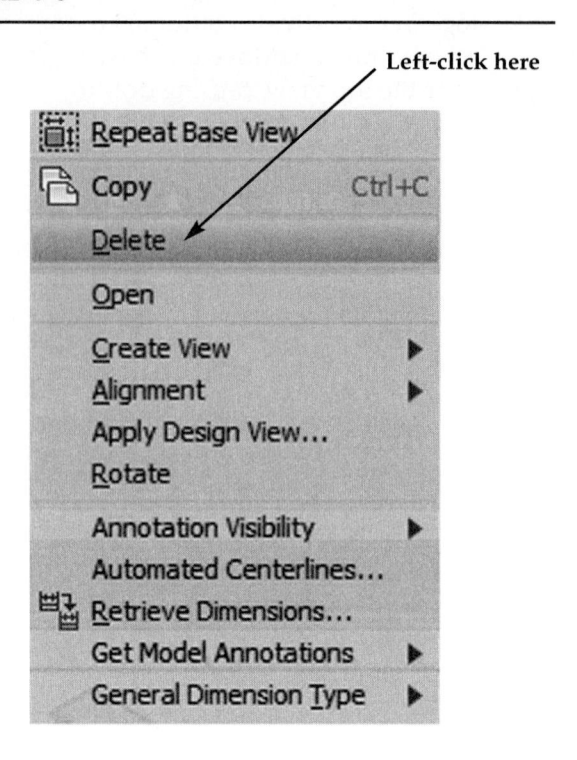

7.   The Delete dialog box will appear. Left-click on **OK** as shown in Figure 4-6.

FIGURE 4-6

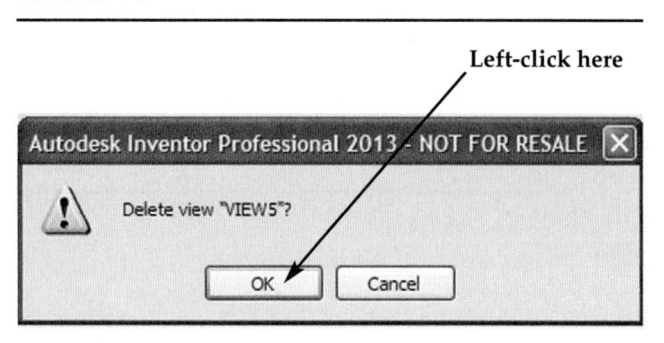

8.   There will now be more room to work. Your screen should look similar to Figure 4-7.

FIGURE 4-7

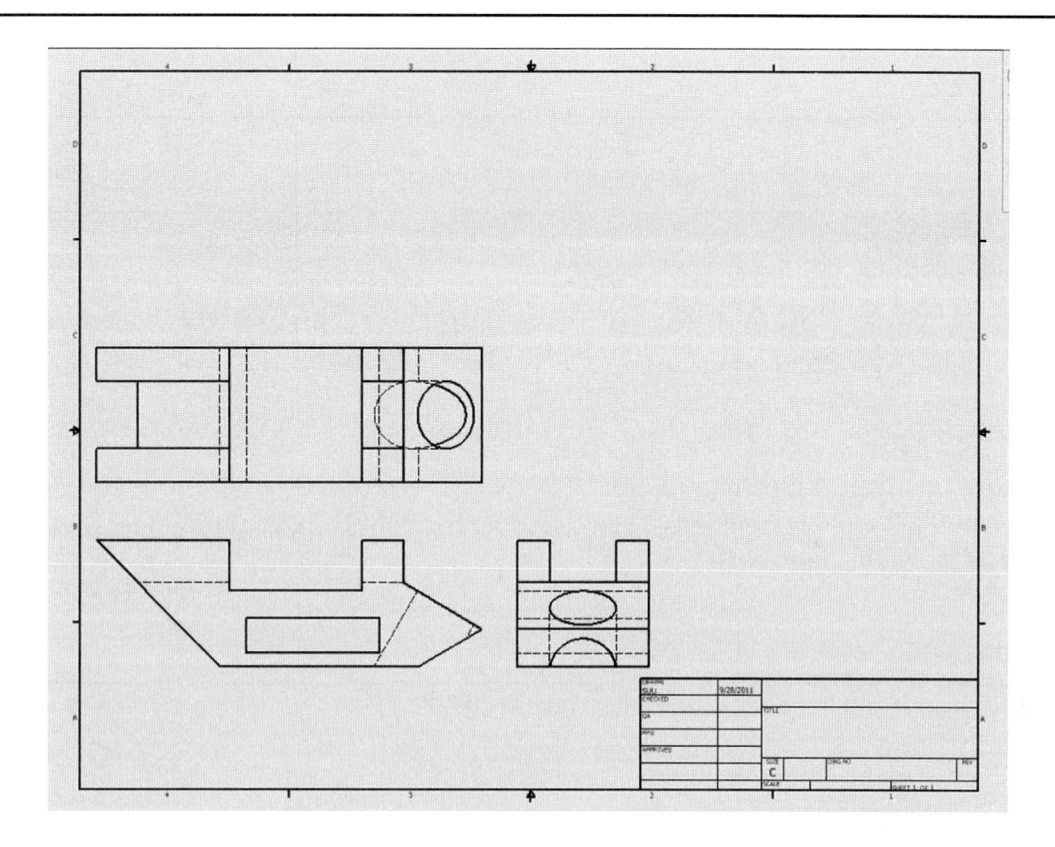

9.   To provide more space on the drawing, the drawing view scale will have to be reduced. Right-click on the front (base) view. A pop-up menu will appear. Left-click on **Edit View** as shown in Figure 4-8.

FIGURE 4-8

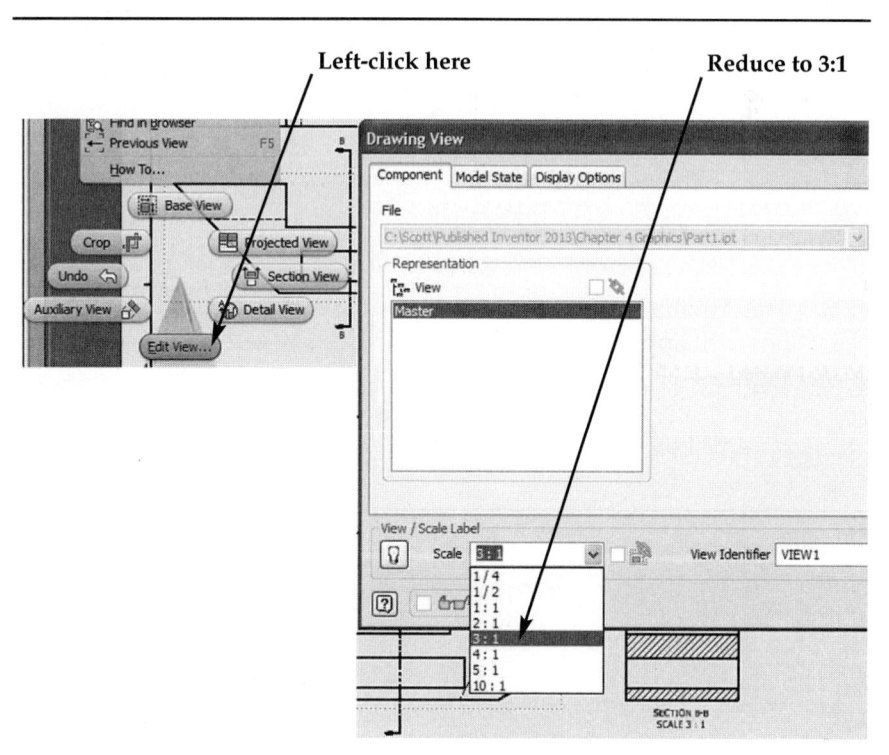

CREATE AN AUXILIARY VIEW USING THE DRAWING VIEWS PANEL

10. Move the cursor to the upper left portion of the screen and left-click on **Auxiliary View** as shown in Figure 4-9.

FIGURE 4-9

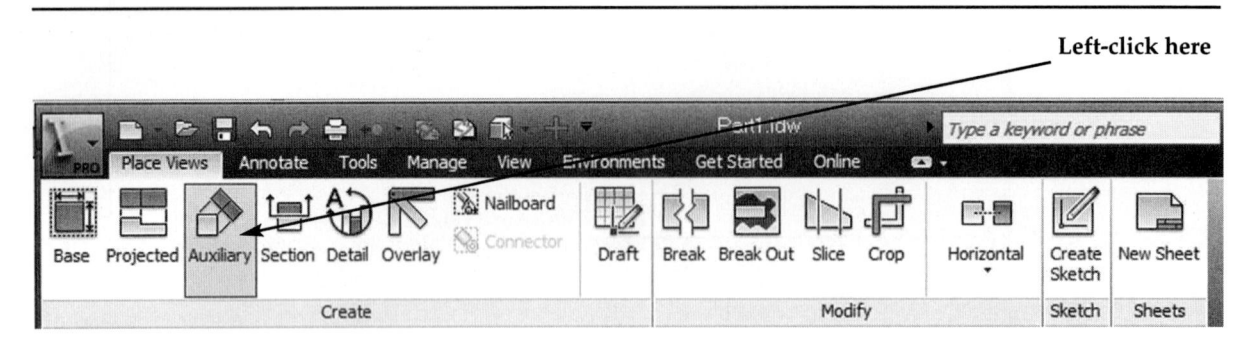

Left-click here

11. Move the cursor to the front (base) view, causing red dots to appear around the view. Left-click once as shown in Figure 4-10.

FIGURE 4-10

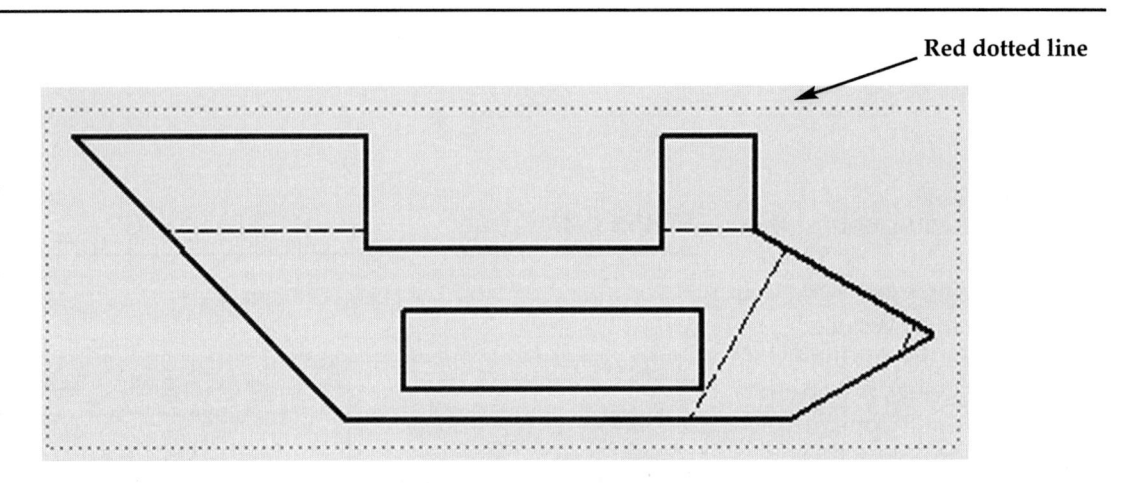

Red dotted line

12. The Auxiliary View dialog box will appear as shown in Figure 4-11.

FIGURE 4-11

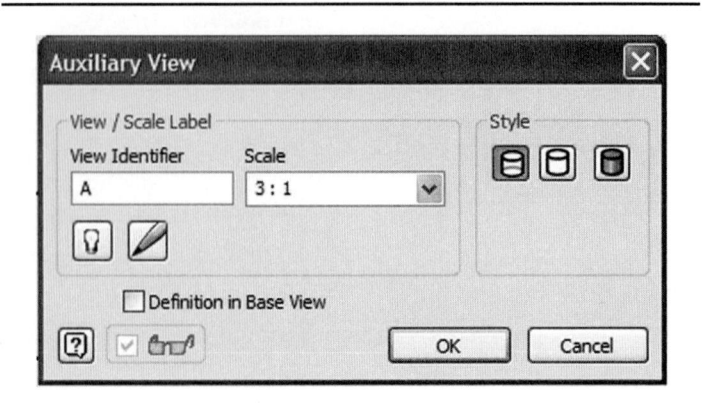

13. Move the cursor over the wedge line, causing it to turn red. Left-click as shown in Figure 4-12.

FIGURE 4-12

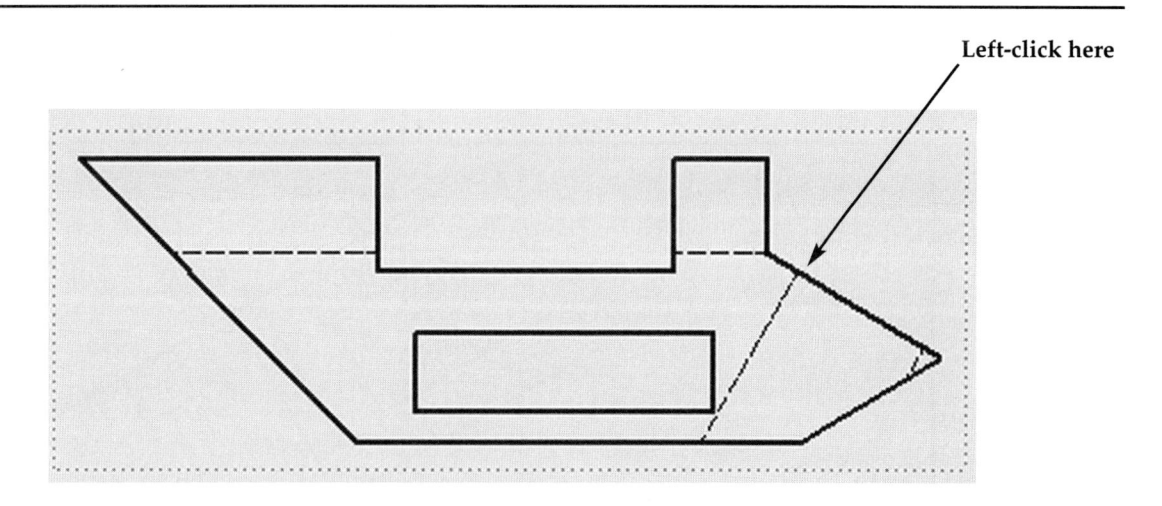

Left-click here

14. Inventor will create an Auxiliary View from the selected surface. The view will be attached to the cursor as shown in Figure 4-13.

FIGURE 4-13

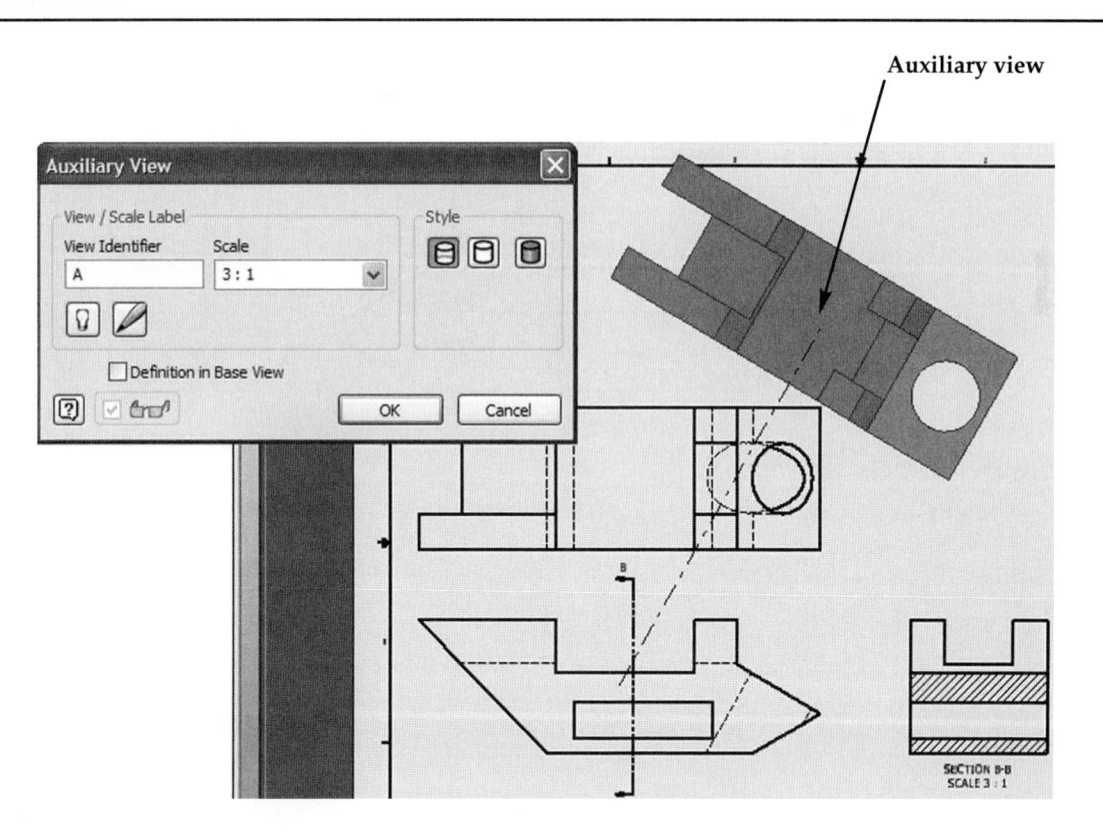

Auxiliary view

15. Move the cursor towards the upper right and left-click. The Auxiliary View dialog box will close as shown in Figure 4-14.

FIGURE 4-14

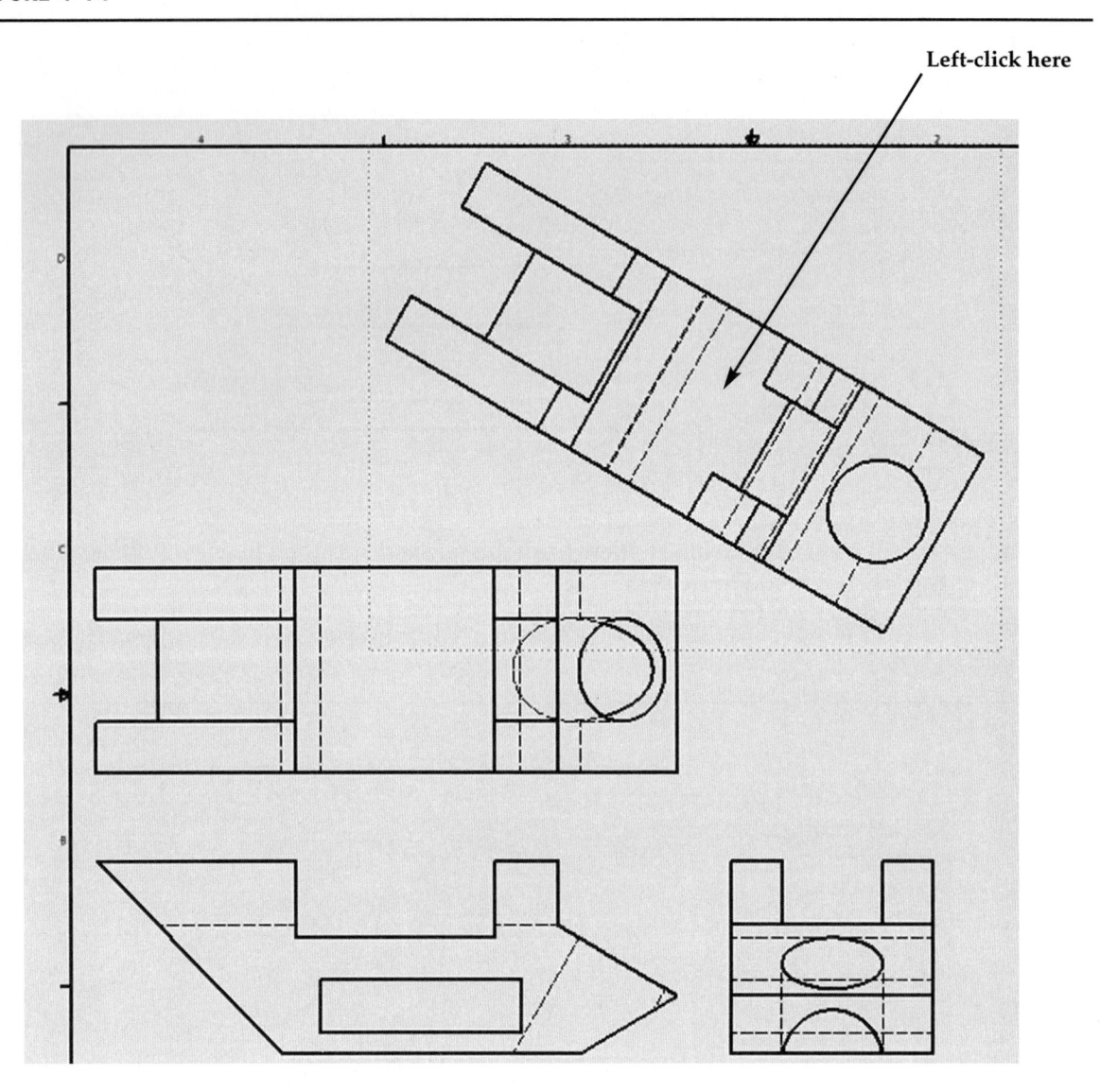

16.   Your screen should look similar to Figure 4-15.

FIGURE 4-15

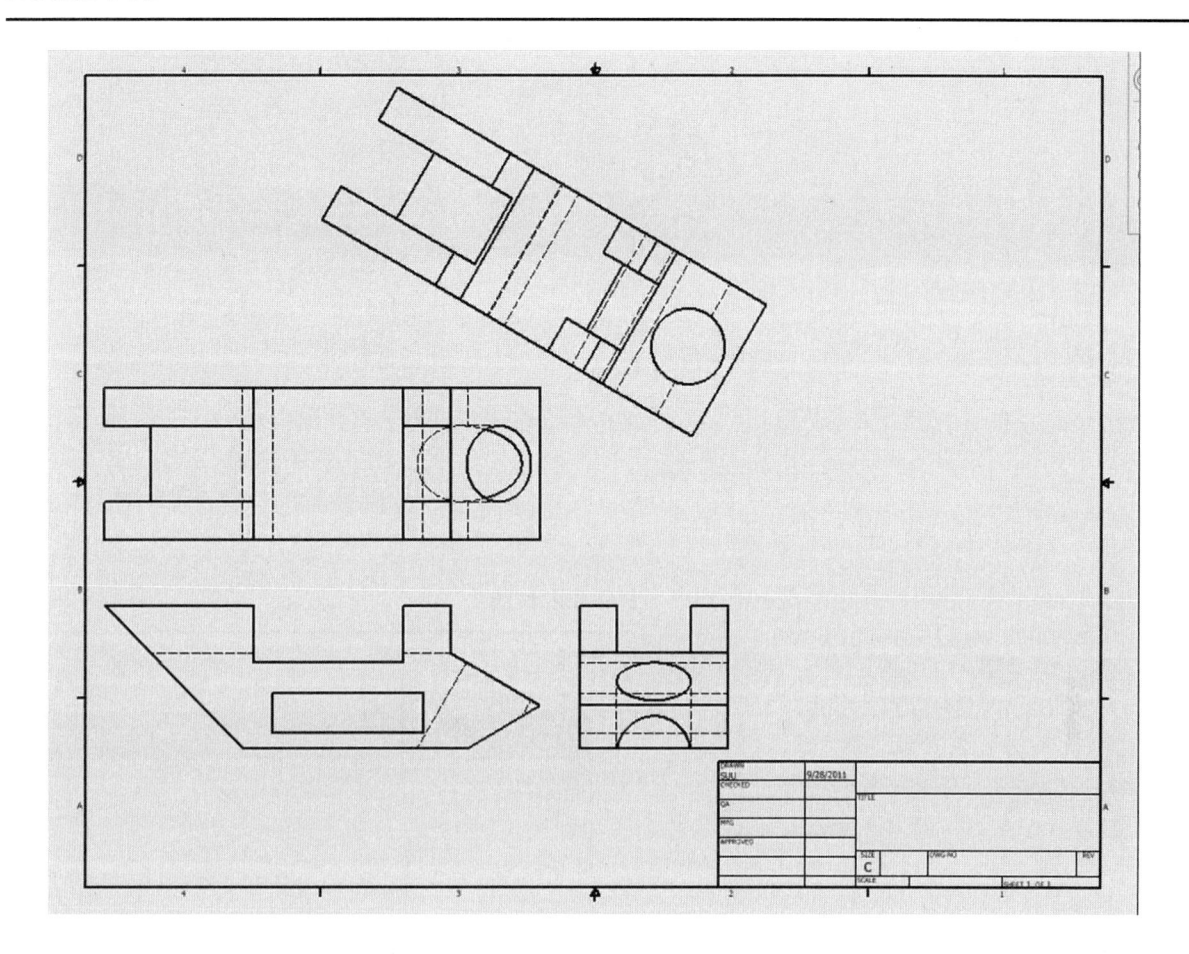

17.   Move the cursor to the side view, causing red dots to appear as shown in Figure 4-16.

FIGURE 4-16

Red dots

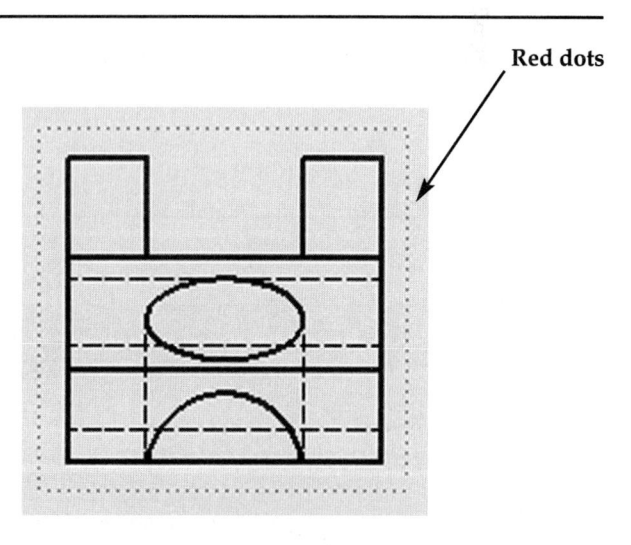

18. Right-click on the view. A pop-up menu will appear. Left-click on **Delete** as shown in Figure 4-17.

FIGURE 4-17

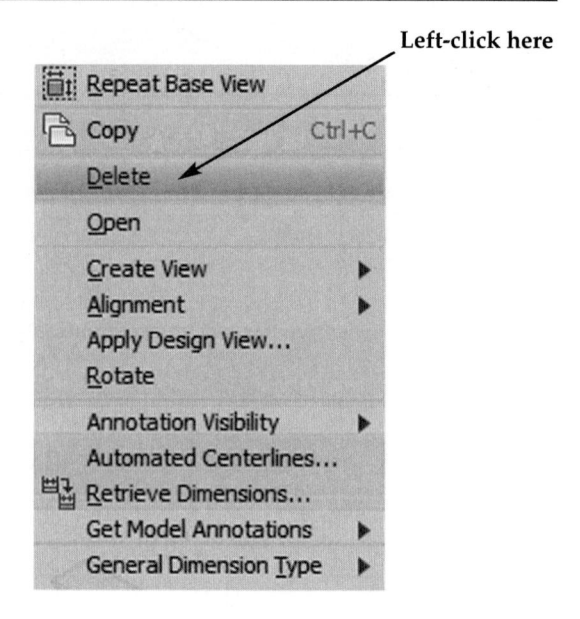

19. A Delete dialog box will appear. Left-click on **OK** as shown in Figure 4-18.

FIGURE 4-18

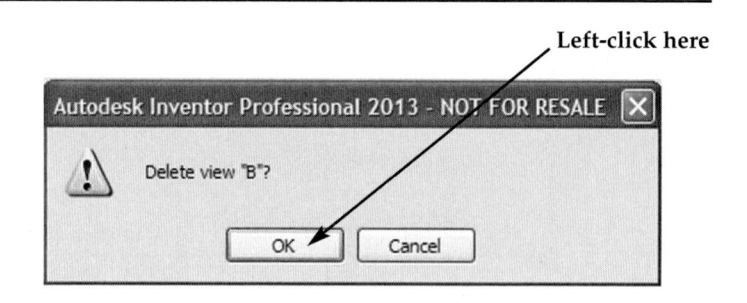

CREATE A SECTION VIEW USING THE DRAWING VIEWS PANEL

20. Move the cursor to the upper left portion of the screen and left-click on **Section View** as shown in Figure 4-19.

FIGURE 4-19

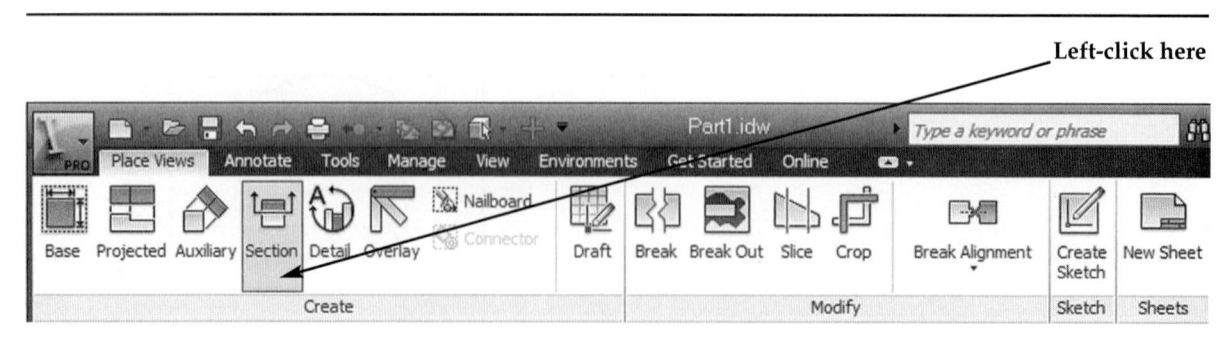

21. Move the cursor over the front view, causing red dots to appear around the view as shown in Figure 4-20.

FIGURE 4-20

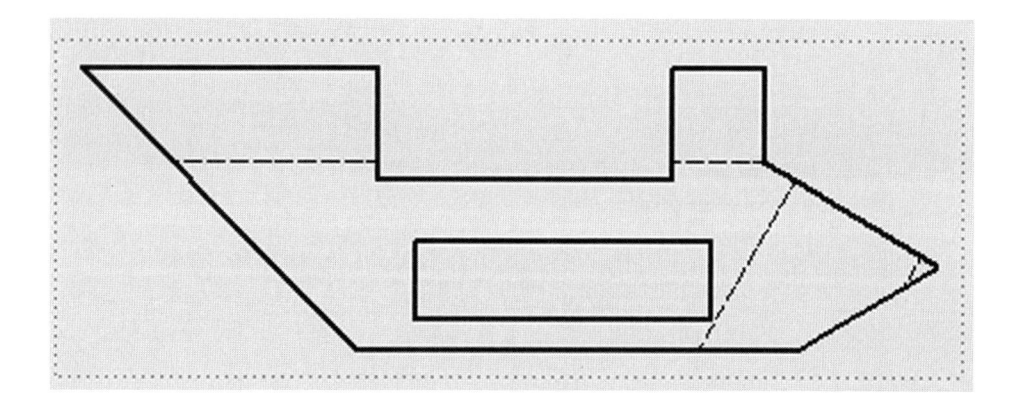

22. Left-click on the view, causing the red dots to turn into a solid red line as shown in Figure 4-21.

FIGURE 4-21

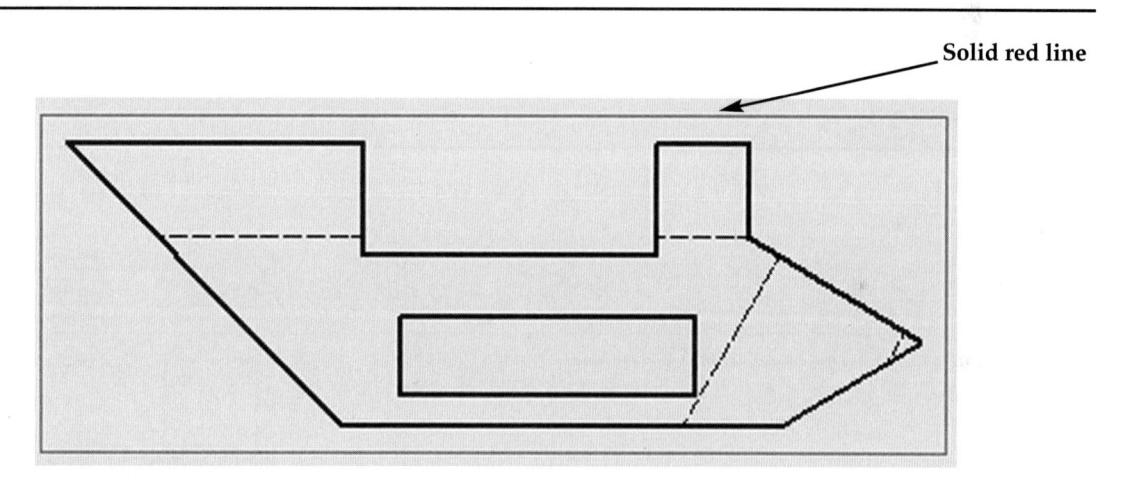

Solid red line

23. Move the cursor around outside the red line and wait for the dotted line to appear as shown in Figure 4-22. It may take a few seconds before the line appears.

FIGURE 4-22

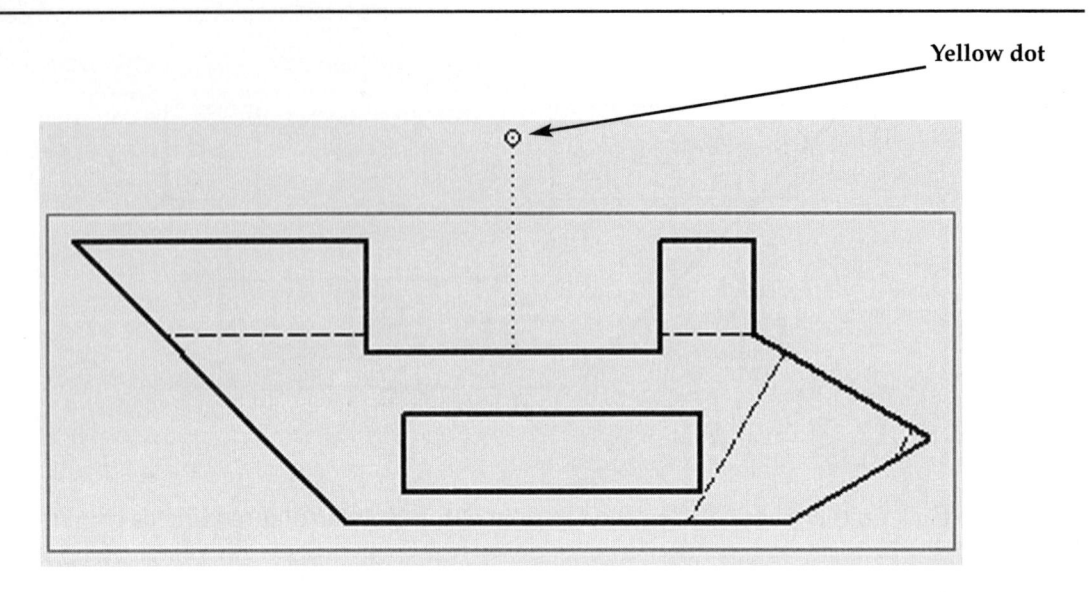

Yellow dot

24. Left-click on the yellow dot, move the line down, and left-click as shown in Figure 4-23.

FIGURE 4-23

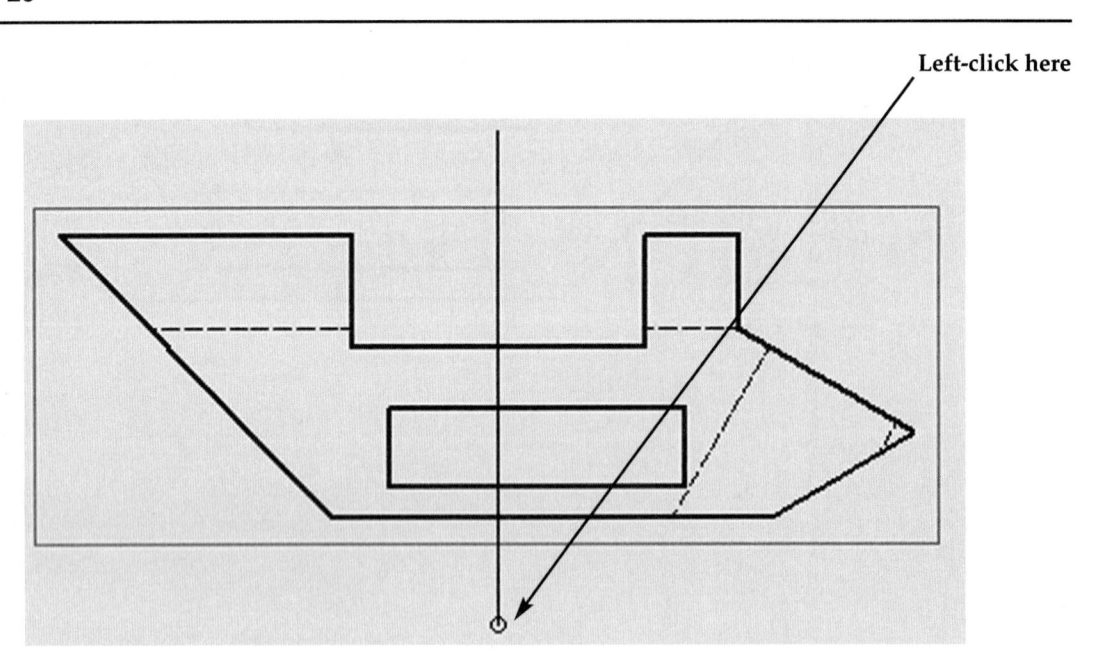

Left-click here

25. Right-click in the same location. A pop-up menu will appear. Left-click on **Continue** as shown in Figure 4-24.

FIGURE 4-24

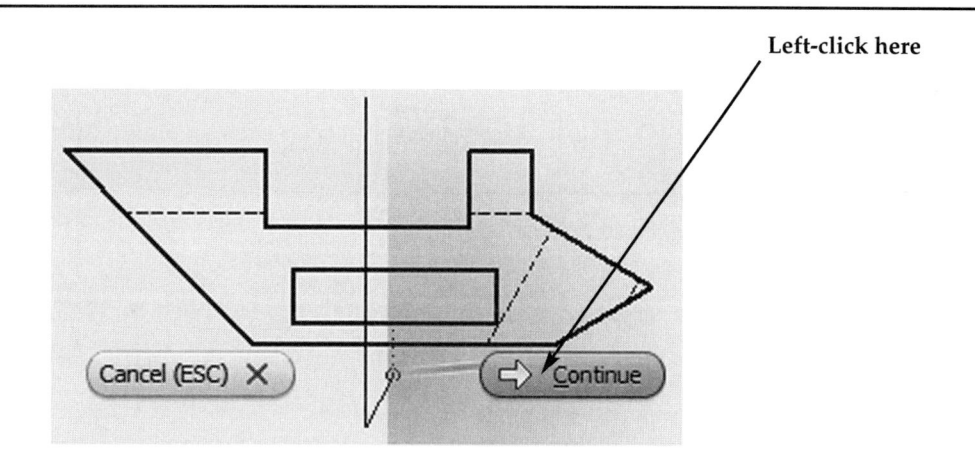

26. The Section View dialog box will appear as shown in Figure 4-25. The Section View will be attached to the cursor. Move the cursor to the right where the side view was located and left-click once. The Section View dialog box will close.

FIGURE 4-25

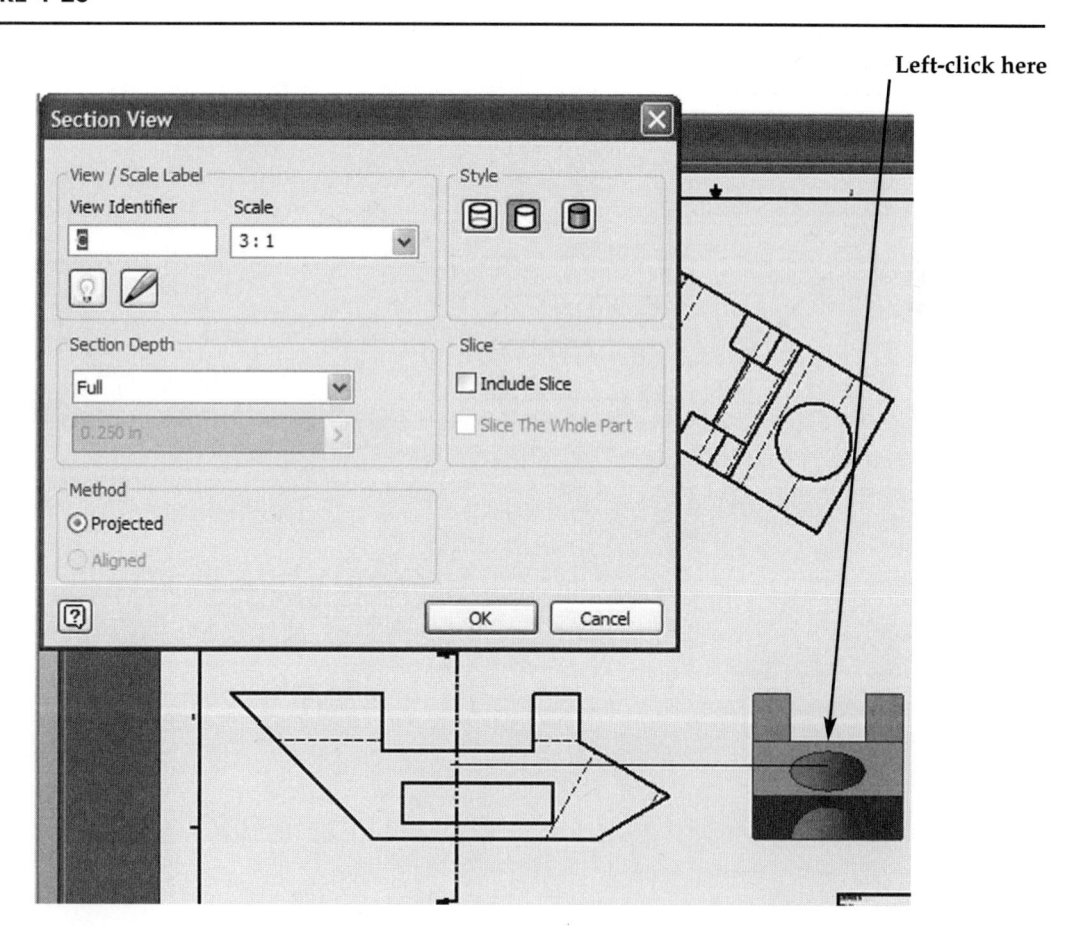

27.   Inventor will create a Section View to the right as shown in Figure 4-26.

FIGURE 4-26

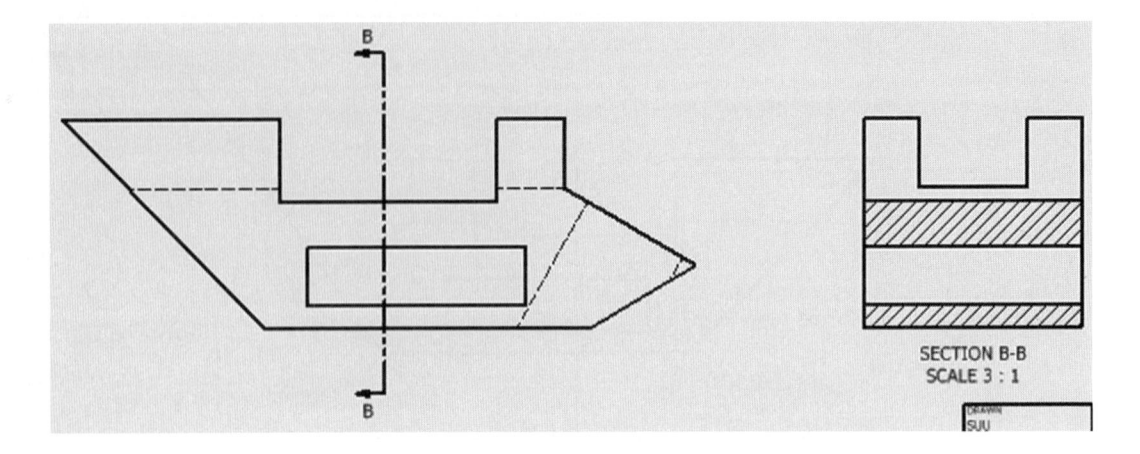

28.   Inventor will create a section view that represents wherever the cutting plane line cuts through the part. Move the cursor over the center of the cutting plane line, causing it to turn red as shown in Figure 4-27.

FIGURE 4-27

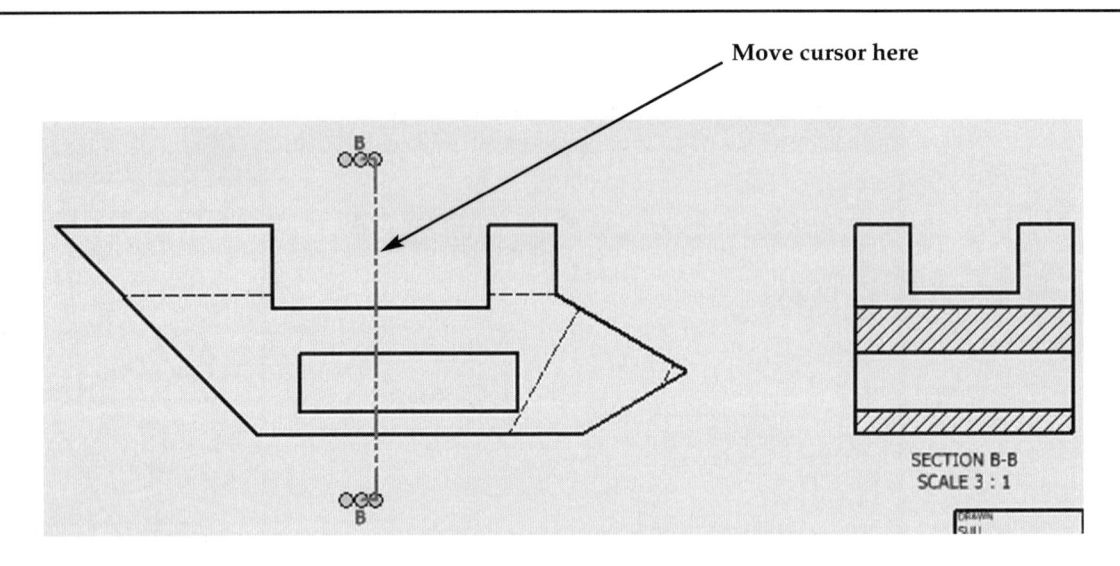

29.   Once the line becomes highlighted (turns red), left-click (holding the left mouse button down) and drag the cursor to the right. The cutting plane line will become a normal looking line while attached to the cursor as shown in Figure 4-28.

FIGURE 4-28

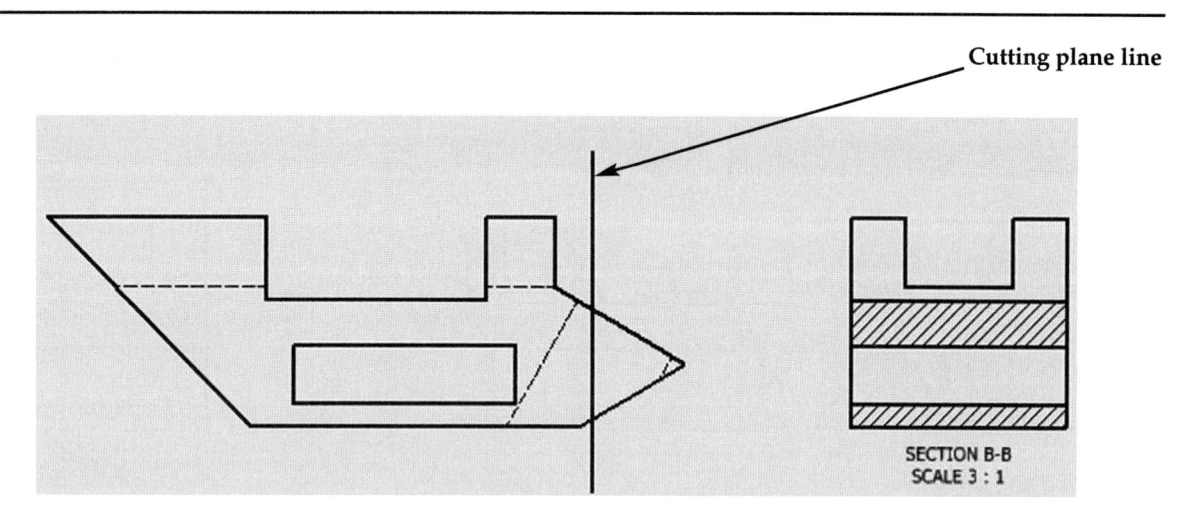

30.   Once the cutting plane line has been moved to a new location, release the left mouse button. The side view now reflects the new location of the cutting plane line as shown in Figure 4-29.

FIGURE 4-29

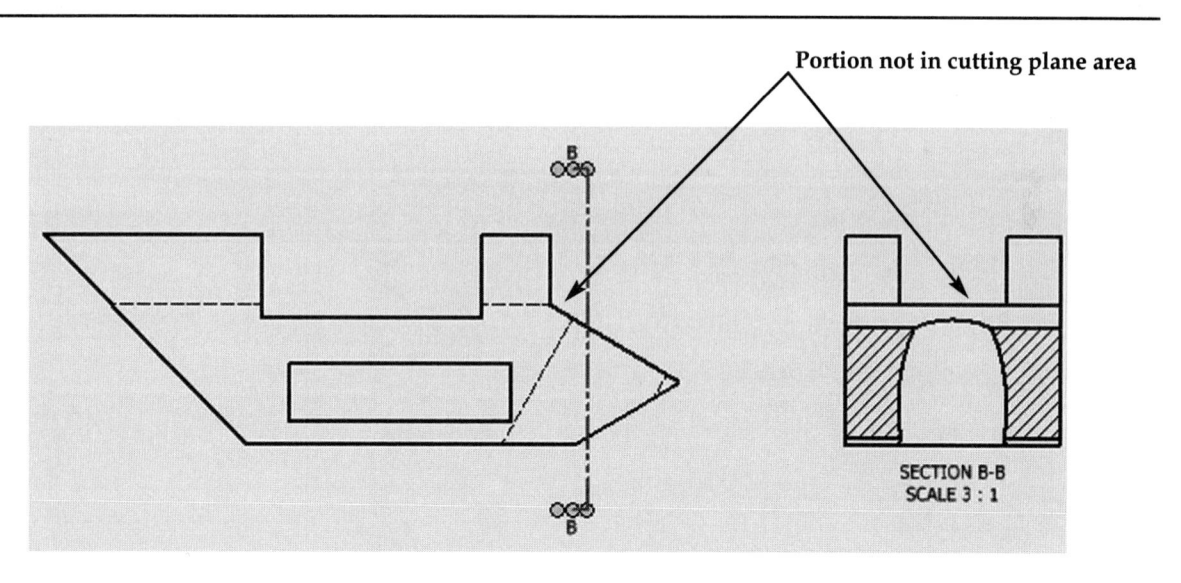

31. Left-click (holding the left mouse button down) on the cutting plane line and move it back to its original location. Notice that the cross hatch in the section view will update to reflect the location of the cutting plane line as shown in Figure 4-30.

FIGURE 4-30

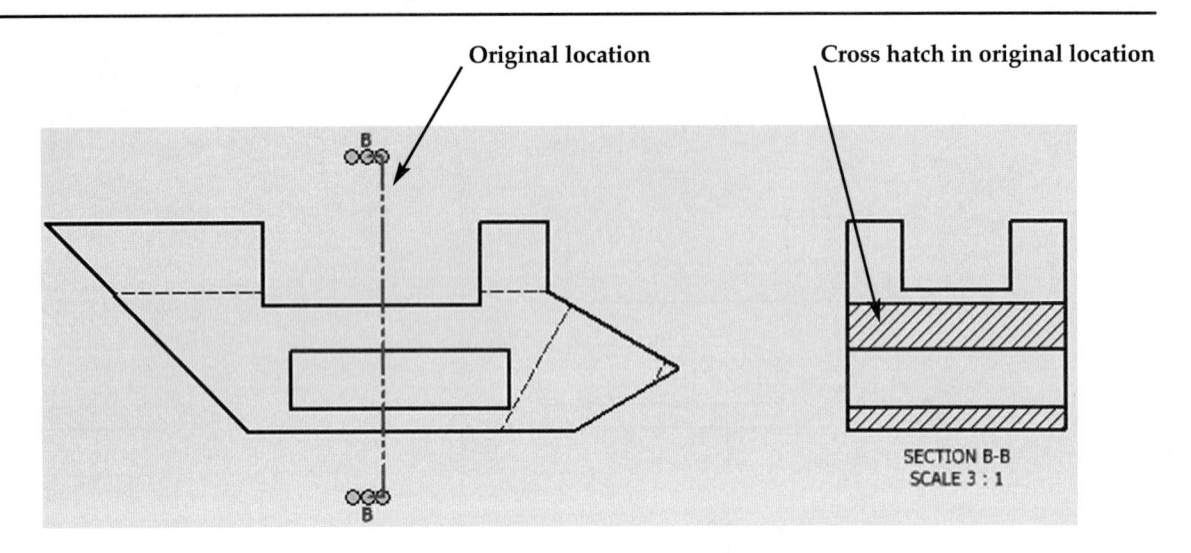

CREATE A BROKEN VIEW USING THE BREAK COMMAND

32. Move the cursor to the middle left portion of the screen and left-click on **Break** as shown in Figure 4-31.

FIGURE 4-31

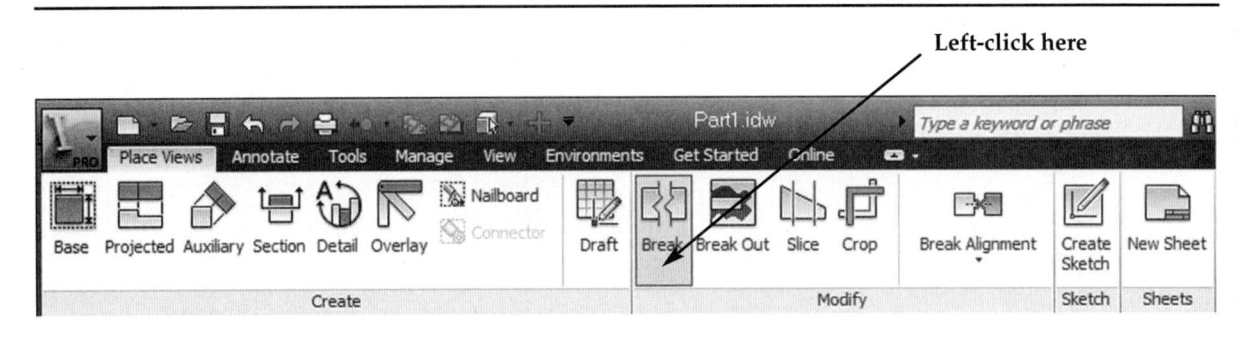

33.   Move the cursor to the front view and left-click once as shown in Figure 4-32.

FIGURE 4-32

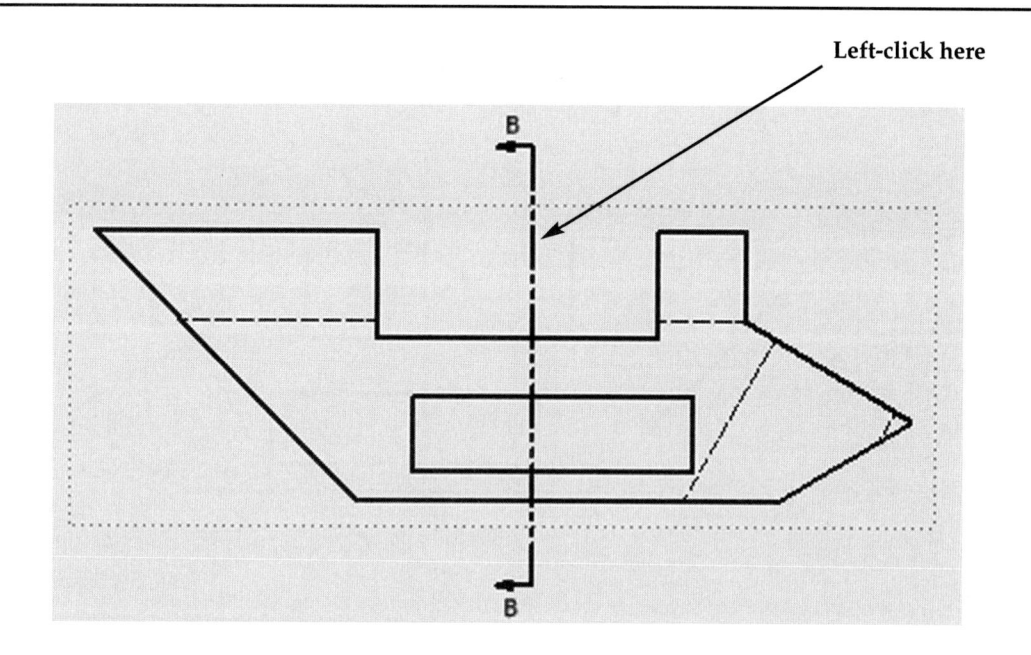

34.   The Break dialog box will appear. The Break symbol will also be attached to the cursor as shown in Figure 4-33.

FIGURE 4-33

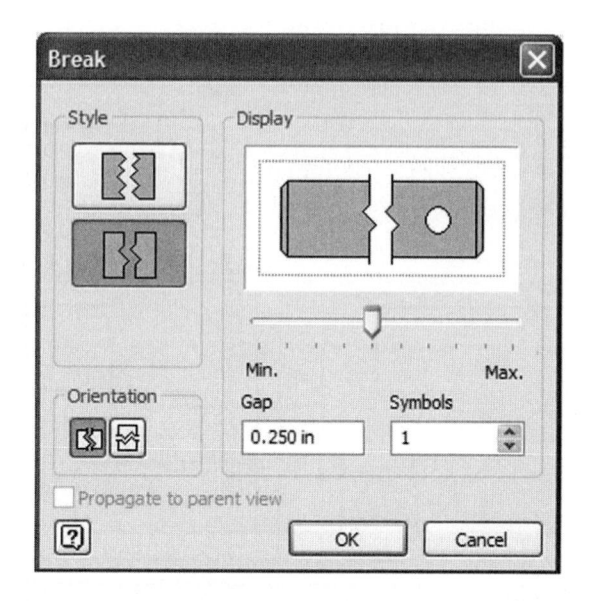

35.   Left-click on the part, causing a red box to appear as shown in Figure 4-34.

FIGURE 4-34

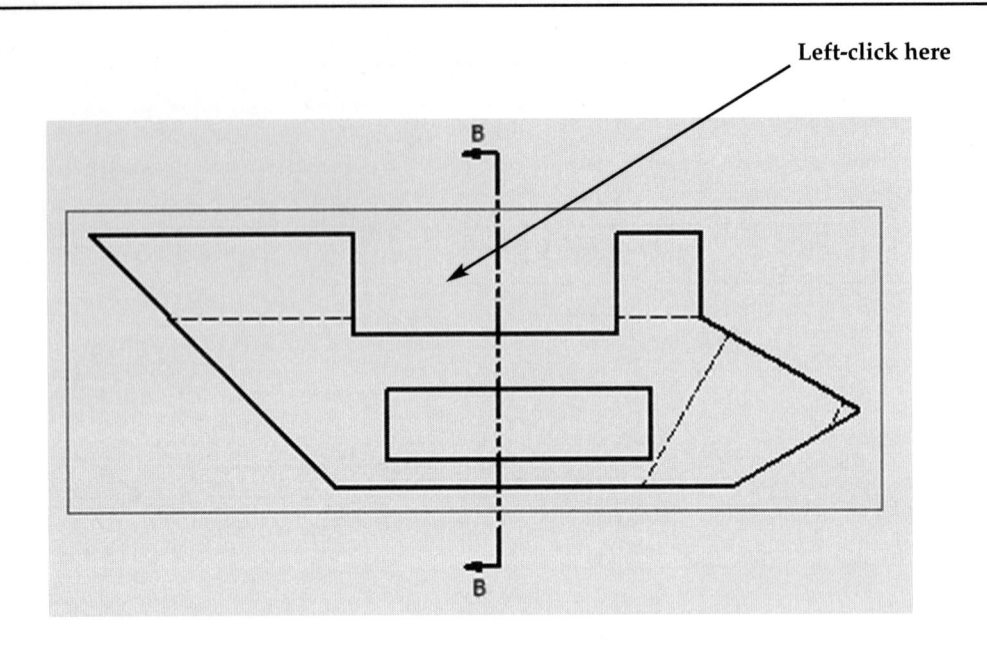

36.   Move the cursor to the left side of the cutting plane line and left-click once. Move the cursor to the left. Another line will appear next to the first line. These two lines represent the size of the gap that Inventor will create in the part. A third line will be attached to the cursor. This line represents how much of the part will be removed from the view. Move the cursor to the far left portion of the part and left-click as shown in Figure 4-35.

FIGURE 4-35

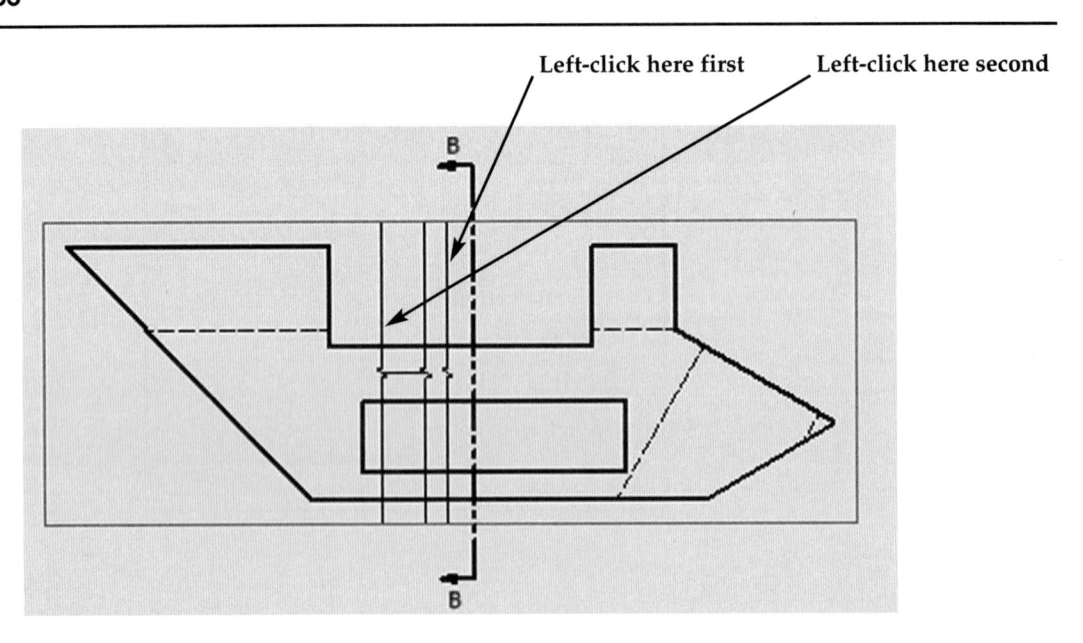

37.   Inventor will remove sections from both the front and top views as shown in Figure 4-36.

FIGURE 4-36

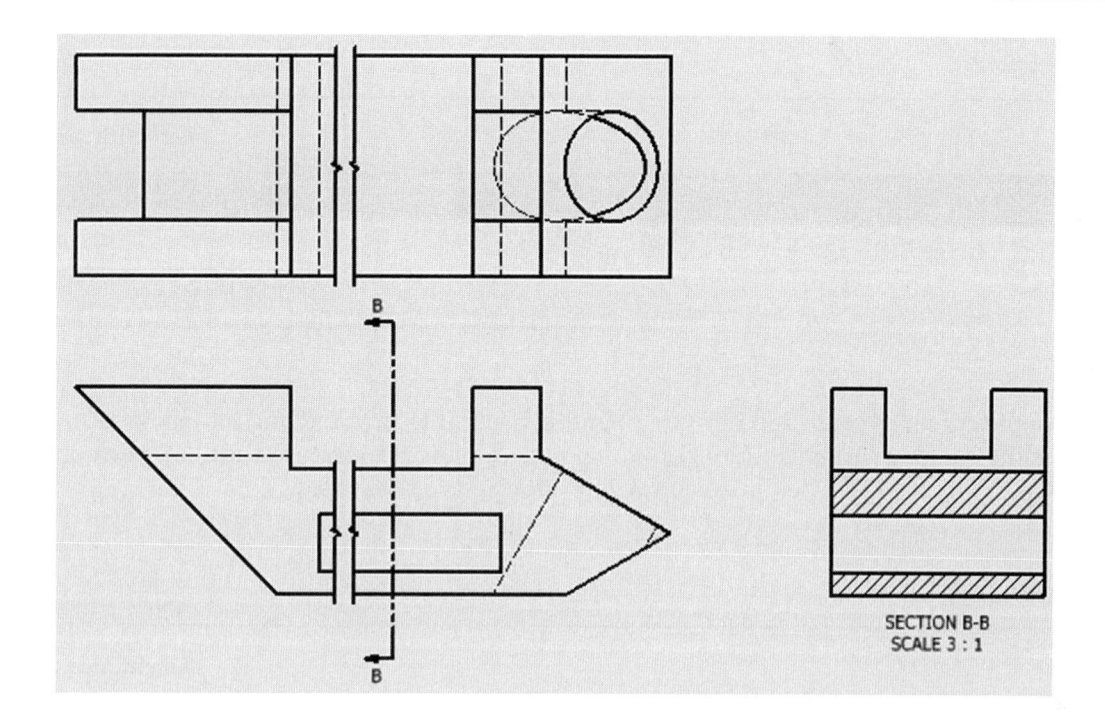

38.   Move the cursor to the upper left portion of the screen and left-click on "Undo" as shown in Figure 4-37.

FIGURE 4-37

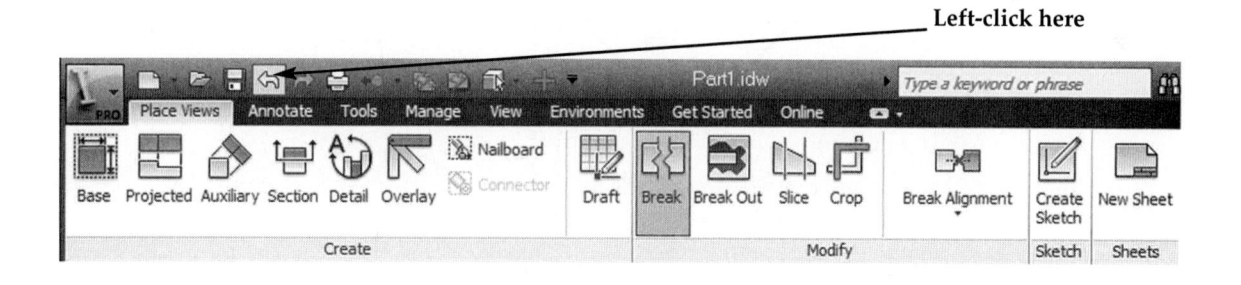

DIMENSION VIEWS USING THE DRAWING ANNOTATION PANEL

39.   Move the cursor to the upper left portion of the screen and left-click on the **Annotation** tab. Left-click on **Dimension** as shown in Figure 4-38.

FIGURE 4-38

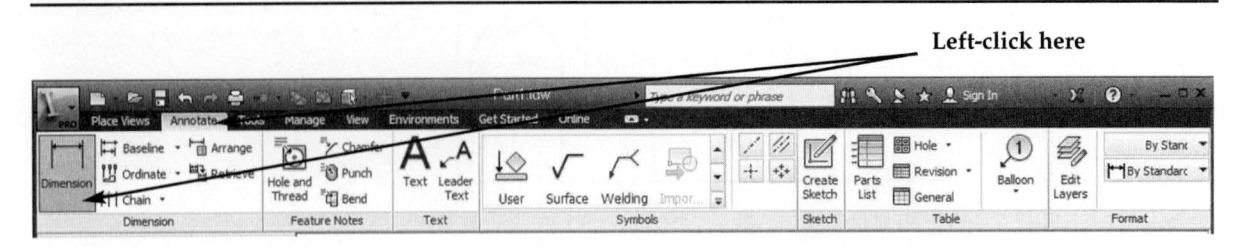

40.   Move the cursor over the top horizontal line, causing it to turn red and left-click once. Then move the cursor over the bottom horizontal line, causing it to turn red and left-click once as shown in Figure 4-39. The dimension will be attached to the cursor.

FIGURE 4-39

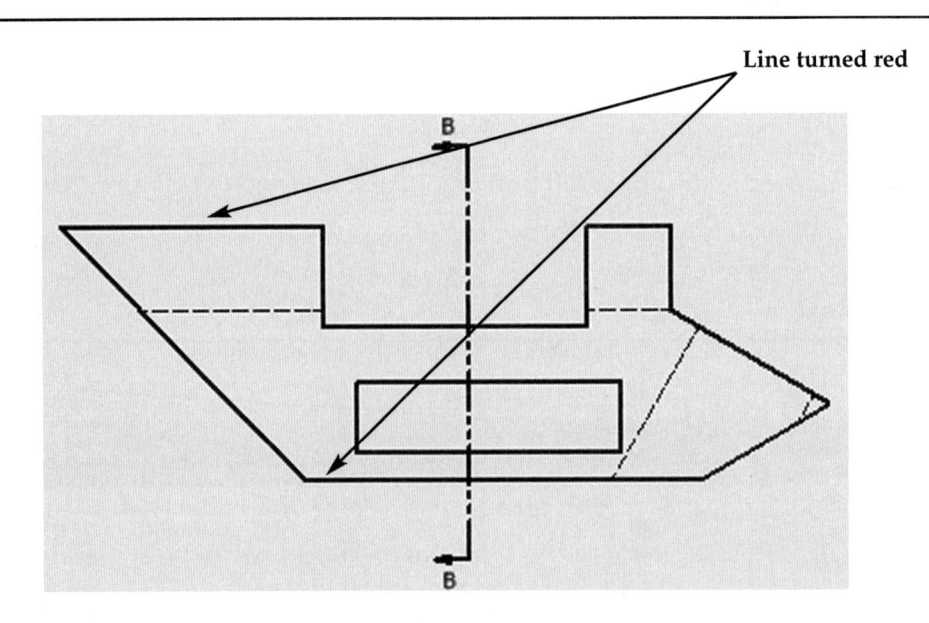

41. Move the cursor to the left and left-click once. The actual dimension of the line will appear as shown in Figure 4-40. The Edit Dimension dialog box (not shown) will appear. Left-click on **OK**.

FIGURE 4-40

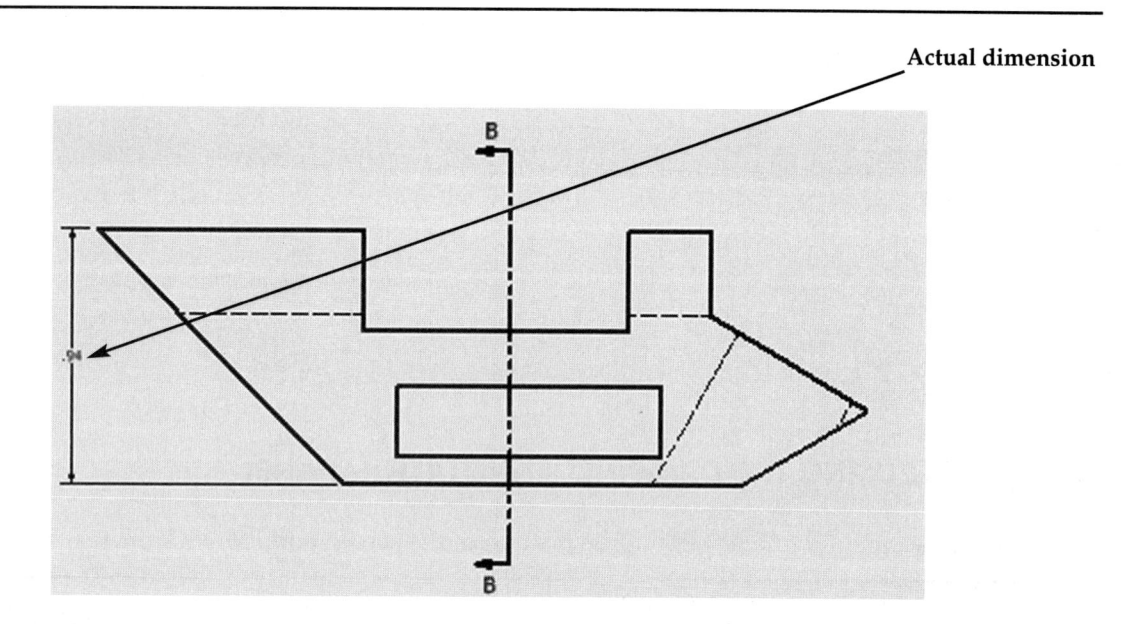

42. Finish dimensioning the part to your own satisfaction. When the part is satisfactorily dimensioned, save the file to a location where it can easily be retrieved.

43. To delete an unwanted dimension, move the cursor over the dimension. The dimension will turn red and several green dots will appear as shown in Figure 4-41.

FIGURE 4-41

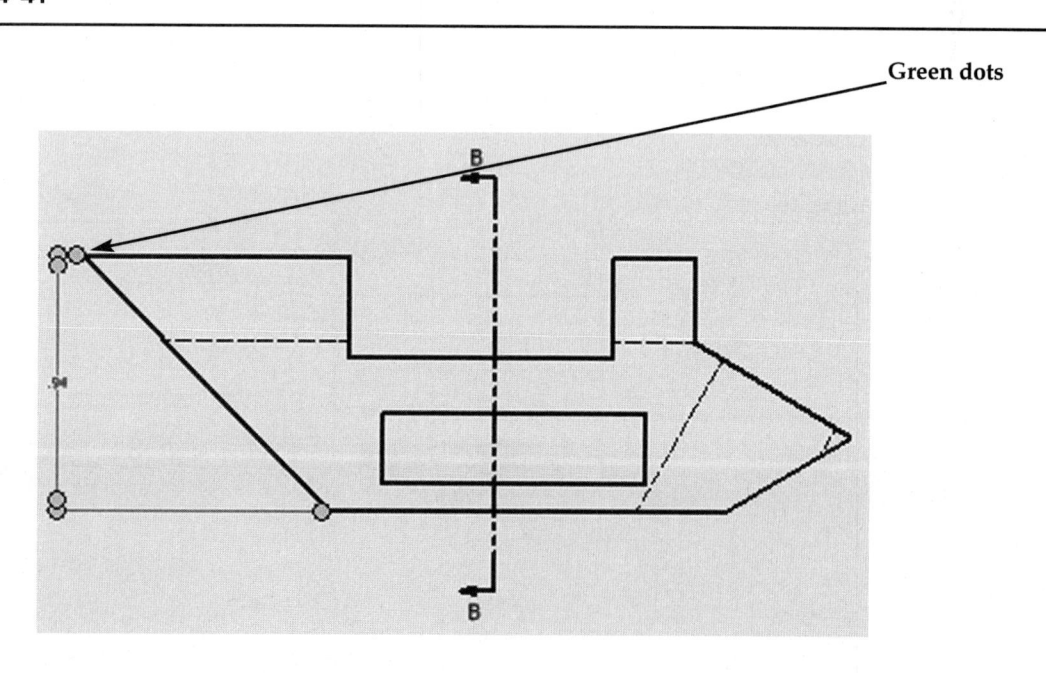

44.   Right-click on the dimension. A pop-up menu will appear. Left-click on **Delete** as shown in Figure 4-42.

FIGURE 4-42

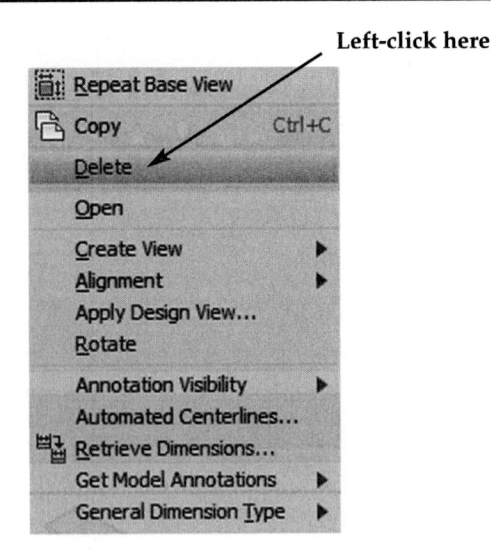

45.   Move the cursor to the upper middle portion of the screen and left-click on **Text** as shown in Figure 4-43.

CREATE TEXT USING THE DRAWING ANNOTATION PANEL

FIGURE 4-43

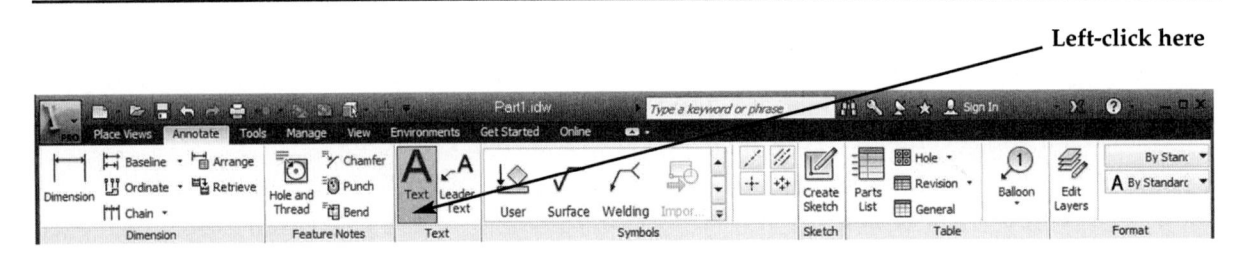

46.   Move the cursor to the title block location as shown in Figure 4-44. Left-click once when the yellow dot appears.

FIGURE 4-44

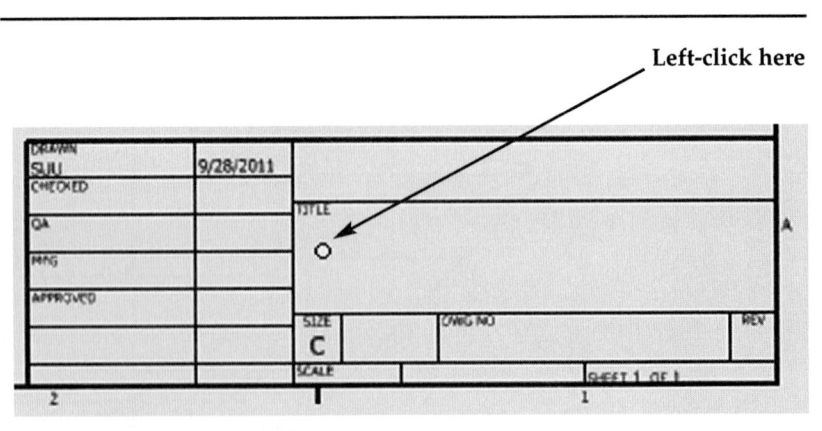

47. The Format Text dialog box will appear. Left-click on the drop-down box and change the text height to **.240** inches as shown in Figure 4-45.

FIGURE 4-45

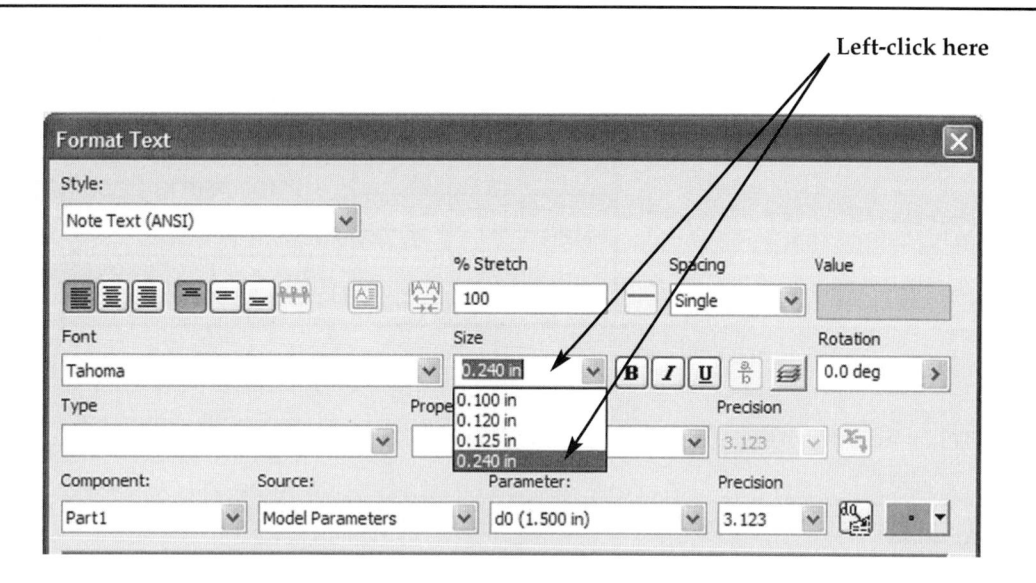

48. Move the cursor to the open area located in the lower half of the Format Text dialog box and enter your first and last name. Text will appear near the flashing cursor as shown in Figure 4-46.

FIGURE 4-46

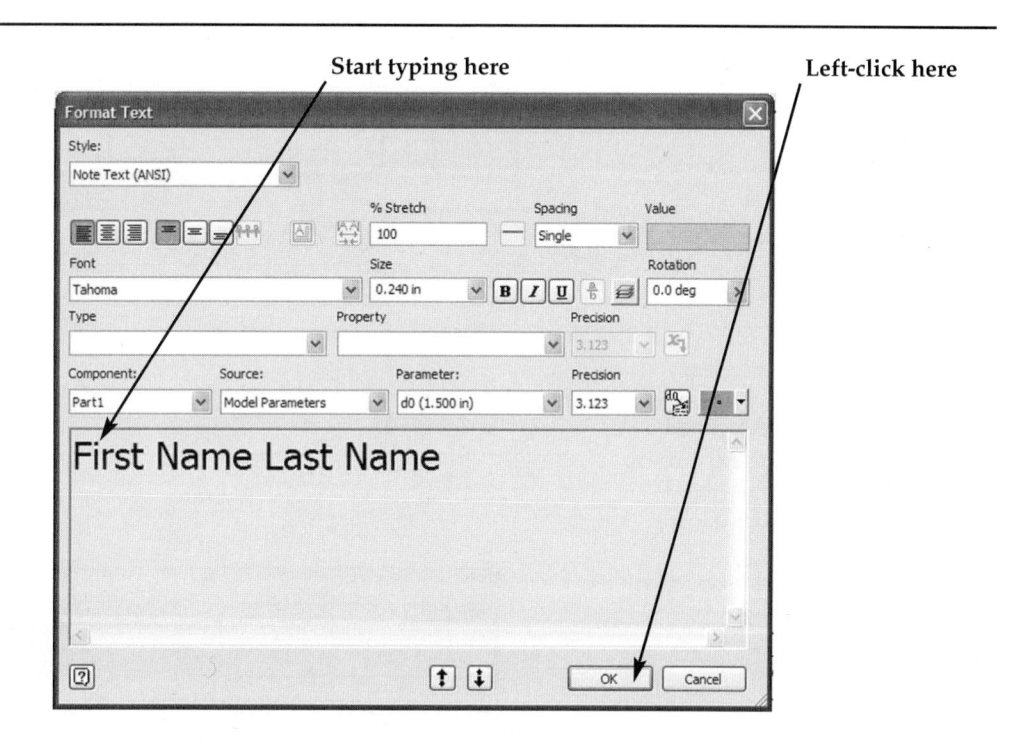

49. After text has been entered, left-click on **OK** as shown in Figure 4-46.

50. The Format Text dialog box will close.

51. Text will appear in the title block as shown in Figure 4-47.

FIGURE 4-47

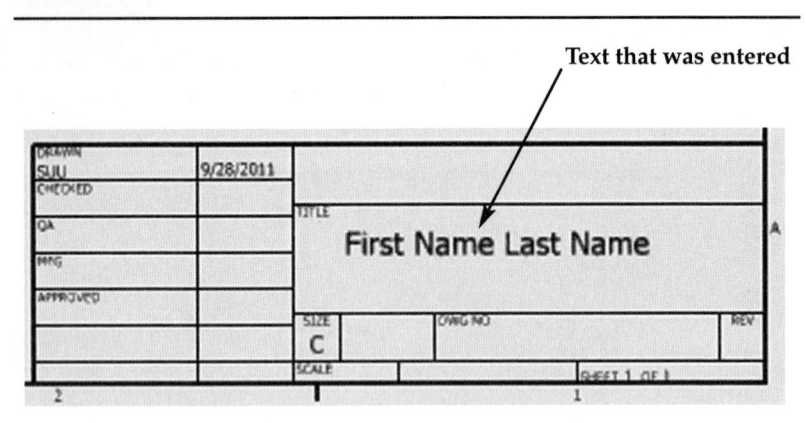

Text that was entered

52. Right-click near the text. A pop-up menu will appear as shown in Figure 4-48.

FIGURE 4-48

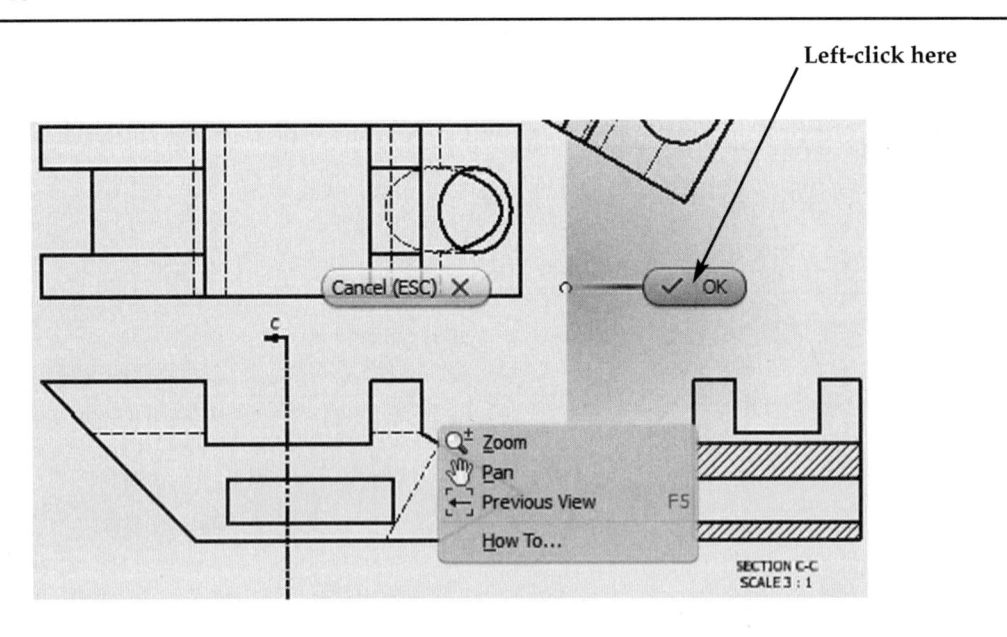

Left-click here

53. If the text needs to be moved, move the cursor over the text, causing several green dots to appear as shown in Figure 4-49.

FIGURE 4-49

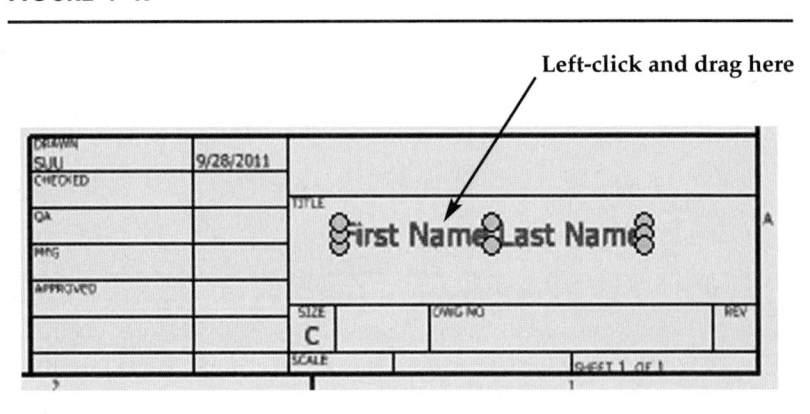

Left-click and drag here

54. While the text is highlighted, left-click (holding the left mouse button down) and drag the text to the desired location. After the text is in the desired location, release the left mouse button, move the cursor away from the text, and left-click once.

55. Move the cursor to the upper left portion of the screen and left-click on the **Place Views** tab as shown in Figure 4-50. This will return Inventor to the Place Views options menu.

FIGURE 4-50

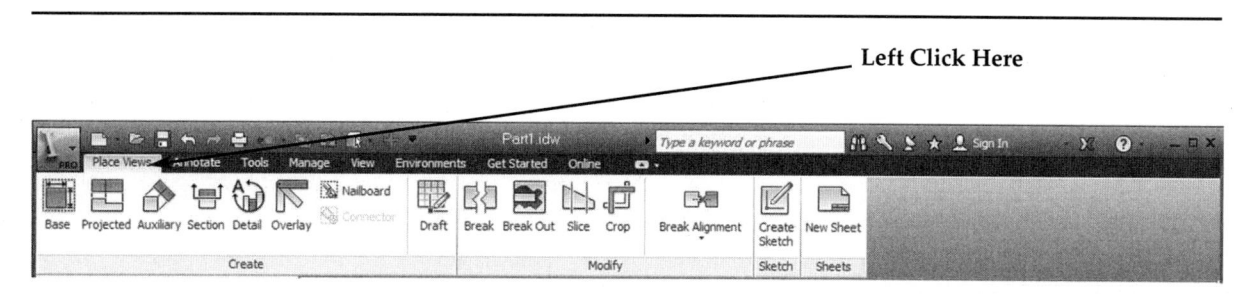

56. Your screen should look similar to Figure 4-51.

FIGURE 4-51

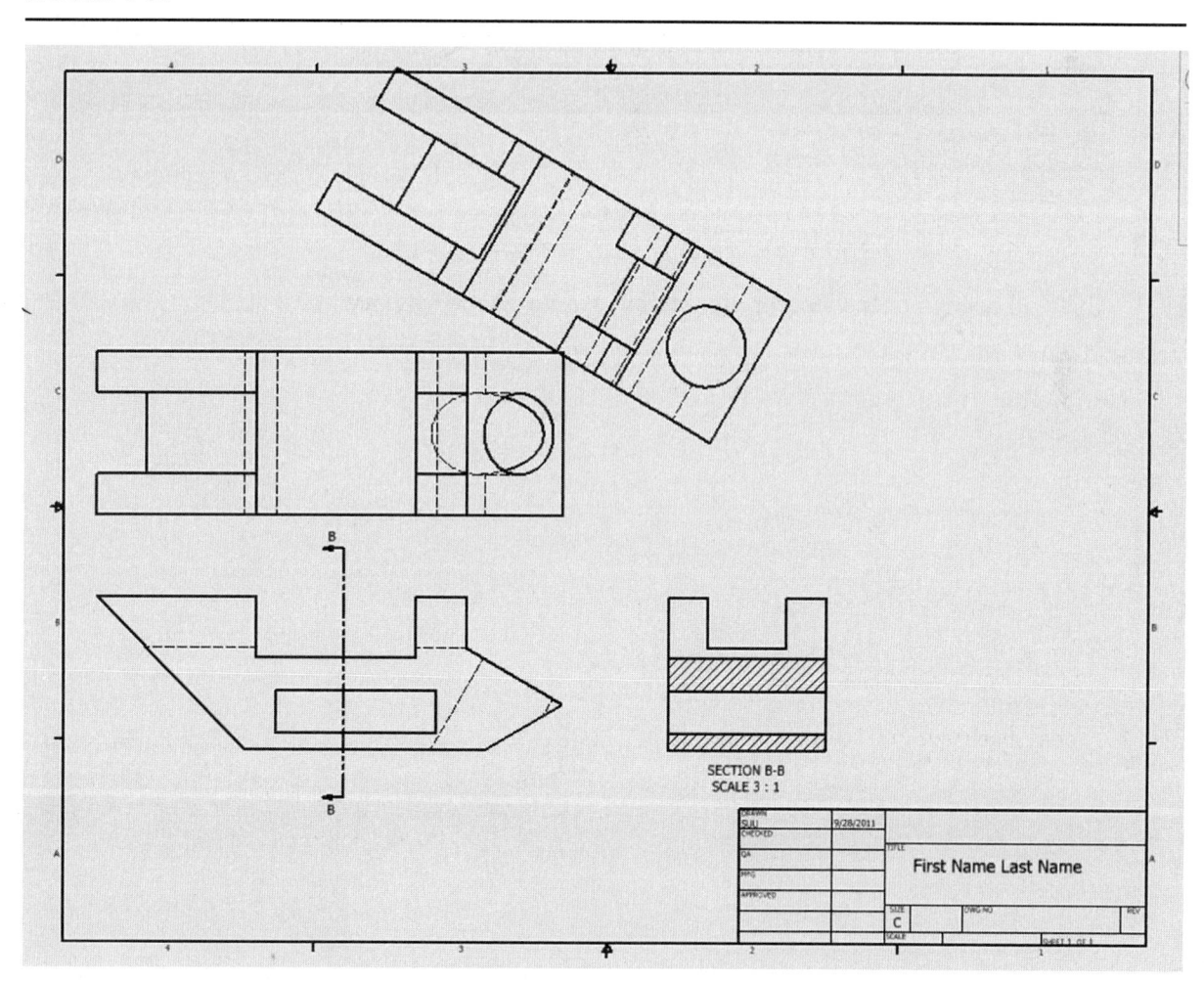

57. Before starting a new sheet of detail drawings, make sure to first save the current sheet. **Caution:** Once a new sheet has been created, the old sheet is not retrievable unless it has been saved. If a new sheet is created before the old sheet was saved, left-click on the "Undo" icon located at the upper left portion of the screen as shown in Figure 4-52.

FIGURE 4-52

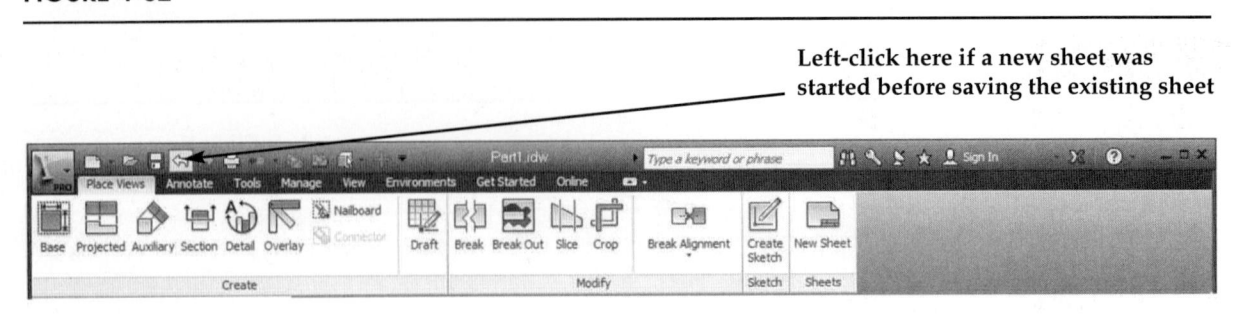

Left-click here if a new sheet was started before saving the existing sheet

58. Move the cursor to the left middle portion of the screen and left-click on **New Sheet** as shown in Figure 4-53.

FIGURE 4-53

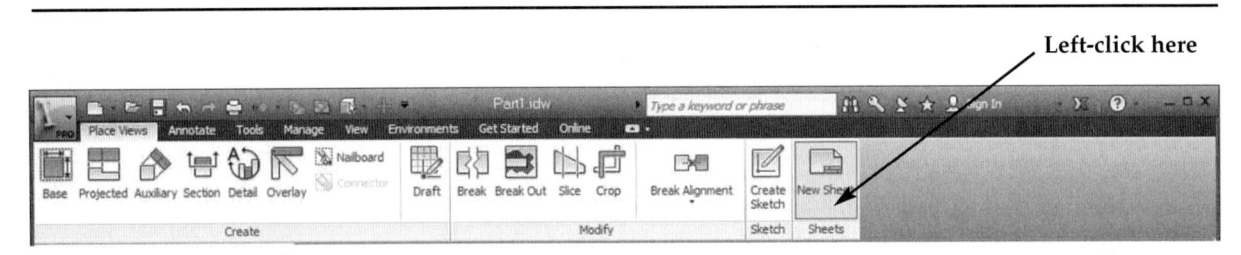

Left-click here

59. This will begin a new sheet for more detail drawings if necessary.

CHAPTER PROBLEMS

Create Section View Drawings for the following problems.

PROBLEM 4-1

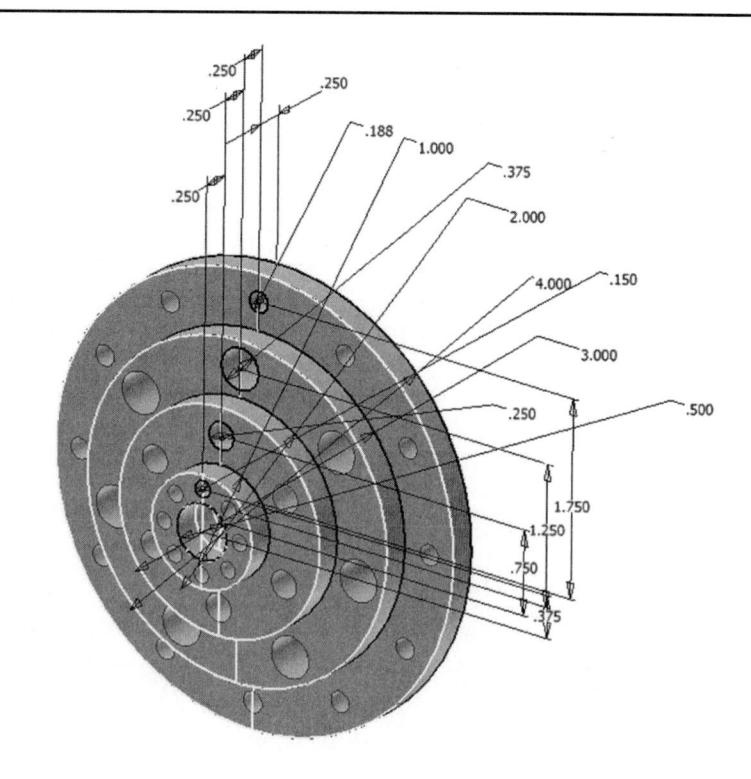

PROBLEM 4-2

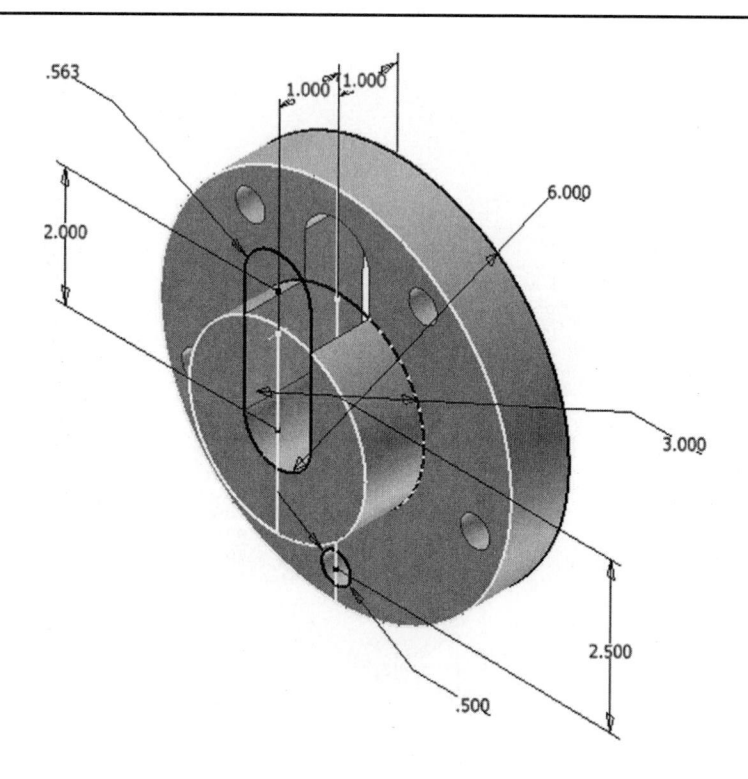

PROBLEM 4-3

Revolve the following sketch, then create a Section View.

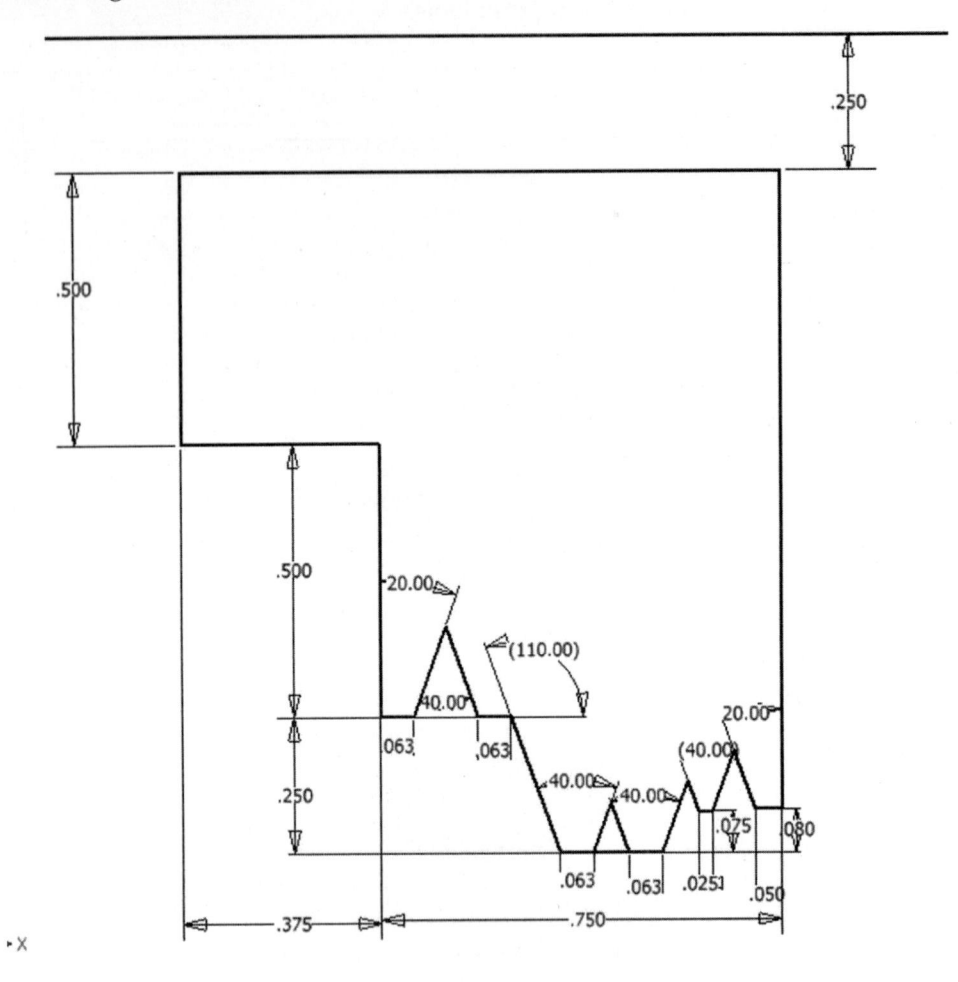

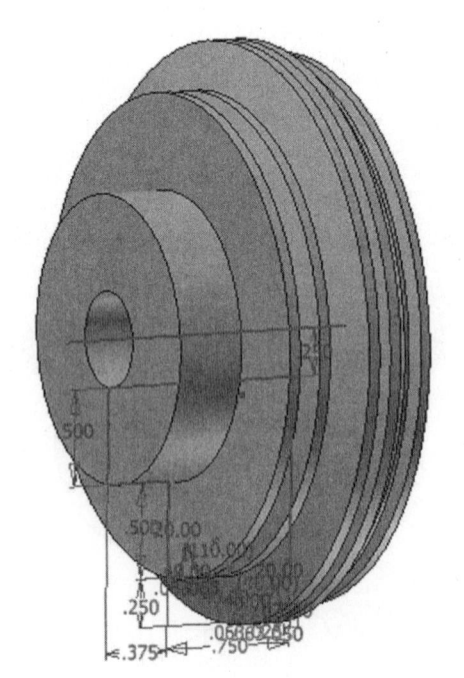

PROBLEM 4-4

Revolve the following sketch, then create a Section View.

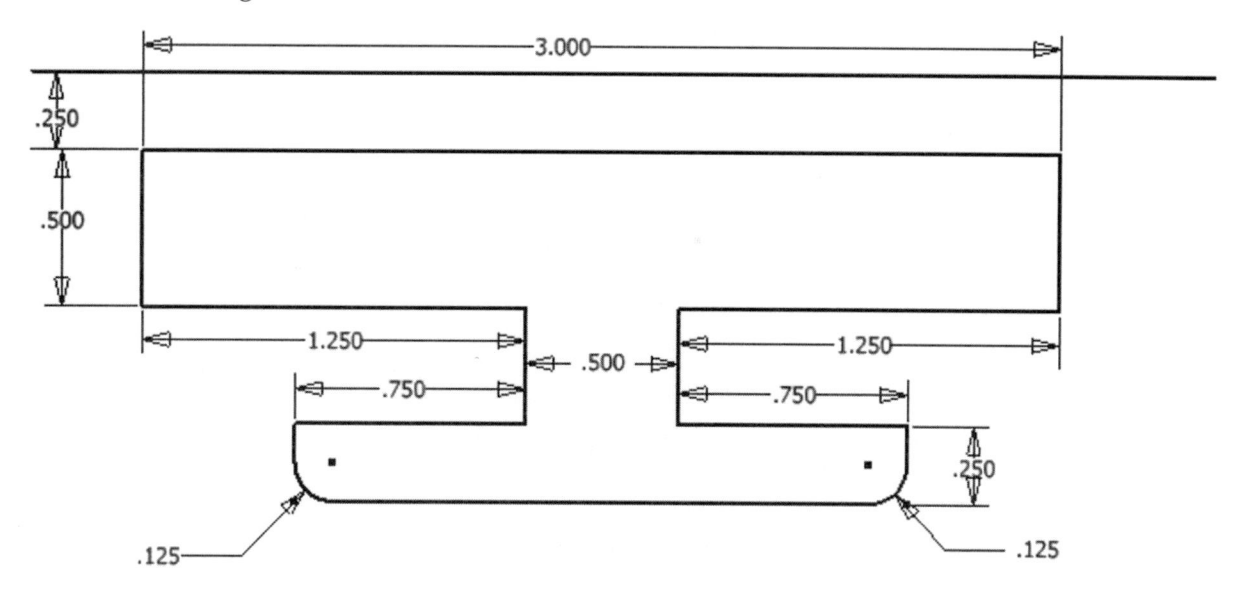

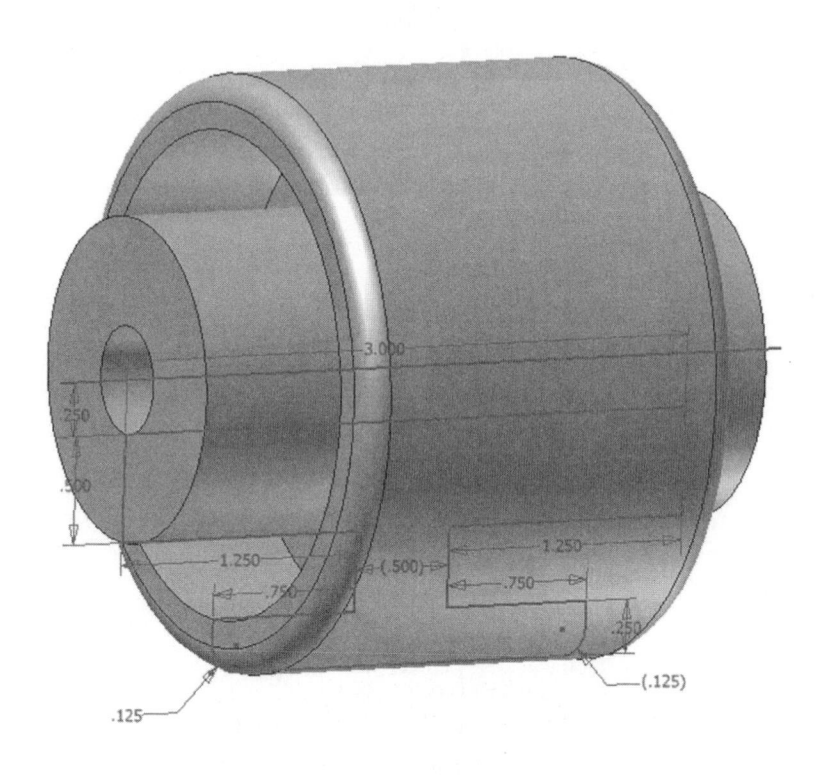

Create Section View Drawings for the following problems.

PROBLEM 4-5

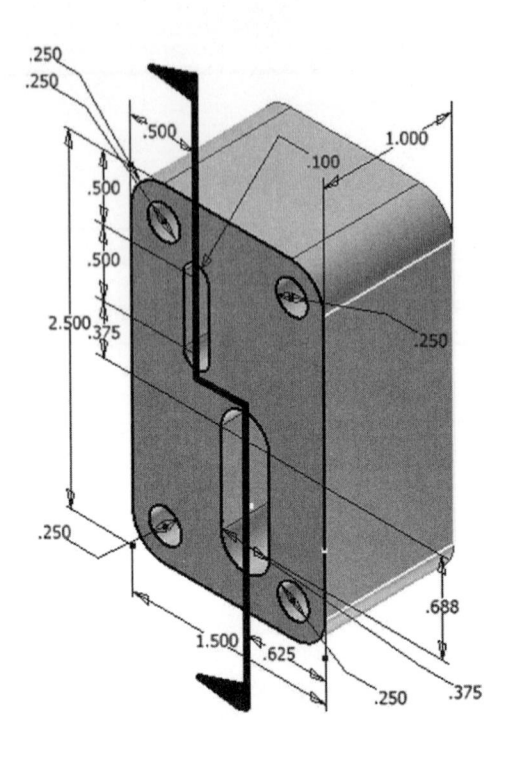

PROBLEM 4-6

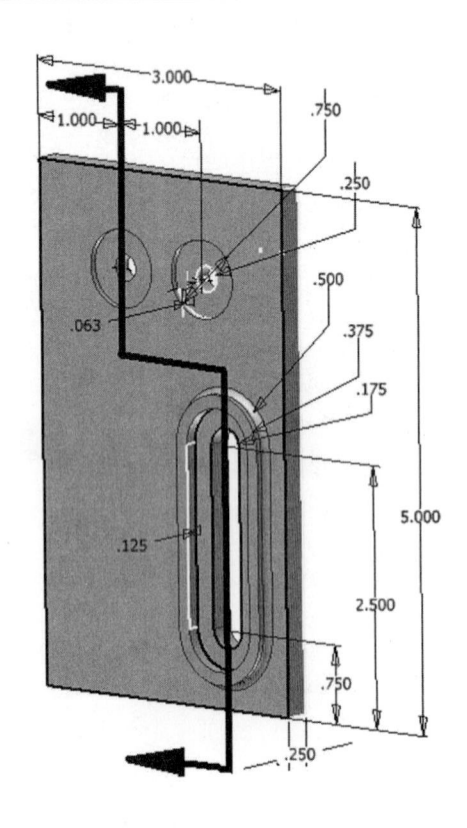

PROBLEM 4-7

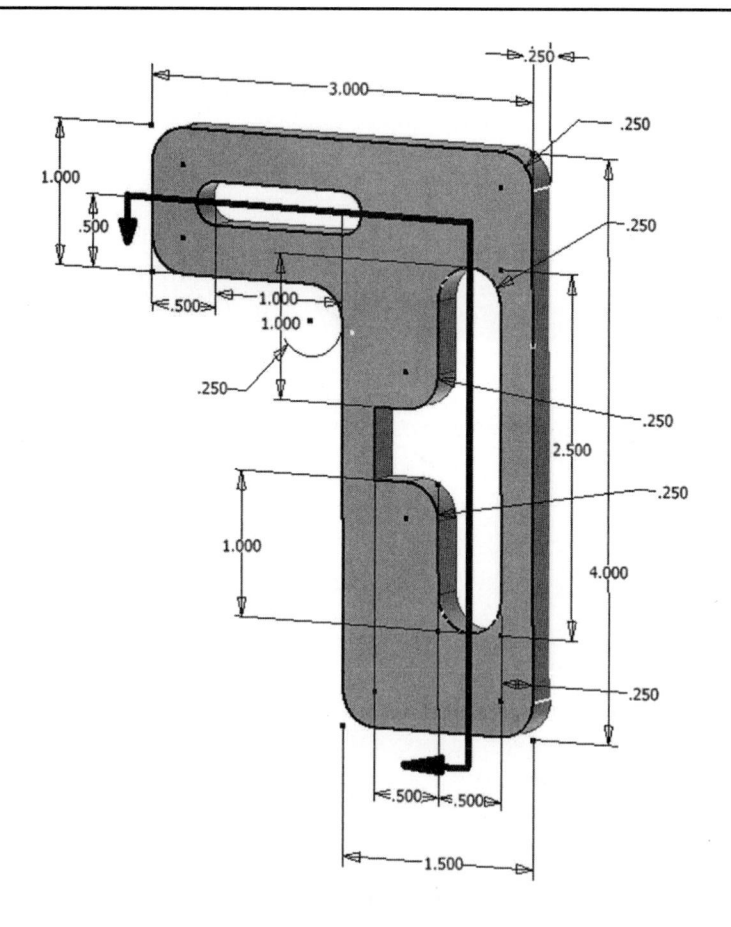

PROBLEM 4-8

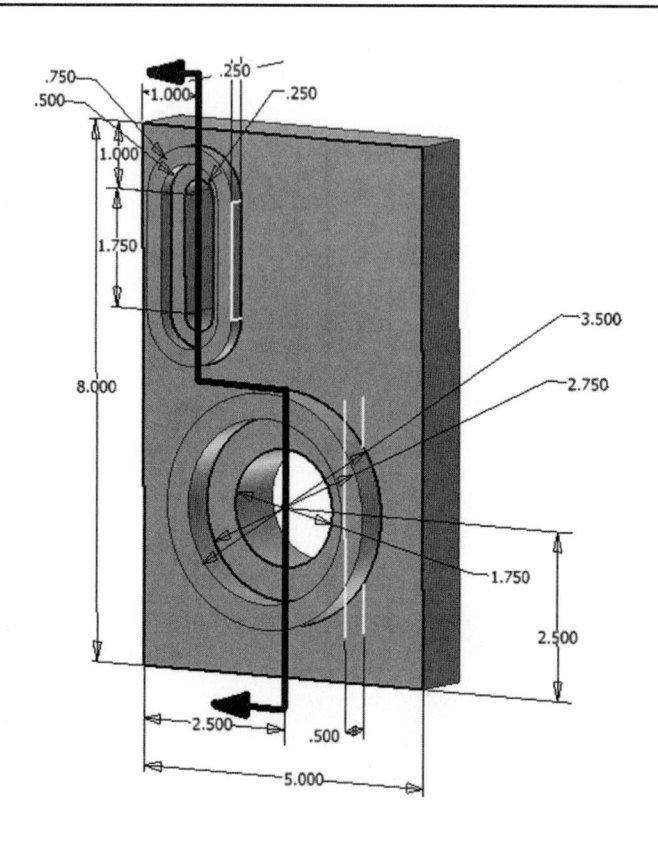

5

LEARNING TO EDIT EXISTING SOLID MODELS

Chapter 5 includes instruction on how to design and edit the part shown above.

OBJECTIVES

1. Design a simple part.
2. Use the Circular Pattern command
3. Edit the part using the Sketch Panel
4. Edit the part using the Extrude command
5. Edit the part using the Fillet command

1.   Start Autodesk Inventor 2013 by referring to Chapter 1.

2.   After Autodesk Inventor 2013 is running, begin a new sketch.

3.   Create the sketch shown in Figure 5-1.   **FIGURE 5-1**

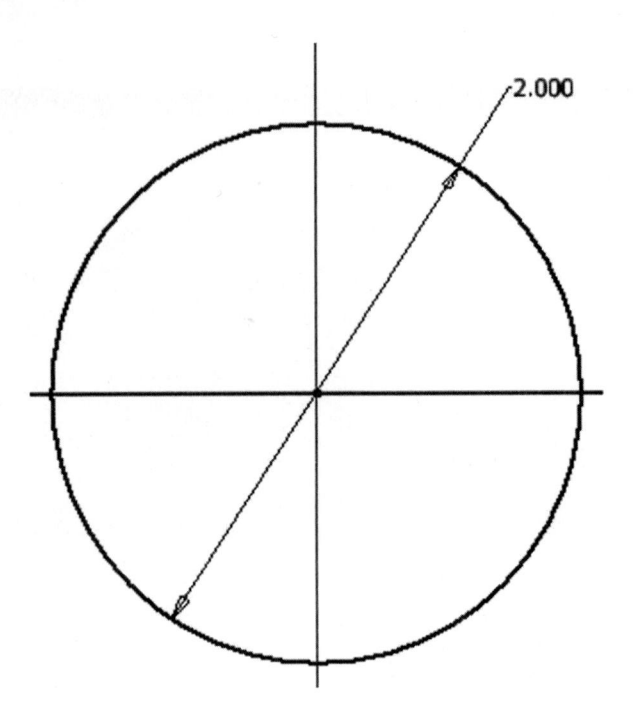

4.   Change the view to Isometric/Home View as shown in Figure 5-2.   **FIGURE 5-2**

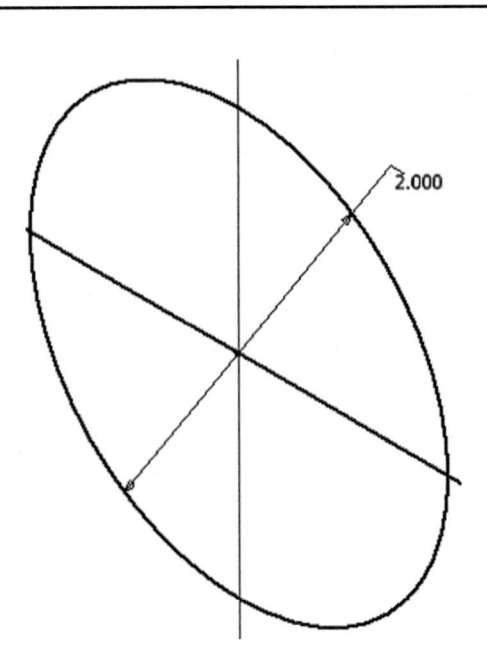

5. Extrude the sketch to a distance of .25 inches as shown in Figure 5-3.

6. Once a solid has been created, begin a New Sketch on the front surface as shown in Figure 5-3.

FIGURE 5-3

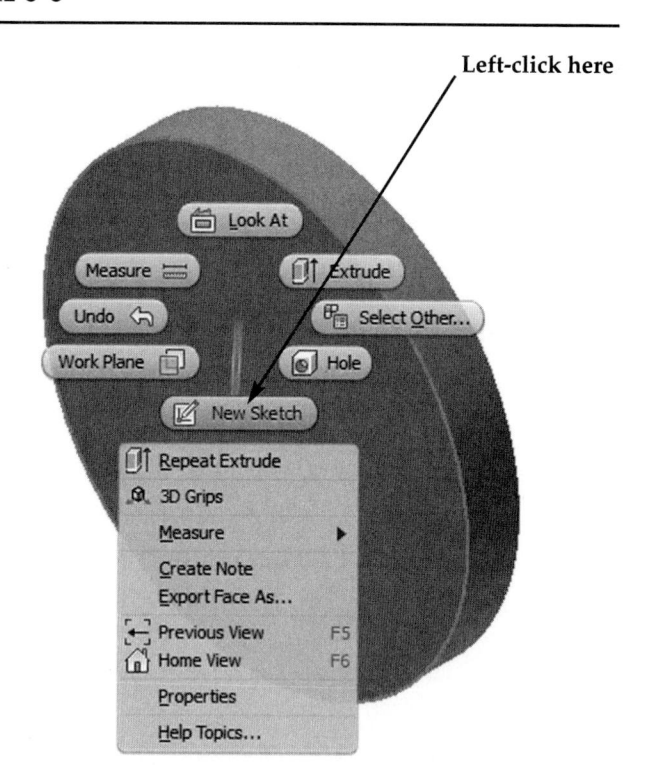

7. Complete the following sketch. Estimate the center location of the hole from the center of the part.

FIGURE 5-4

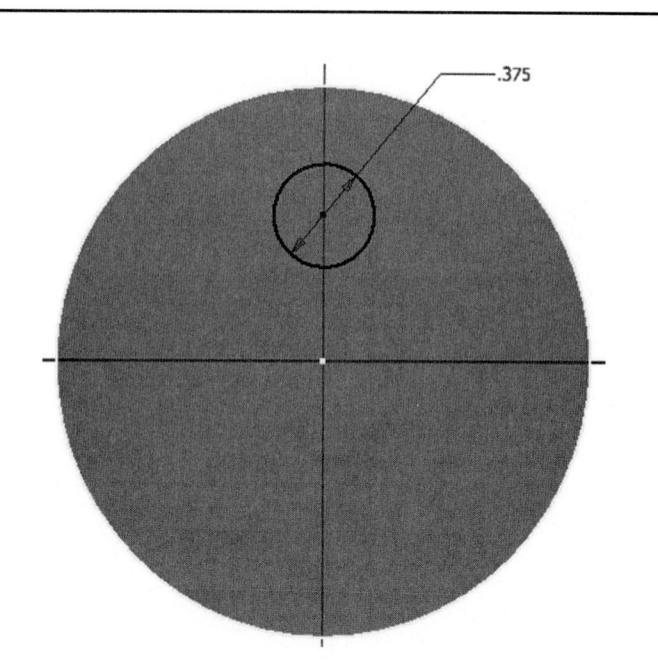

8.   Exit out of the Sketch Panel.
      Change the view to Isometric/
      Home View as shown in
      Figure 5-5.

FIGURE 5-5

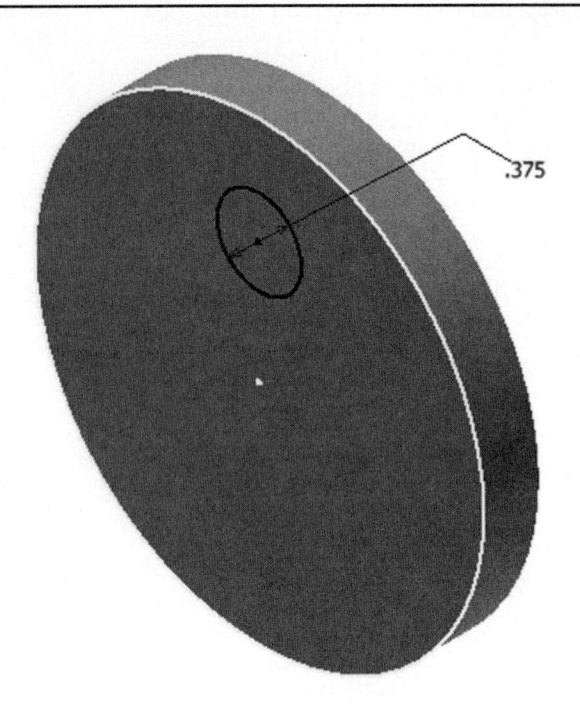

.375

9.   Use the "Extrude-Cut" command
      to cut a hole in the part as shown
      in Figure 5-6.

FIGURE 5-6

10. Use the "Circular Pattern" command to create 3 holes in the part as shown in Figure 5-7.

FIGURE 5-7

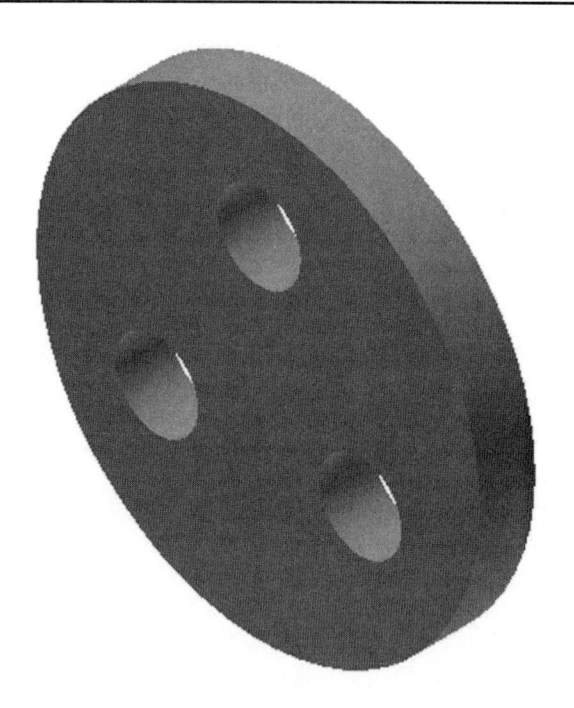

11. Use the **Fillet** command to create a fillet with .0625 radius as shown in Figure 5-8.

FIGURE 5-8

EDIT THE PART USING THE SKETCH PANEL

12.   If for some reason a change needs to be made to this part, it can be accomplished by editing either a sketch or a feature located in the Part Tree at the upper left corner of the screen as shown in Figure 5-9.

FIGURE 5-9

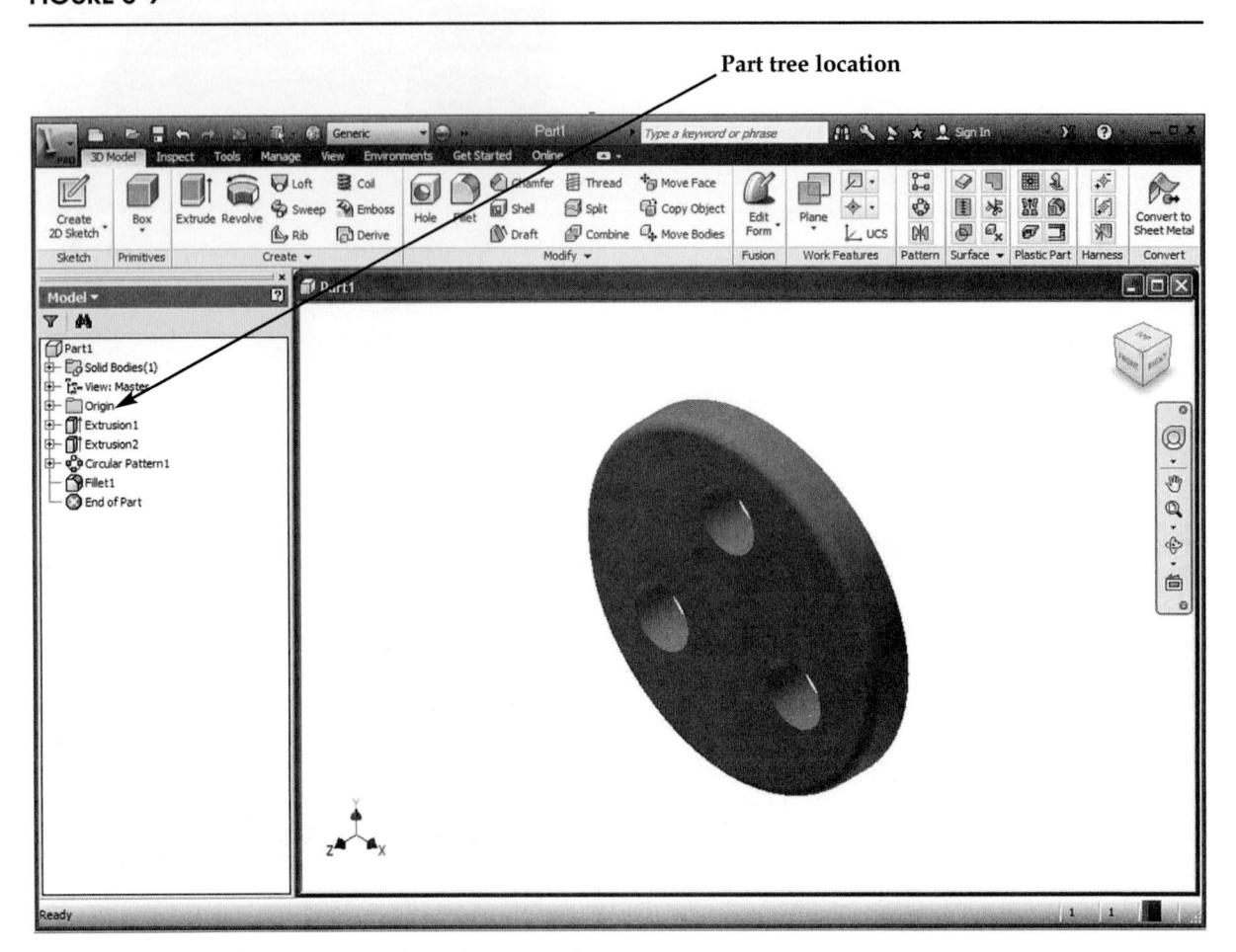

13. A close-up of the Part Tree is shown in Figure 5-10. Left-click on each of the "plus" signs in the part tree. The tree will expand, showing more details for part construction.

FIGURE 5-10

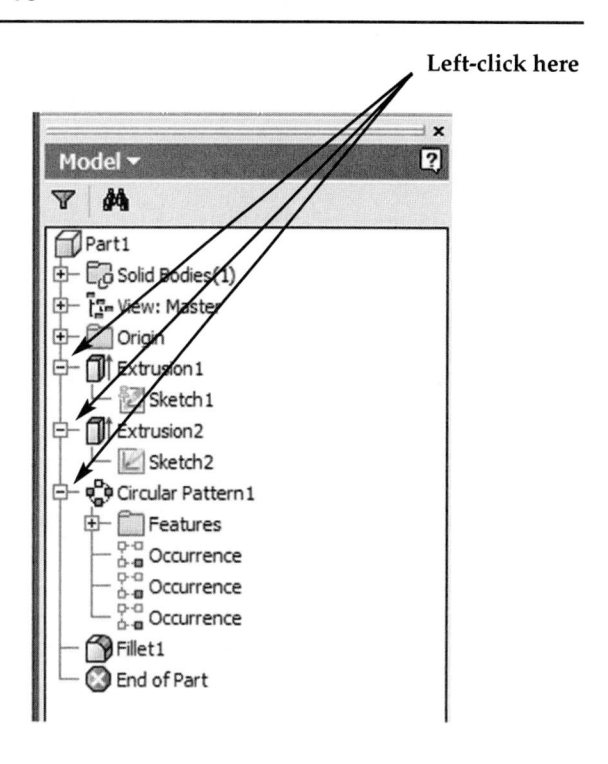

14. If a change needs to be made to any portion of the part that was constructed using a "Sketch1," the change can be made here.

15. Move the cursor over Sketch1. A red box will appear around the text "Sketch1" as shown in Figure 5-11.

FIGURE 5-11

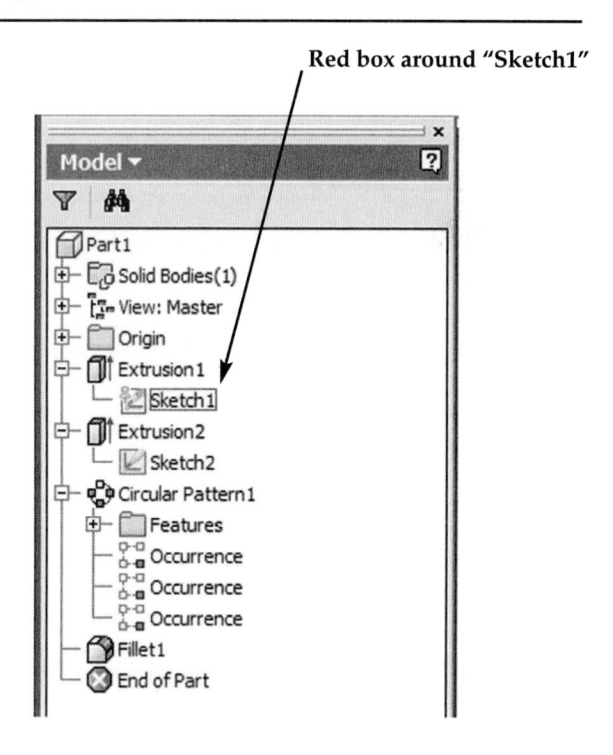

16. The original sketch will also appear as shown in Figure 5-12.

FIGURE 5-12

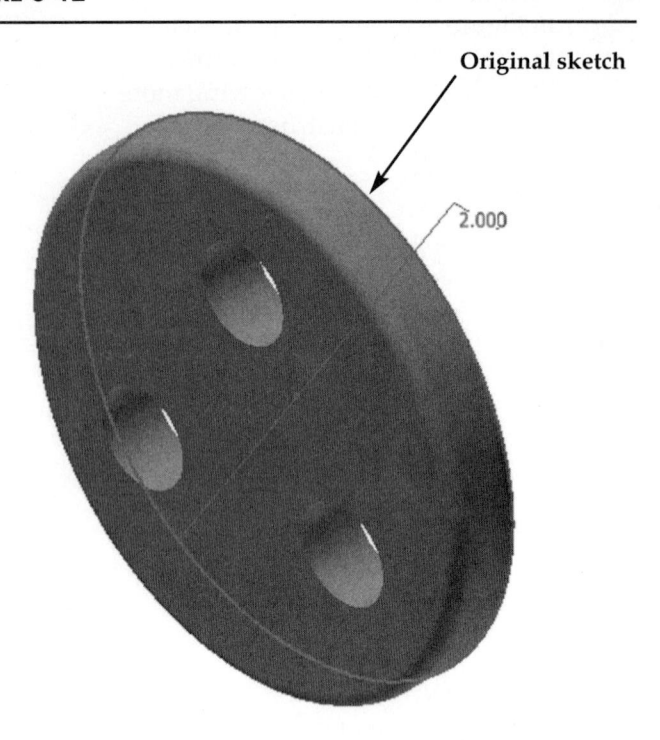

Original sketch

2.000

17. Right-click on **Sketch1**. The text "Sketch1" will become highlighted. A pop-up menu will appear. Left-click on **Edit Sketch** as shown in Figure 5-13.

FIGURE 5-13

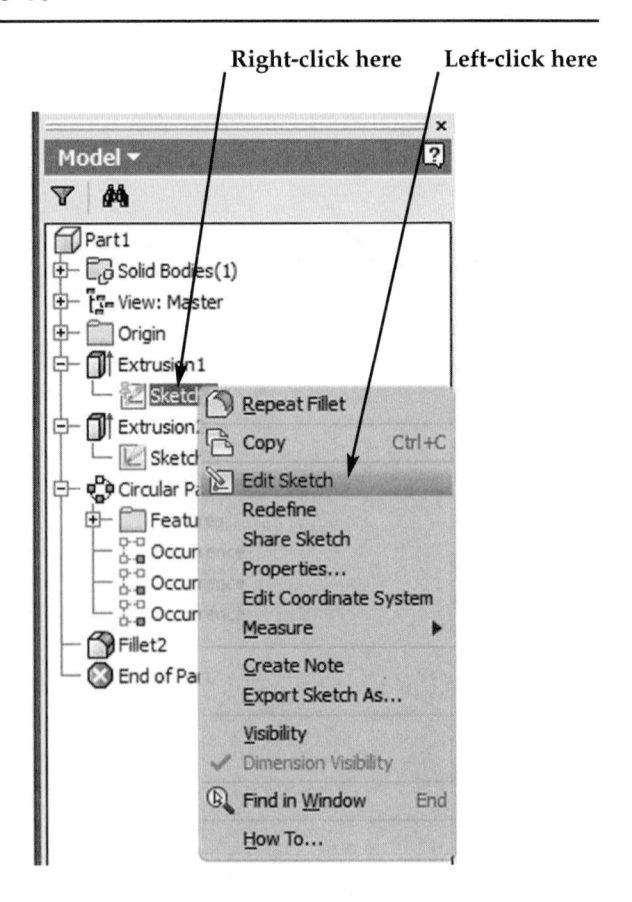

Right-click here   Left-click here

18. The original sketch will appear as shown in Figure 5-14.

FIGURE 5-14

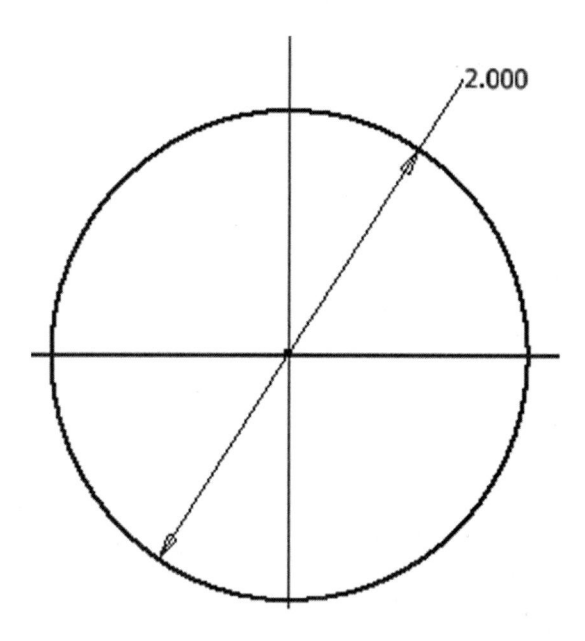

19. If Inventor rotated the part to provide a perpendicular view, then skip to Number 22. If the sketch is not a perpendicular view, move the cursor to the upper right portion of the screen and left-click on the "Face View/Look At" icon as shown in Figure 5-15.

FIGURE 5-15

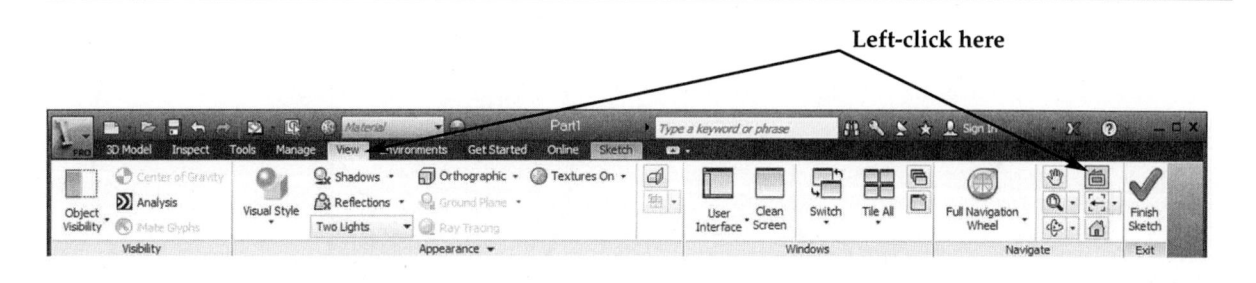

20. Move the cursor over to the part tree and left-click on the "plus" sign to the left of Origin. The part tree will expand displaying all three work planes. Move the cursor over the text, "XY Plane." A red box will appear around XY Plane and the sketch itself. After the red box appears, left-click once on the **XY Plane** as shown in Figure 5-16.

FIGURE 5-16

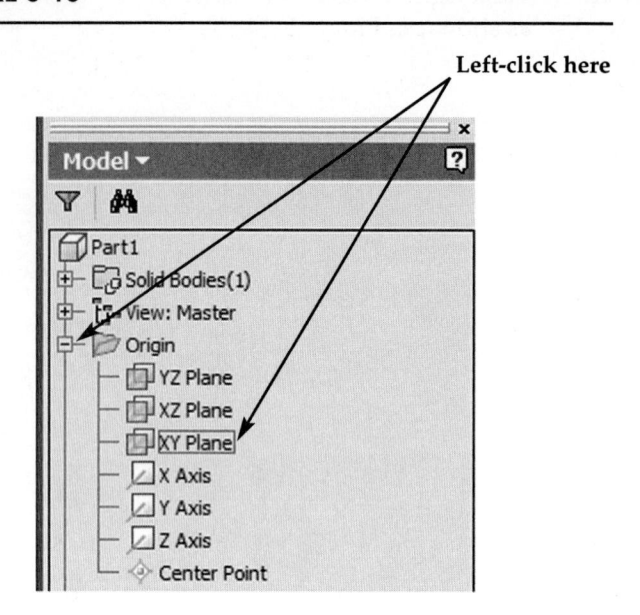

21. Inventor will provide a perpendicular view of the sketch similar to when the sketch was first constructed. Your screen should look similar to Figure 5-17.

FIGURE 5-17

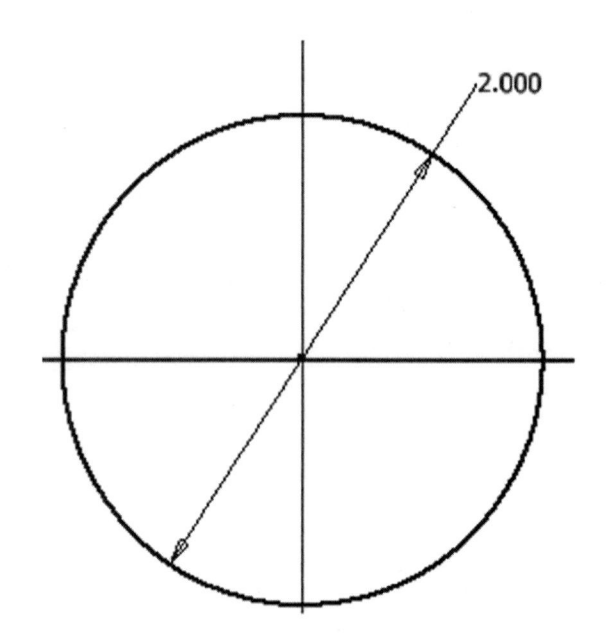

22. Start by modifying the diameter of the part. First, double-click on the overall dimension. The Edit Dimension dialog box will appear as shown in Figure 5-18.

FIGURE 5-18

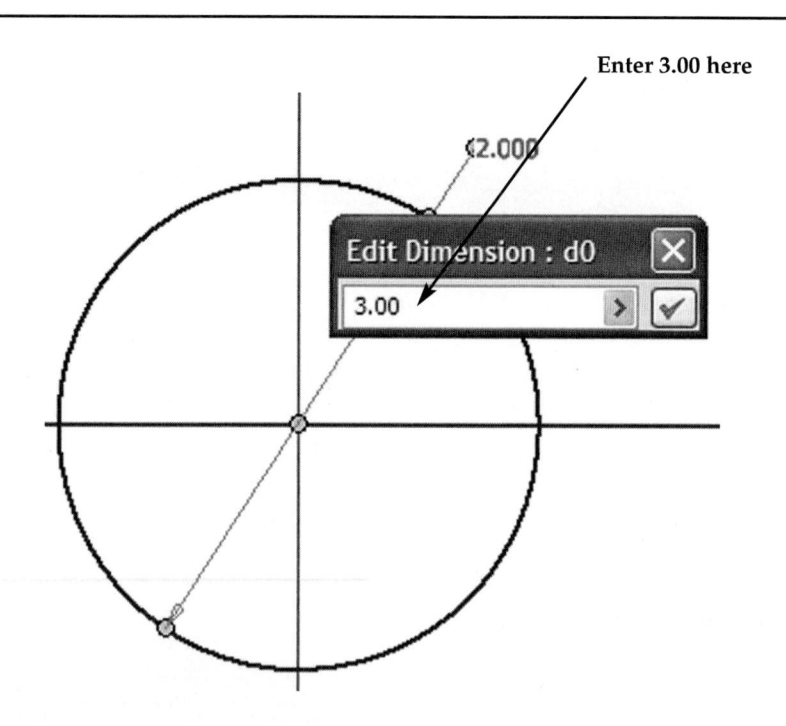

23. Enter **3.00** as shown in Figure 5-18. Press **Enter** on the keyboard.

24. The diameter of the part will increase to 3.00 as shown in Figure 5-19.

FIGURE 5-19

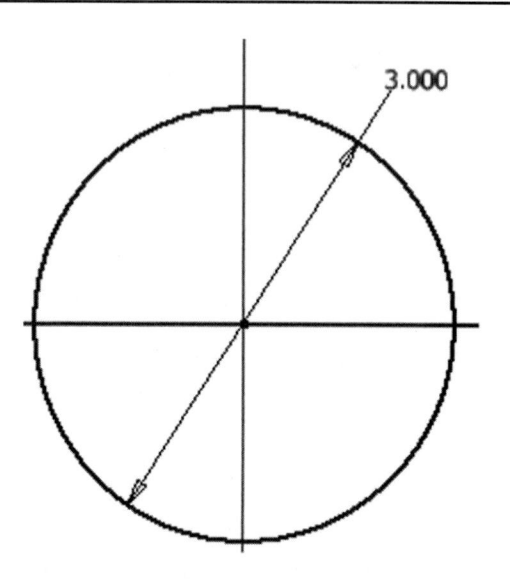

25.   Move the cursor to the upper left portion of the screen and left-click on the **Manage** tab. Left-click on the drop-down arrow below **Update**. A drop-down menu will appear. Left-click on **Update** as shown in Figure 5-20.

FIGURE 5-20

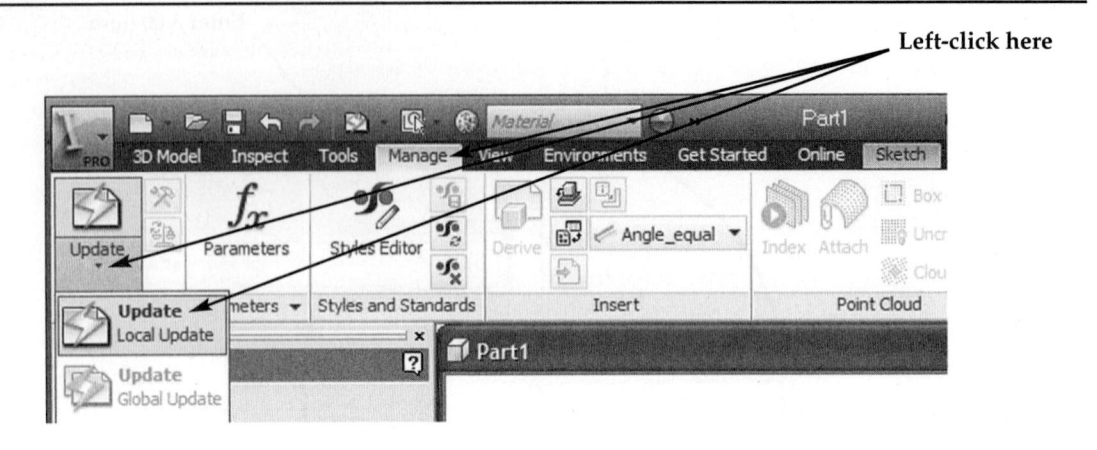

26.   Inventor will automatically update the part as shown in Figure 5-21. The part will be updated without the need to repeat any of the steps that created the original part.

FIGURE 5-21

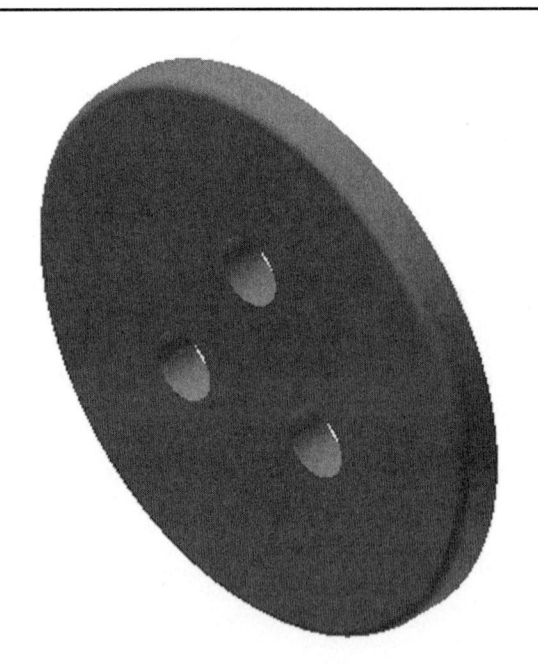

EDIT THE PART USING THE EXTRUDE COMMAND

27. Move the cursor over the text, "Extrusion1." A red box will appear around the text. After the red box appears, right-click once on **Extrusion1** as shown in Figure 5-22.

FIGURE 5-22

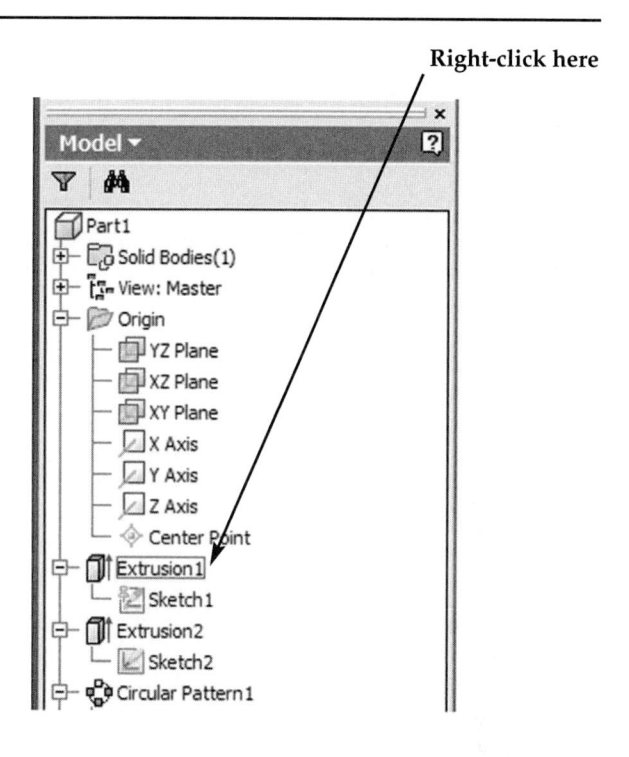

28. A pop-up menu will appear. Left-click on **Edit Feature** as shown in Figure 5-23.

FIGURE 5-23

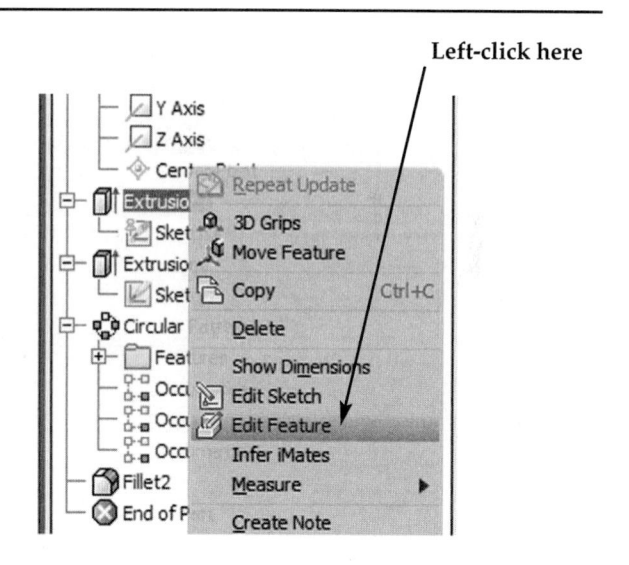

29. The Extrusion dialog box will appear. Enter **.500** for the extrusion distance and left-click on **OK** as shown in Figure 5-24.

FIGURE 5-24

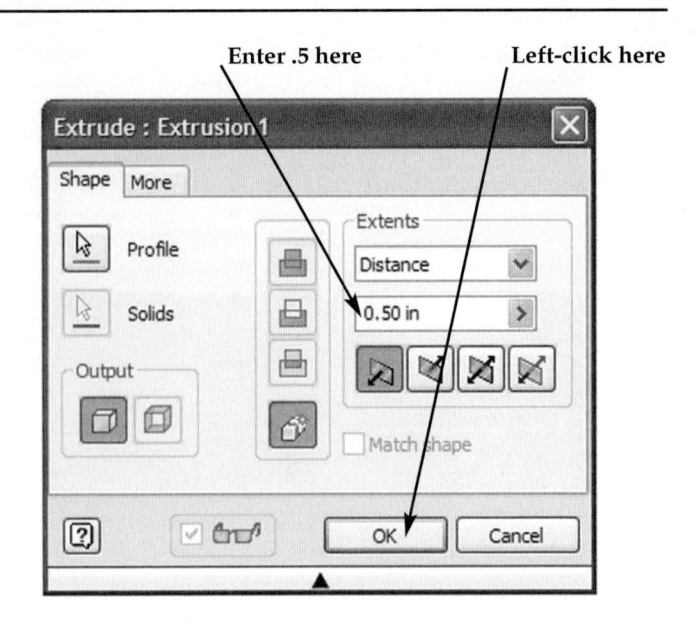

30. Move the cursor to the upper left portion of the screen and left-click on the **Manage** tab. Left-click on the drop-down arrow below **Update**. A drop-down menu will appear. Left-click on **Update** as shown in Figure 5-25.

FIGURE 5-25

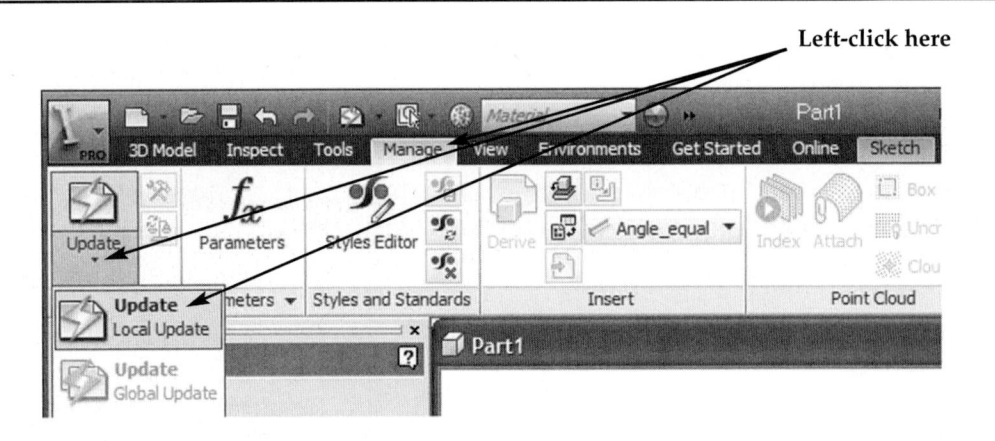

31. Inventor will automatically update the part. Notice that the holes are no longer thru holes as shown in Figure 5-26.

FIGURE 5-26

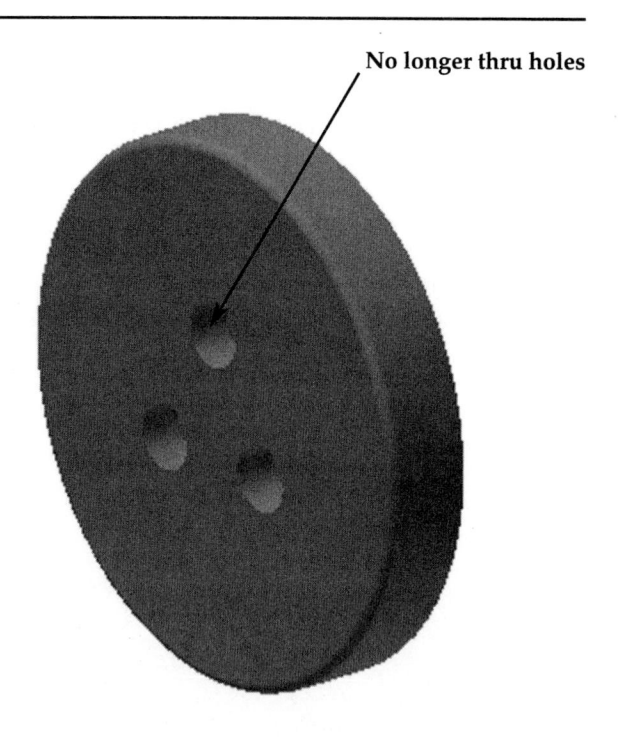

No longer thru holes

32. Rotate the view around close to perpendicular using "Face View/Look At" command to see that the holes are no longer thru as shown in Figure 5-27.

FIGURE 5-27

33. Move the cursor over the text, "Sketch2." A red box will appear around the text. After the red box appears, right-click once on **Sketch2** as shown in Figure 5-28.

FIGURE 5-28

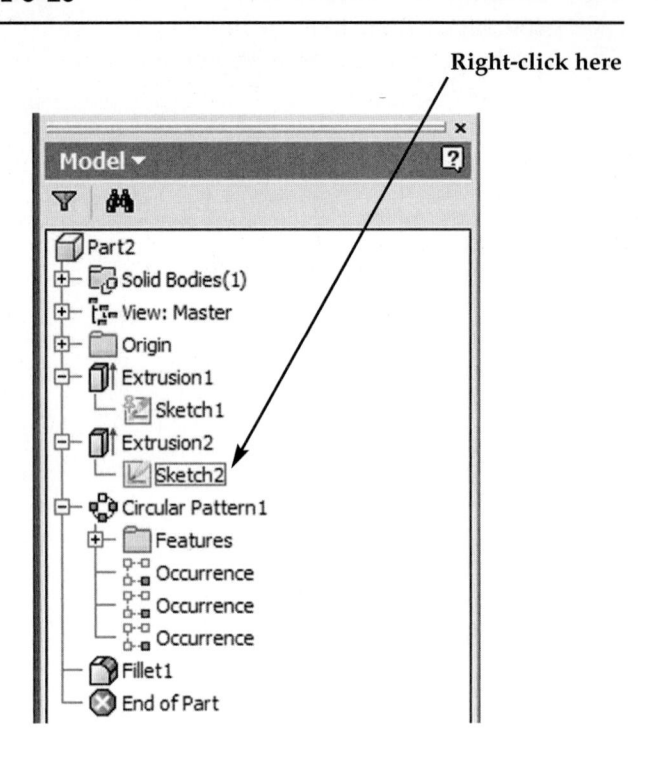

34. A pop-up menu will appear. Left-click on **Edit Sketch** as shown in Figure 5-29.

FIGURE 5-29

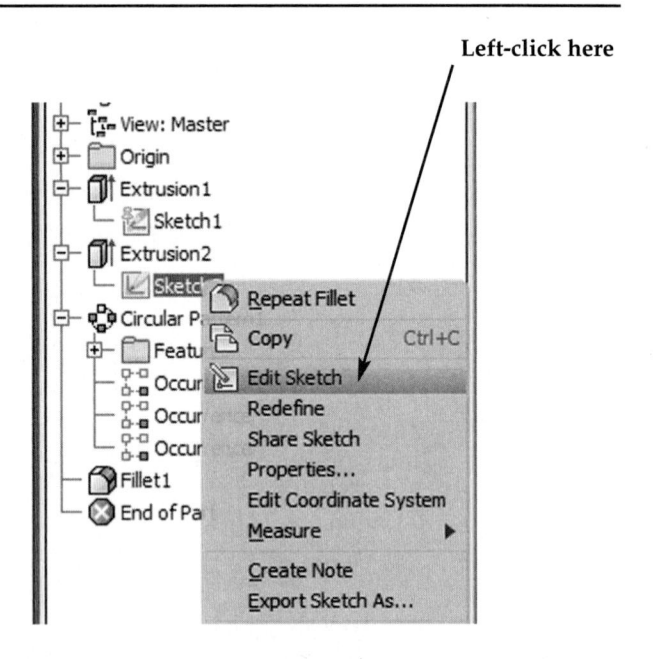

35.   The original sketch will appear as shown in Figure 5-30.

FIGURE 5-30

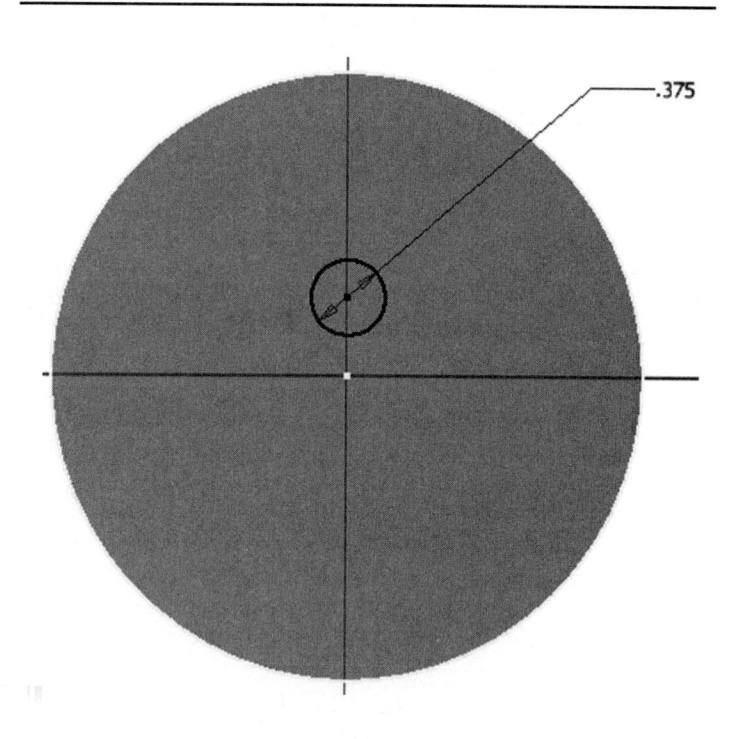

36.   Modify the diameter of the holes by double-clicking on the overall dimension. The Edit Dimension dialog box will appear as shown in Figure 5-31.

37.   Enter **.125** and press **Enter** on the keyboard as shown in Figure 5-31.

FIGURE 5-31

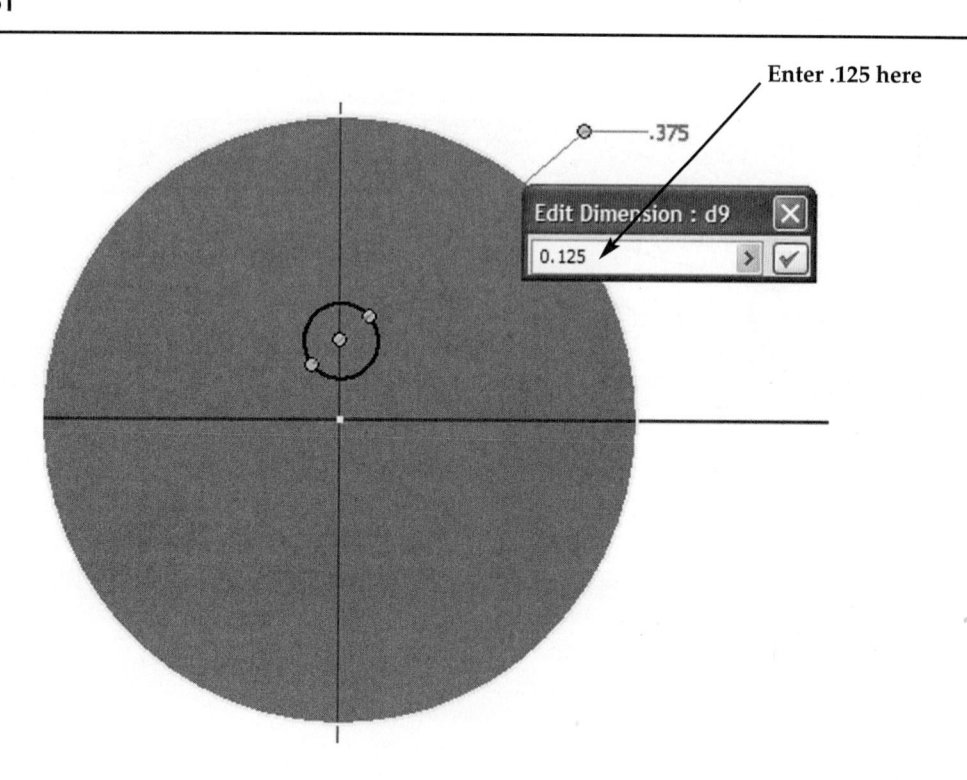

38. The diameter of all the holes will be reduced to .125 as shown in Figure 5-32.

FIGURE 5-32

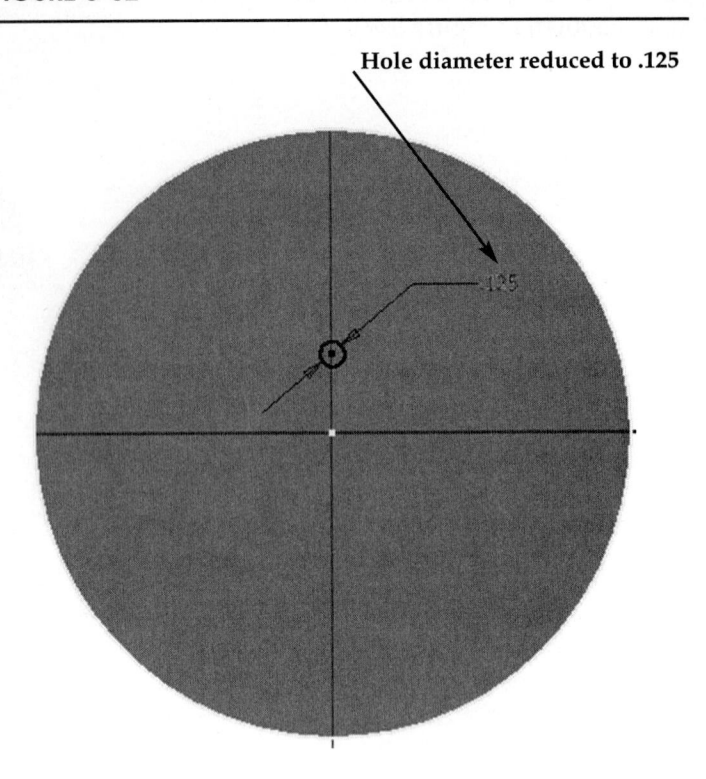

Hole diameter reduced to .125

39. Move the cursor to the upper left portion of the screen and left-click on the **Manage** tab. Left-click on the drop-down arrow below **Update**. A drop-down menu will appear. Left-click on **Update** as shown in Figure 5-33.

FIGURE 5-33

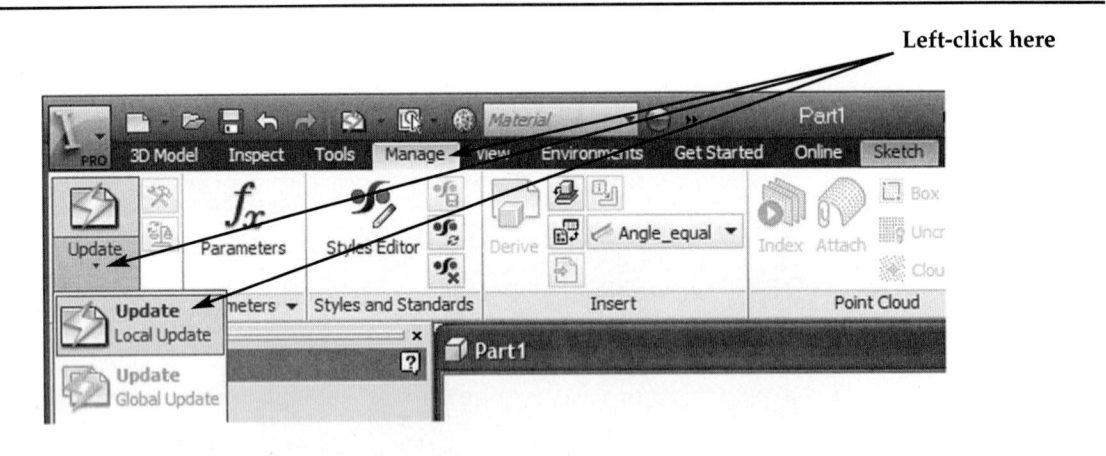

Left-click here

40. Inventor will automatically update the part as shown in Figure 5-34.

FIGURE 5-34

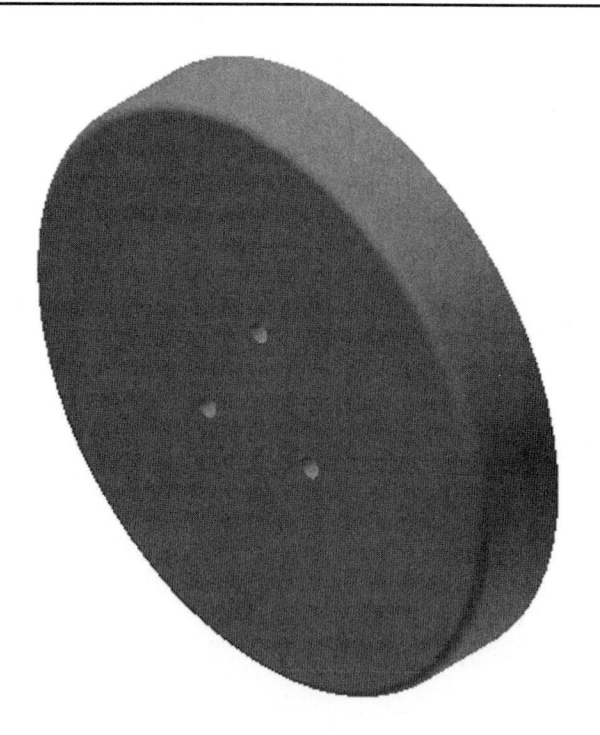

41. Move the cursor over the text, "Extrusion2." A red box will appear around the text. After the red box appears, right-click once on **Extrusion2** as shown in Figure 5-35.

FIGURE 5-35

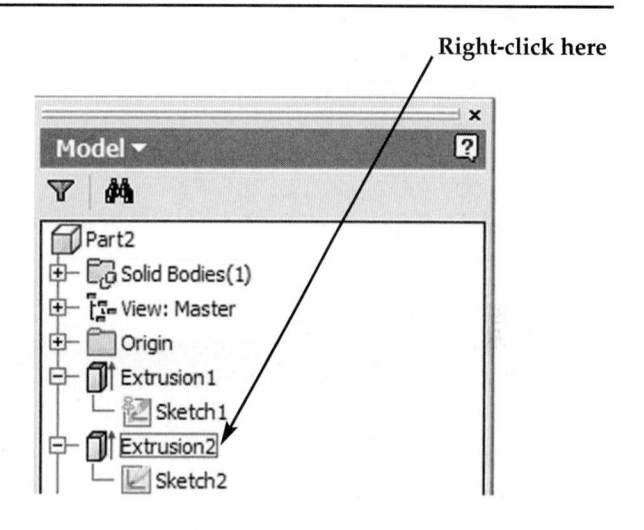

42. A pop-up menu will appear. Left-click on **Edit Feature** as shown in Figure 5-36.

FIGURE 5-36

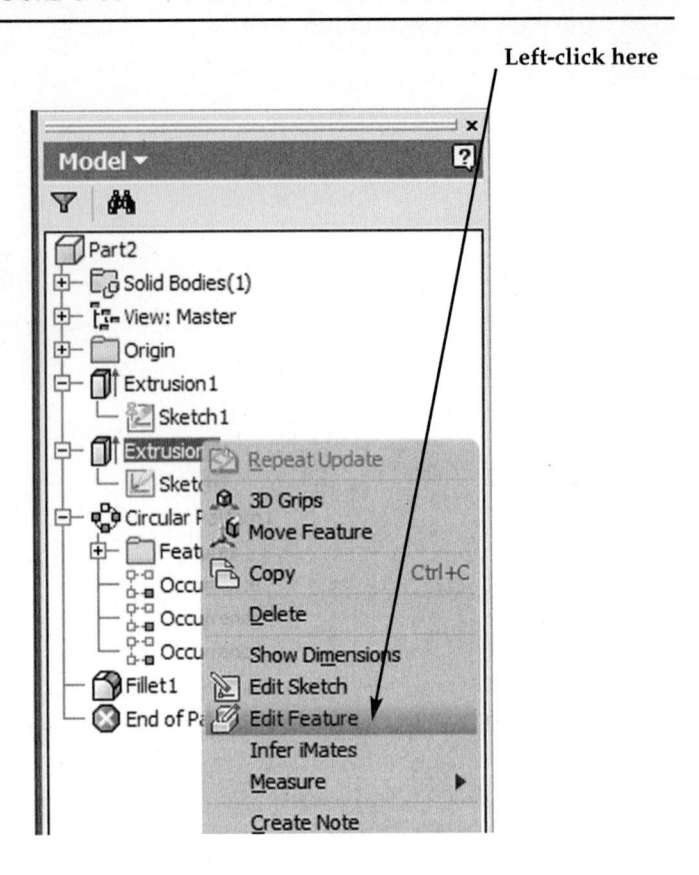

43. The Extrusion dialog box will appear. Enter **.5** for the extrusion distance and left-click on **OK** as shown in Figure 5-37.

FIGURE 5-37

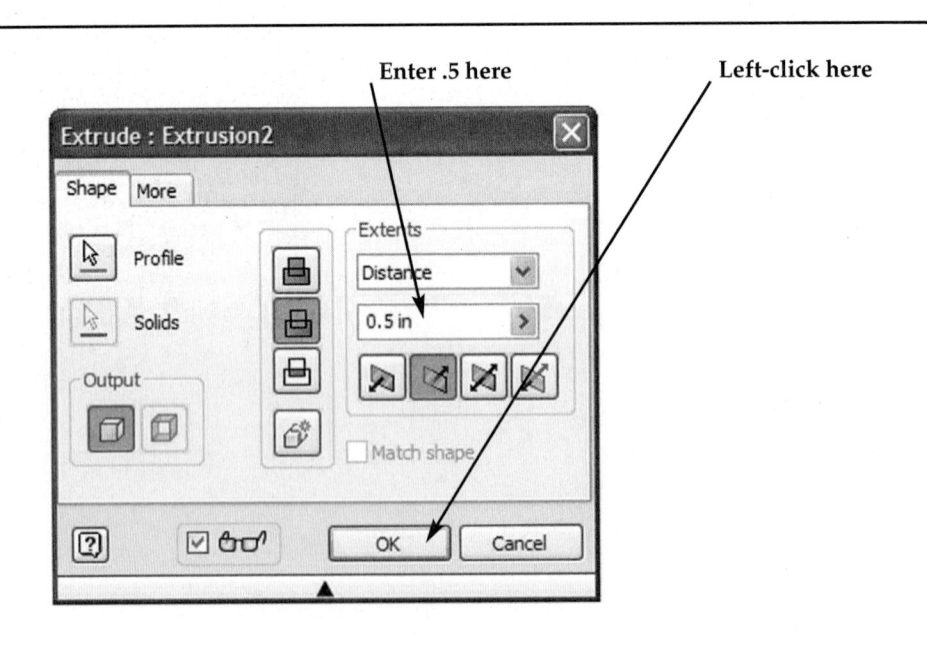

44. Move the cursor to the upper left portion of the screen and left-click on the **Manage** tab. Left-click on the drop-down arrow below **Update**. A drop-down menu will appear. Left-click on **Update** as shown in Figure 5-38.

FIGURE 5-38

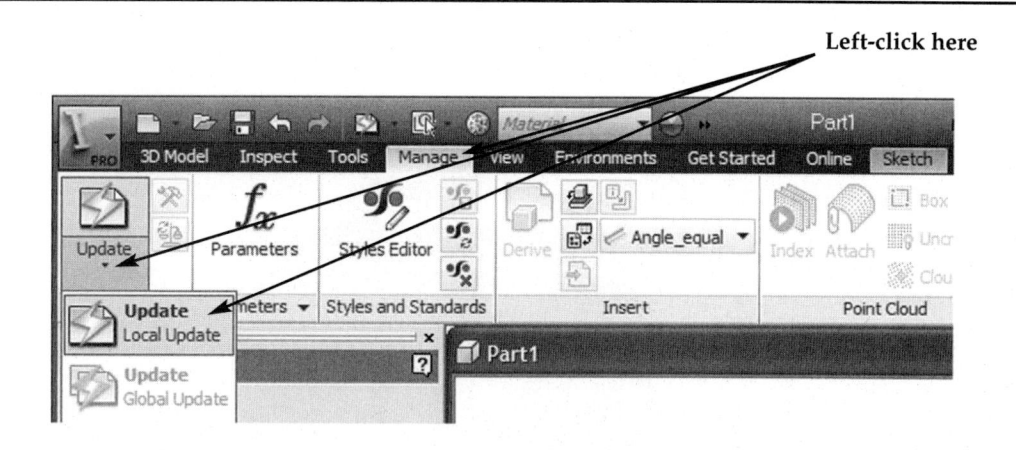

45. Inventor will automatically update the part. Notice that the holes are now thru holes as shown in Figure 5-39.

FIGURE 5-39

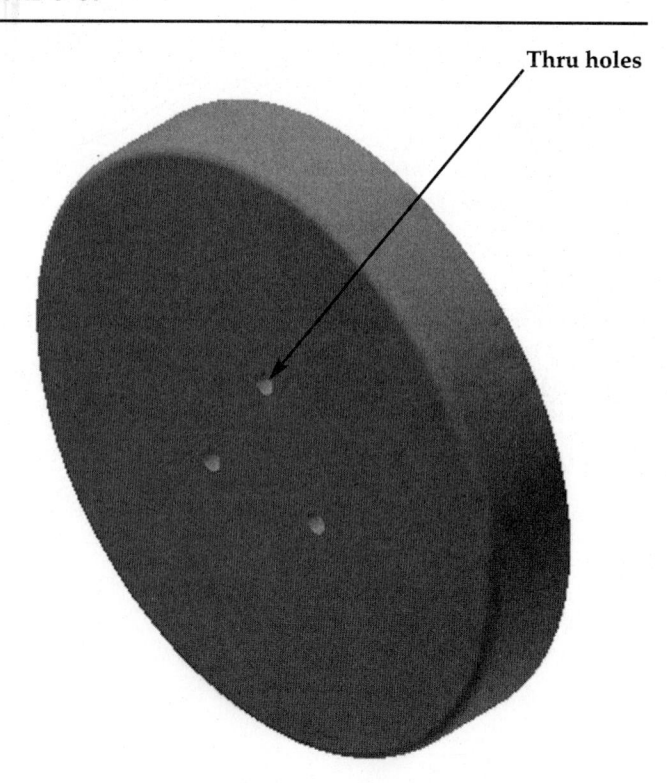

46.   Use the "Free Orbit/Rotate" command to rotate the part as shown in Figure 5-40.

FIGURE 5-40

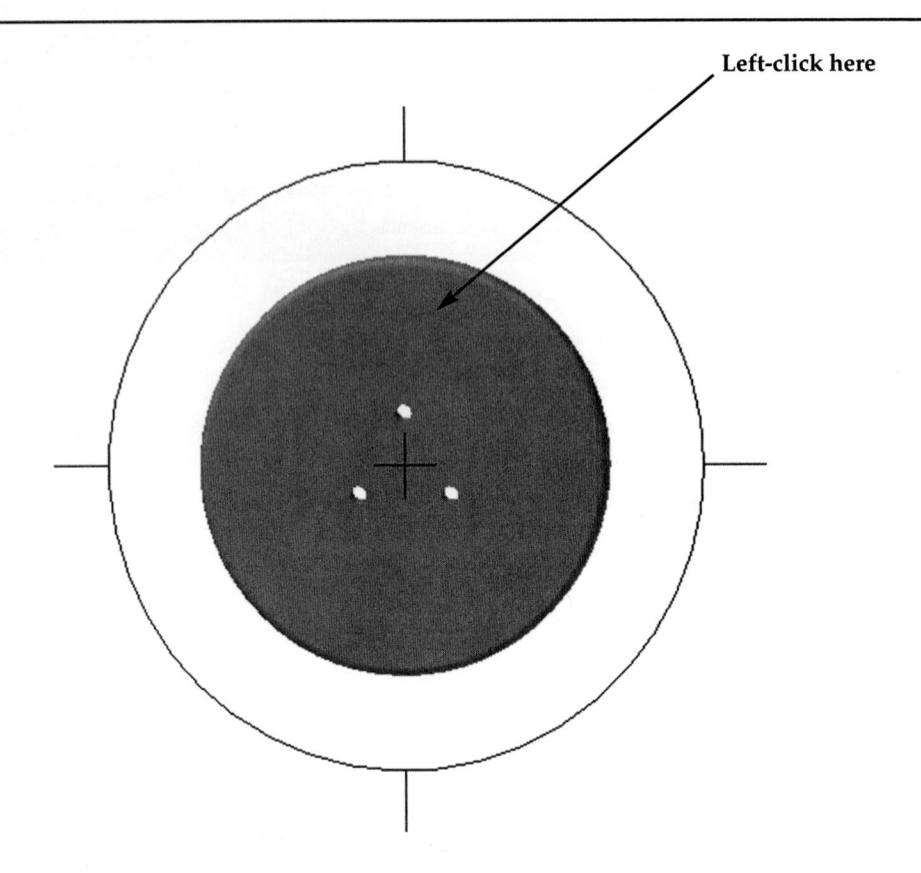

Left-click here

EDIT THE PART USING THE CIRCULAR PATTERN COMMAND

47.   Move the cursor over the text, "Circular Pattern1." A red box will appear around the text. After the red box appears, left-click once on **Circular Pattern1** as shown in Figure 5-41.

FIGURE 5-41

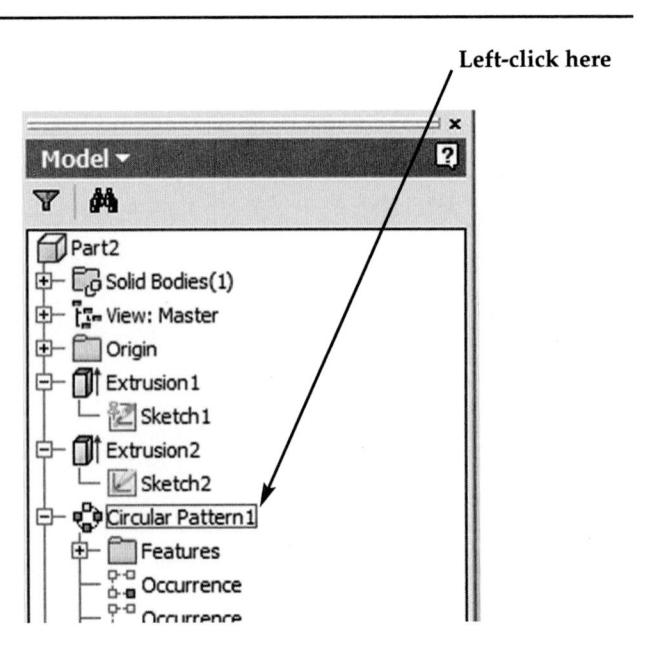

Left-click here

48. Right-click on **Circular Pattern1**. A pop-up menu will appear. Left-click on **Edit Feature** as shown in Figure 5-42.

FIGURE 5-42

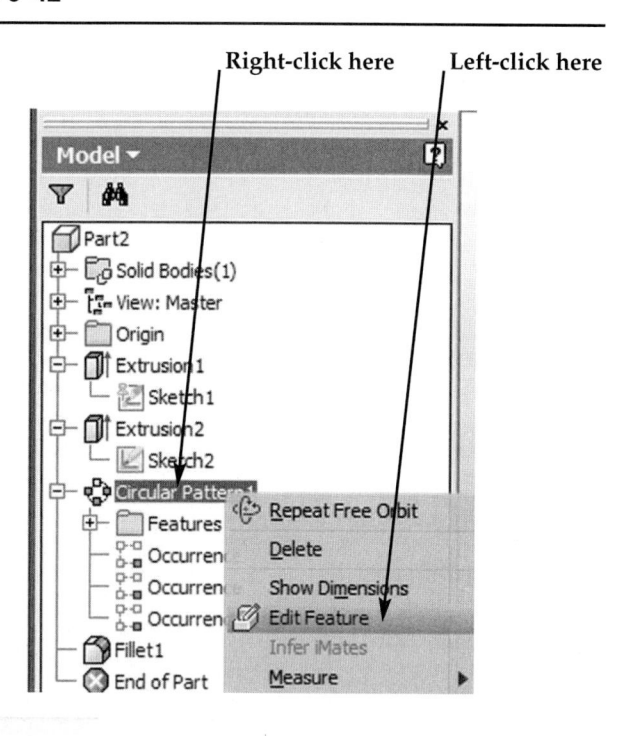

49. The Circular Pattern dialog box will appear. Enter **6** under **Placement** as shown in Figure 5-43.

FIGURE 5-43

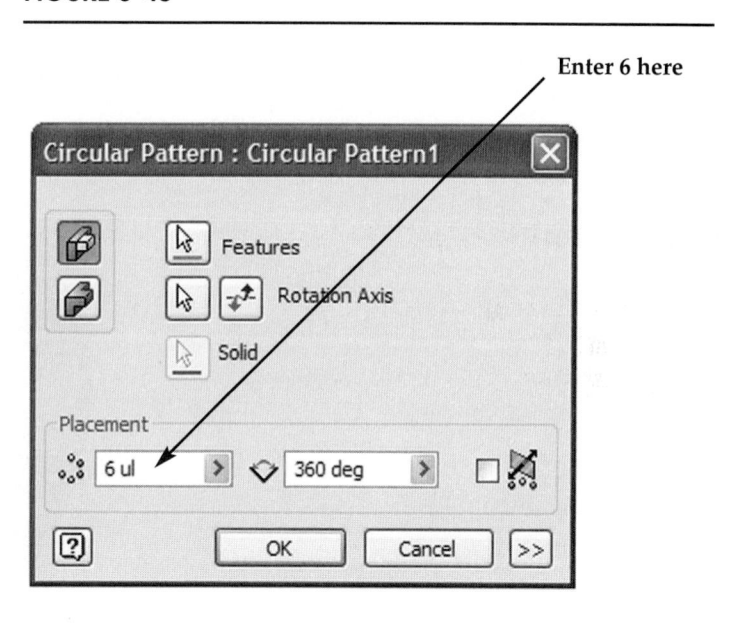

50. Inventor will provide a preview as shown in Figure 5-44.

FIGURE 5-44

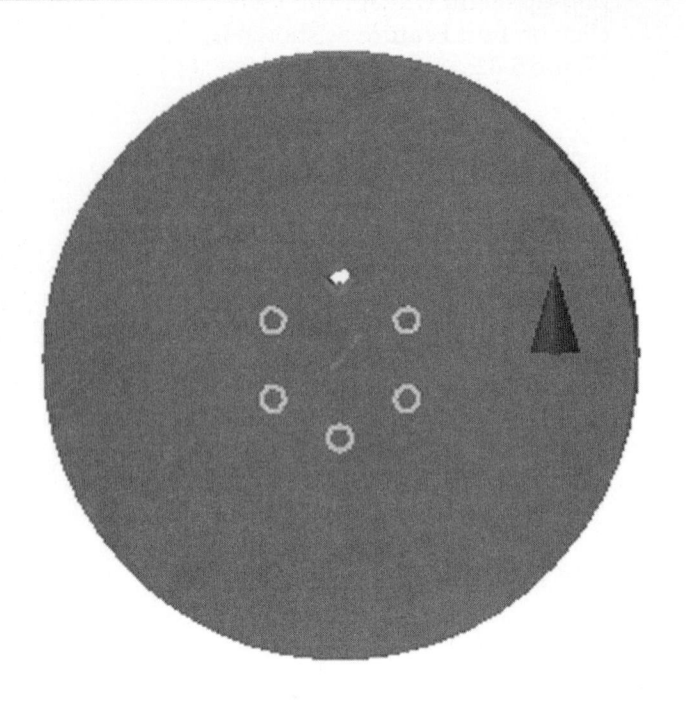

51. Left-click on **OK** in the Circular Pattern dialog box. Your screen should look similar to Figure 5-45.

FIGURE 5-45

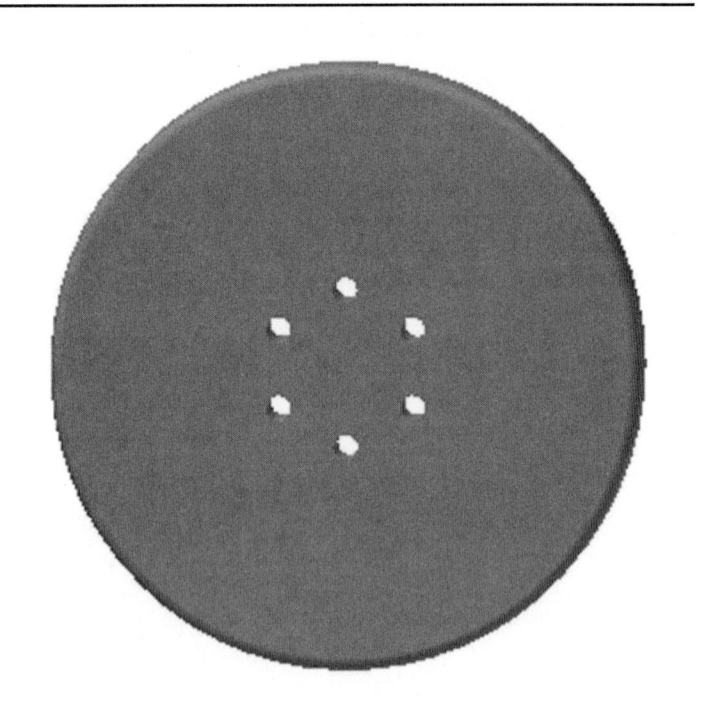

52. Change the view to **Home/Isometric View** as shown in Figure 5-46.

FIGURE 5-46

EDIT THE PART USING THE FILLET COMMAND

53. Move the cursor over the text, "Fillet1." A red box will appear around the text. After the red box appears, left-click once on **Fillet1** as shown in Figure 5-47.

FIGURE 5-47

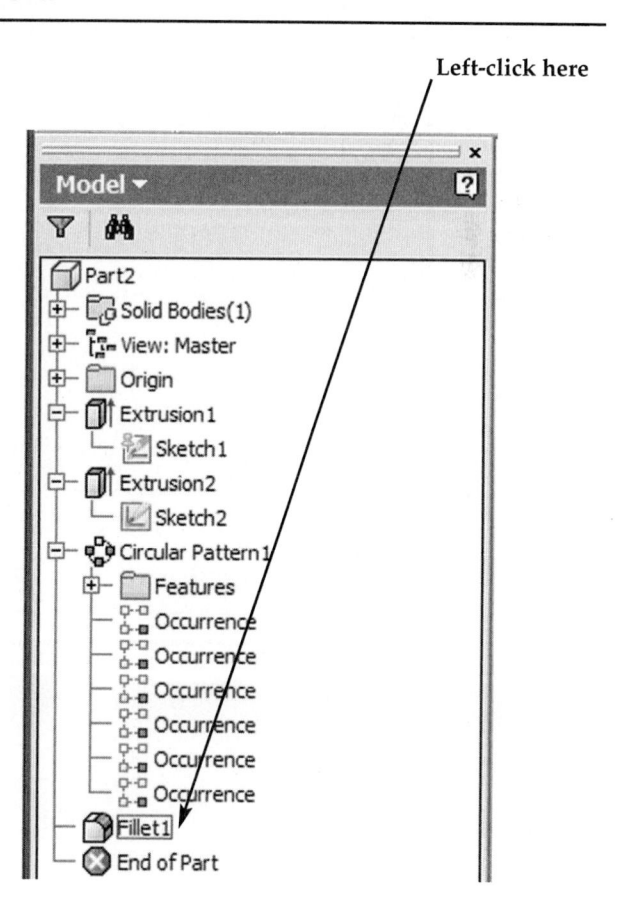

54.   Right-click on **Fillet1**. A pop-up menu will appear. Left-click on **Edit Feature** as shown in Figure 5-48.

FIGURE 5-48

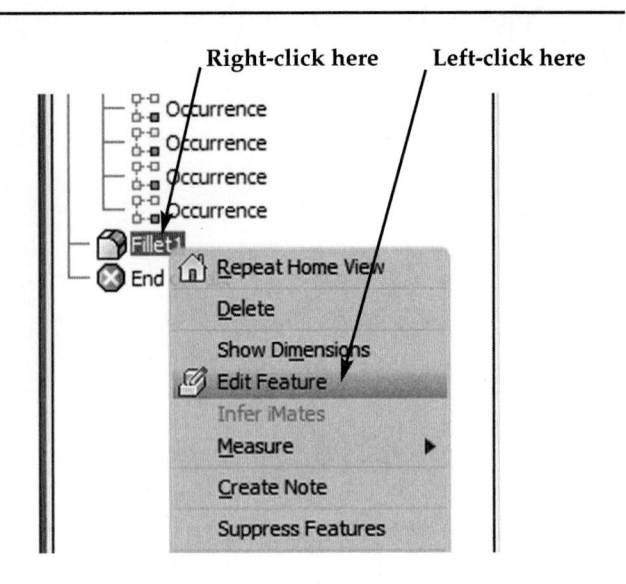

55.   The Fillet dialog box will appear. Enter **.250** for the Radius and left-click on **OK** as shown in Figure 5-49.

FIGURE 5-49

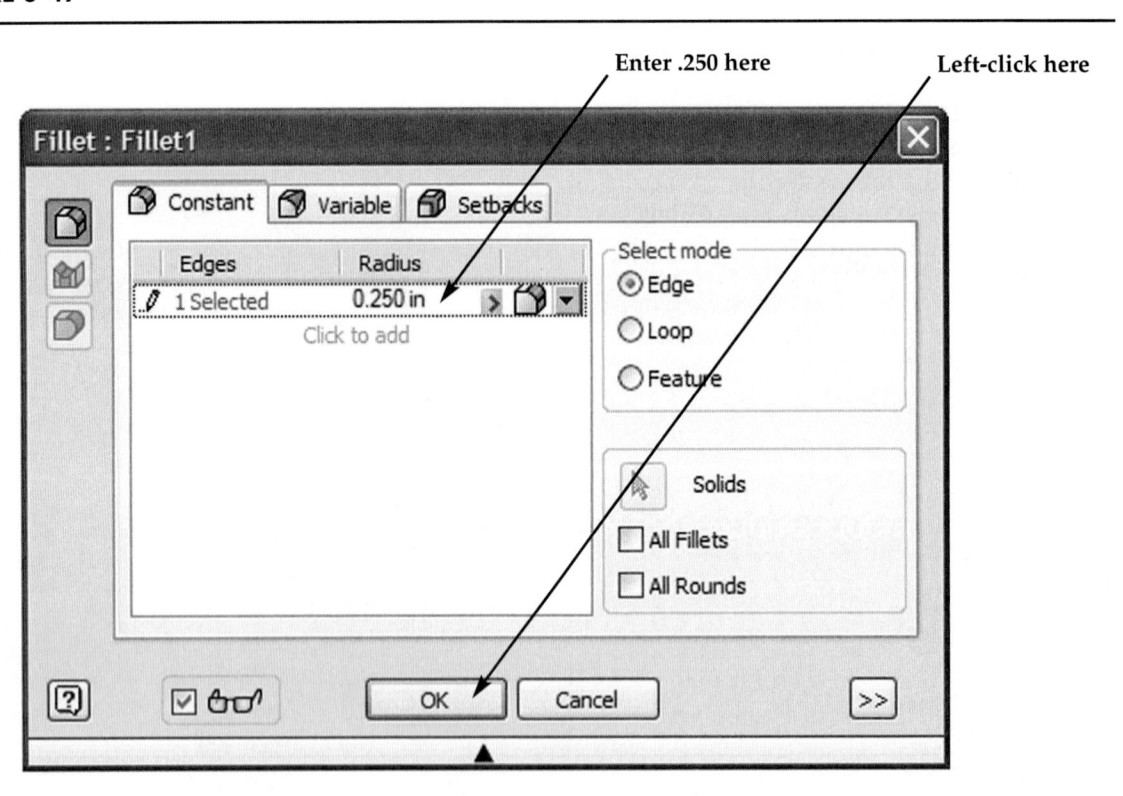

56. Your screen should look similar to Figure 5-50.

FIGURE 5-50

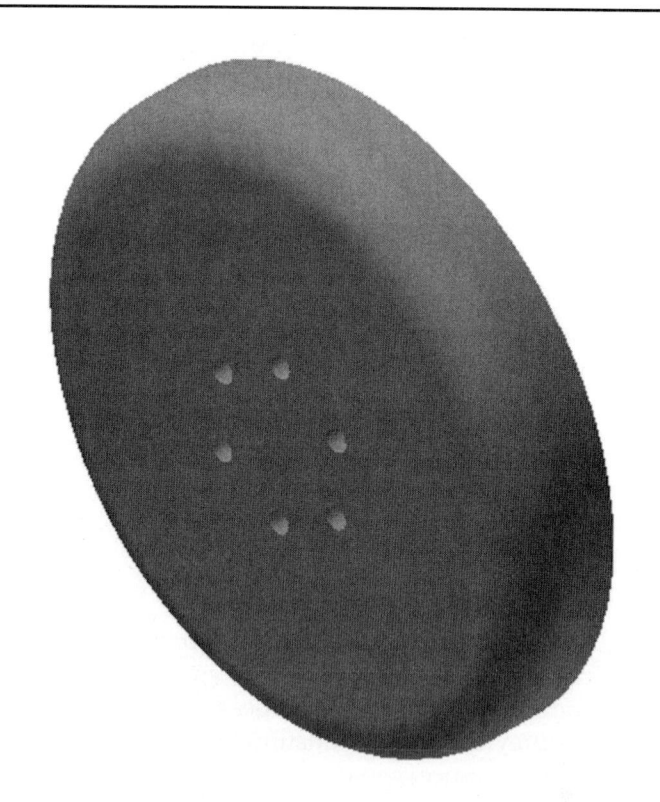

57. Move the cursor over the second listed text, "Occurrence." A red box will appear around the text. After the red box appears, left-click once on **Occurrence** as shown in Figure 5-51.

FIGURE 5-51

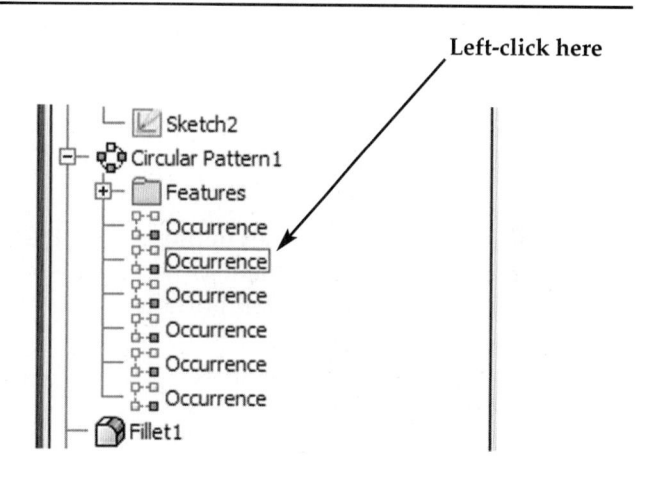

58.   Right-click once. A pop-up menu will appear. Left-click on **Suppress** as shown in Figure 5-52. Inventor will suppress that particular occurrence while leaving all others active.

FIGURE 5-52

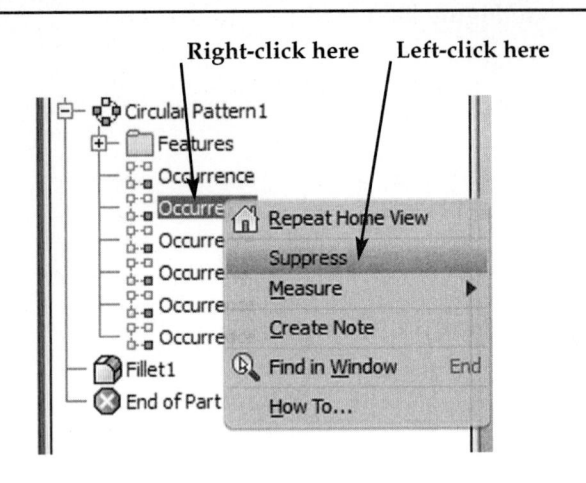

59.   Inventor will draw a line through and gray the text as shown in Figure 5-53. You will notice that the second hole created using the "Circular Pattern" command is not visible. Repeat the previous steps to unsuppress the occurrence.

FIGURE 5-53

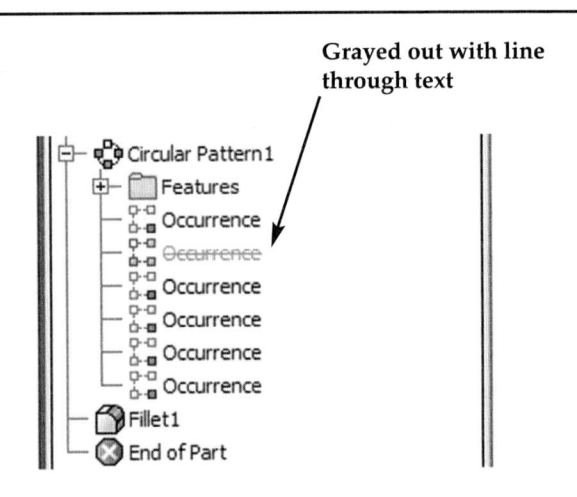

60.   The names of all branches in the part tree can also be edited. Move the cursor to the lower left portion of the screen where the part tree is located. Move the cursor over **Extrusion1** and left-click once, causing the text to become highlighted. After the text is highlighted, left-click one time. The text may be edited as shown in Figure 5-54.

FIGURE 5-54

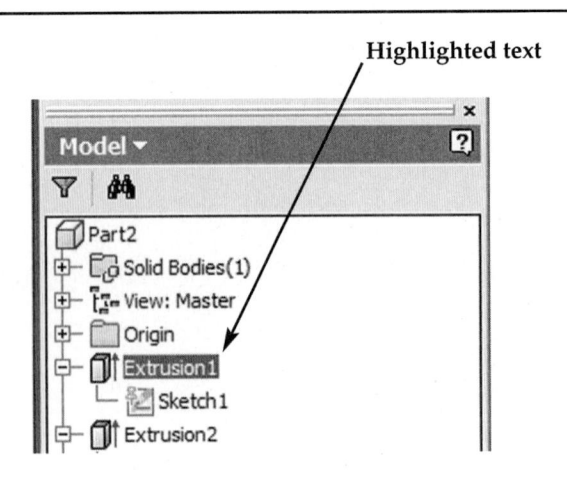

61. Enter the text, **Base Extrusion** as shown in Figure 5-55. Press **Enter** on the keyboard. Text for each individual operation can be edited if desired.

FIGURE 5-55

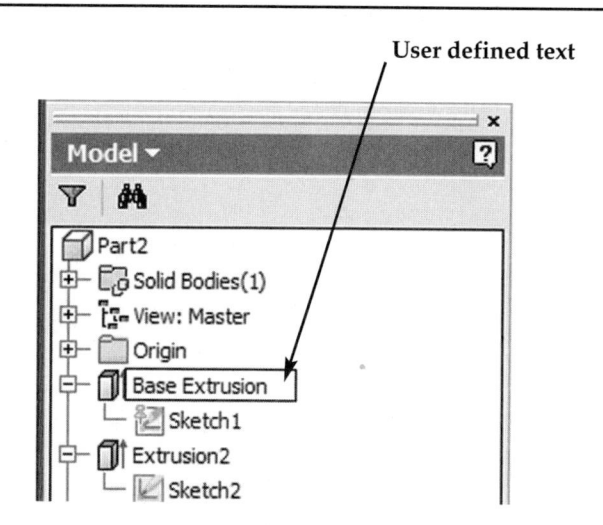

62. Notice that the final design looks significantly different than the original design. The part was redesigned by modifying the existing part as shown in Figure 5-56.

FIGURE 5-56

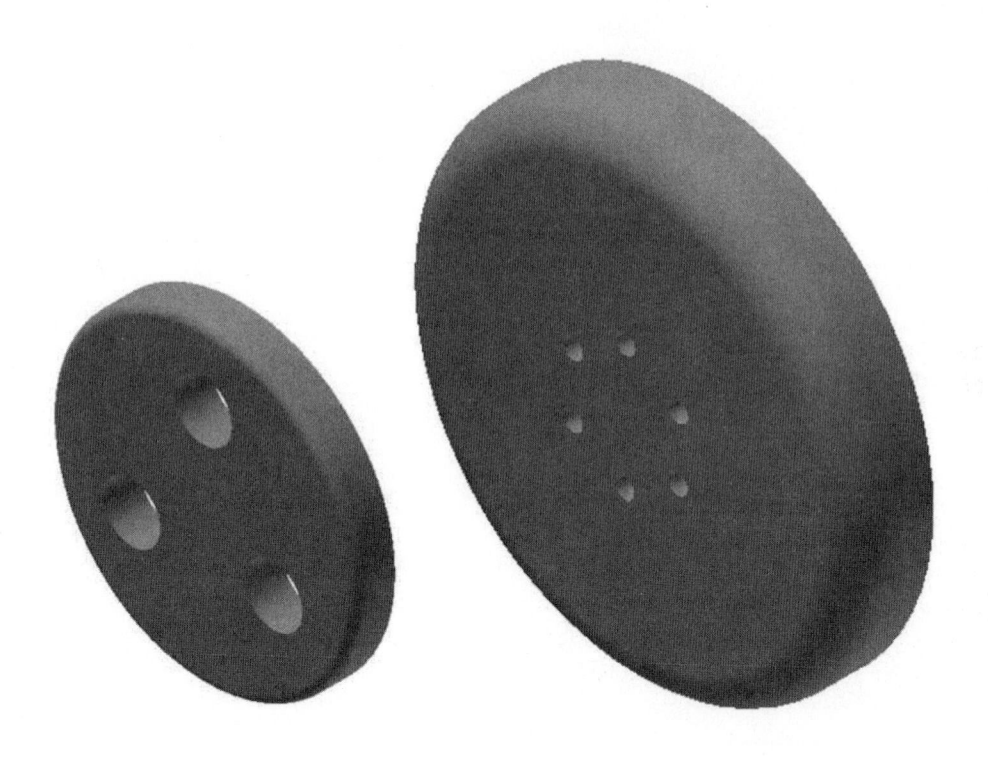

CHAPTER PROBLEMS

Create the following parts as they are shown. Then use the directions to the right and modify the part accordingly. Use the Sketch Panel and Features Panel to modify/edit each part.

PROBLEM 5-1

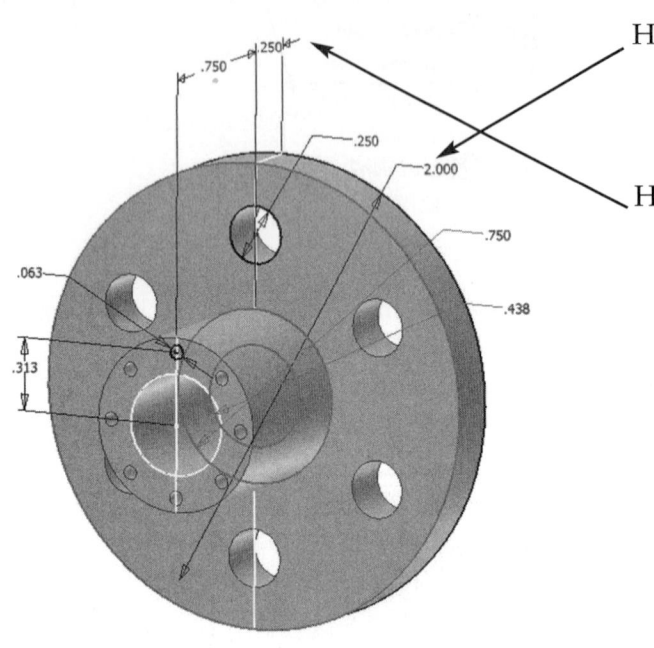

Hint: Use the Edit Sketch command to modify the sketch to 3.00 inches

Hint: Use the Extrude command to modify the extrusion thickness to 0.125 inches

PROBLEM 5-2

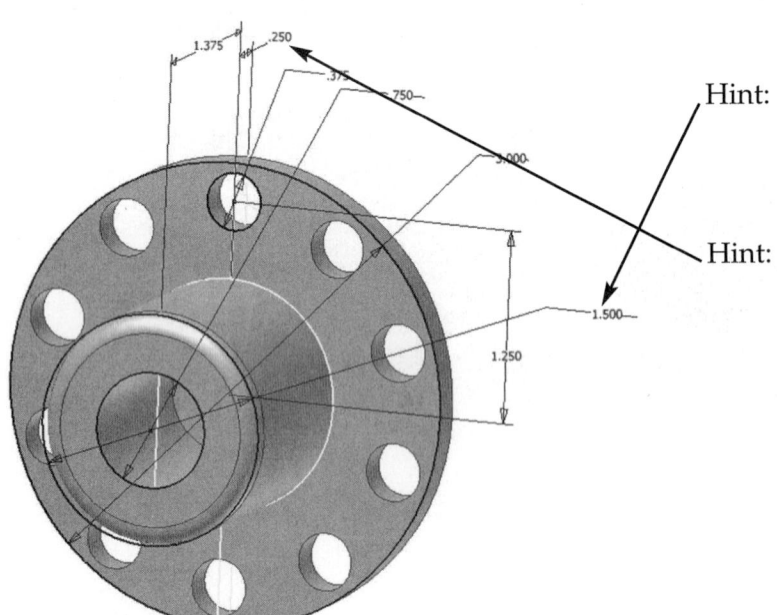

Hint: Use the Edit Sketch command to modify the sketch to 1.00 inch

Hint: Use the Extrude command to modify the Extrusion distance to 2.00 inches. This will cause the hub to protrude out of the back of the part.

PROBLEM 5-3

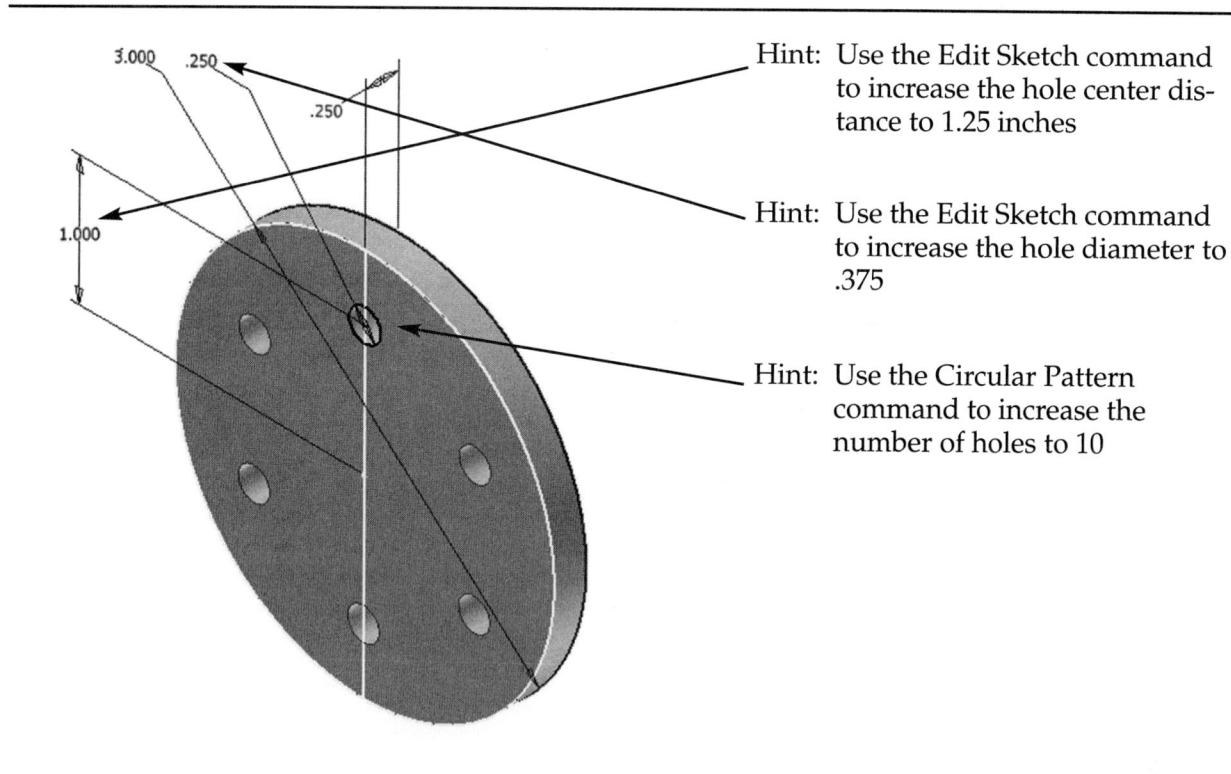

3.000   .250

.250

1.000

Hint: Use the Edit Sketch command to increase the hole center distance to 1.25 inches

Hint: Use the Edit Sketch command to increase the hole diameter to .375

Hint: Use the Circular Pattern command to increase the number of holes to 10

PROBLEM 5-4

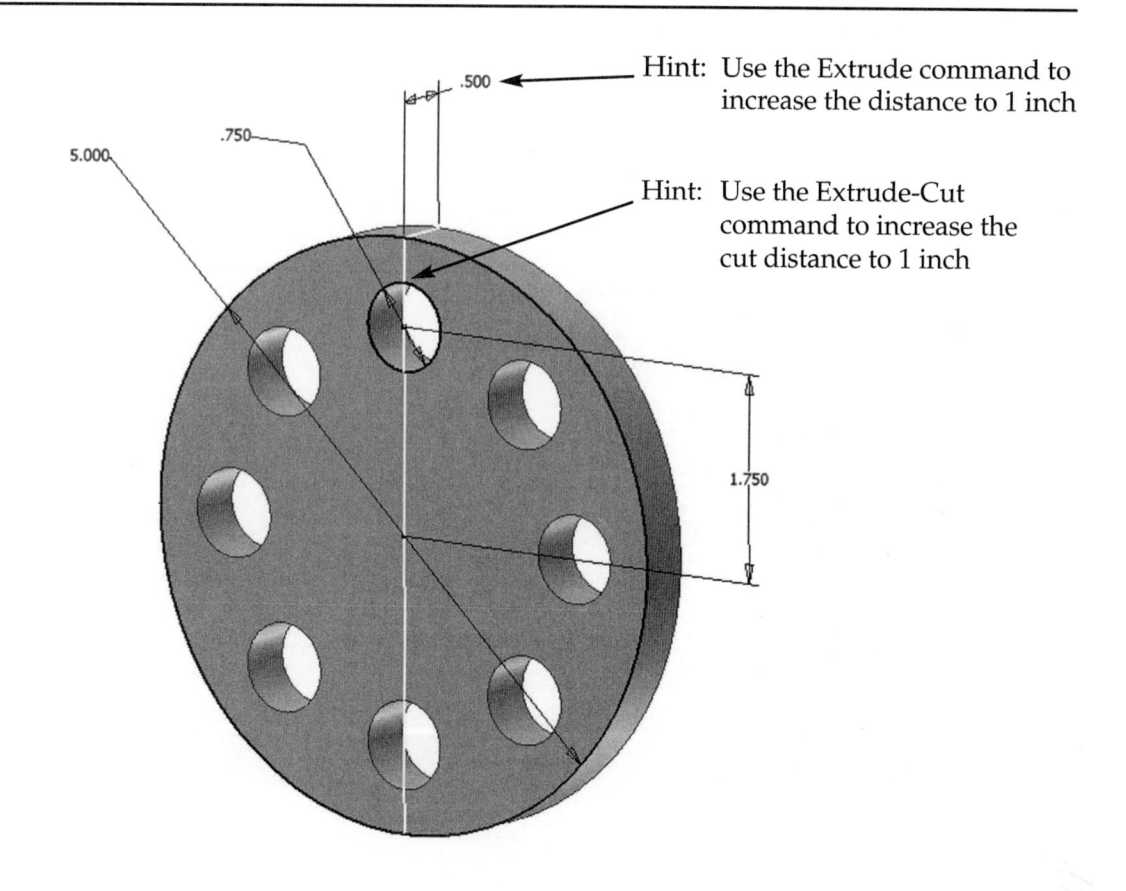

.500

.750

5.000

1.750

Hint: Use the Extrude command to increase the distance to 1 inch

Hint: Use the Extrude-Cut command to increase the cut distance to 1 inch

PROBLEM 5-5

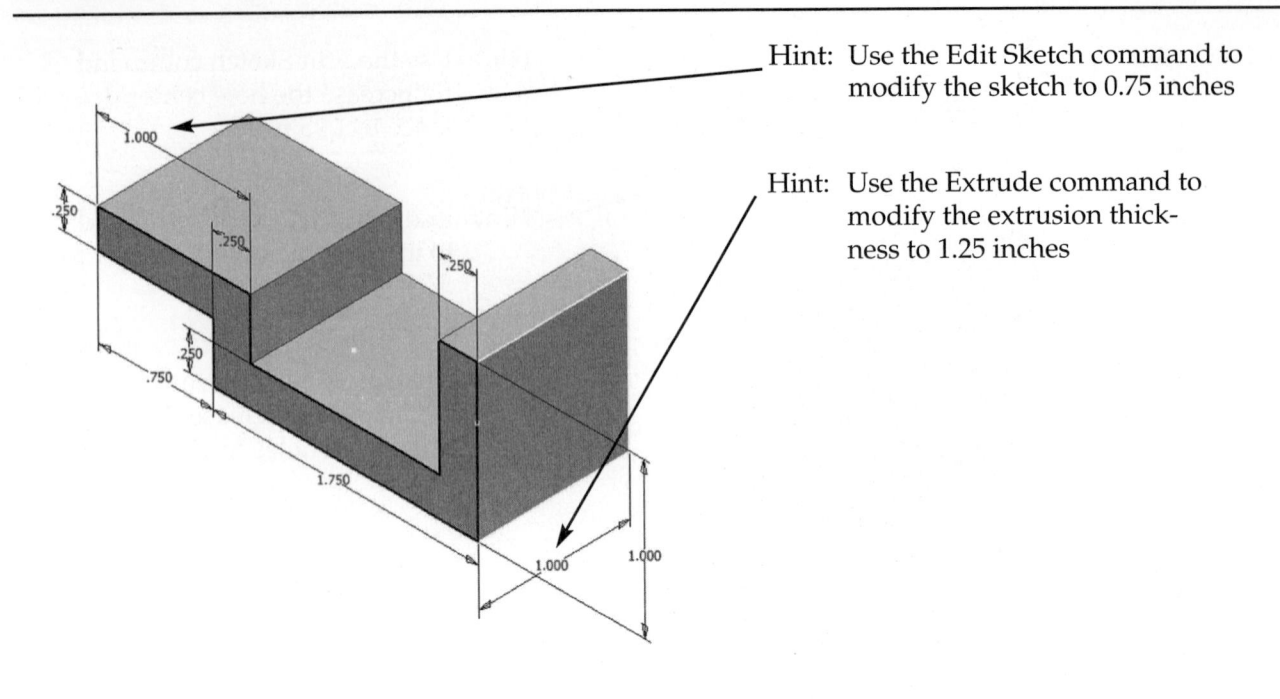

Hint: Use the Edit Sketch command to modify the sketch to 0.75 inches

Hint: Use the Extrude command to modify the extrusion thickness to 1.25 inches

PROBLEM 5-6

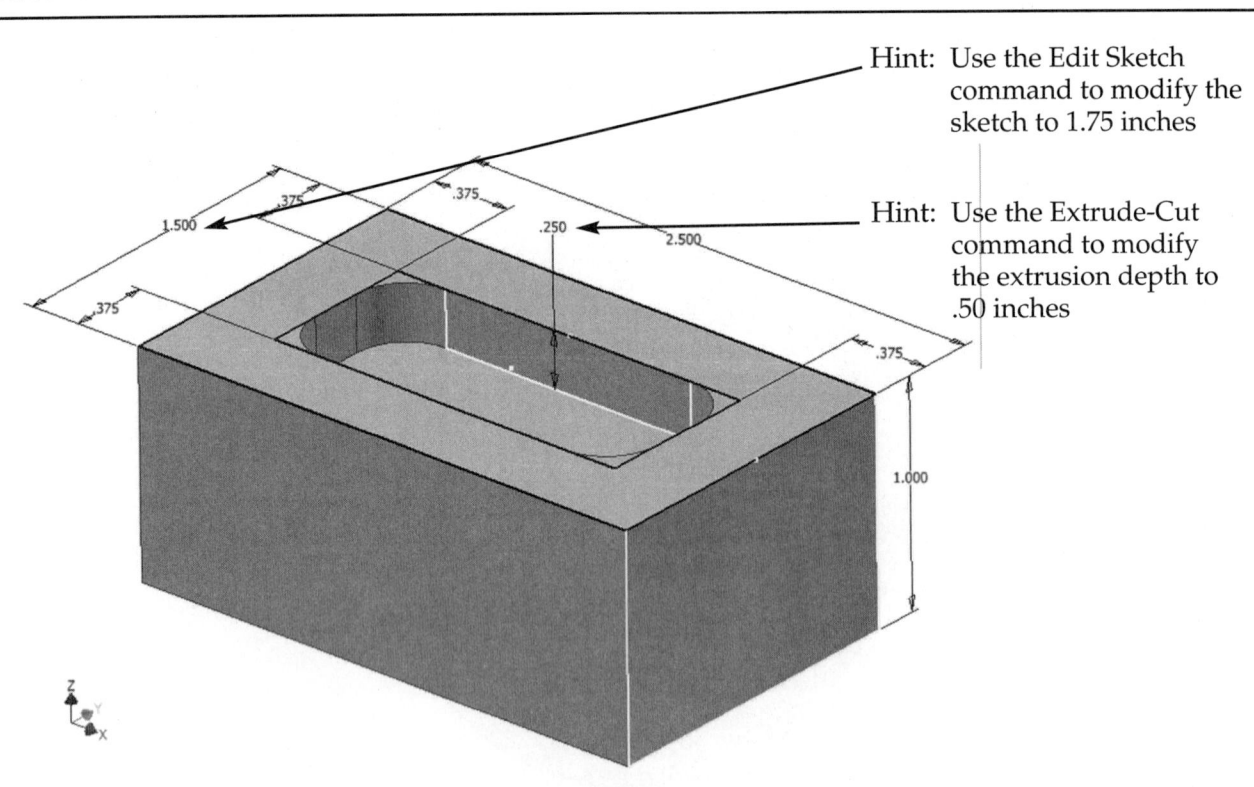

Hint: Use the Edit Sketch command to modify the sketch to 1.75 inches

Hint: Use the Extrude-Cut command to modify the extrusion depth to .50 inches

PROBLEM 5-7

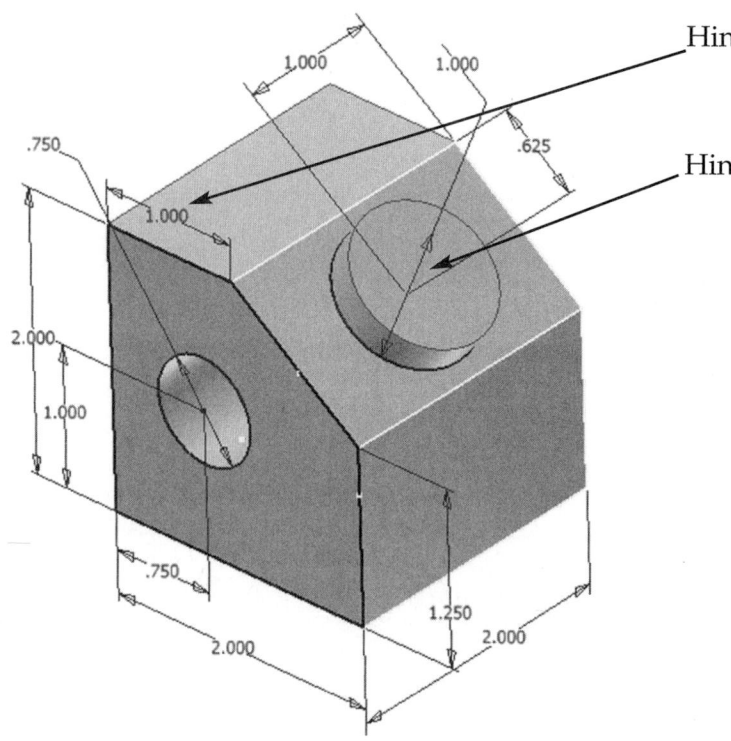

Hint: Use the Edit Sketch command to modify the sketch to 0.875 inches

Hint: Use the Extrude-Cut command to modify the Extrusion to a Reverse Direction Cut

PROBLEM 5-8

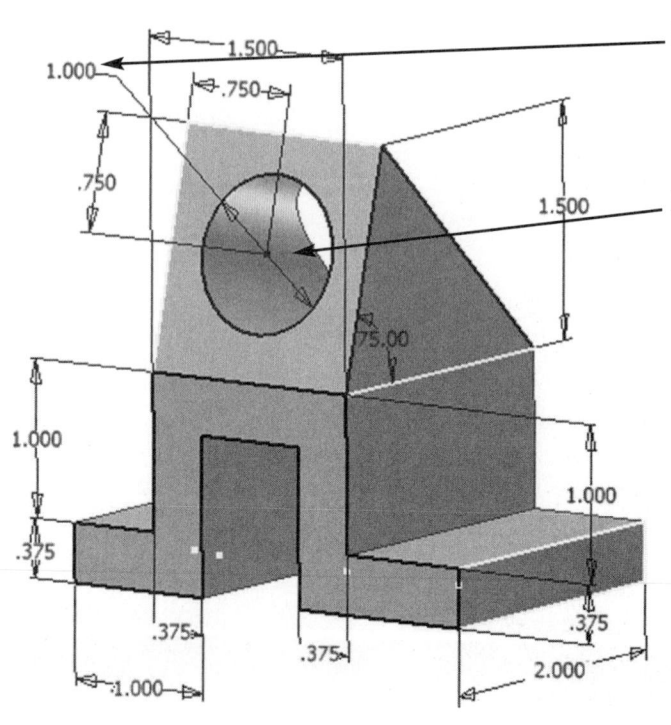

Hint: Use the Edit Sketch command to modify the sketch to 1.275 inches

Hint: Use the Extrude-Cut command to modify the Extruded Cut Depth to .25 inches

CHAPTER 6

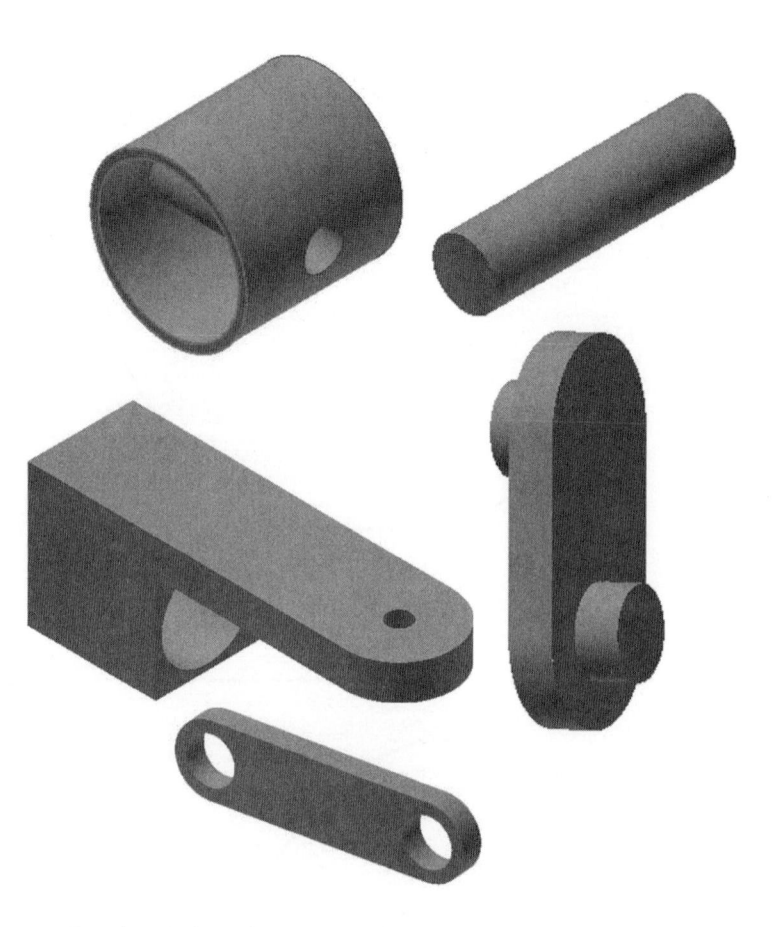

Chapter 6 includes instruction on how to design the parts shown above.

DESIGNING PART MODELS FOR ASSEMBLY

OBJECTIVES

1. Design multiple sketch parts
2. Use the X, Y, and Z Planes
3. Use the Wireframe viewing command
4. Project geometry onto a new sketch
5. Use the Shell command
6. Use Constraints while constructing a sketch

1. Start Autodesk Inventor 2013 by referring to Chapter 1.

2. After Autodesk Inventor 2013 is running, begin a new sketch.

3. Complete the sketch shown in Figure 6-1.

FIGURE 6-1

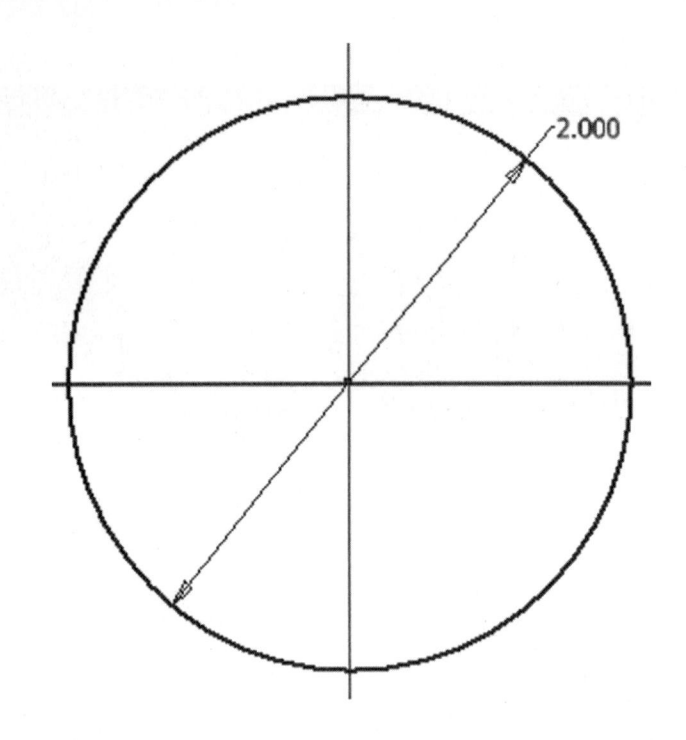

4. Exit out of the Sketch Panel and change the view to Isometric as shown in Figure 6-2.

FIGURE 6-2

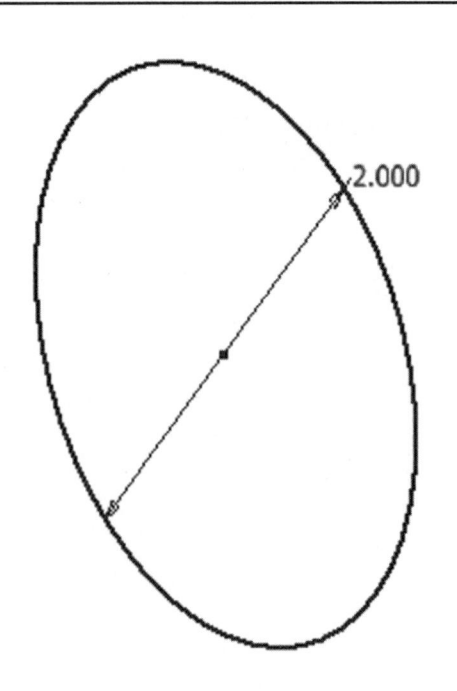

5. Extrude the sketch a distance of 2 inches as shown in Figure 6-3.

FIGURE 6-3

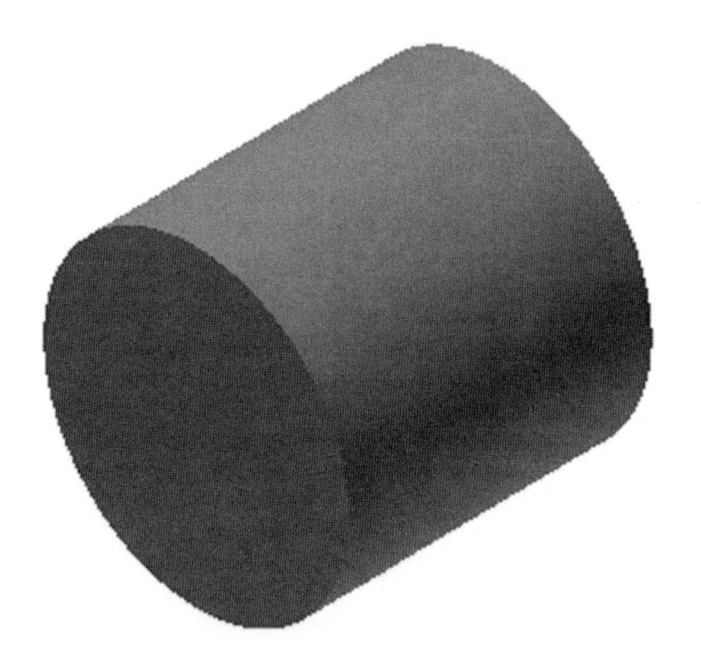

6. Move the cursor to the upper left portion of the screen and left-click on the "plus" sign next to the text, "Origin" as shown in Figure 6-4.

FIGURE 6-4

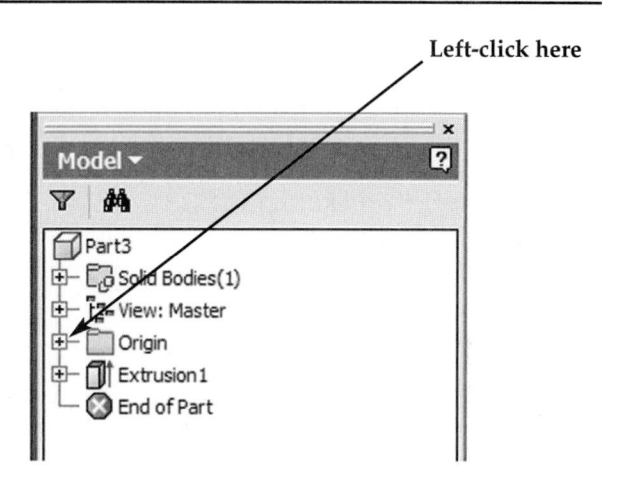

USE THE X, Y, AND Z PLANES

7. The part tree will expand. Move the cursor over the text, "YZ Plane," causing a red box to appear around the text as shown in Figure 6-5.

FIGURE 6-5

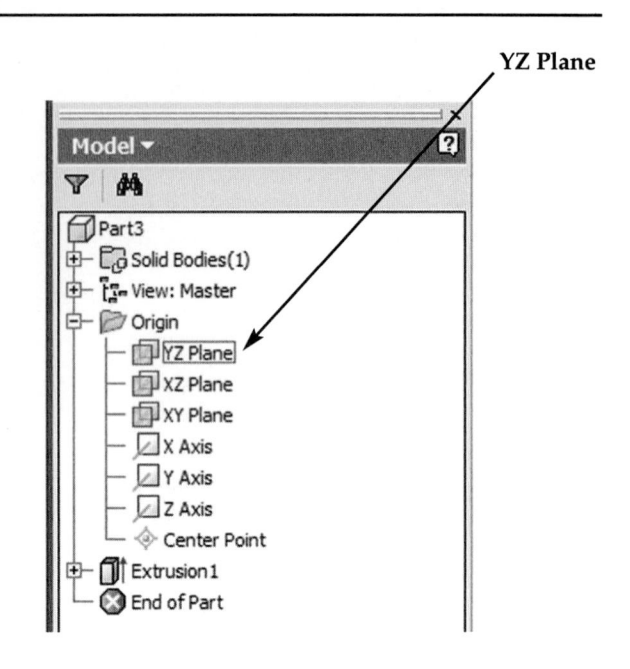

YZ Plane

8. The YZ Plane will become visible as shown in Figure 6-6.

FIGURE 6-6

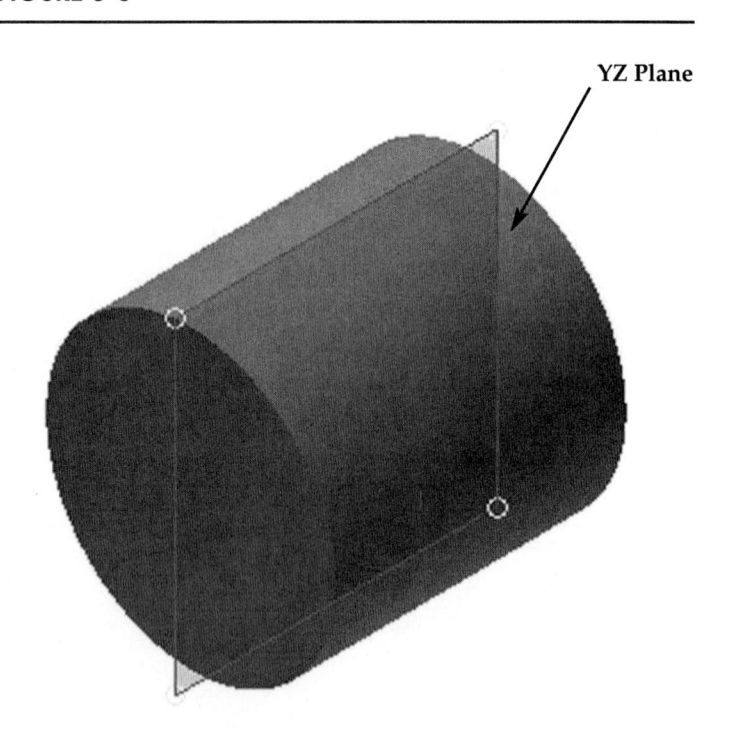

YZ Plane

9.   Right-click on the text, **YZ Plane**. A pop-up menu will appear. Left-click on **New Sketch** as shown in Figure 6-7.

FIGURE 6-7

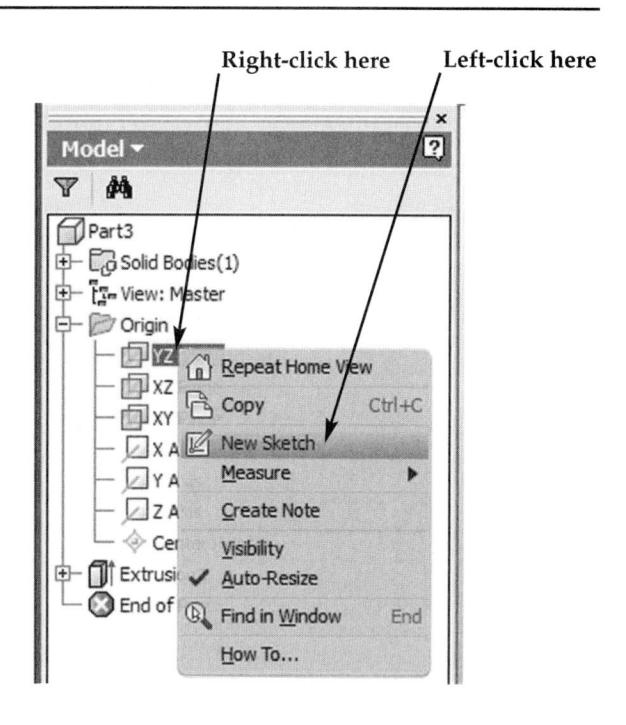

10. Use the **Home View** command to rotate the part as shown. Your screen should look similar to Figure 6-8.

FIGURE 6-8

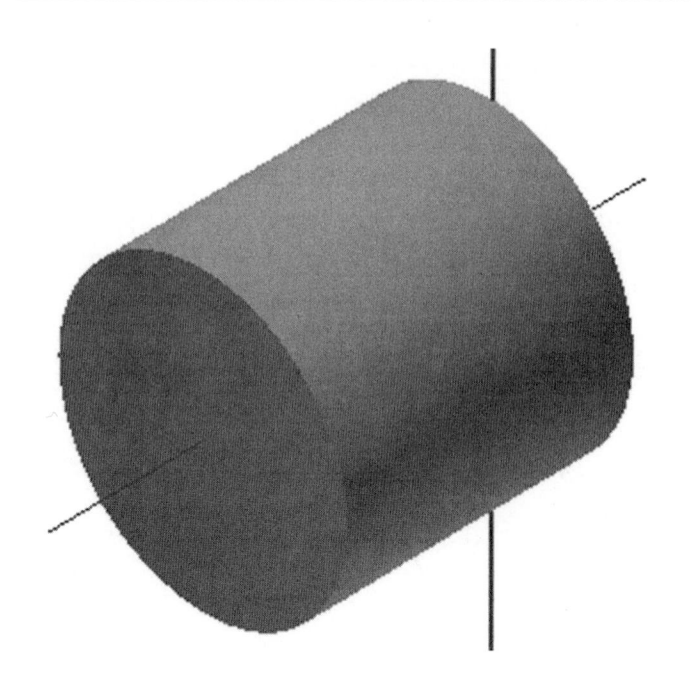

USE THE WIREFRAME VIEWING COMMAND

11.   Move the cursor to the upper left portion of the screen and left-click on the **View** tab. Left-click on the drop-down arrow below the **Visual Style** icon. A drop-down menu will appear. Left-click on **Wireframe** as shown in Figure 6-9.

FIGURE 6-9

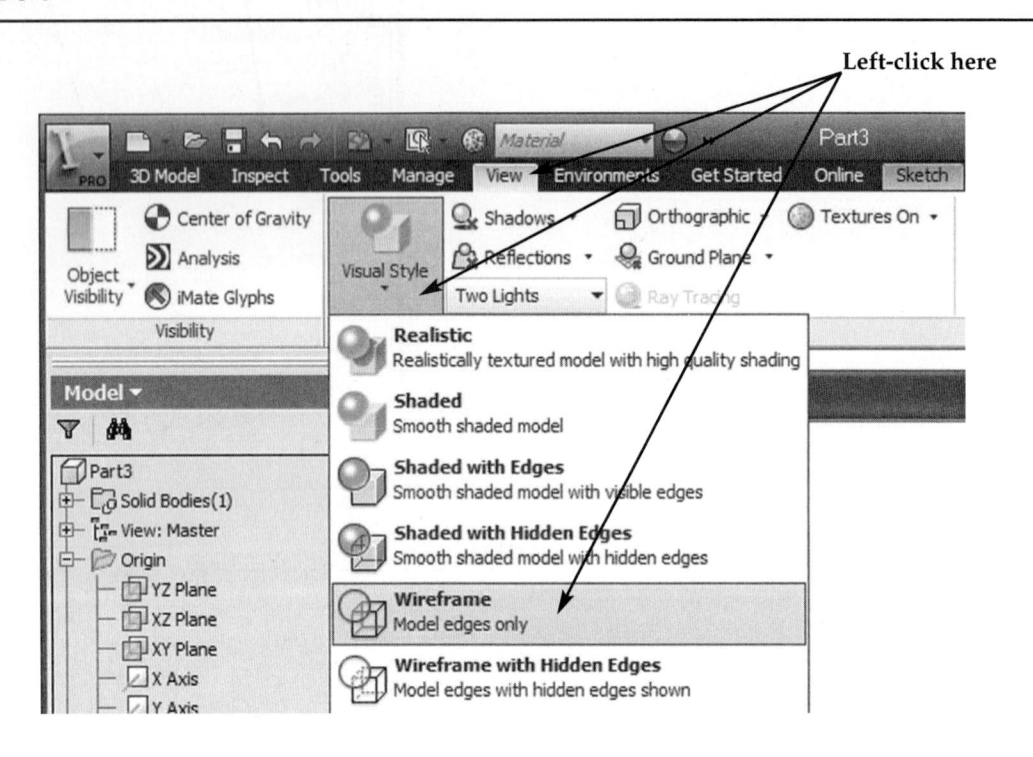

12.   Your screen should look similar to Figure 6-10.

FIGURE 6-10

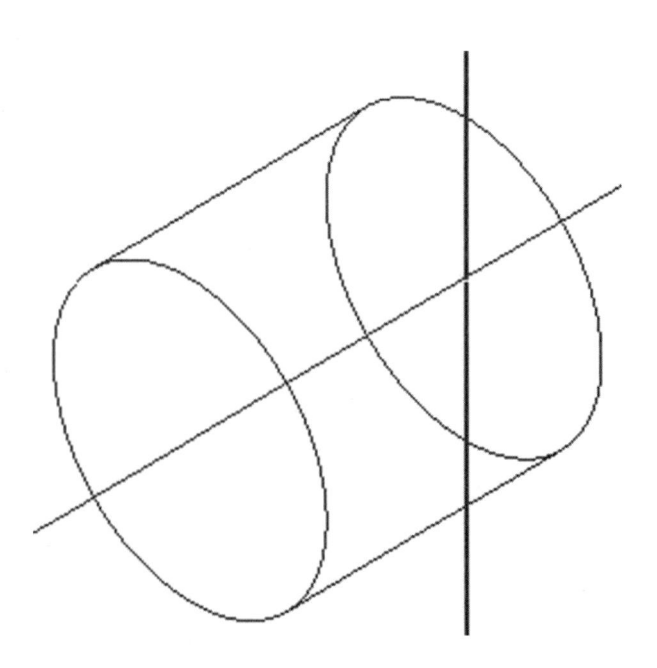

13. Move the cursor to the upper right portion of the screen and left-click on the "Face View/Look At" icon as shown in Figure 6-11.

FIGURE 6-11

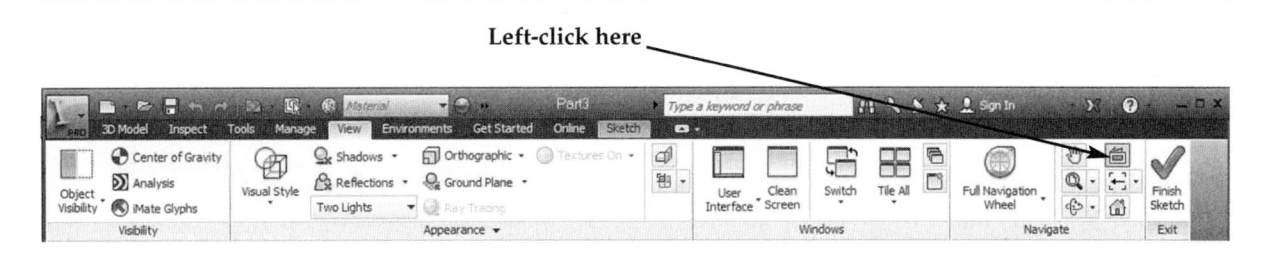

Left-click here

14. Move the cursor to the upper left portion of the screen and left-click on the text, **YZ Plane** in the part tree as shown in Figure 6-12.

FIGURE 6-12

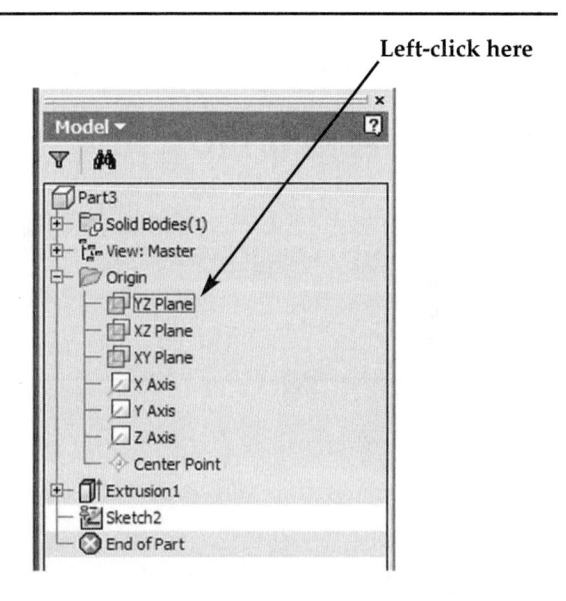

Left-click here

15. Notice the YZ Plane becoming visible through the part as shown in Figure 6-13.

FIGURE 6-13

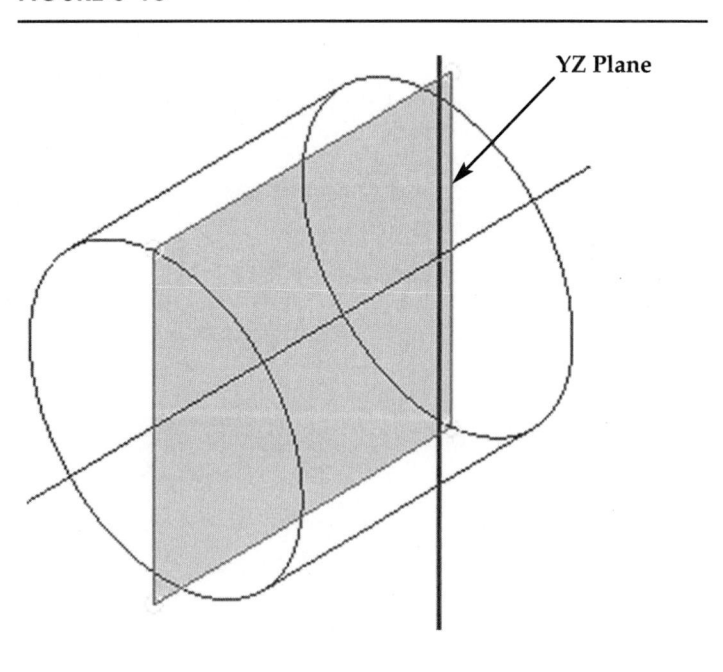

YZ Plane

16. Inventor will rotate the YZ Plane to provide a perpendicular view as shown in Figure 6-14.

FIGURE 6-14

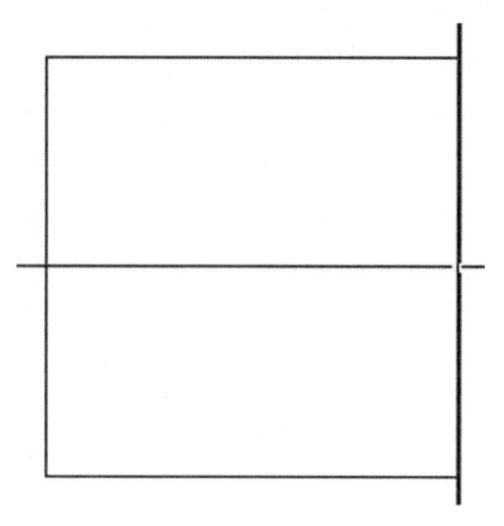

PROJECT GEOMETRY TO A NEW SKETCH

17. Move the cursor to the upper middle portion of the screen and left-click on the **Sketch** tab. Left-click on **Project Geometry** as shown in Figure 6-15.

FIGURE 6-15

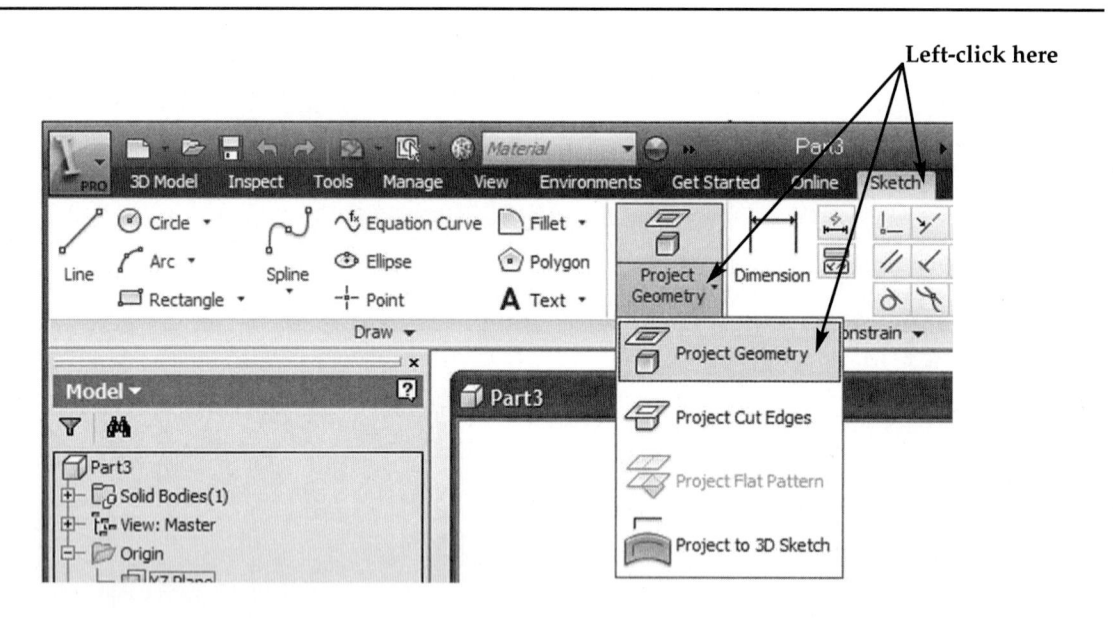

18.   Move the cursor over the top and bottom lines and left-click. Inventor will project these lines onto the sketch for reference purposes as shown in Figure 6-16. They will need to be deleted before exiting the sketch.

FIGURE 6-16

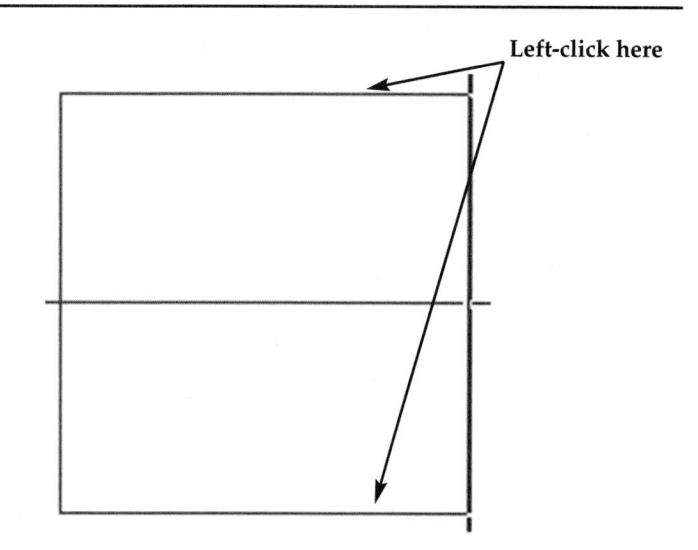

19.   Use the **Line** command to draw a line from the midpoint of the top line to the midpoint of the bottom line as shown in Figure 6-17.

FIGURE 6-17

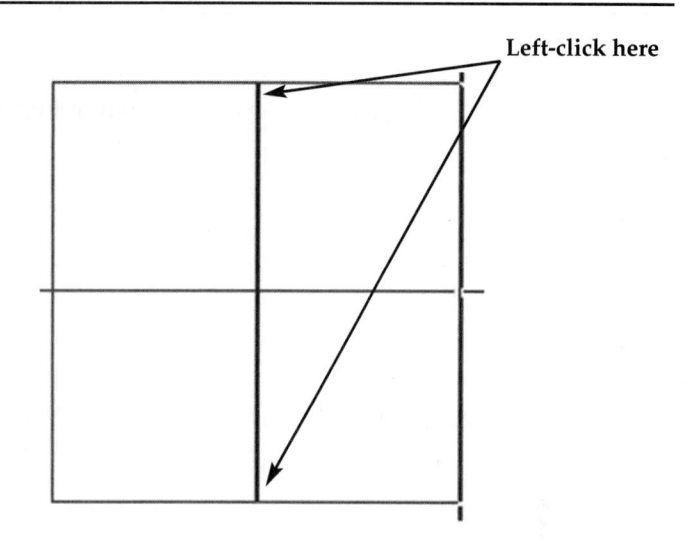

20.   Create a .500 inch circle at the midpoint as shown in Figure 6-18.

FIGURE 6-18

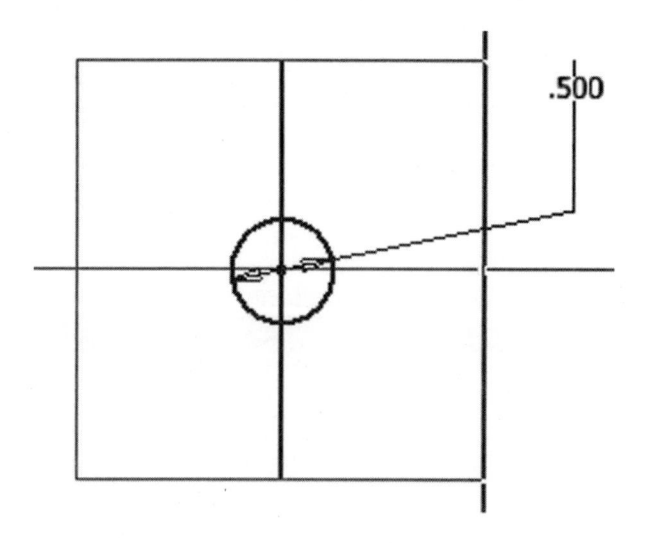

21.   Delete the lines that were projected onto the sketch along with the center line that was used to create the circle as shown in Figure 6-19.

FIGURE 6-19

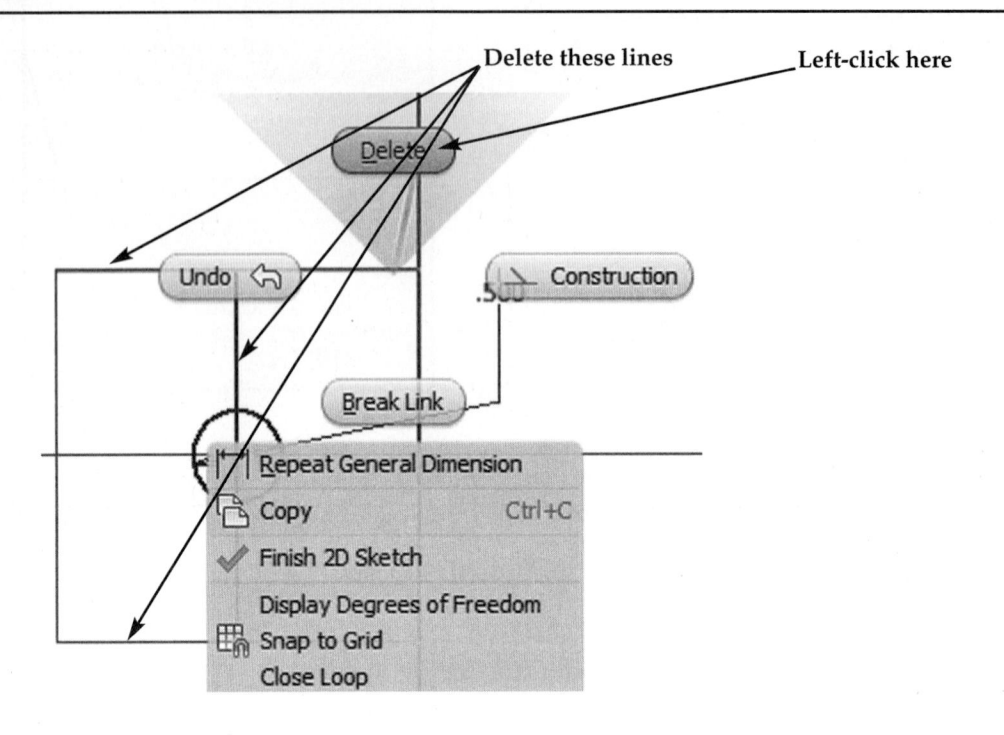

22.   Exit out of the Sketch Panel and change the view to Isometric as shown in Figure 6-20.

FIGURE 6-20

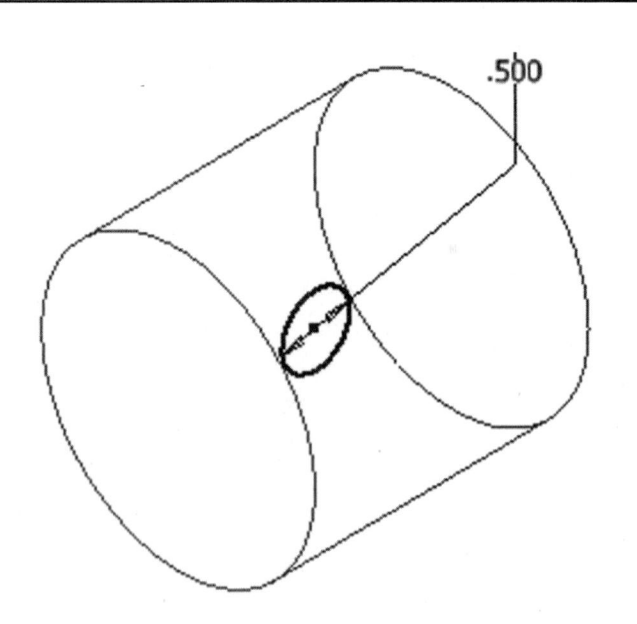

23. Use the "Extrude" command to create a hole in the part using the newly created circle. Left-click on the "Cut" icon. Enter **2.00** for the distance. Left-click on the "Bi-directional" icon. Inventor will provide a preview of the extrusion. Left-click on **OK** as shown in Figure 6-21.

FIGURE 6-21

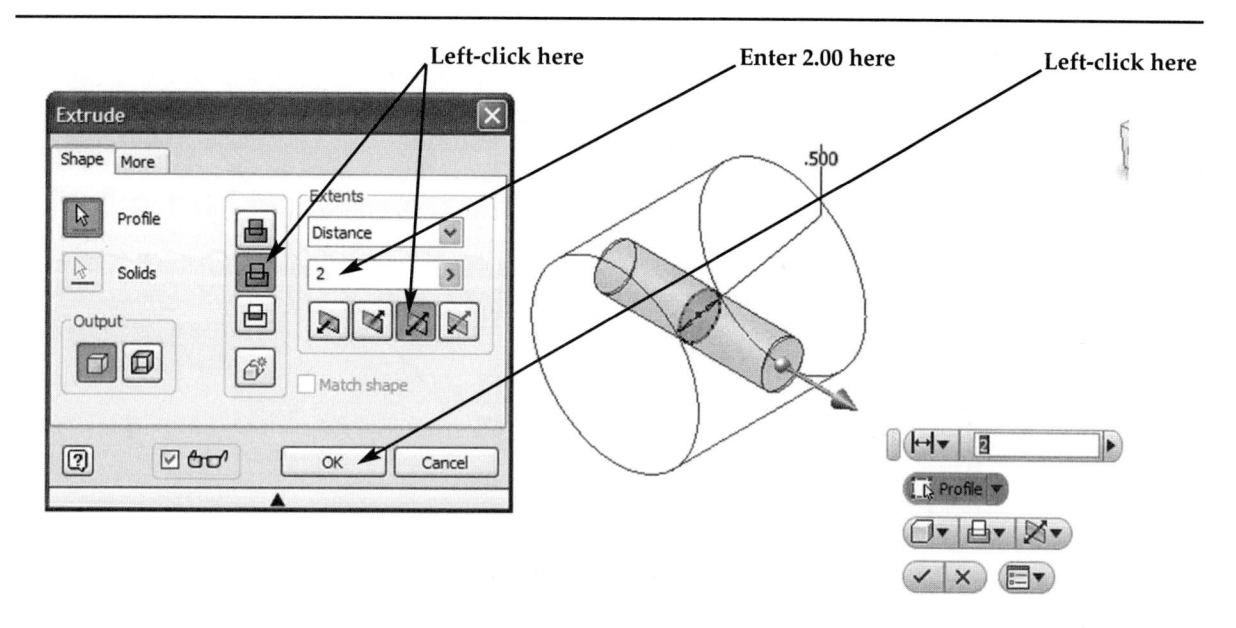

24. Your screen should look similar to Figure 6-22.

FIGURE 6-22

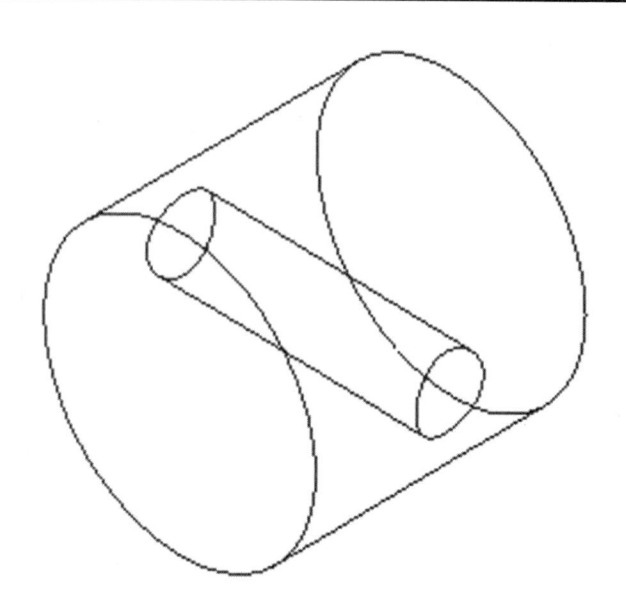

25. Move the cursor to the upper middle portion of the screen and left-click on the **View** tab. Left-click on the drop-down arrow below the **Visual Style** icon. Left-click on **Shaded with Edges** as shown in Figure 6-23.

FIGURE 6-23

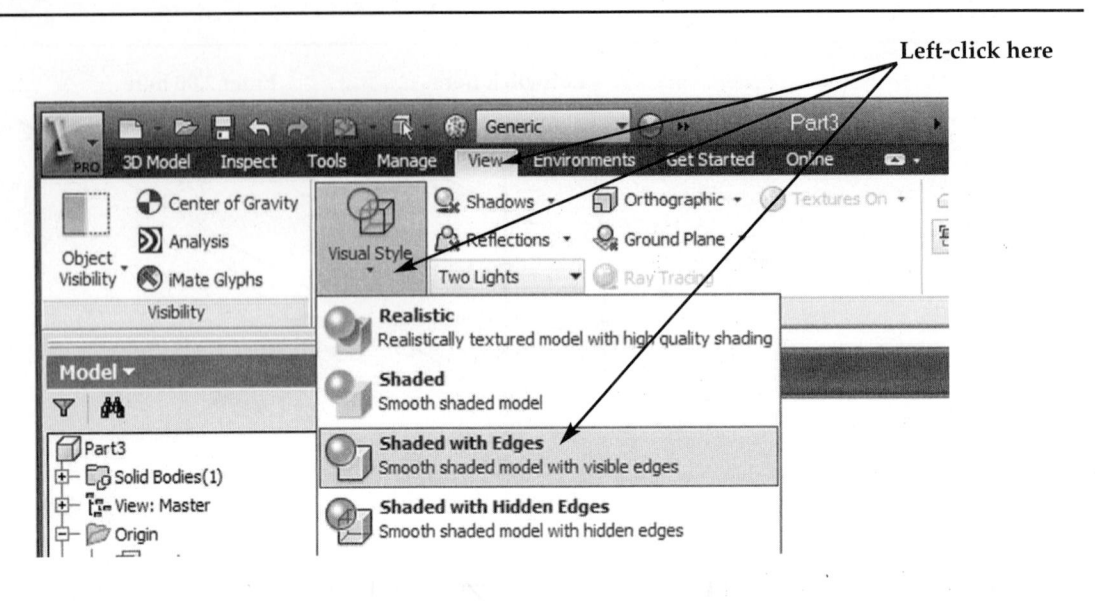

26. Your screen should look similar to Figure 6-24.

FIGURE 6-24

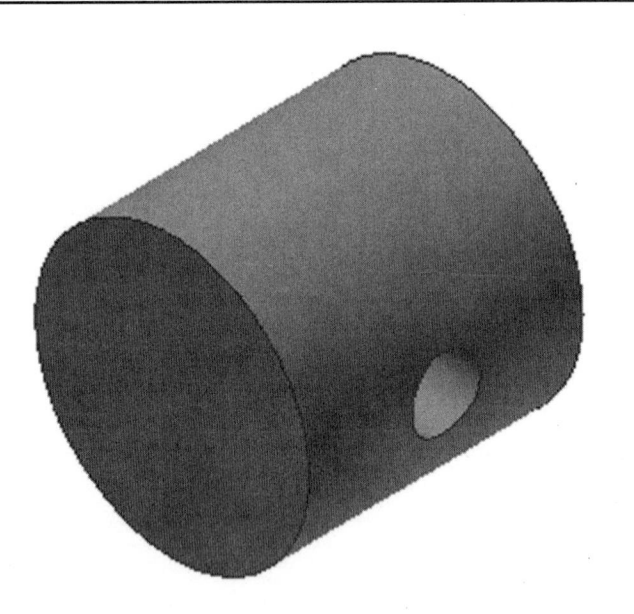

USE THE SHELL COMMAND

27. Move the cursor to the upper left portion of the screen and left-click on the **3D Model** tab. Left-click on the **Shell** icon. The Shell dialog box will appear as shown in Figure 6-25.

FIGURE 6-25

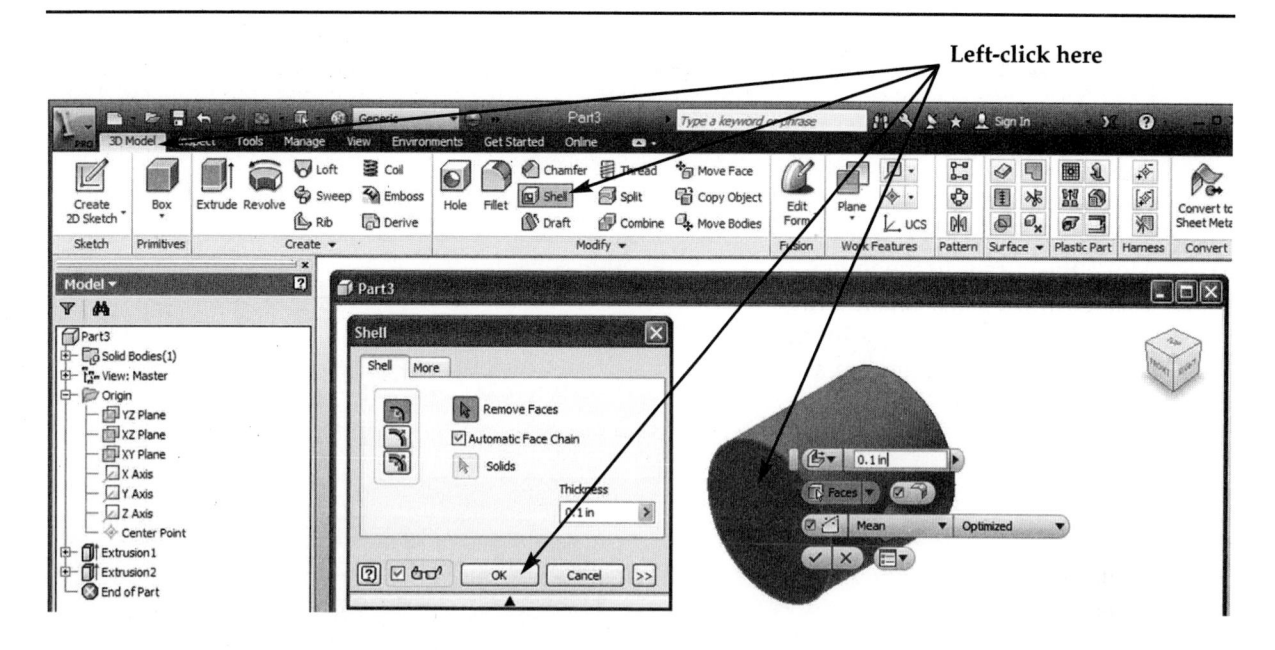

28. Left-click on the lower surface of the part and left-click on **OK** as shown in Figure 6-25.

29. Your screen should look similar to Figure 6-26.

FIGURE 6-26

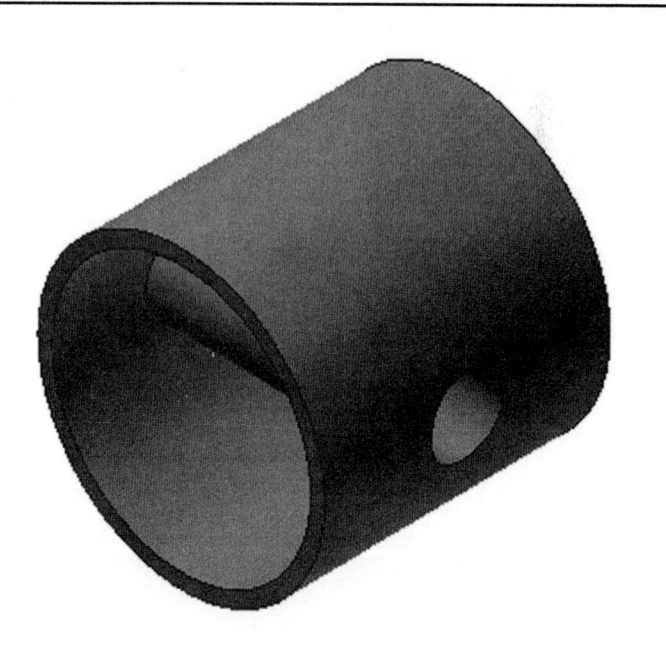

30.   Move the cursor to the upper middle portion of the screen and left-click on the **View** tab. Left-click on the "Face View/Look At" icon as shown in Figure 6-27.

FIGURE 6-27

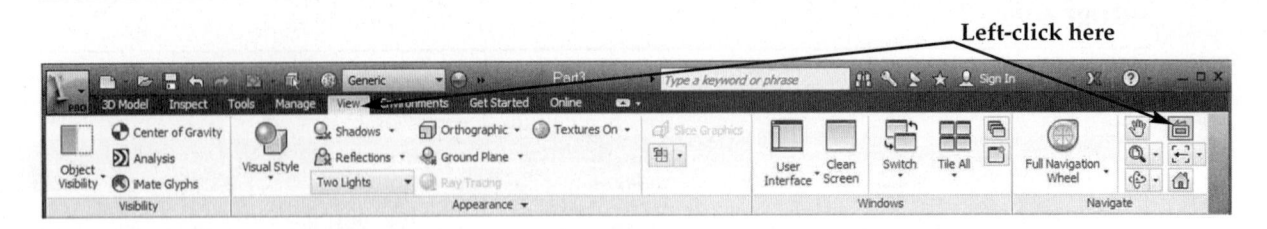

31.   Move the cursor to the lower surface of the part, causing the inside and outside edges to turn red and left-click as shown in Figure 6-28.

FIGURE 6-28

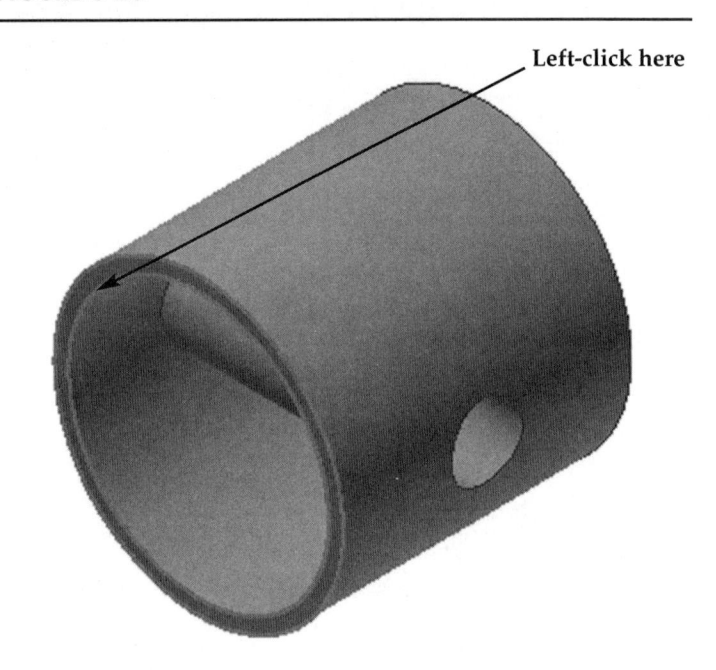

32.   After both edges turn red, left-click once. Inventor will rotate the part providing a perpendicular view of the inside as shown in Figure 6-29.

FIGURE 6-29

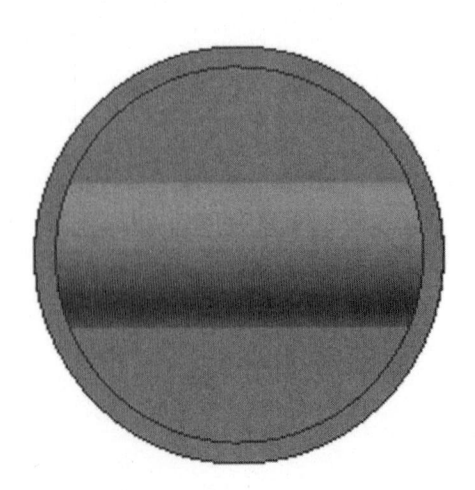

33.   Begin a new sketch on the surface shown in Figure 6-30.

FIGURE 6-30

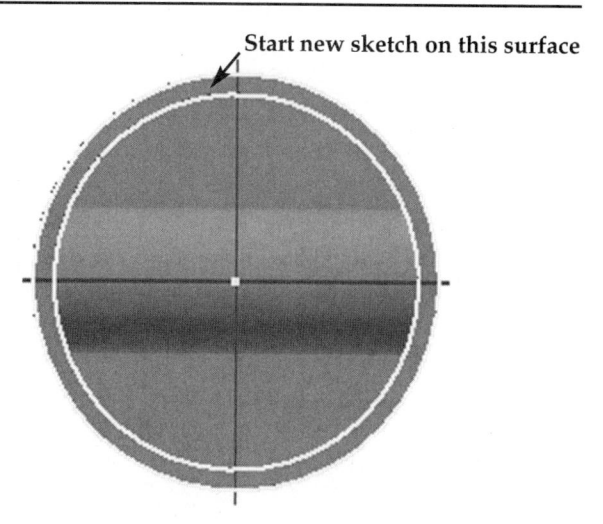

Start new sketch on this surface

34.   Use the "Rectangle" command to complete the following sketch. You will need to dimension the rectangle from the origin as shown in Figure 6-31.

FIGURE 6-31

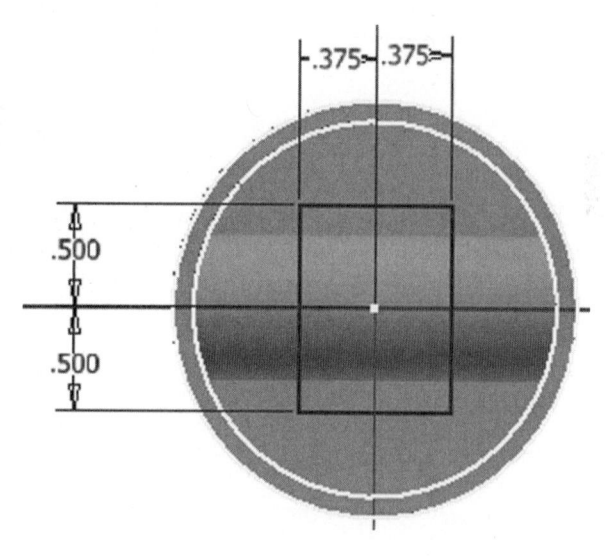

35.   Exit the Sketch Panel and change the view to Isometric as shown in Figure 6-32.

FIGURE 6-32

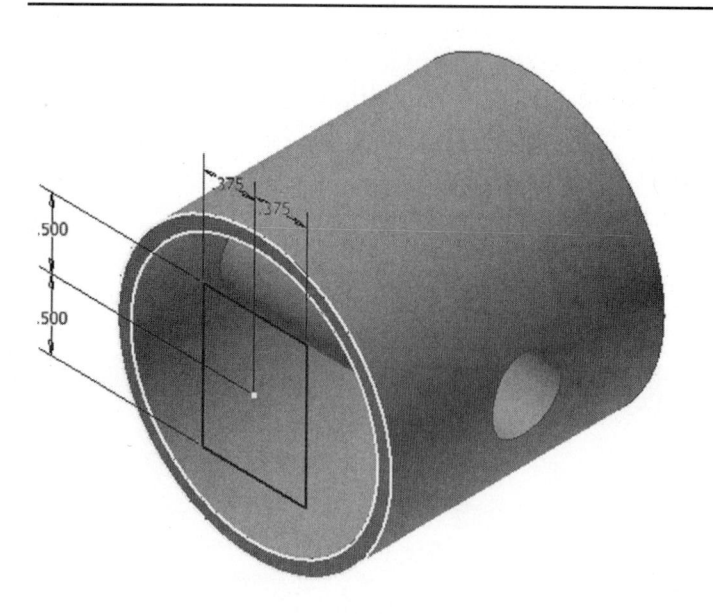

36. Use the "Extrude" command to "Cut" back into the part a distance of **1.875** as shown in Figure 6-33.

FIGURE 6-33

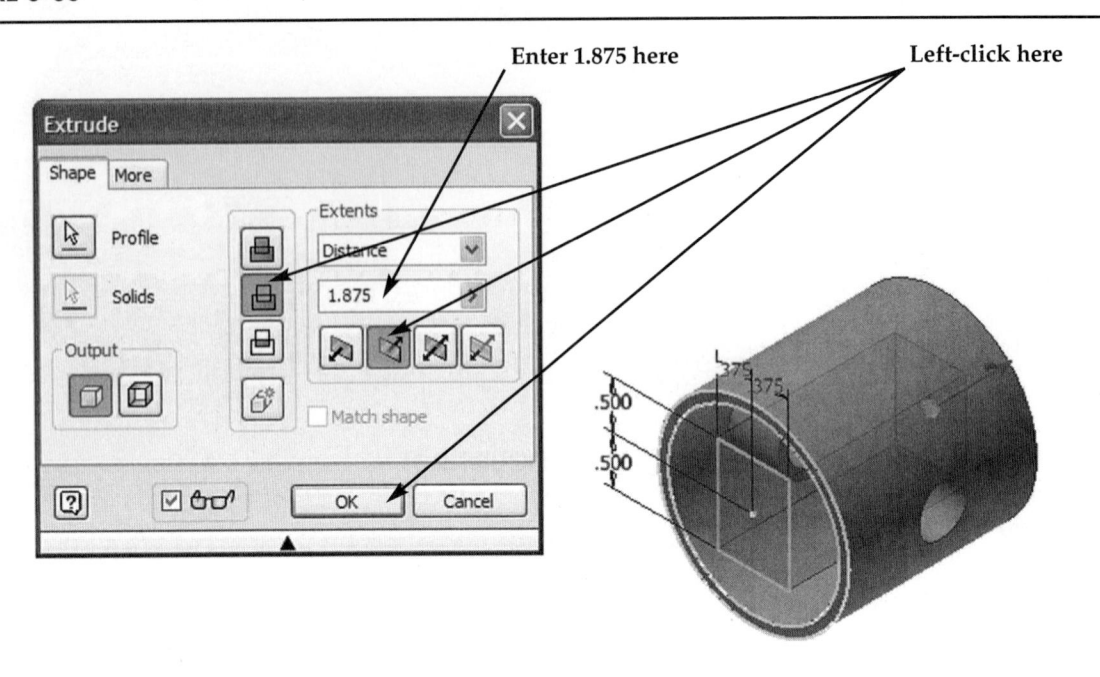

37. Save the part as Piston1.ipt where it can be easily retrieved later.

38. Begin a new drawing as described in Chapter 1.

39. Draw a circle in the center of the grid as shown in Figure 6-34.

FIGURE 6-34

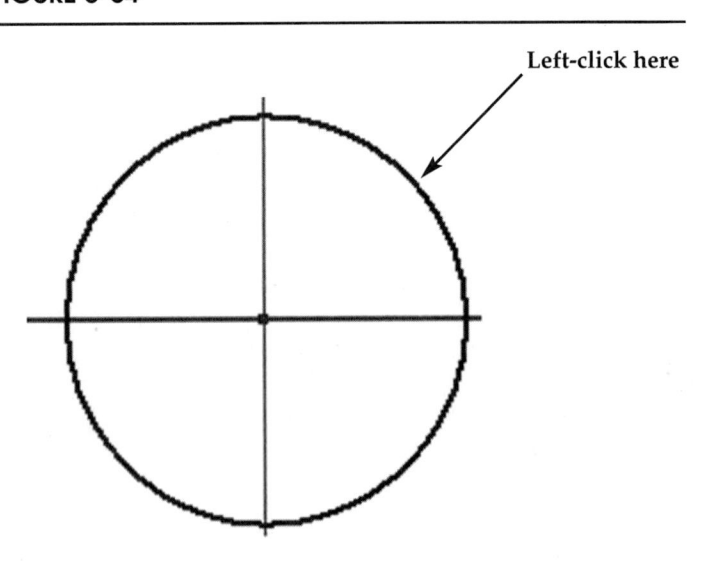

40. Use the **Dimension** command to dimension the circle to **.5** inches as shown in Figure 6-35.

FIGURE 6-35

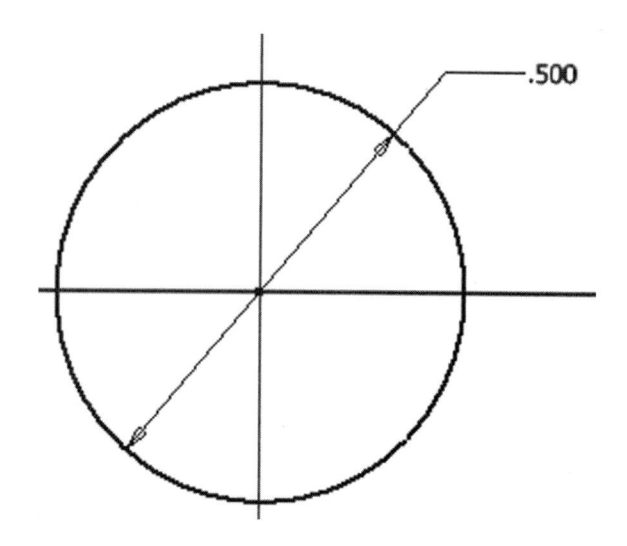

41. Use the **Home View** command to view the sketch in Isometric. Exit the Sketch Panel and Extrude the circle to a length of **1.875** inches as shown in Figure 6-36.

FIGURE 6-36

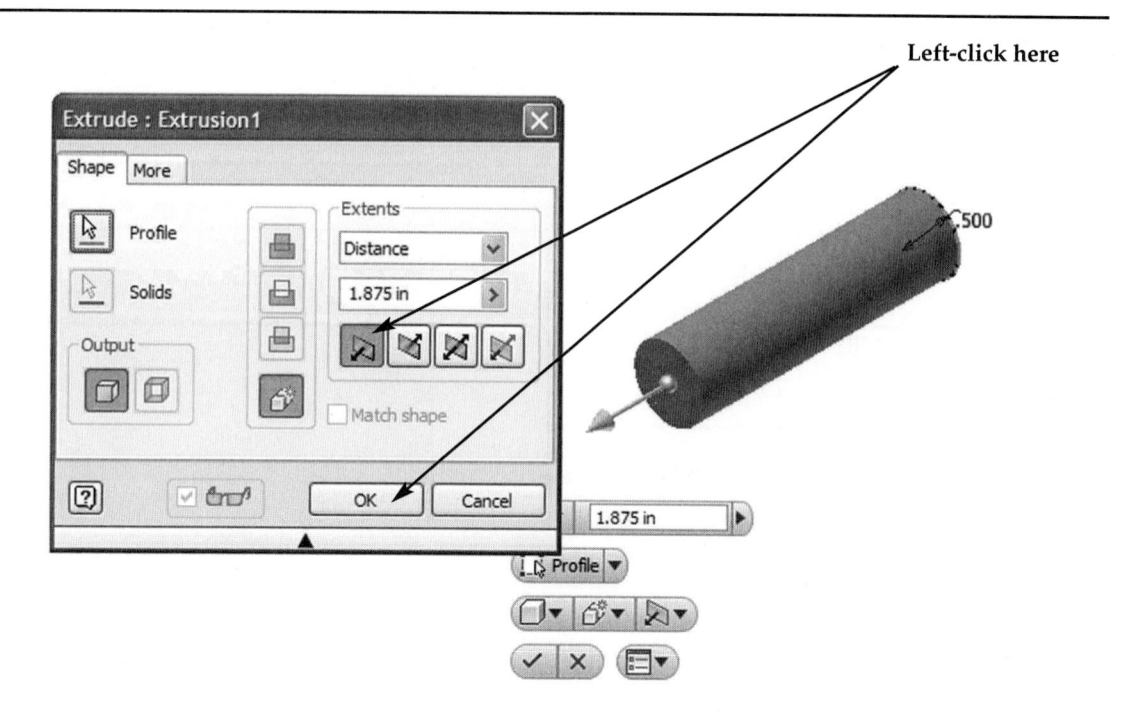

42.   Your screen should look similar
      to Figure 6-37.

FIGURE 6-37

43.   Save the part as Wristpin1.ipt where it can be easily retrieved later.

44.   Begin a new sketch as described in Chapter 1.

45.   Complete the sketch shown in Figure 6-38.

FIGURE 6-38

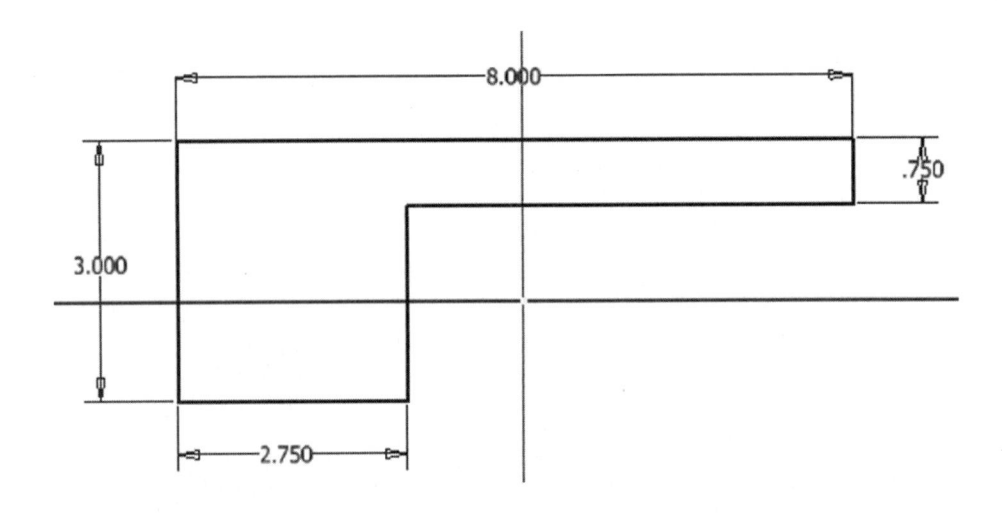

46.   Exit the Sketch Panel and change the view to Isometric as shown in Figure 6-39.

FIGURE 6-39

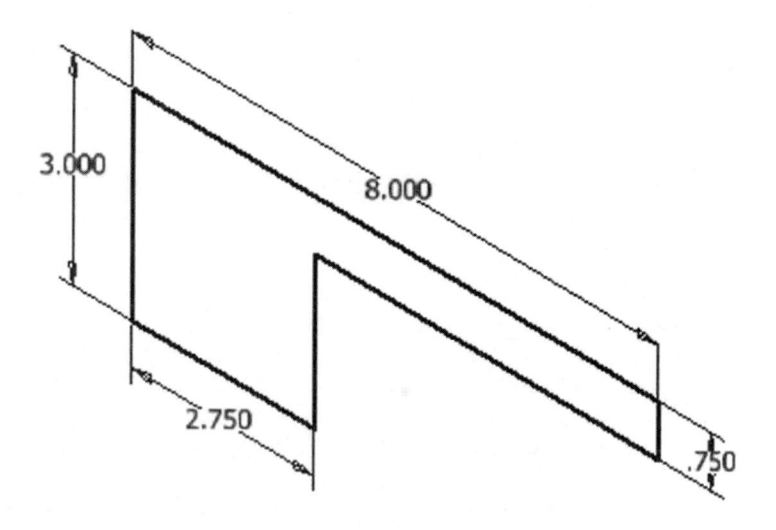

47.   Extrude the sketch to a distance of **2.25** inches. Your screen should look similar to what is shown in Figure 6-40.

FIGURE 6-40

48.   Use the **Fillet** command to create **1.125** inch fillets on the front portion of the part as shown in Figure 6-41.

FIGURE 6-41

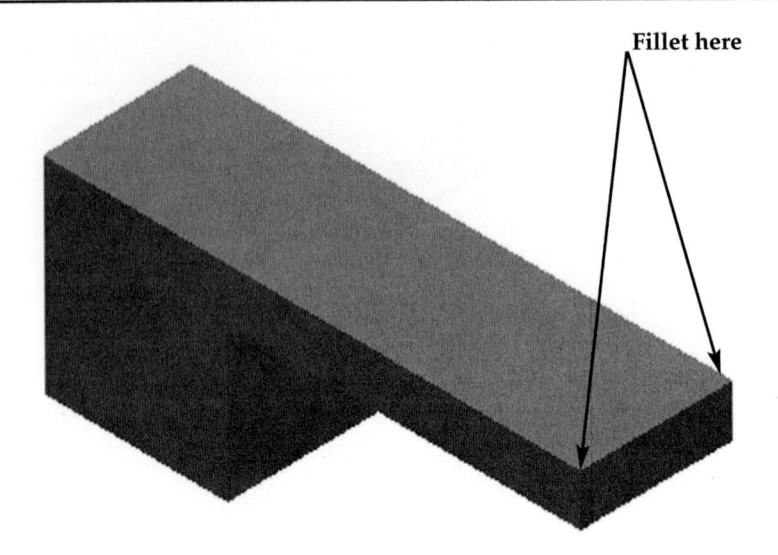

49.   Your screen should look similar to Figure 6-42.

FIGURE 6-42

50. Move the cursor to the upper middle portion of the screen and left-click on the **View** tab. Left-click on the "Face View/Look At" icon as shown in Figure 6-43.

FIGURE 6-43

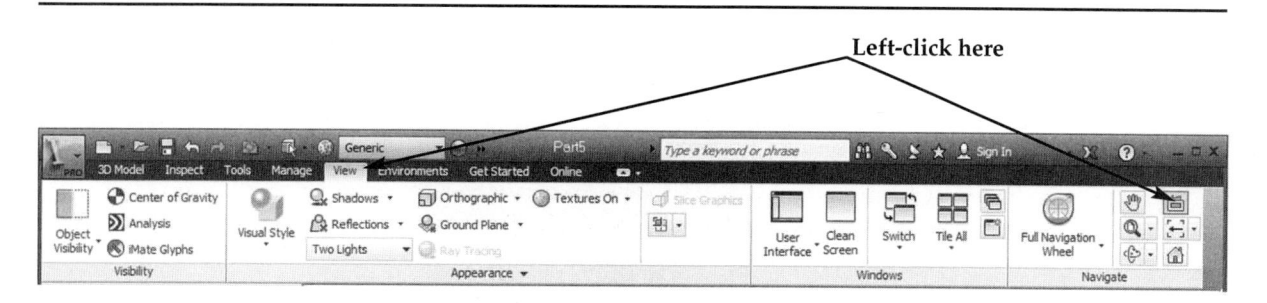

51. Move the cursor to the surface shown in Figure 6-44, causing it to turn red. Left-click once.

FIGURE 6-44

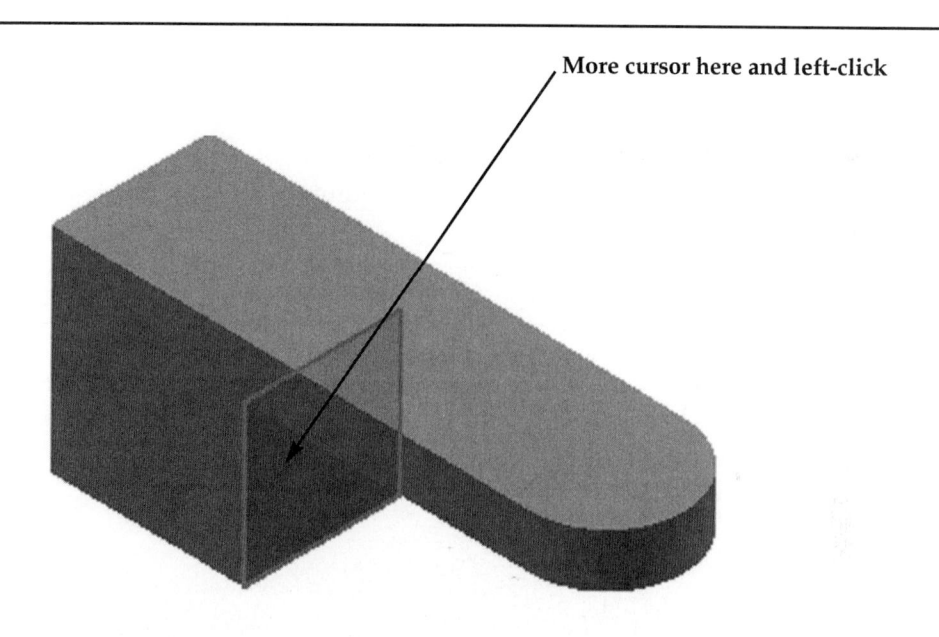

52. Complete the sketch shown in Figure 6-45.

FIGURE 6-45

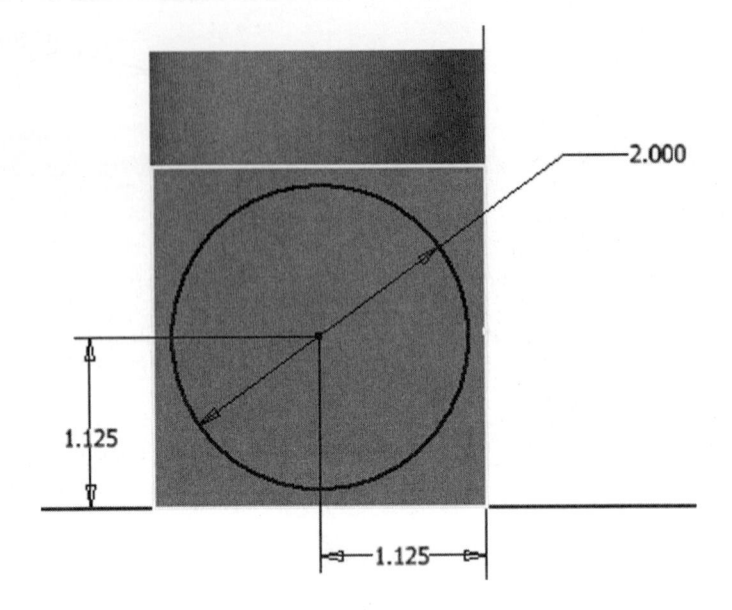

53. Change the view to Isometric as shown in Figure 6-46.

FIGURE 6-46

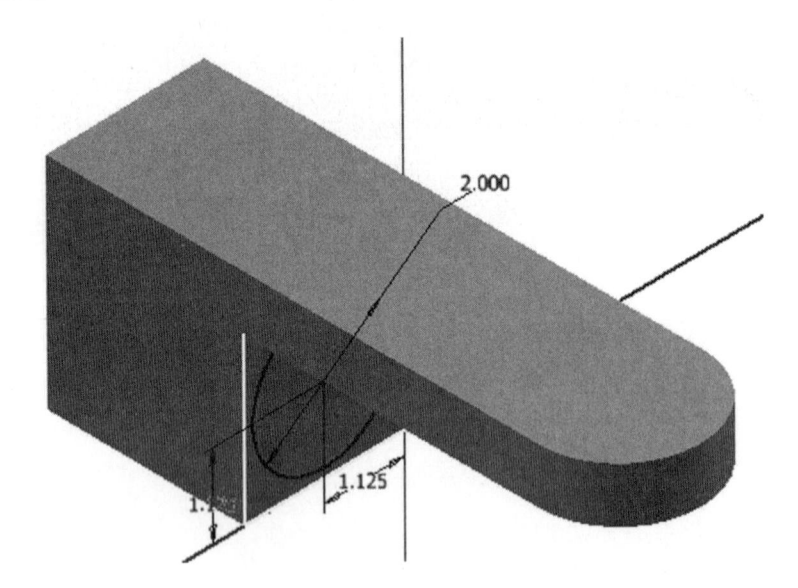

54. Use the "Extrude" command to extrude or cut out the circle that was just completed creating a thru hole. Your screen should look similar to Figure 6-47.

FIGURE 6-47

55. Move the cursor to the upper middle portion of the screen and left-click on the **View** tab. Left-click on the "Face View/Look At" icon as shown in Figure 6-48.

FIGURE 6-48

Left-click here

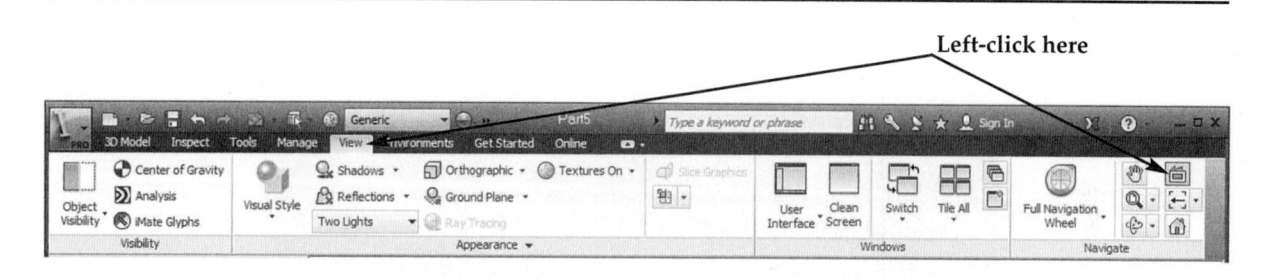

56. Left-click on the surface shown in Figure 6-49.

FIGURE 6-49

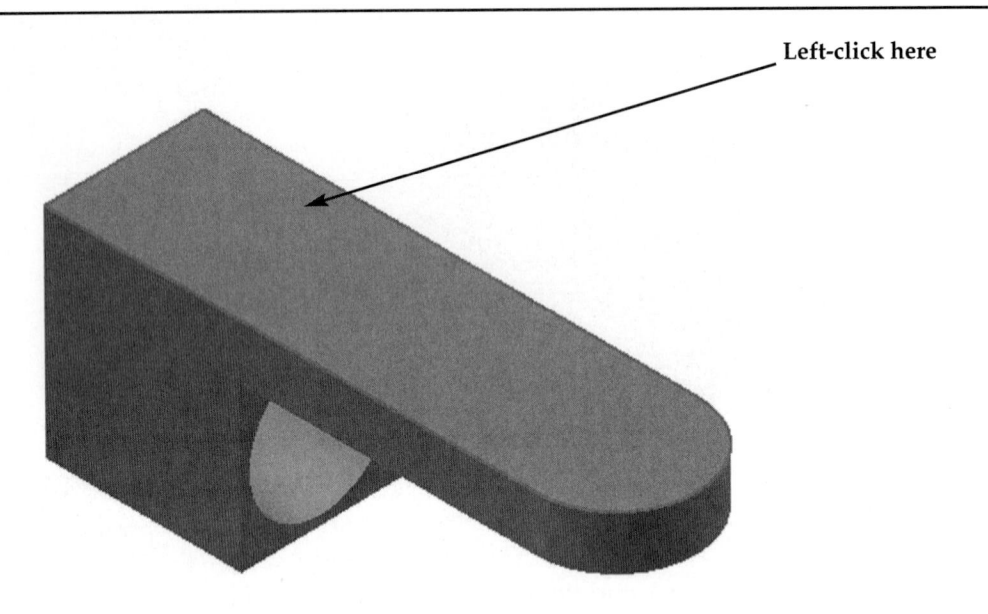

57. Complete the sketch shown in Figure 6-50.

FIGURE 6-50

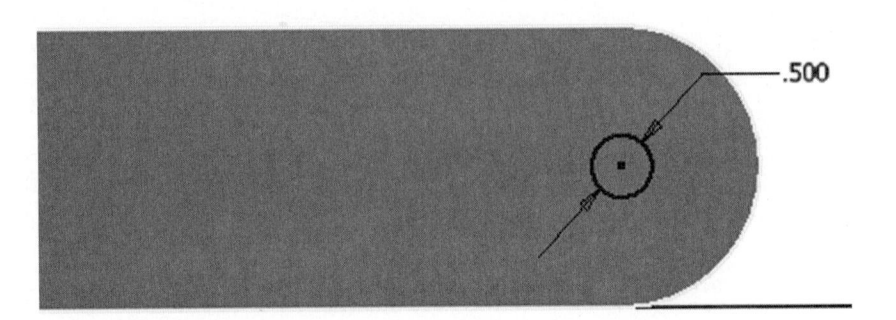

58. Use the "Extrude" command to extrude or cut out the circle that was just completed. Change the view to Isometric as shown in Figure 6-51.

FIGURE 6-51

59. Save the part as Pistoncase1.ipt where it can be easily retrieved later.

60. Begin a new drawing as described in Chapter 1.

61. Begin a sketch as shown in Figure 6-52. Make sure that the circles are located on the endpoint of a line. Also make sure the circles are <u>not</u> the same diameter.

FIGURE 6-52

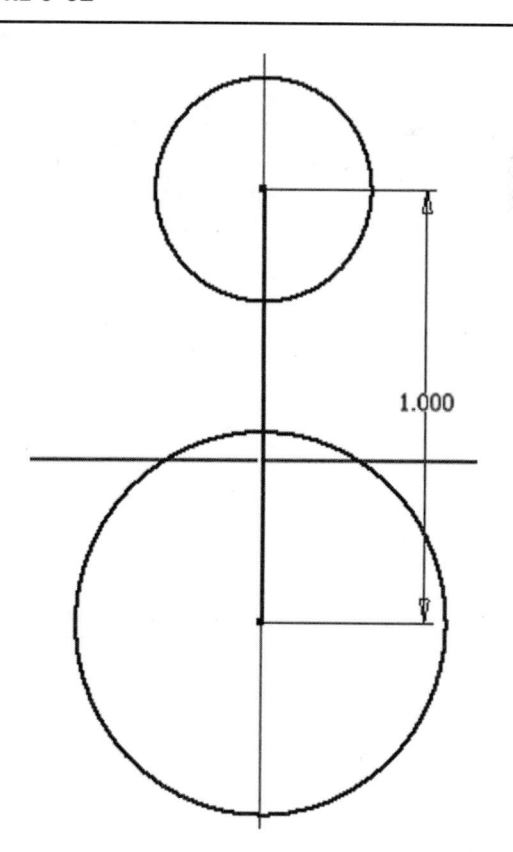

1.000

62. Move the cursor to the upper middle portion of the screen and left-click on the "Equal" constraints icon as shown in Figure 6-53.

FIGURE 6-53

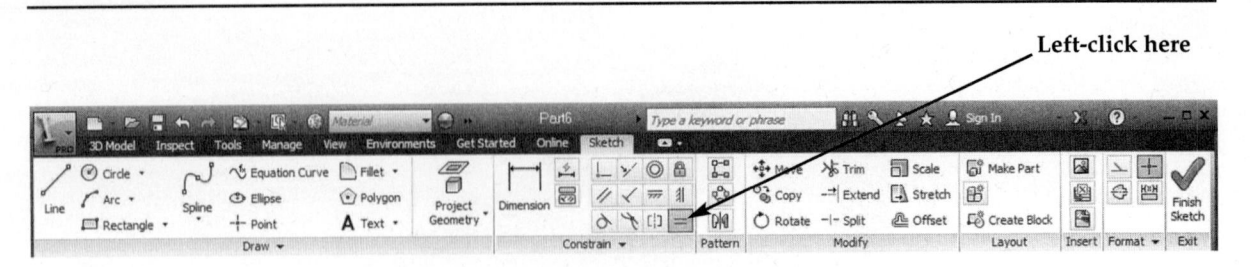

63. Left-click on each of the circles as shown in Figure 6-54.

FIGURE 6-54

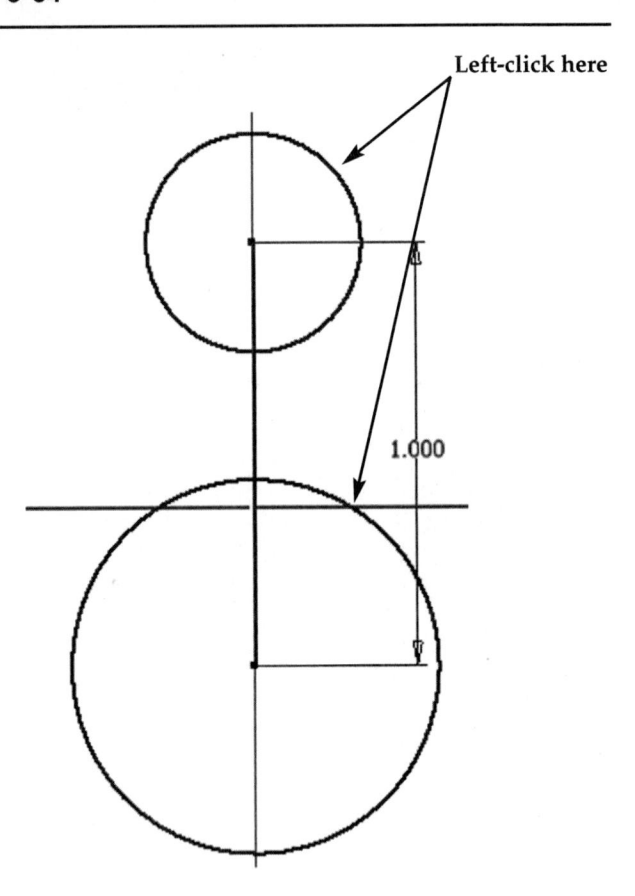

64.   Inventor will create two circles of the same size as shown in Figure 6-55. When one circle is dimensioned, Inventor will automatically update the size of the other circle.

FIGURE 6-55

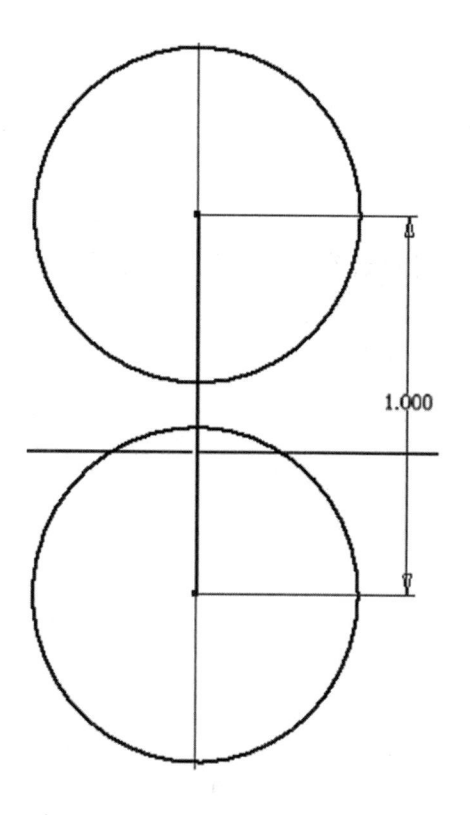

1.000

65.   Finish completing the sketch shown in Figure 6-56.

FIGURE 6-56

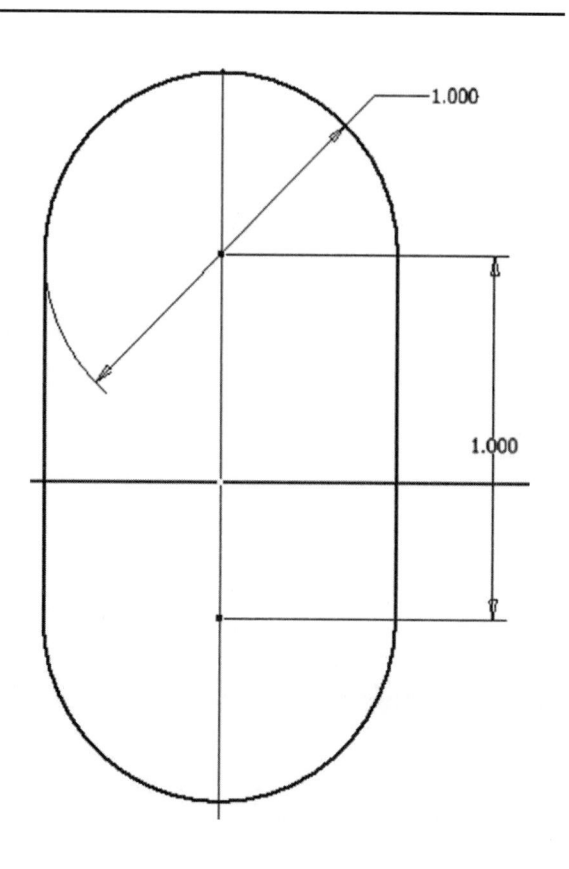

1.000

1.000

66.   Extrude the sketch into a solid with a thickness of **.25** as shown in Figure 6-57.

FIGURE 6-57

67.   Complete the following sketch. Use the center of the outside Fillet radius as the center of the circle as shown in Figure 6-58.

FIGURE 6-58

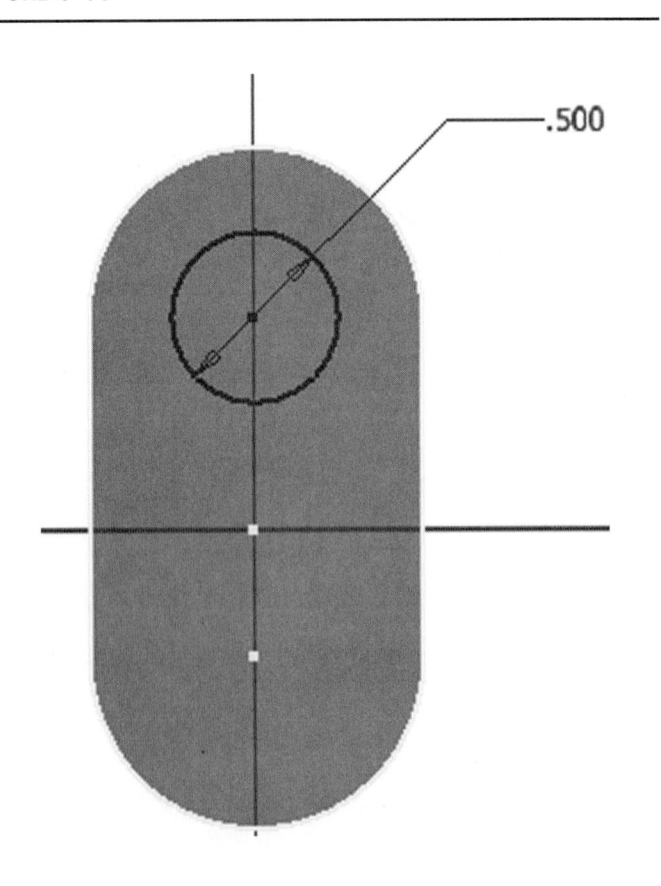

.500

68. Extrude the sketch into a solid with a thickness of **.25** as shown in Figure 6-59.

FIGURE 6-59

69. Rotate the part around to gain access to the opposite side as shown in Figure 6-60.

FIGURE 6-60

70. Complete the following sketch as shown in Figure 6-61.

FIGURE 6-61

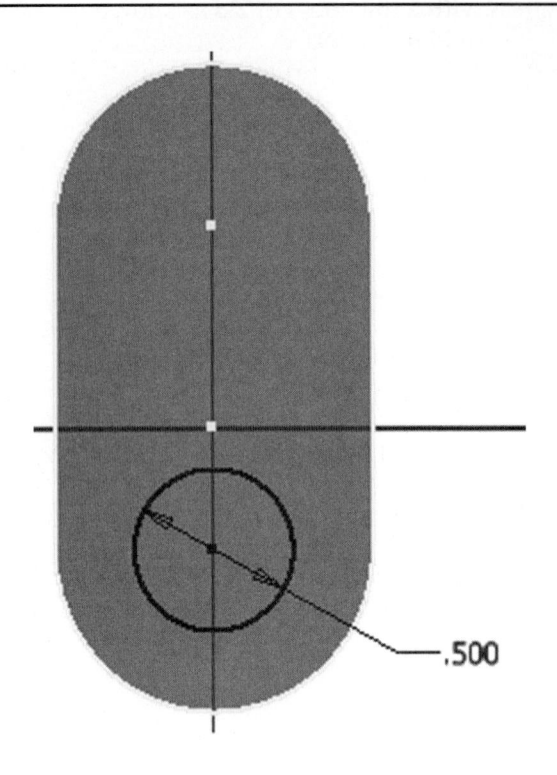

.500

71. Extrude the sketch into a solid with a thickness of **.25** as shown in Figure 6-62.

FIGURE 6-62

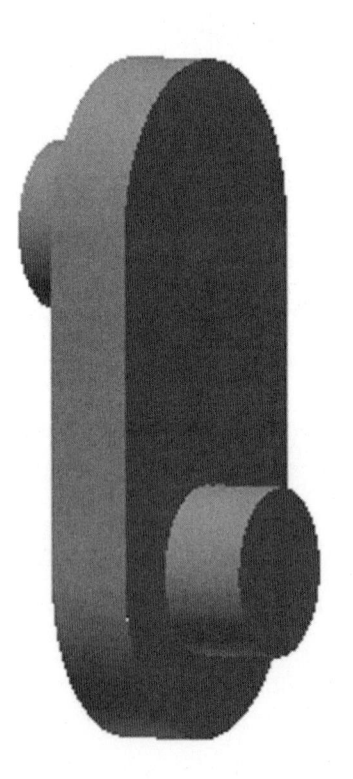

72.   Save the part as Crankshaft1.ipt where it can be easily retrieved later.

73.   Begin a new drawing as described in Chapter 1.

74.   Create a new sketch as shown. Make sure each of the circles are <u>not</u> sharing the same center and are <u>not</u> in line with each other as shown in Figure 6-63.

FIGURE 6-63

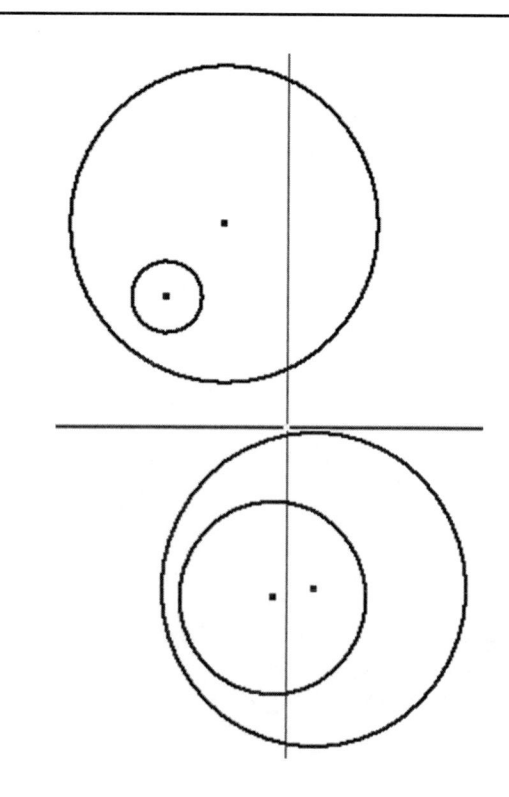

75.   Move the cursor to the upper middle portion of the screen and left-click on the "Concentric Constraint" icon as shown in Figure 6-64.

FIGURE 6-64

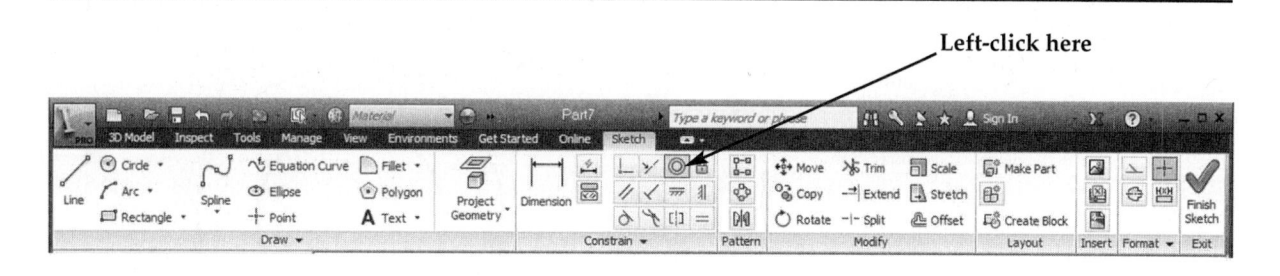

76. Holding the Shift key down, left-click on each circle as shown in Figure 6-65.

FIGURE 6-65

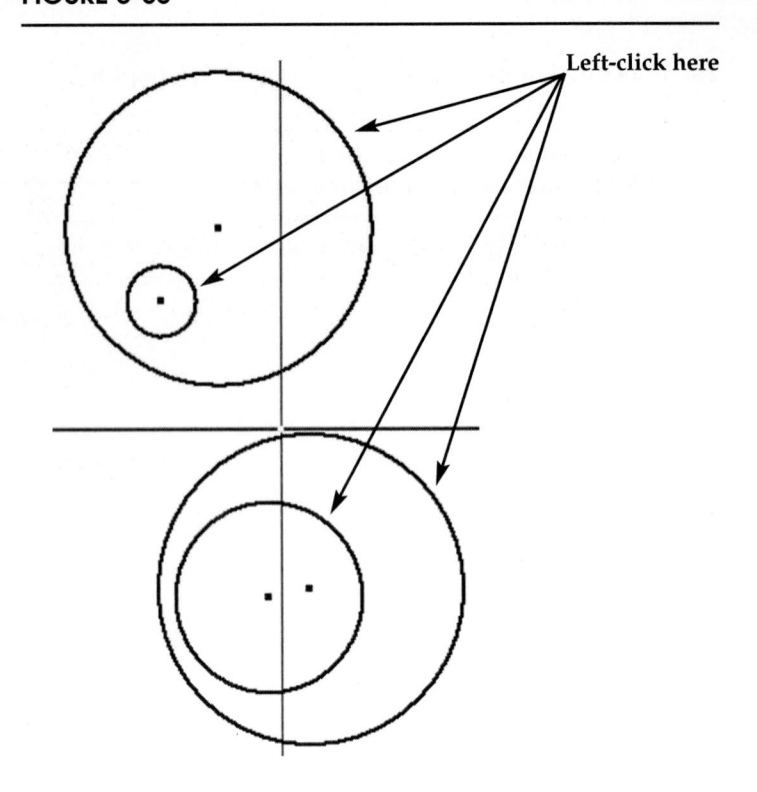

Left-click here

77. Inventor will create concentric circles as shown in Figure 6-66.

FIGURE 6-66

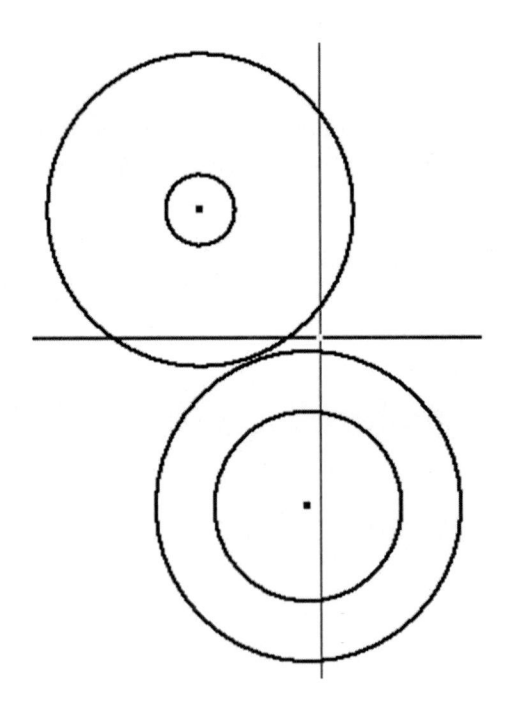

78.   Move the cursor to the upper middle portion of the screen and left-click on the "Vertical Constraint" icon as shown in Figure 6-67.

FIGURE 6-67

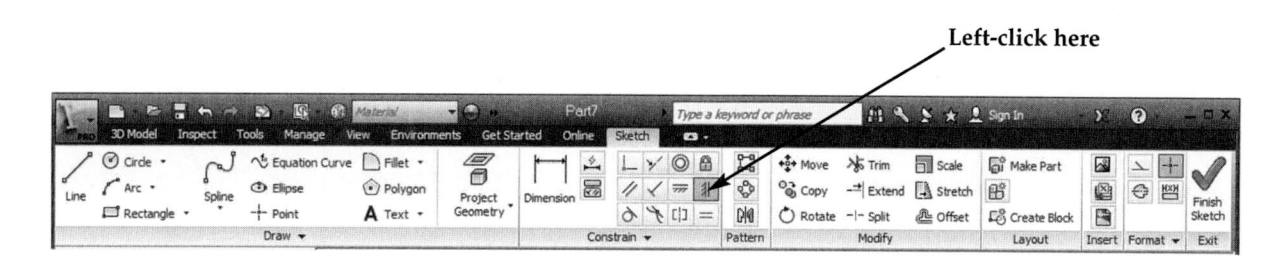

79.   Left-click on the centers of the circles as shown in Figure 6-68.

FIGURE 6-68

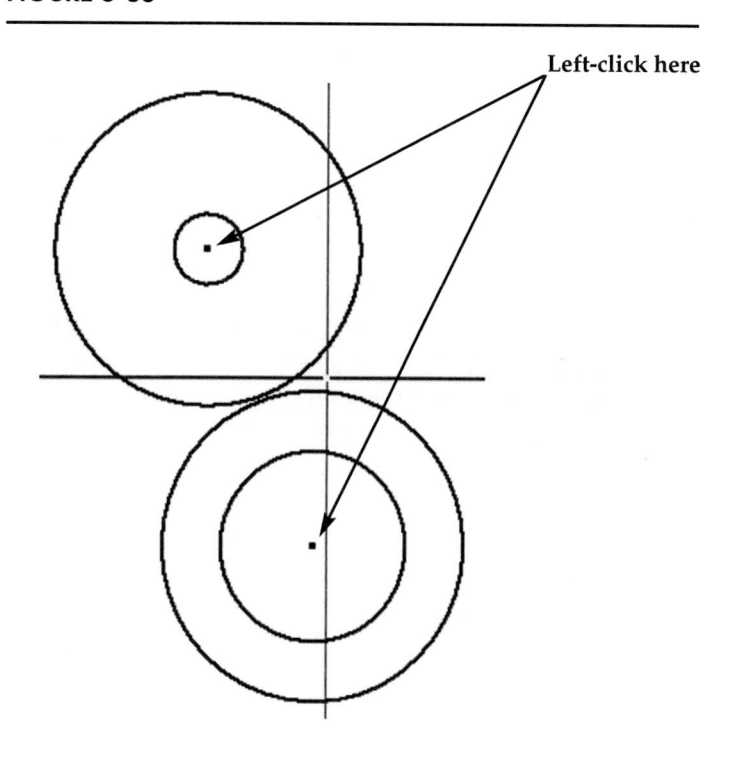

80.   Inventor will create a vertical constraint between the centers of the circles as shown in Figure 6-69.

FIGURE 6-69

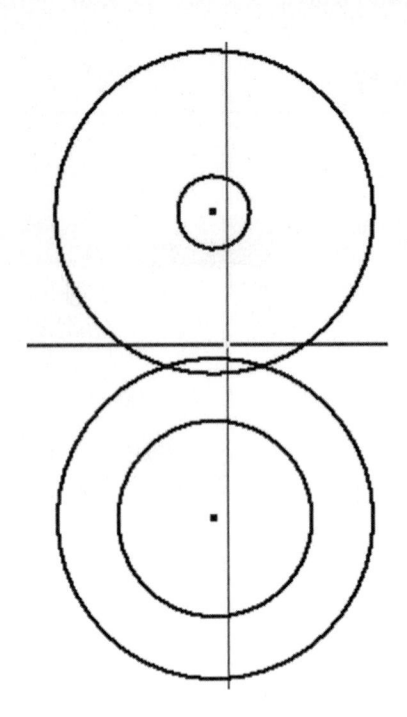

81.   Use the "Free Orbit/Rotate" command to rotate the sketch around onto its side as shown in Figure 6-70.

FIGURE 6-70

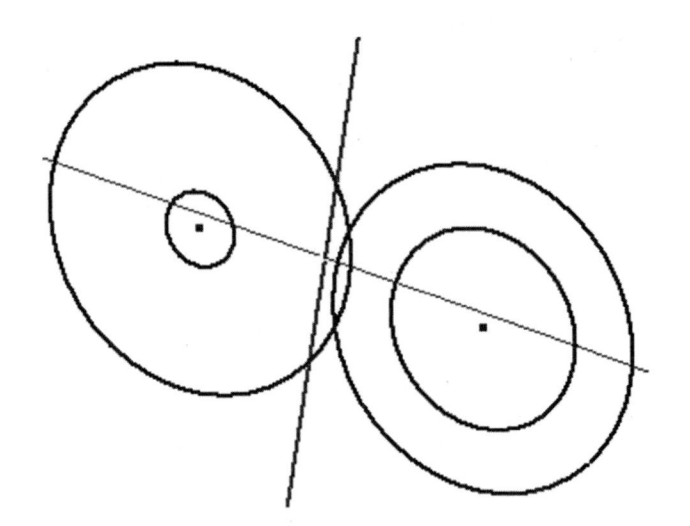

82. Using the geometry created on the previous page, complete the sketch shown. Extrude the sketch to a thickness of **.25** inches as shown in Figure 6-71.

FIGURE 6-71

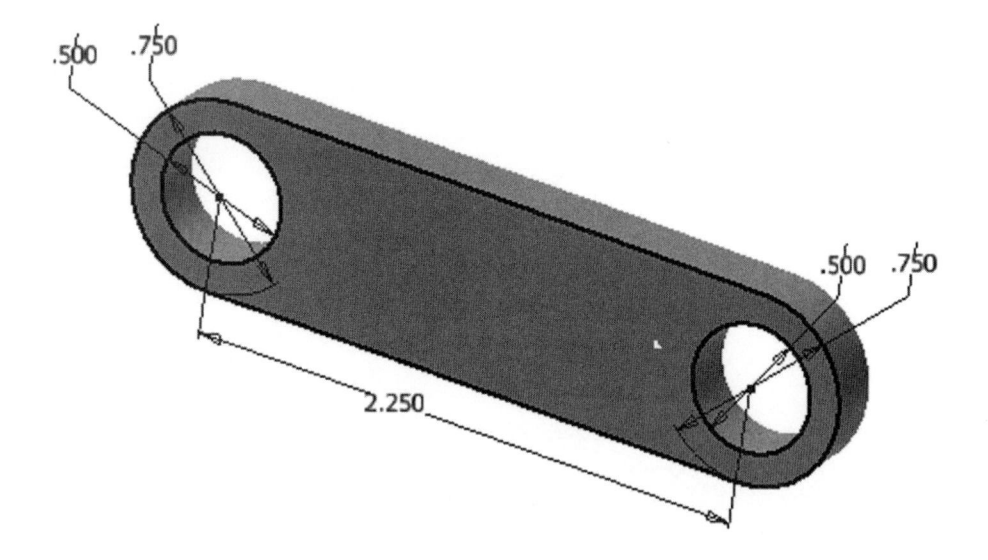

83. Save the part as Conrod1.ipt where it can be easily retrieved later.

84. All of these parts will be used in the next chapter.

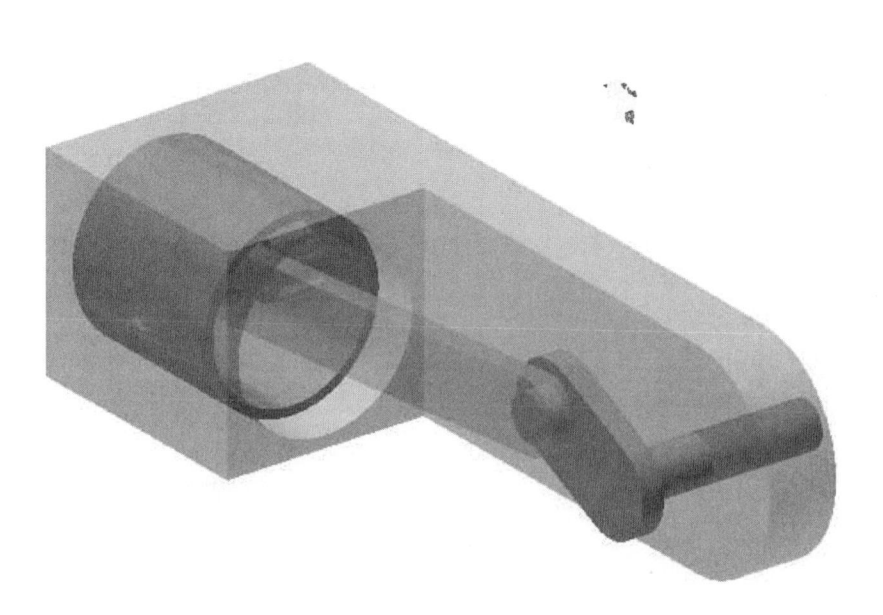

Chapter 7 includes instruction on how to construct the assembly shown above.

INTRODUCTION TO ASSEMBLY VIEW PROCEDURES

OBJECTIVES

1. Import existing solid models into the Assembly Panel

2. Constrain all parts in the Assembly Panel

3. Edit/modify parts while in the Assembly Panel

4. Assign colors to different parts in the Assembly Panel

5. Animate/simulate motion

6. Create an .avi or .wmv file while in the Assembly Panel

1.   Start Inventor 2013 by referring to Chapter 1.

2.   After Autodesk Inventor 2013 is running, begin an Assembly Drawing. First, move the cursor to the upper left corner of the screen and left-click on **New**. The Create New File dialog box will appear. Left-click on the **English** folder. Left-click on **Standard (in).iam** as shown in Figure 7-1.

FIGURE 7-1

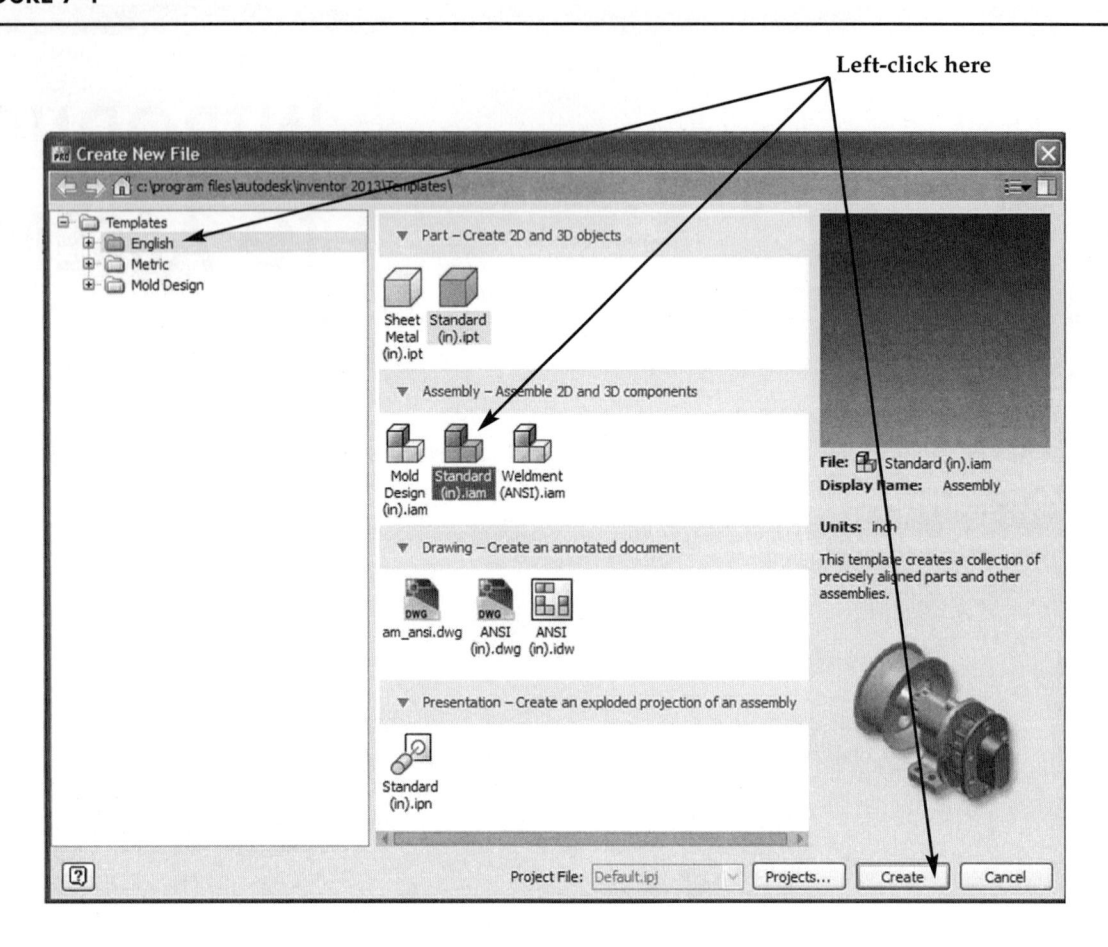

3.   Left-click on **Create**.

IMPORT EXISTING SOLID MODELS INTO THE ASSEMBLY PANEL

4.   The Assembly Panel will open. Your screen should look similar to Figure 7-2.

FIGURE 7-2

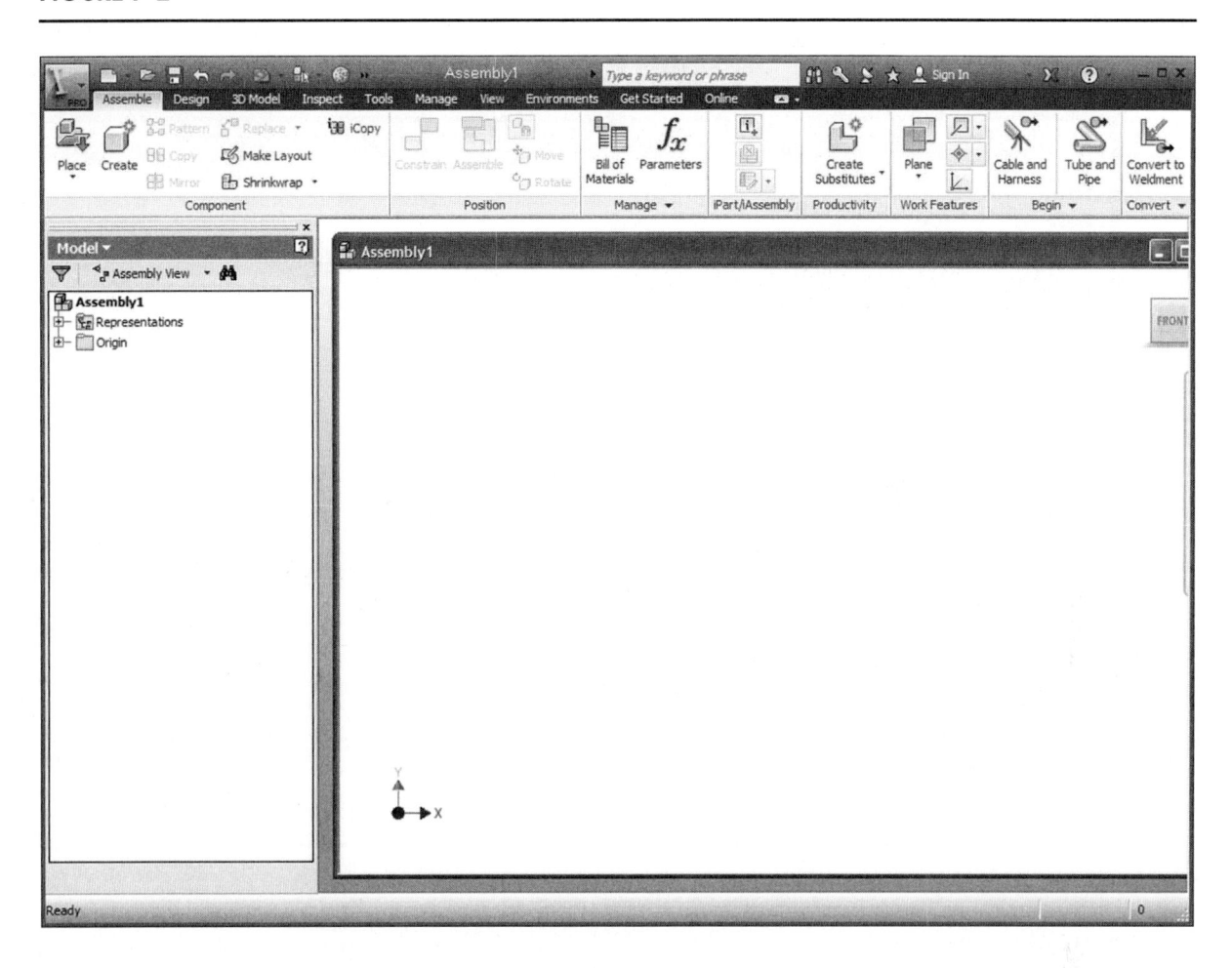

5.   Move the cursor to the upper left portion of the screen and left-click on **Place** as shown in Figure 7-3.

FIGURE 7-3

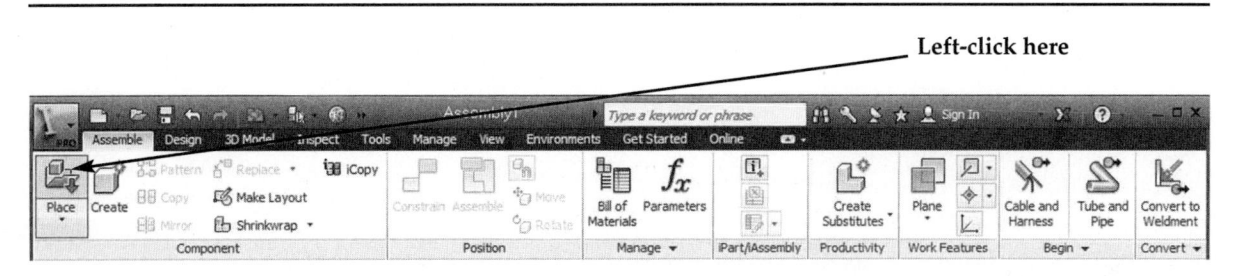

6. The Place Component dialog box will appear. Locate the PistonCase1.ipt file and left-click on **Open** as shown in Figure 7-4.

FIGURE 7-4

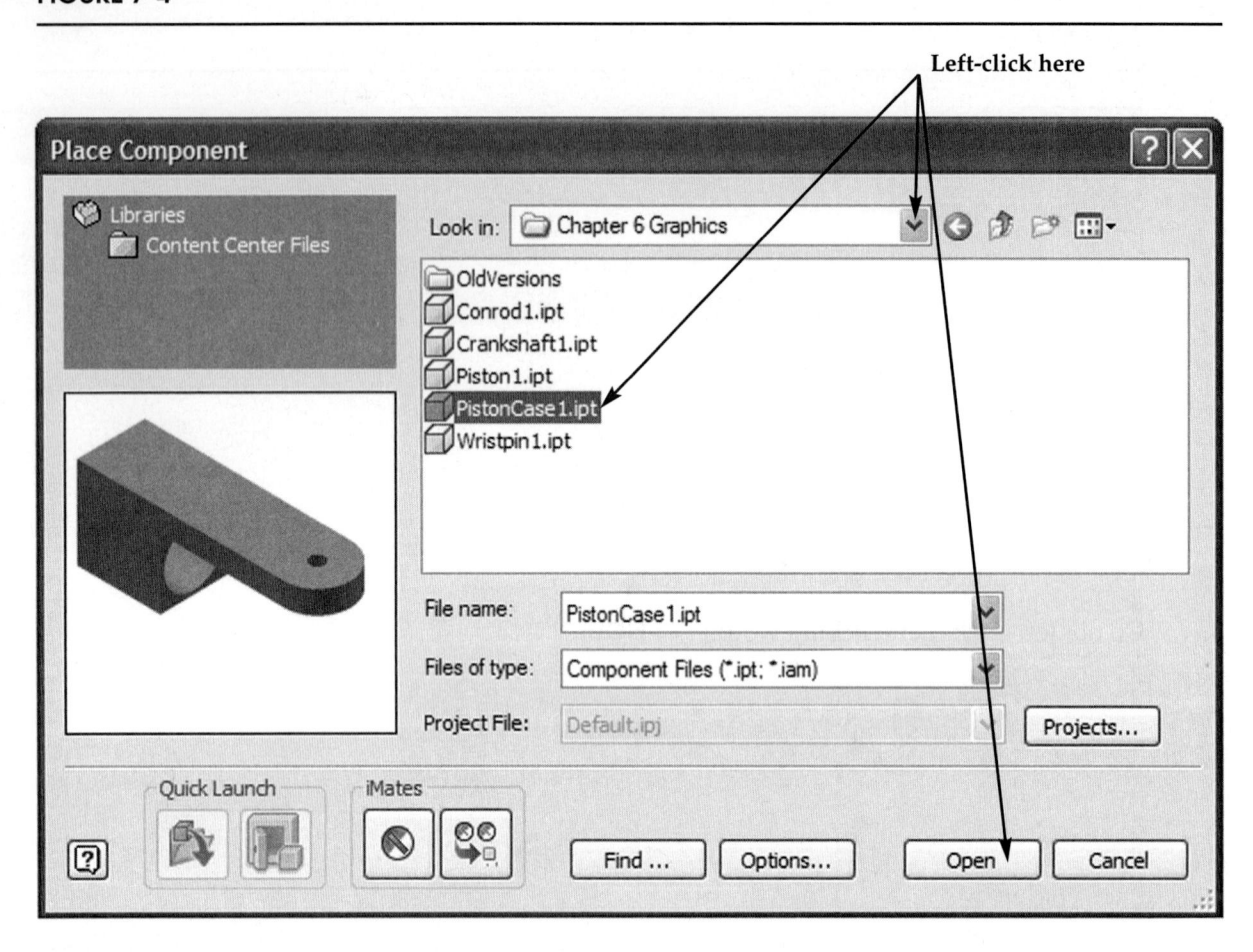

7. Inventor will place one piston case in the drawing space while another piston case will be attached to the cursor as shown in Figure 7-5. A dialog box may appear indicating that "The location of the selected file drawing is not in the active project." Left-click on **Yes**.

FIGURE 7-5

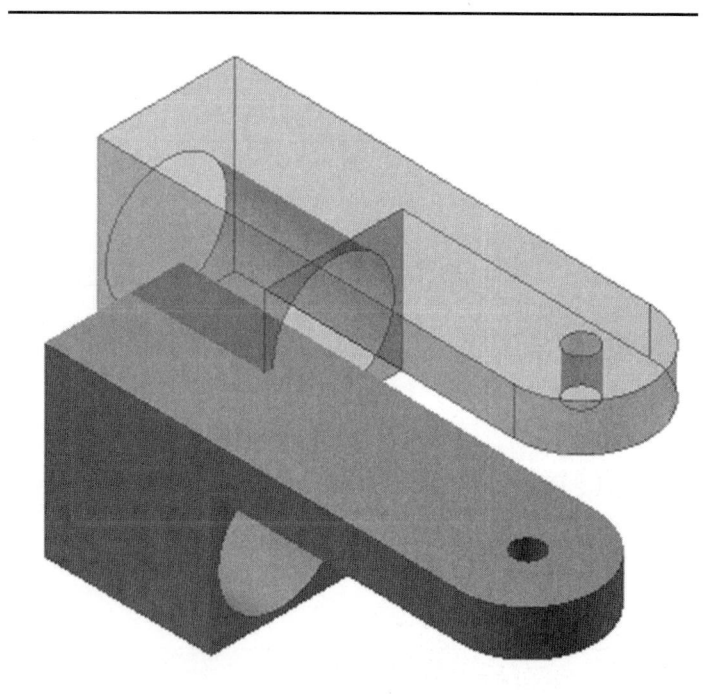

8. Do <u>not</u> left-click. Left-clicking would cause Inventor to place two piston cases in the Assembly area. Press the **Esc** key on the keyboard. Your screen should look similar to Figure 7-6.

FIGURE 7-6

9. Move the cursor to the upper left portion of the screen and left-click on **Place** as shown in Figure 7-7.

FIGURE 7-7

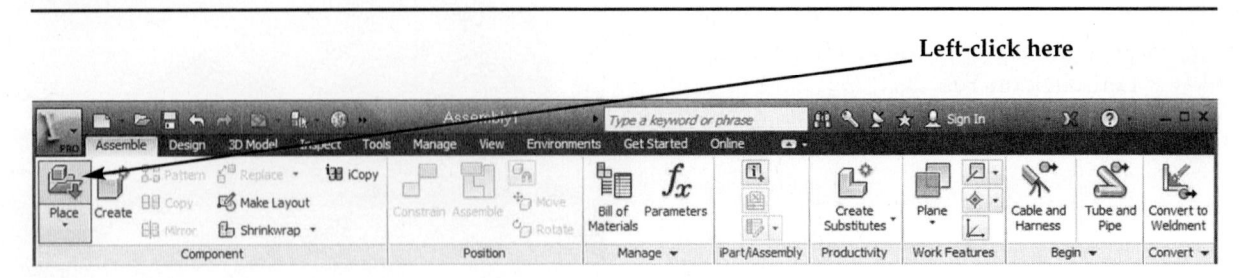

Left-click here

10. The Place Component dialog box will appear. Locate the Piston1.ipt file and left-click on **Open** as shown in Figure 7-8.

FIGURE 7-8

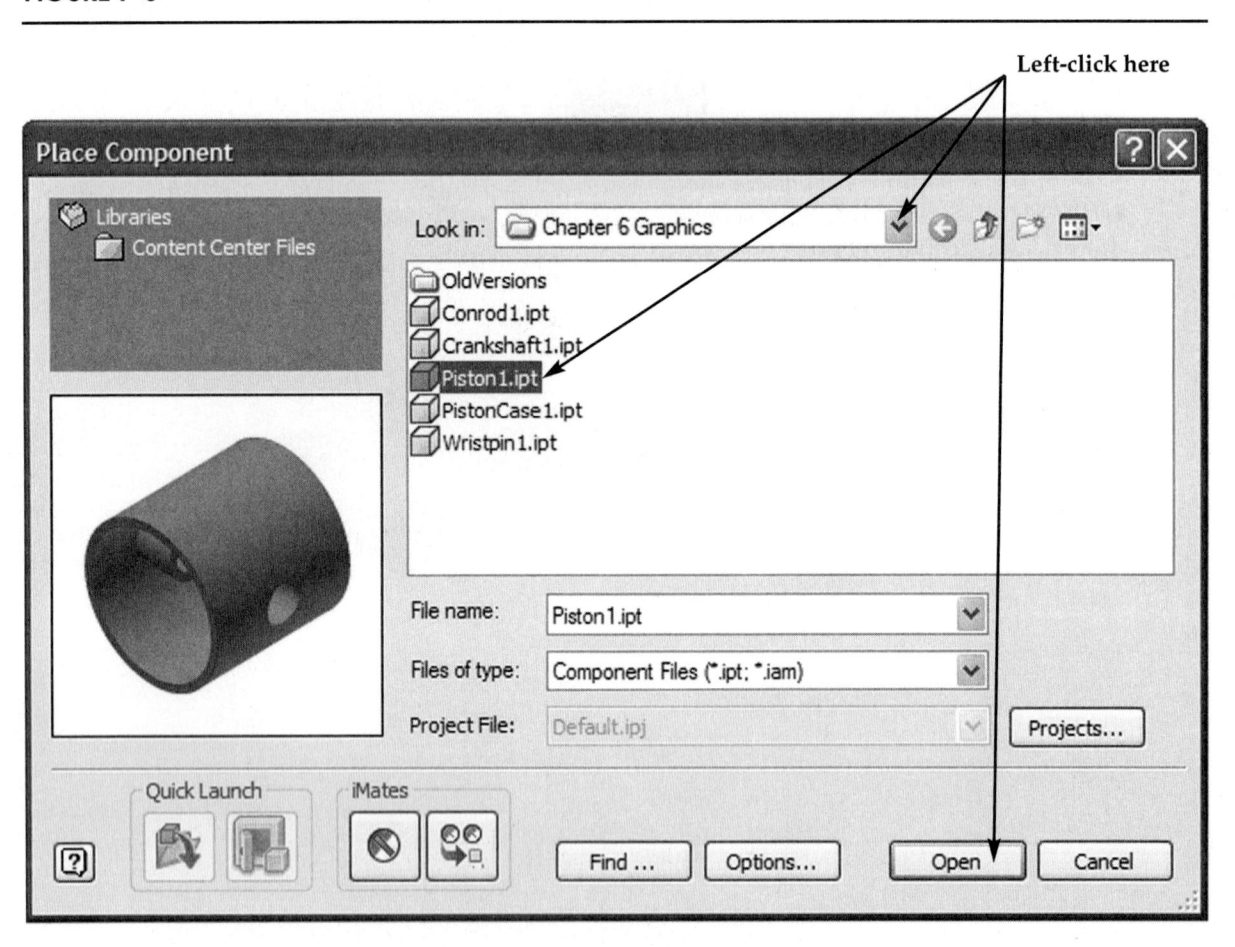

Left-click here

11.   The piston will be attached to the cursor. Place the piston anywhere near the piston case and left-click once. Another piston will be attached to the cursor in case another will be used. In this drawing there is no need to import the same part multiple times. Press the **Esc** button on the keyboard once. Your screen should look similar to Figure 7-9.

FIGURE 7-9

12.   Continue to "Place" the remaining parts into the Assembly area as shown in Figure 7-10.

FIGURE 7-10

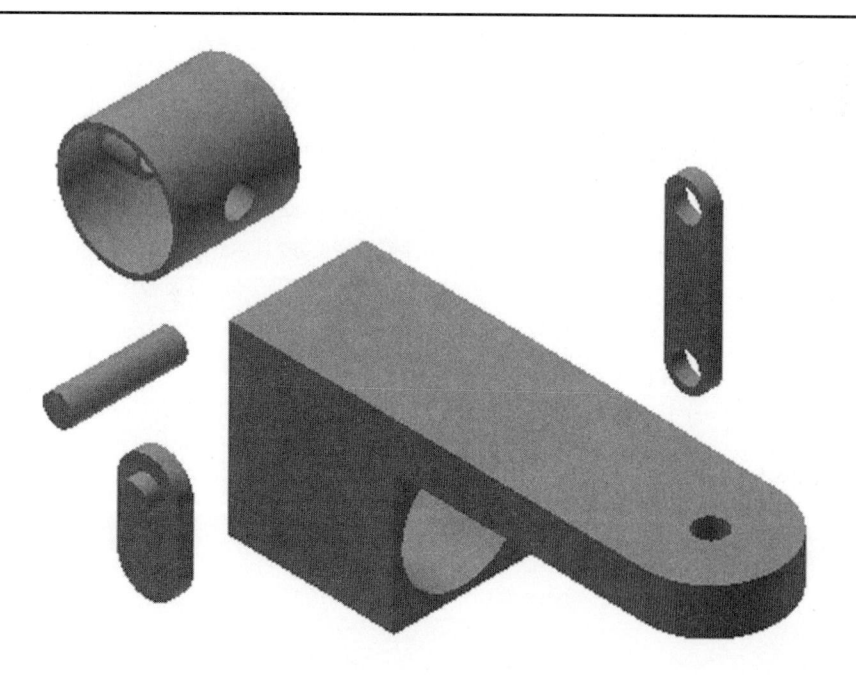

13.   Move the cursor to the lower left portion of the screen in the part tree. Notice the picture of a push pin that appears next to the piston case text in the branch of the part tree. Move the cursor over the words "Pistoncase:1" and left-click once. The text will turn blue. Right-click once. A pop-up menu will appear. Left-click on **Grounded** as shown in Figure 7-11. Inventor will Unground the case allowing it to be moved using the Rotate Component command as shown in Figure 7-12. **The first part placed into any assembly is automatically grounded.**

FIGURE 7-11

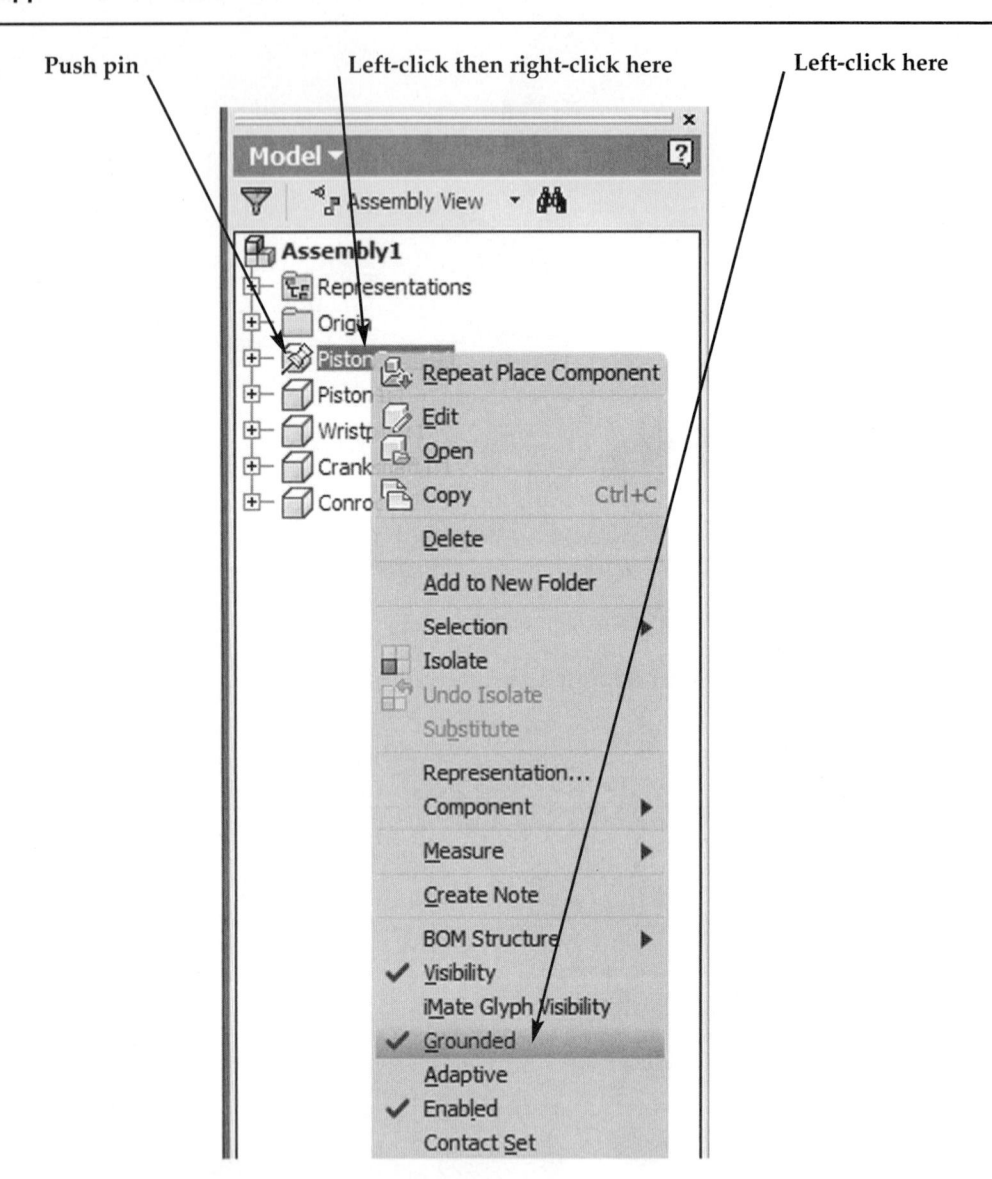

14. Move the cursor to the upper middle portion of the screen and left-click on the **Rotate Component** icon as shown in Figure 7-12.

FIGURE 7-12

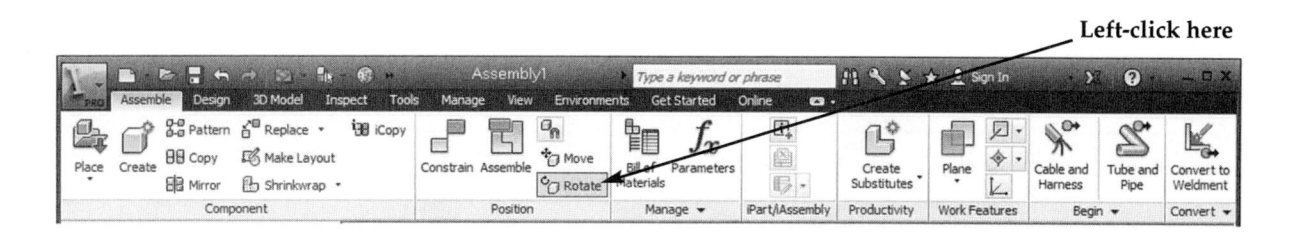

15. Move the cursor to the piston case and left-click once. A white circle will appear around the piston case. Rotate the piston case upward as shown in Figure 7-13.

FIGURE 7-13

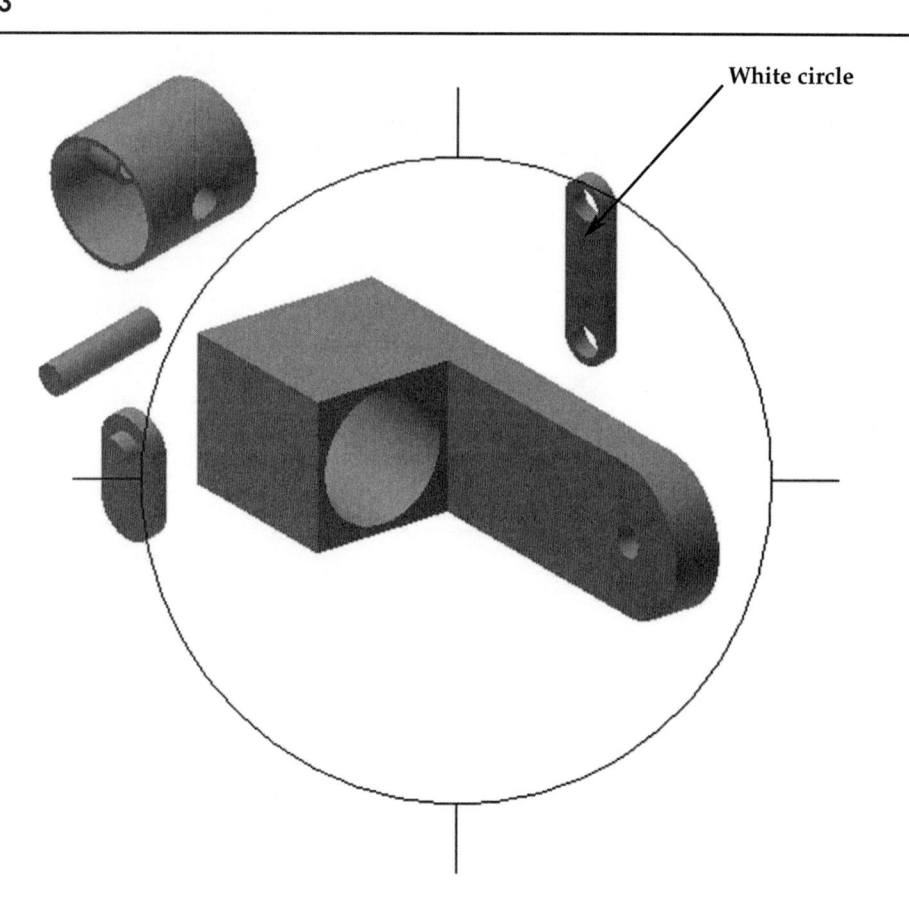

16. Your screen should look similar to Figure 7-13.

17. After the piston case is rotated as shown in Figure 7-14, right-click once. A pop-up menu will appear. Left-click on **Done** as shown in Figure 7-14.

FIGURE 7-14

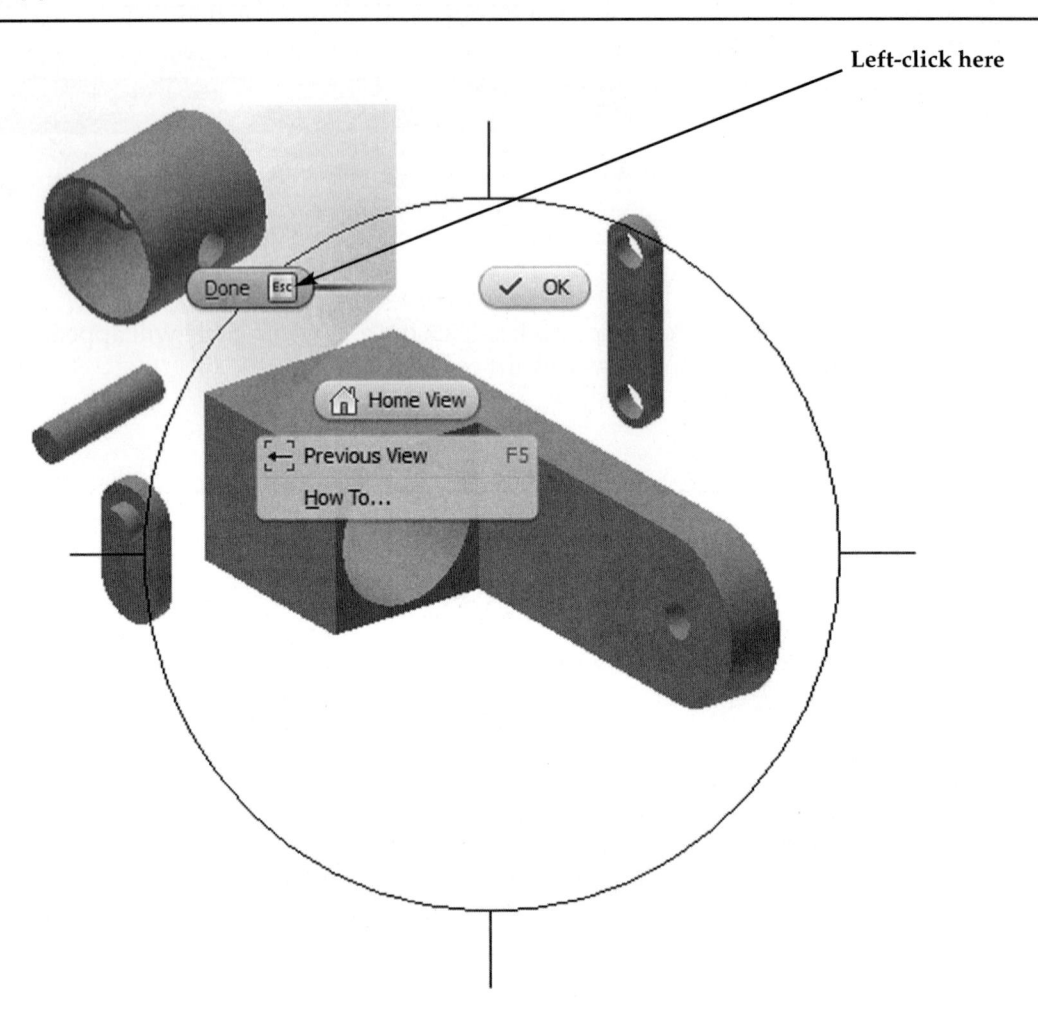

18.   Move the cursor to the lower left portion of the screen in the part tree. Notice the picture of the push pin that appeared next to the piston case text is gone. This means the piston case is <u>not</u> grounded. Move the cursor over the text, "Pistoncase1:1" and left-click once. The text will turn blue. Right-click once. A pop-up menu will appear. Left-click on **Grounded** as shown in Figure 7-15. Inventor will ground the case, preventing it from being moved while the rest of the assembly is constructed. **Caution:** Only ground the Piston Case. If any other parts inadvertently become grounded it will be impossible to assemble parts together.

FIGURE 7-15

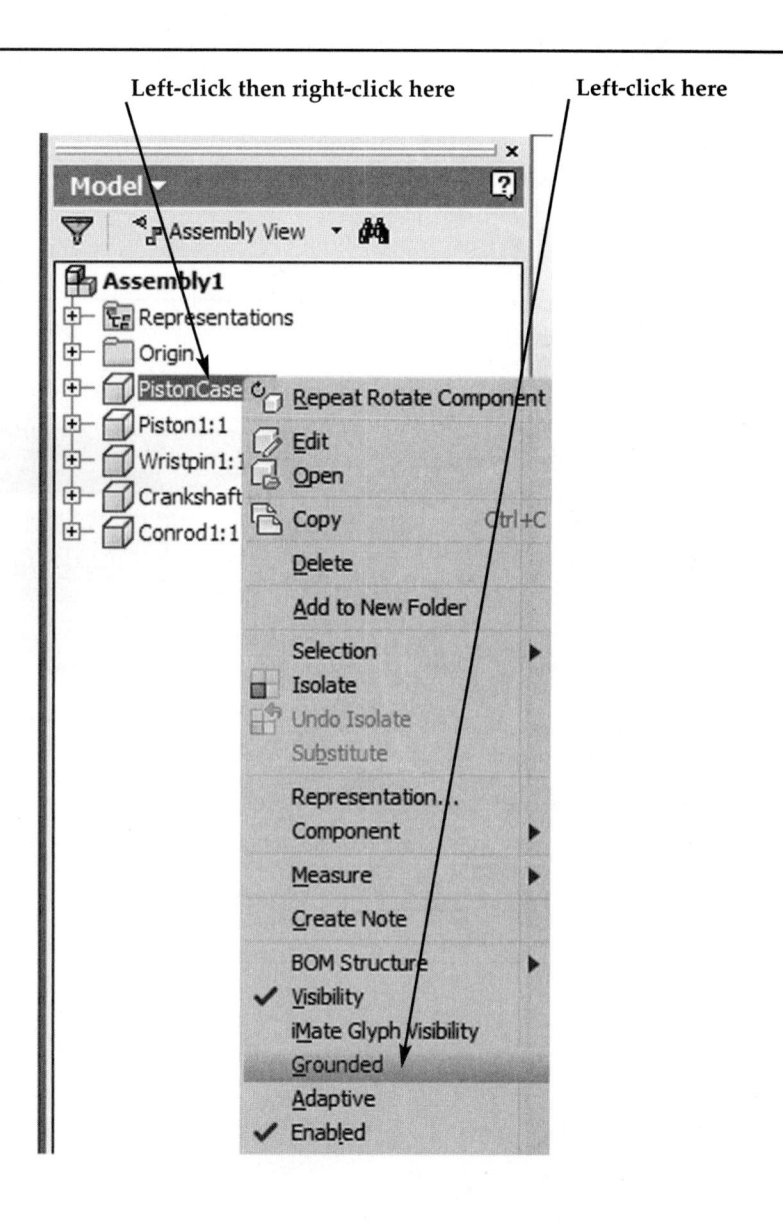

CONSTRAIN ALL PARTS IN THE ASSEMBLY PANEL

19.   Move the cursor to the upper middle portion of the screen and left-click on **Constrain** as shown in Figure 7-16.

FIGURE 7-16

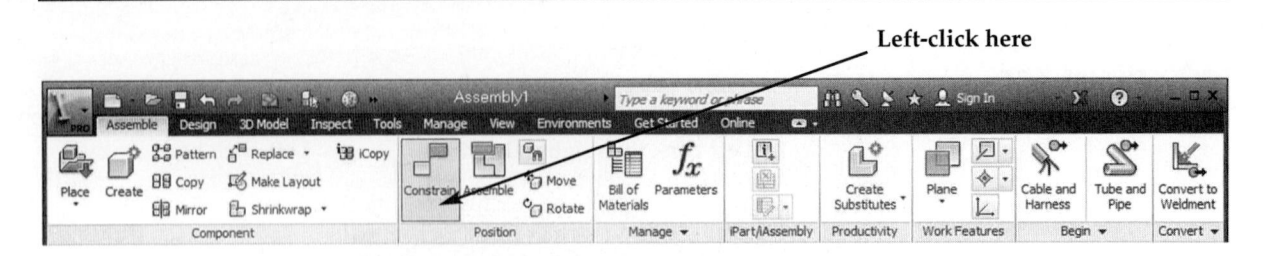

20.   Move the cursor over the piston until a red center line appears as shown in Figure 7-17. Left-click once.

FIGURE 7-17

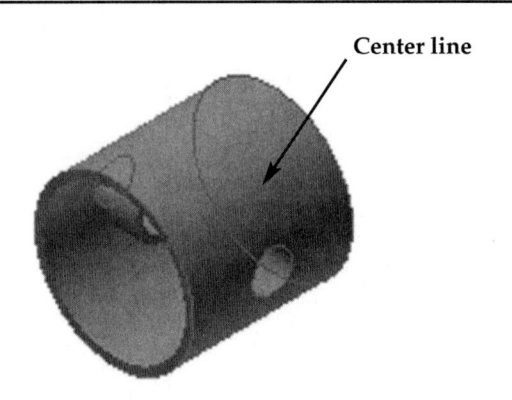

21.   Move the cursor over the piston case until a red center line appears as shown in Figure 7-18. Left-click once.

FIGURE 7-18

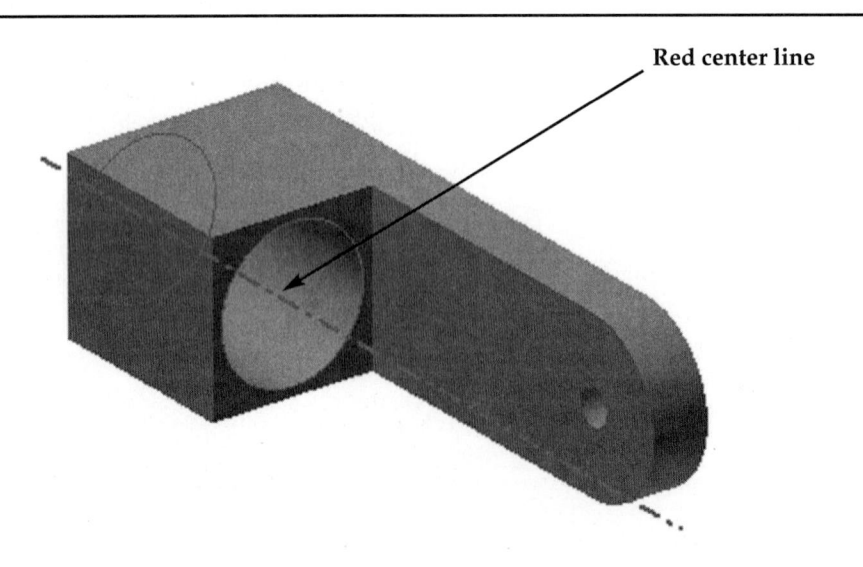

22.   Inventor will align the centers of the piston and the piston case. Your screen should look similar to Figure 7-19.

FIGURE 7-19

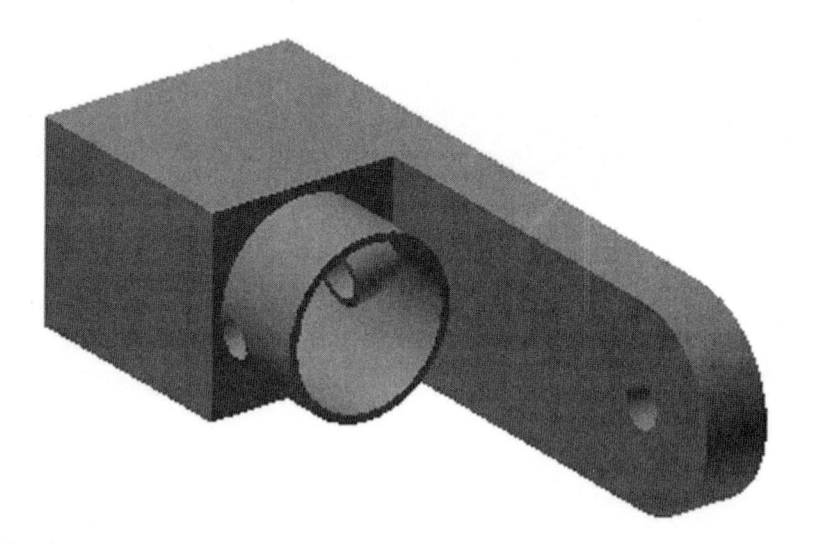

23.   If Inventor installed the piston upside down, click on the "Undo" icon. Use the **Rotate Component** command to rotate the piston so that Inventor has to rotate it less than 180 degrees to install it.

24.   Left-click on **OK** as shown in Figure 7-20.

FIGURE 7-20

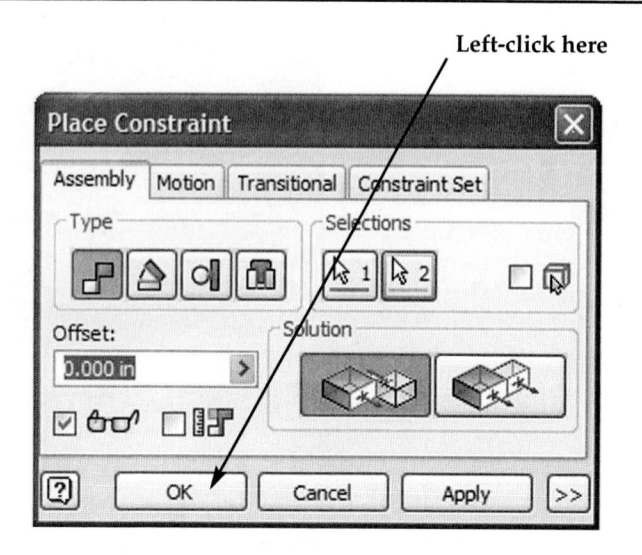

25.   Your screen should look similar to Figure 7-21.

FIGURE 7-21

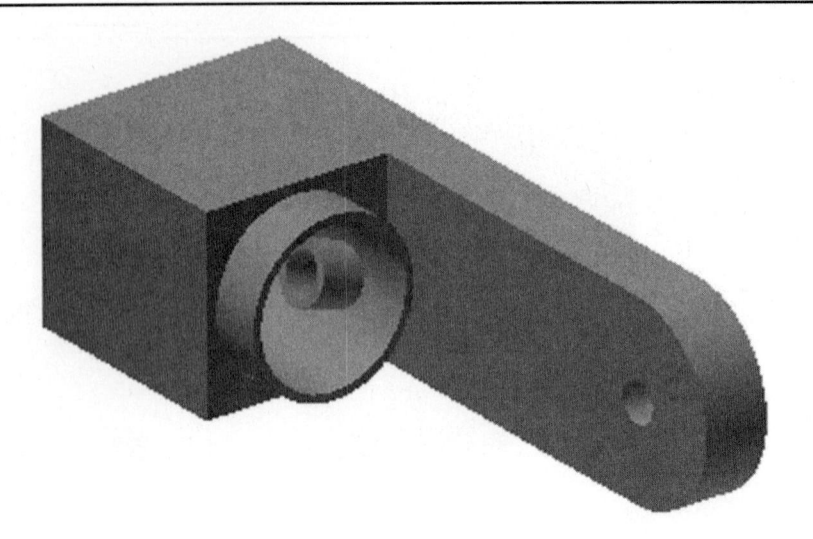

26.   Move the cursor to the lower left portion of the piston. Left-click (holding the left mouse button down) and slide the piston down out below the bore as shown in Figure 7-22.

FIGURE 7-22

27. Move the cursor to the upper middle portion of the screen and left-click on **Constrain**. The Place Constraint dialog box will appear as shown in Figure 7-23.

FIGURE 7-23

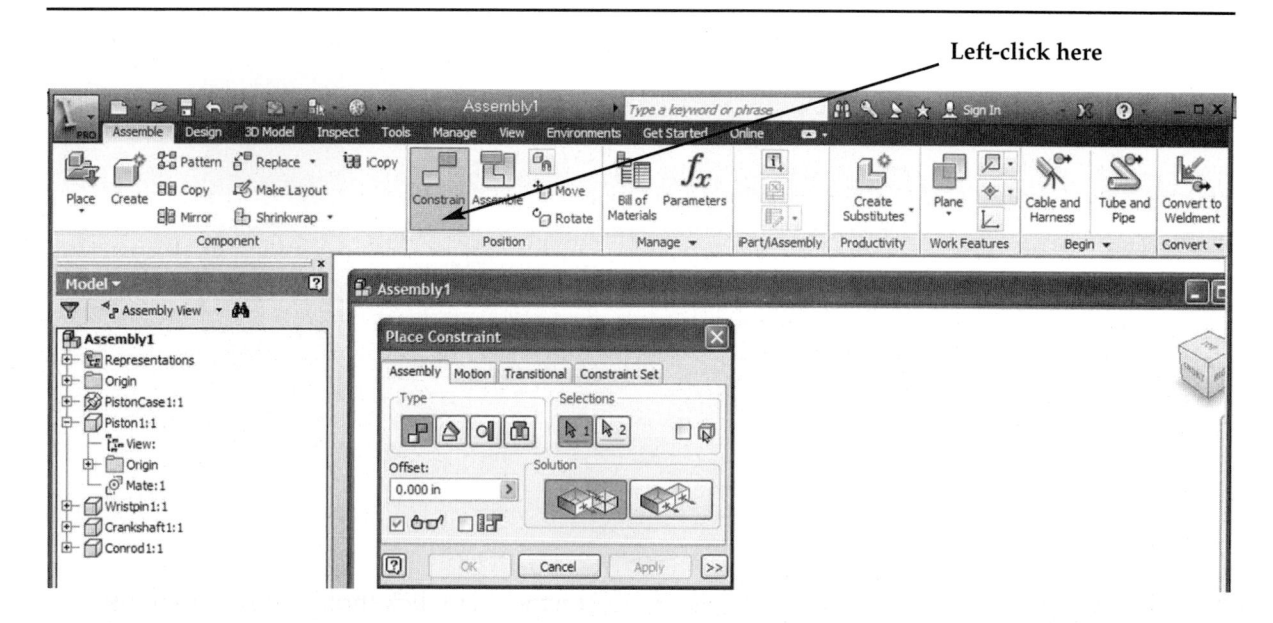

28. Move the cursor to the wristpin hole on the piston. A red center line will appear. Left-click once as shown in Figure 7-24.

FIGURE 7-24

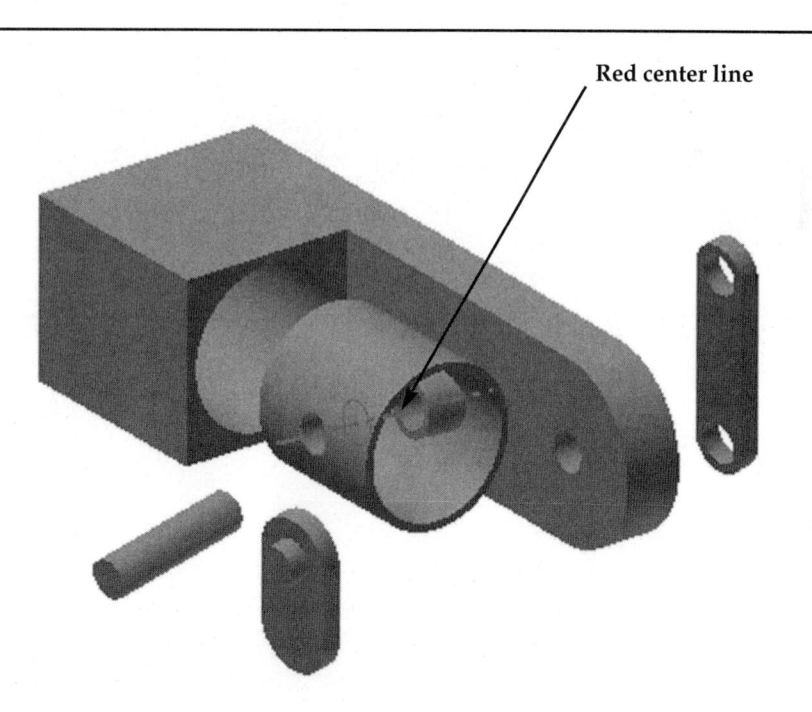

29. Move the cursor to the upper
     portion of the connecting rod.
     A red center line will appear.
     Left-click once as shown in
     Figure 7-25. You may have to
     zoom in to accomplish this.

FIGURE 7-25

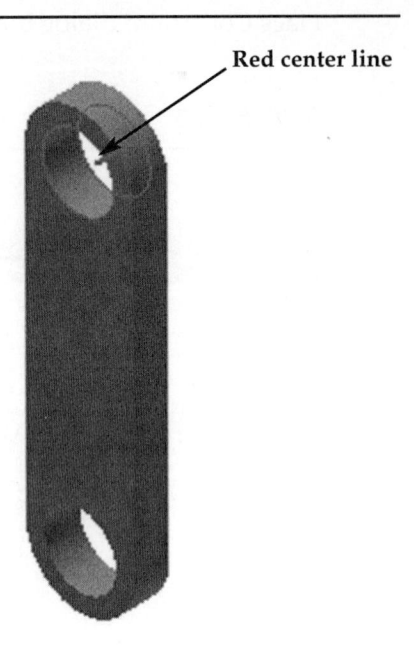

Red center line

30. Left-click on **OK** as shown in
     Figure 7-26.

FIGURE 7-26

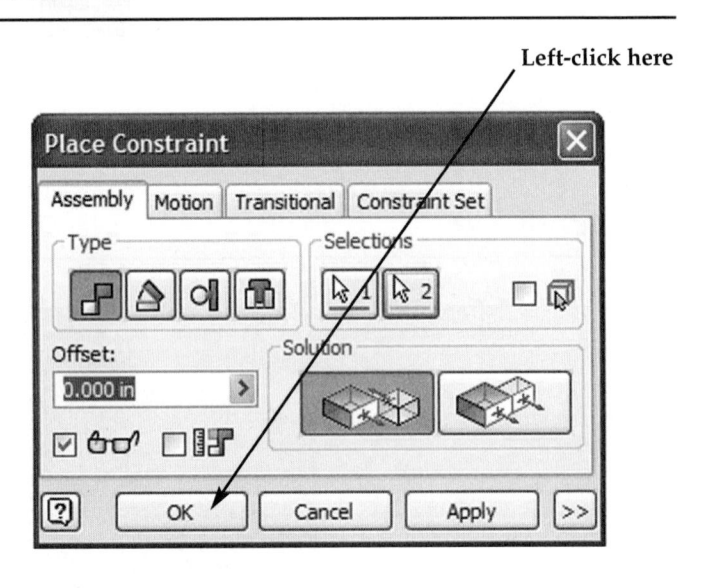

Left-click here

31.   Your screen should look similar to Figure 7-27.

FIGURE 7-27

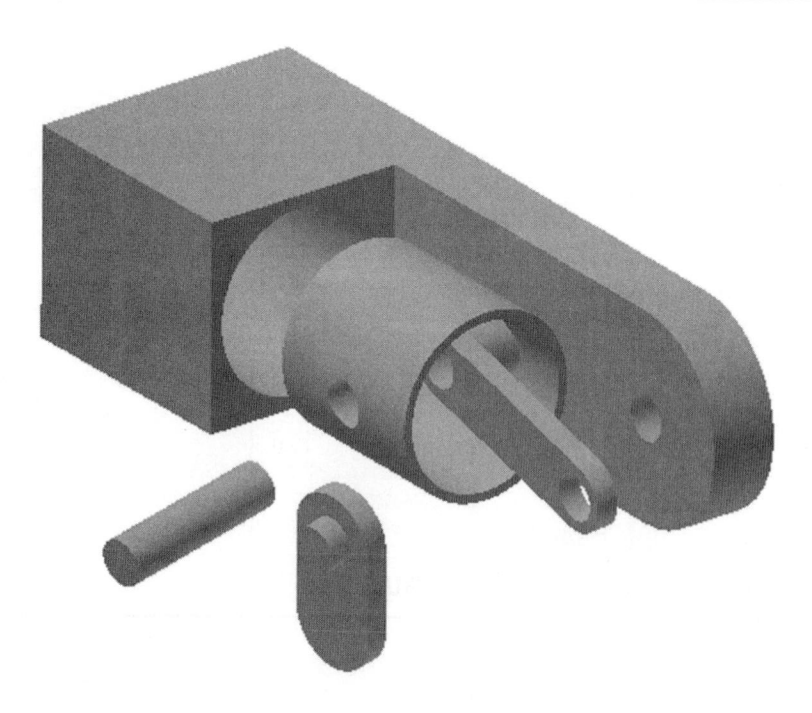

32.   Use the "Free Orbit/Rotate" command to rotate the entire assembly to gain access to the underside of the piston as shown in Figure 7-28.

FIGURE 7-28

33. Move the cursor to the upper middle portion of the screen and left-click on **Constrain**. The Place Constraint dialog box will appear as shown in Figure 7-29.

FIGURE 7-29

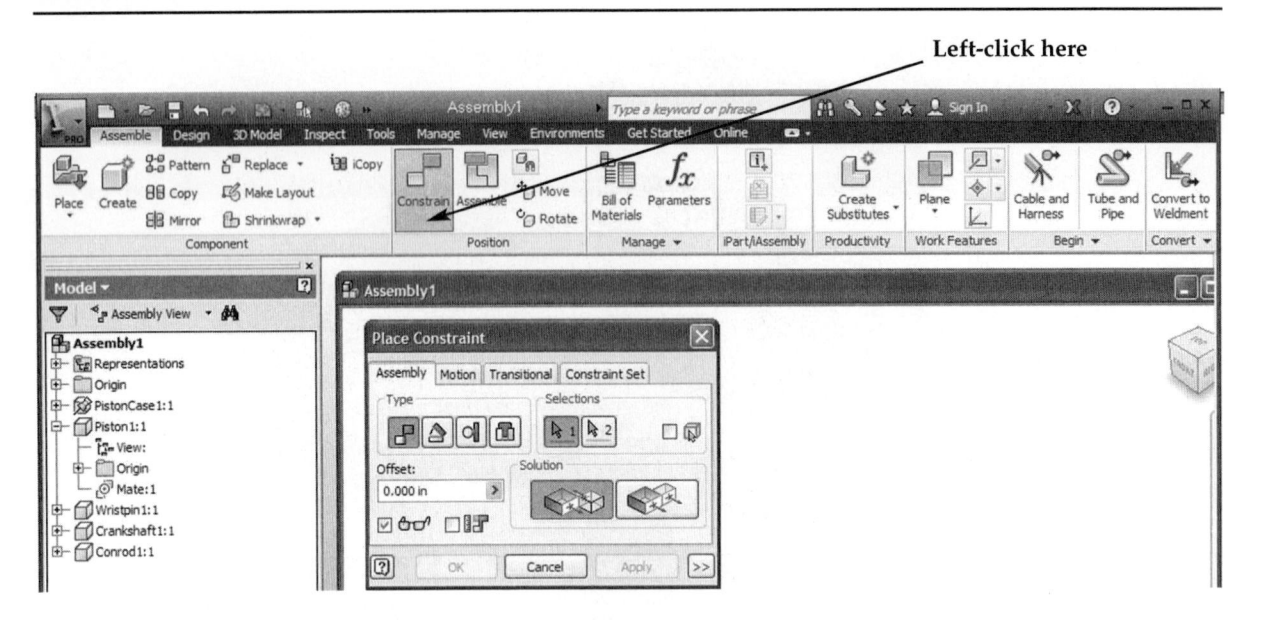

34. Move the cursor to the left side of the connecting rod, causing a red arrow to appear. Left-click as shown in Figure 7-30. You may have to zoom in so that Inventor will find the proper surface.

FIGURE 7-30

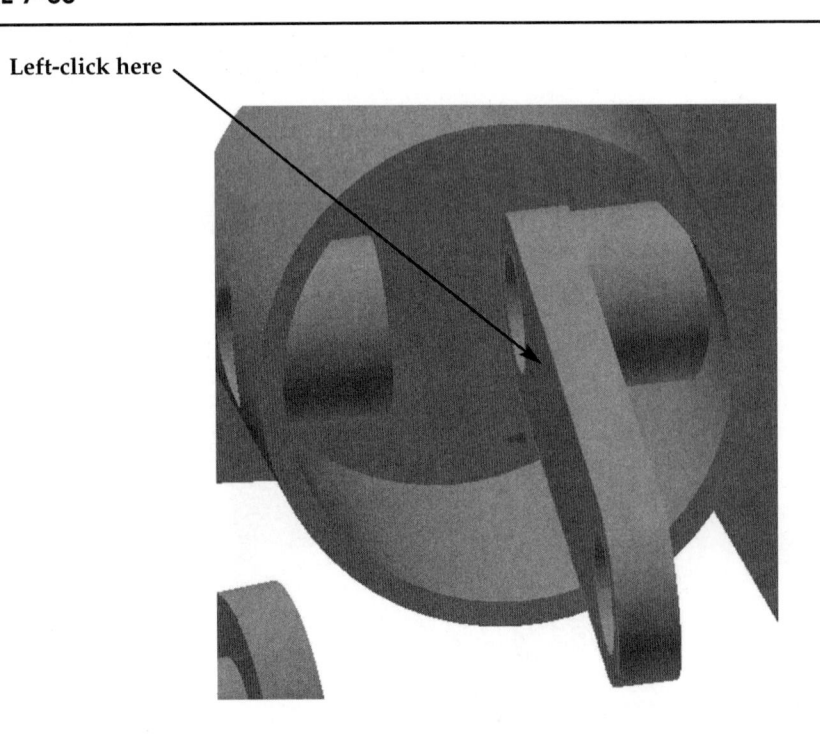

35.   Use the **Rotate** command to turn the piston in order to gain access to the surface opposite the previously selected surface. Hit the **Esc** key once or right-click and select **Done** to get out of the Rotate command. Left-click on the surface opposite the previously selected surface as shown in Figure 7-31.

FIGURE 7-31

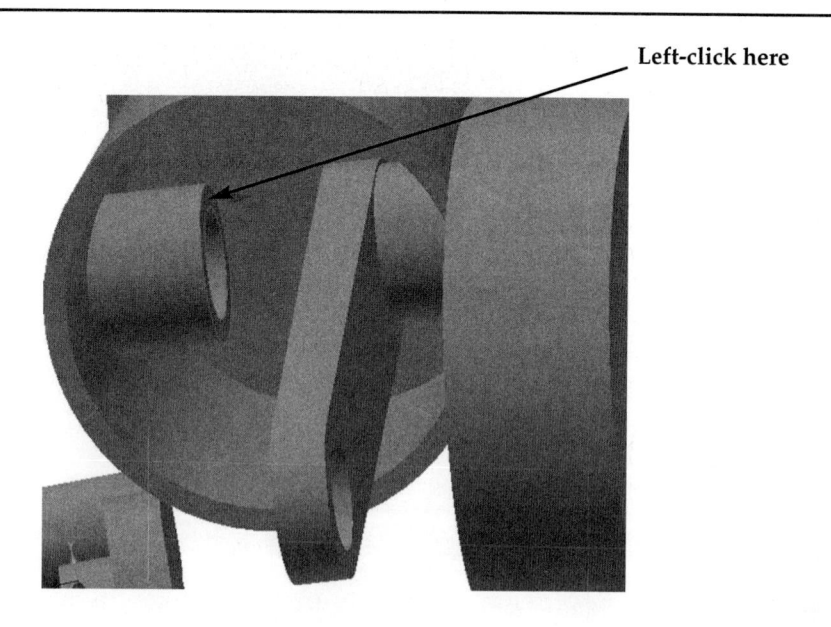

Left-click here

36.   Enter **.250** for the offset as shown in Figure 7-32.

37.   Left-click on **OK**.

FIGURE 7-32

Enter .250 here   Left-click here

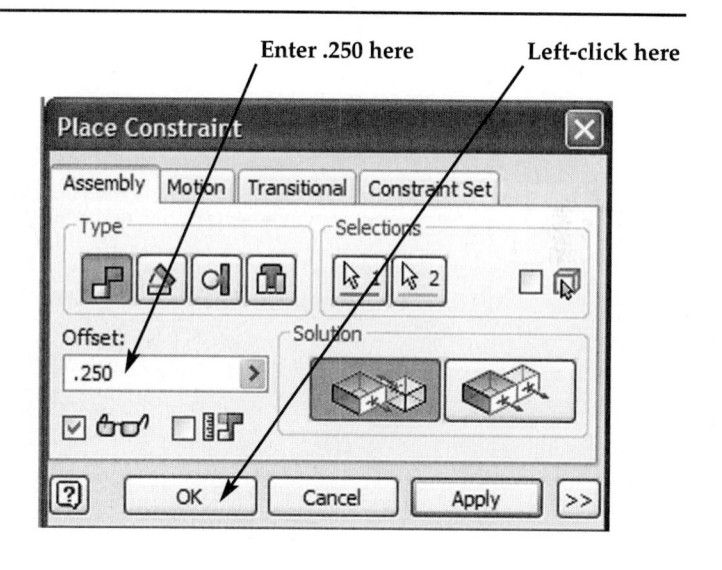

38.   The connecting rod should be centered in the piston. Your screen should look similar to Figure 7-33.

FIGURE 7-33

39.   Right-click anywhere around the drawing. A pop-up menu will appear. Left-click on **Home View** as shown in Figure 7-34.

FIGURE 7-34

Left-click here

40.   Inventor will provide an isometric view of the assembly as shown in Figure 7-35.

FIGURE 7-35

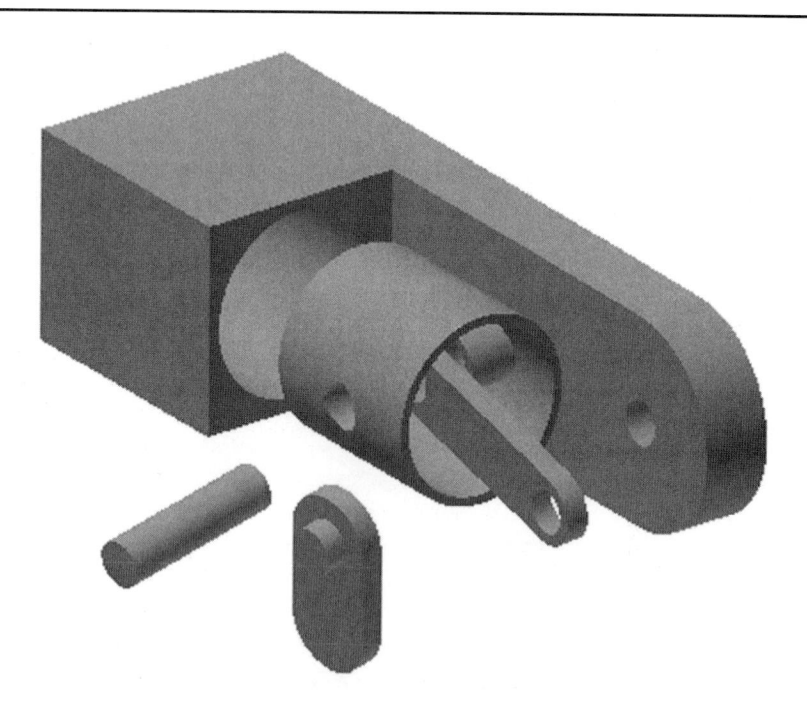

41.   Move the cursor to the upper middle portion of the screen and left-click on **Constrain**. The Place Constraint dialog box will appear as shown in Figure 7-36.

FIGURE 7-36

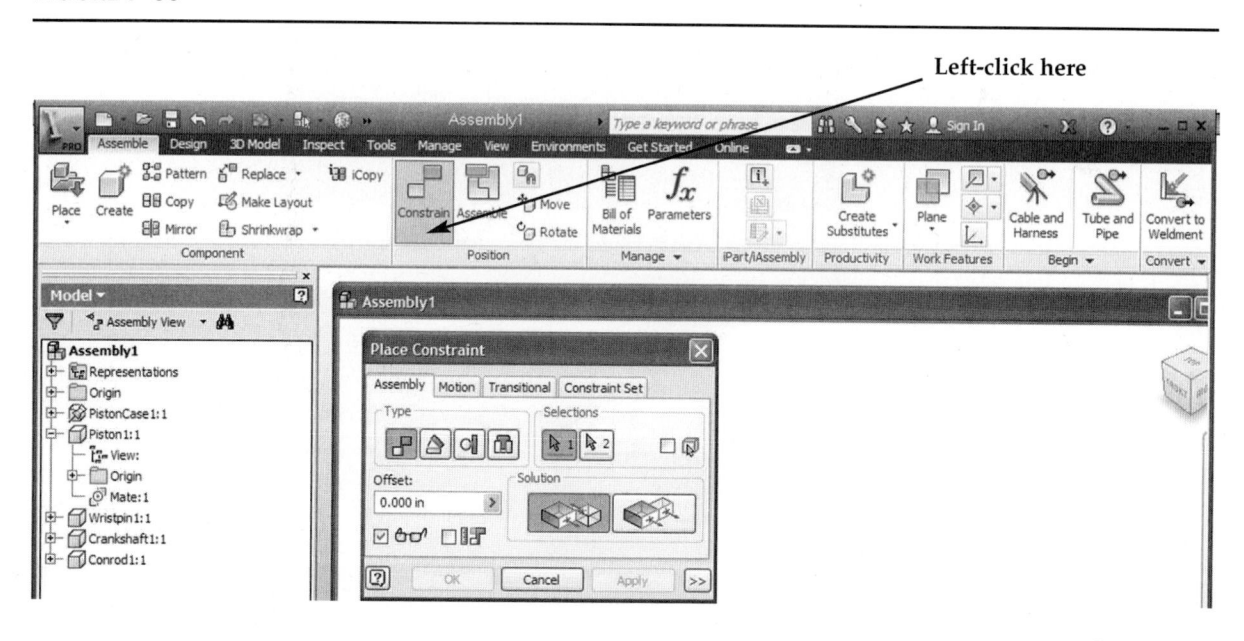

42.   Move the cursor to the wrist pin, causing a red center line to appear. After a red center line appears, left-click once as shown in Figure 7-37.

FIGURE 7-37

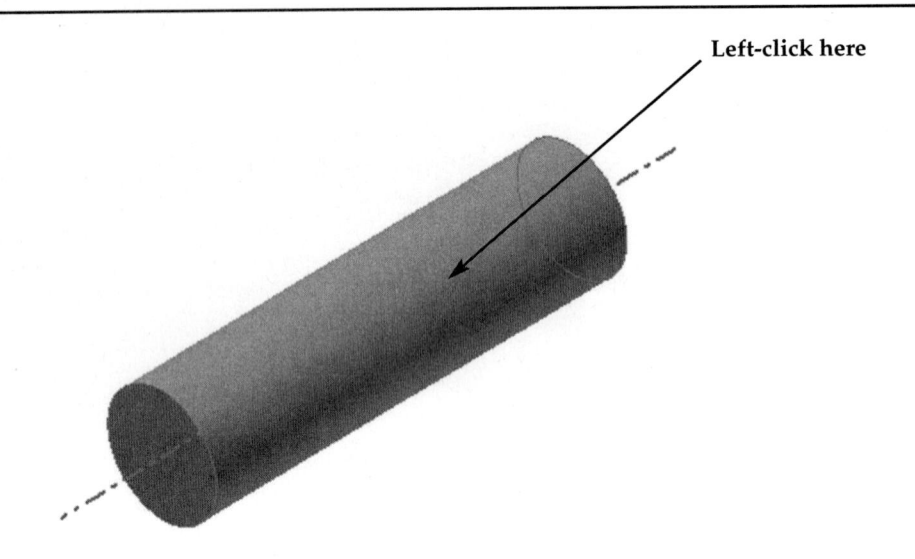

43.   Move the cursor to the piston, causing a red center line to appear. After a red center line appears, left-click once as shown in Figure 7-38.

FIGURE 7-38

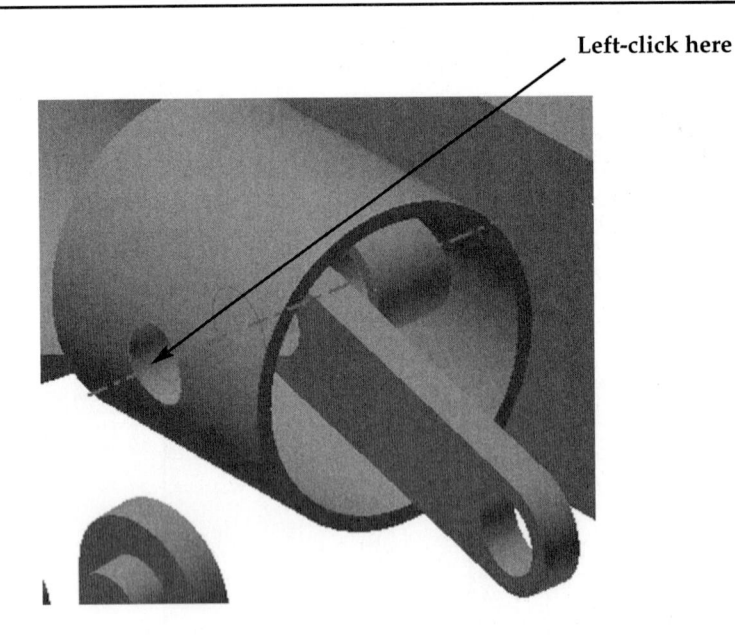

44. Left-click on **OK** as shown in Figure 7-39.

FIGURE 7-39

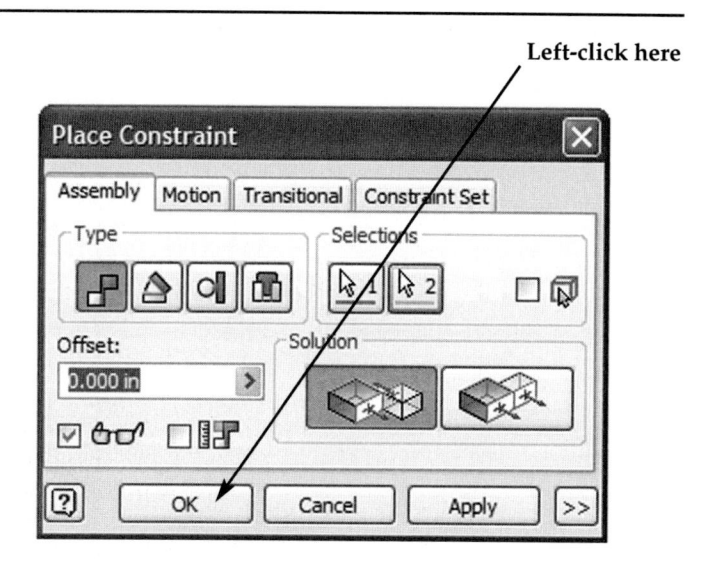

Left-click here

45. Move the cursor to the upper middle portion of the screen and left-click on **Constrain**. The Place Constraint dialog box will appear as shown in Figure 7-40.

FIGURE 7-40

Left-click here

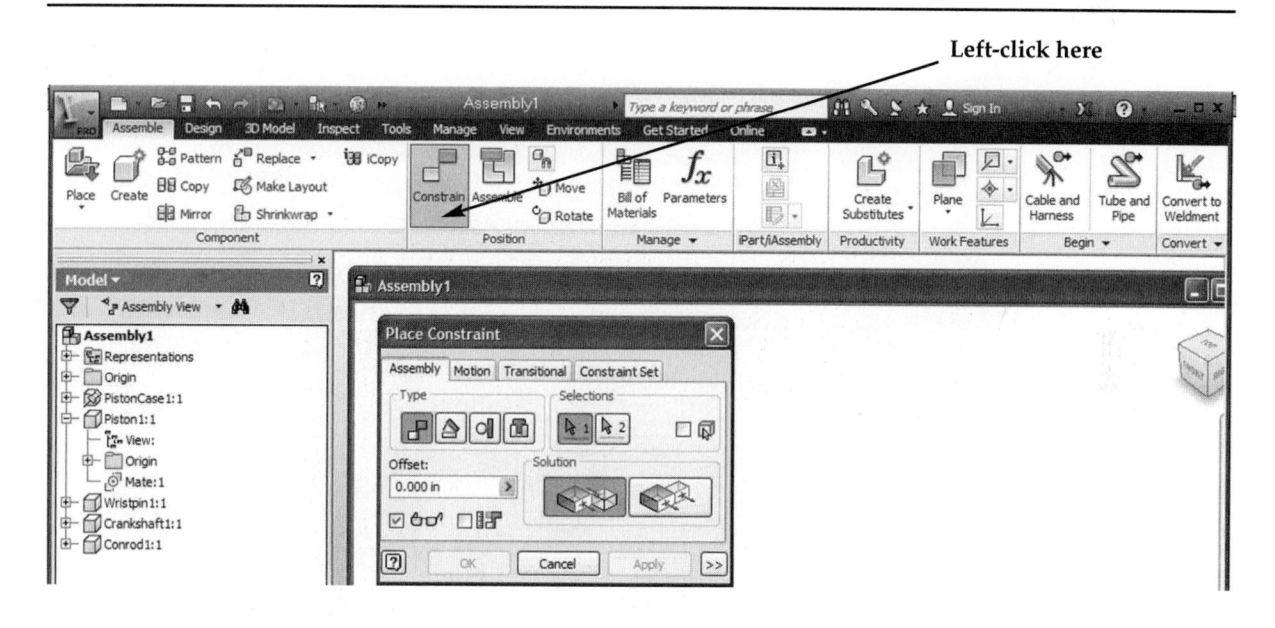

46.   Left-click on the "Flush" icon as shown in Figure 7-41.

FIGURE 7-41

FIGURE 7-41

Left-click here

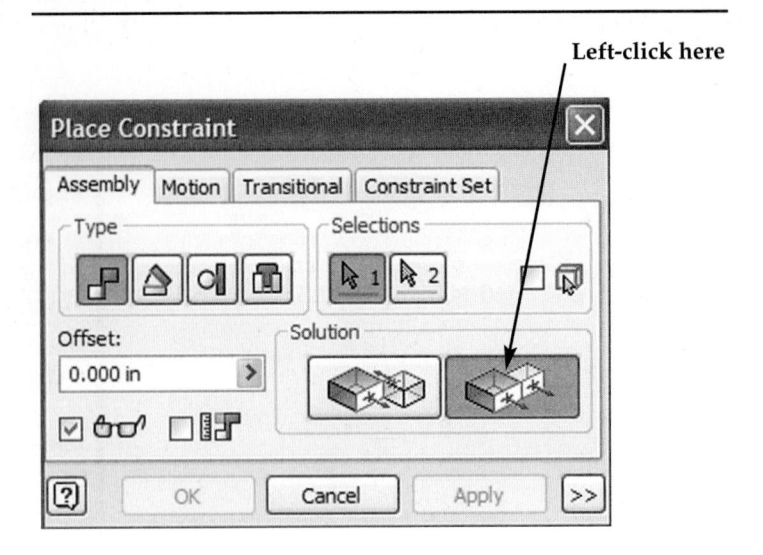

47.   Move the cursor to the side of the wrist pin, causing a red arrow to appear. After a red arrow appears, left-click once as shown in Figure 7-42.

FIGURE 7-42

Left-click here

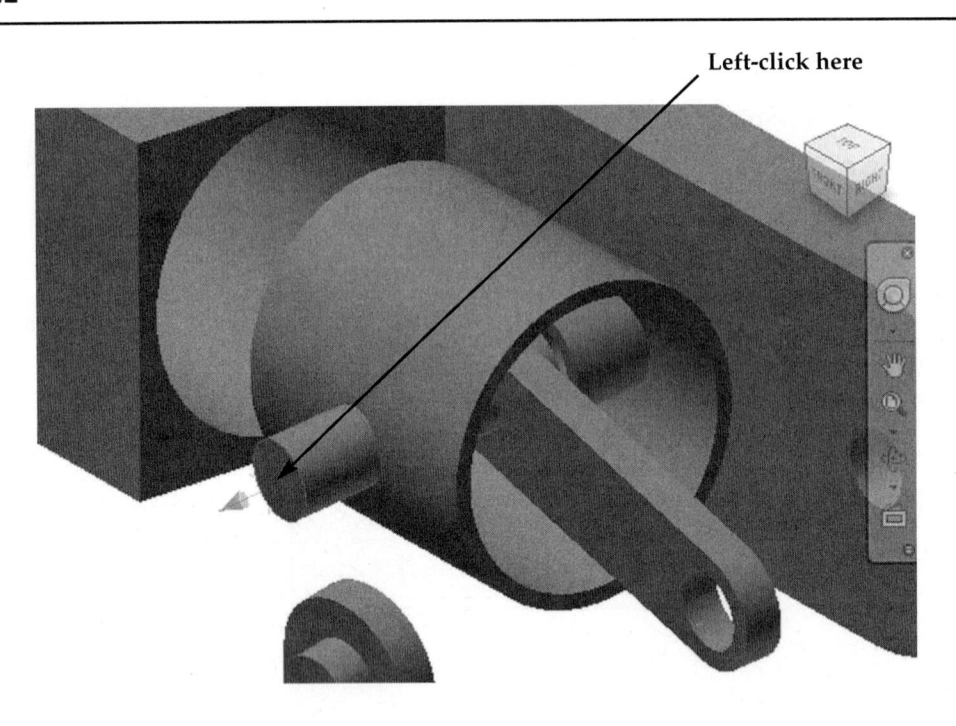

48. Move the cursor to the side of the connecting rod, causing a red arrow to appear. After a red arrow appears, left-click once as shown in Figure 7-43.

FIGURE 7-43

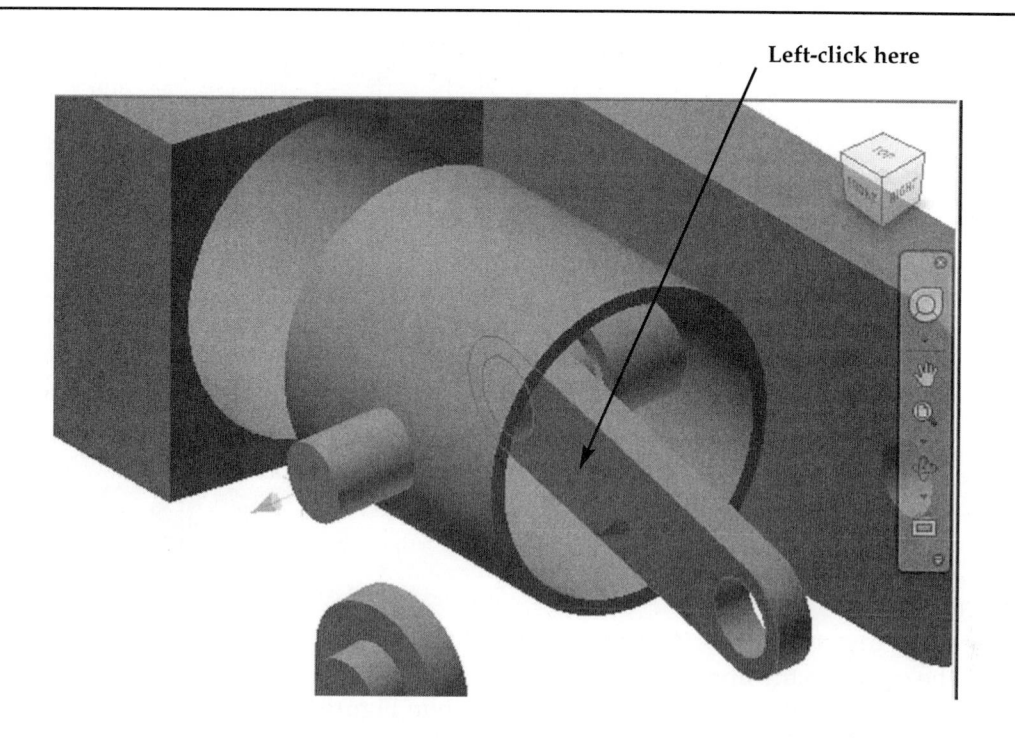

49. Enter **-.7825** under Offset. Left-click on **OK** as shown in Figure 7-44.

FIGURE 7-44

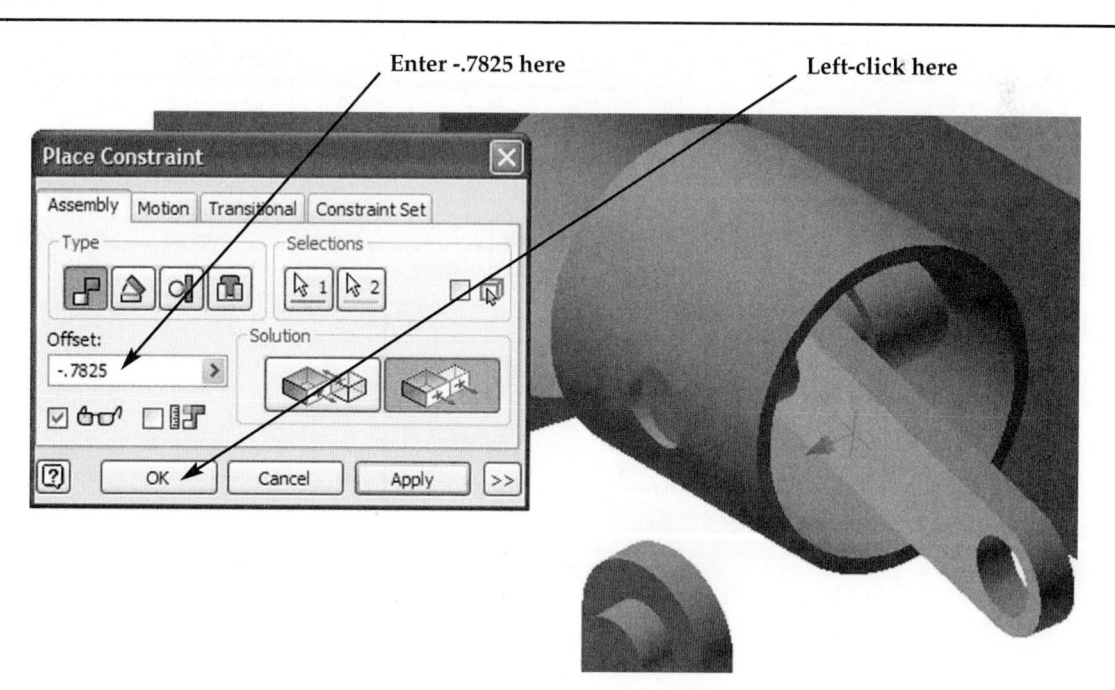

50.   Your screen should look similar to Figure 7-45.

FIGURE 7-45

51.   Move the cursor to the upper middle portion of the screen and left-click on **Constrain**. The Place Constraint dialog box will appear as shown in Figure 7-46.

FIGURE 7-46

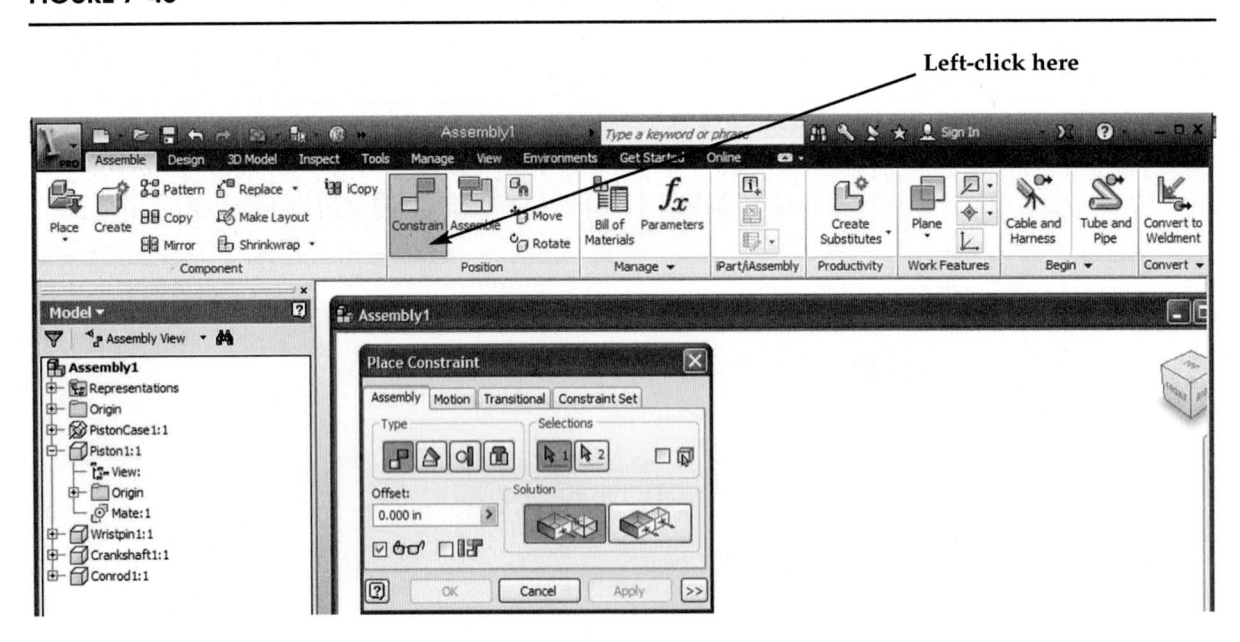

52. Move the cursor to the crankshaft pin, causing a red center line to appear. After a red center line appears, left-click once as shown in Figure 7-47. The crankshaft pin will be secured to the connecting rod.

FIGURE 7-47

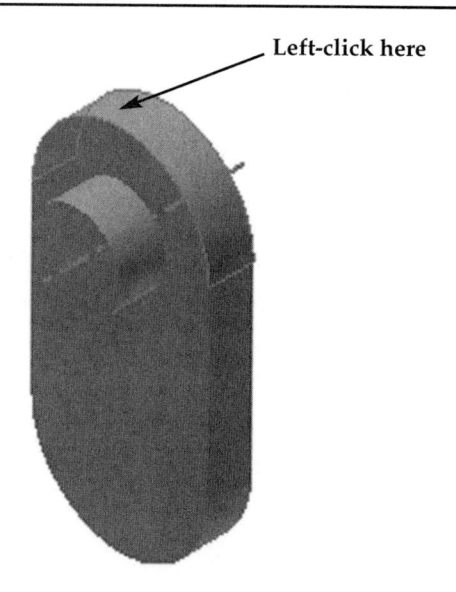

Left-click here

53. Move the cursor to the connecting rod end, causing the red center line to appear. The connecting rod will be secured to the crankshaft. After the red center line appears, left-click once as shown in Figure 7-48.

FIGURE 7-48

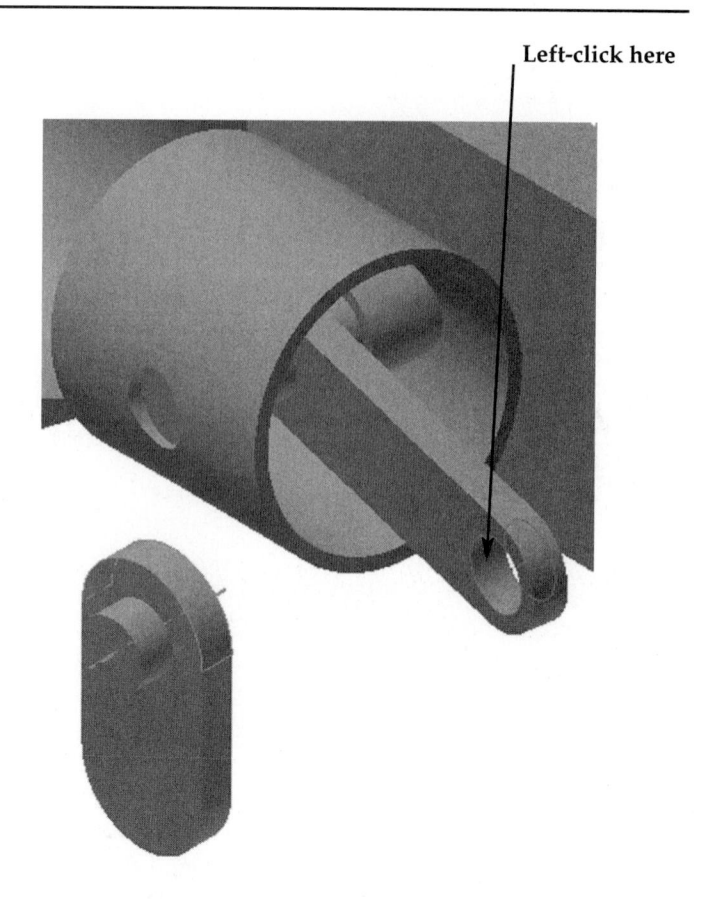

Left-click here

54. Inventor will place the connecting rod and crankshaft together as shown in Figure 7-49.

FIGURE 7-49

55. Left-click on **OK** as shown in Figure 7-50.

FIGURE 7-50

Left-click here

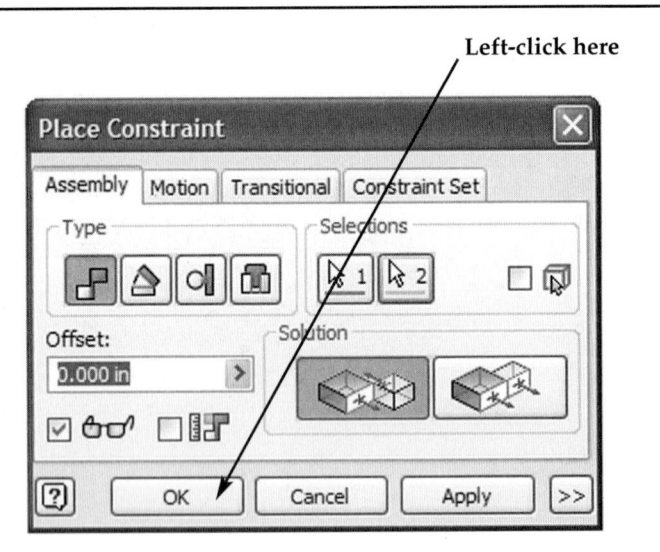

56.   Move the cursor over the piston. Left-click (holding the left mouse button down) and drag the piston upward toward the bottom of the bore as shown in Figure 7-51.

FIGURE 7-51

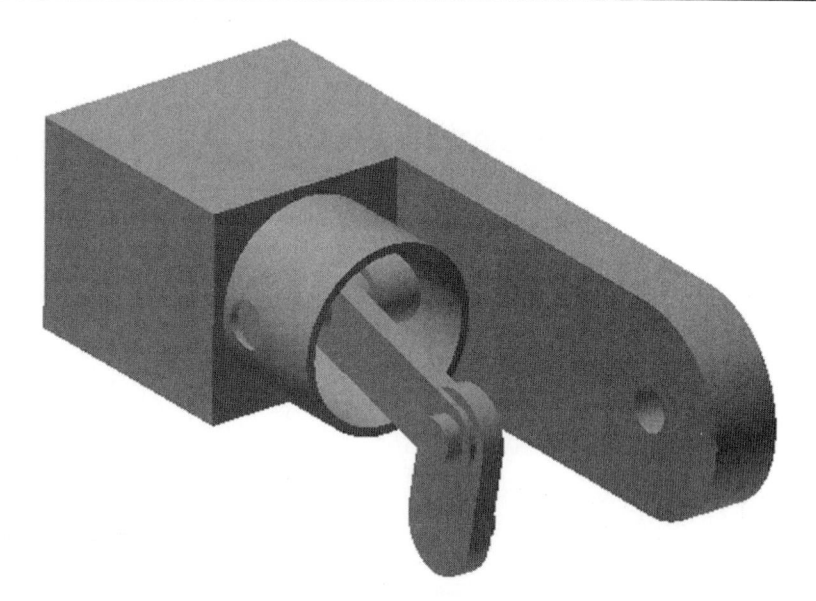

57.   Use the **Rotate** command and roll the assembly around to gain access to the opposite side as shown in Figure 7-52.

FIGURE 7-52

58. Move the cursor to the upper middle portion of the screen and left-click on **Constrain**. The Place Constraint dialog box will appear as shown in Figure 7-53.

FIGURE 7-53

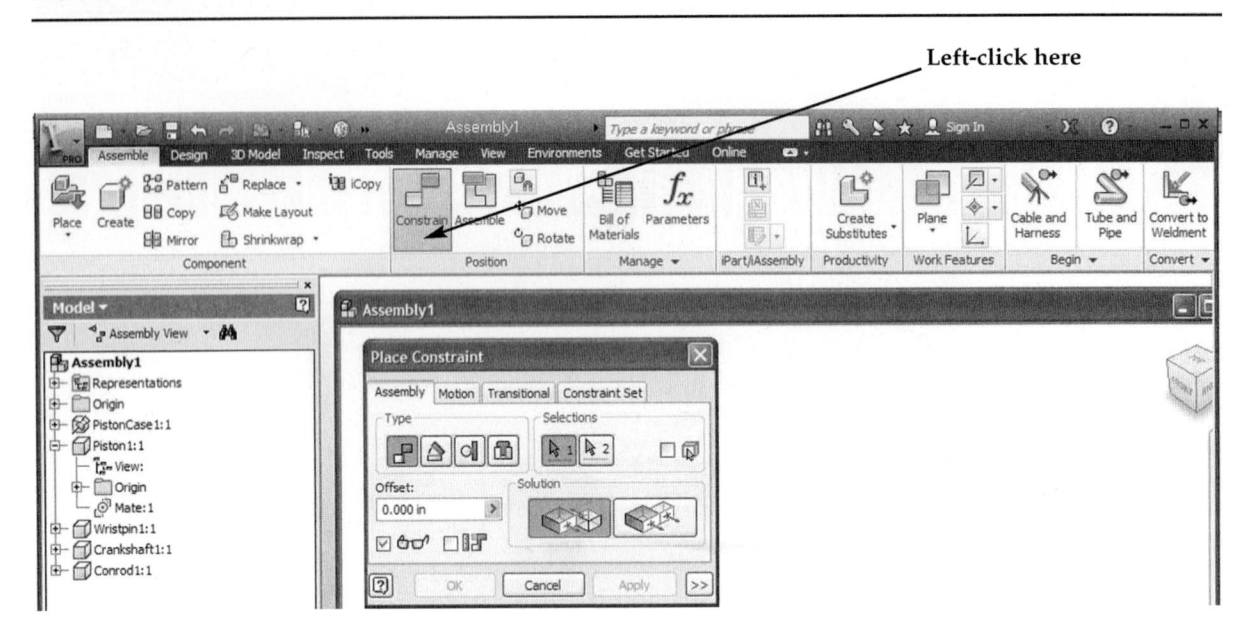

Left-click here

59. Move the cursor to the crankshaft pin, which will be secured in the piston case, causing a red center line to appear. After the red center line appears, left-click once as shown in Figure 7-54.

FIGURE 7-54

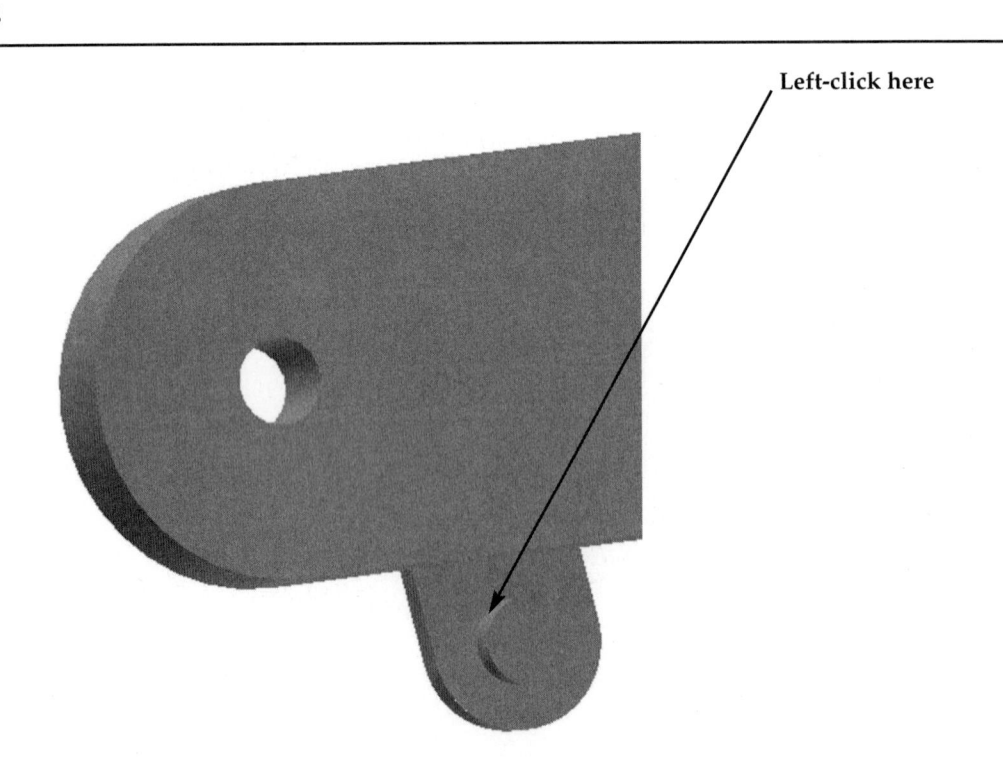

Left-click here

60. Move the cursor to the piston case hole that will secure the crankshaft, causing a red center line appear. After the red center line appears, left-click once as shown in Figure 7-55.

FIGURE 7-55

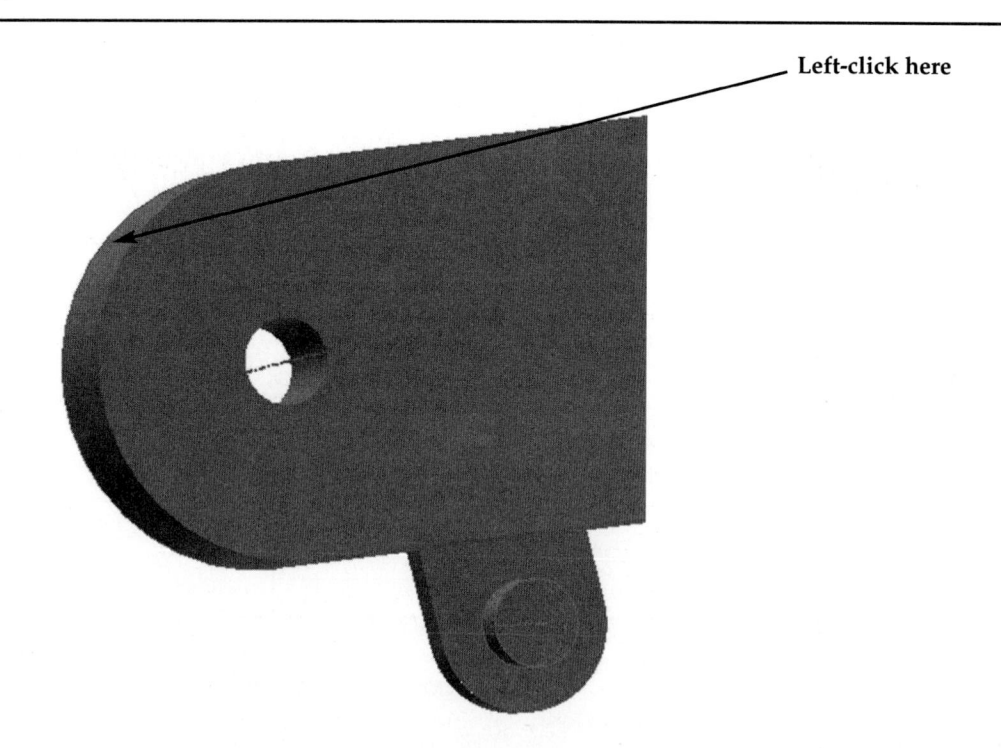

Left-click here

61. Inventor will place the crankshaft pin into the piston case as shown in Figure 7-56.

FIGURE 7-56

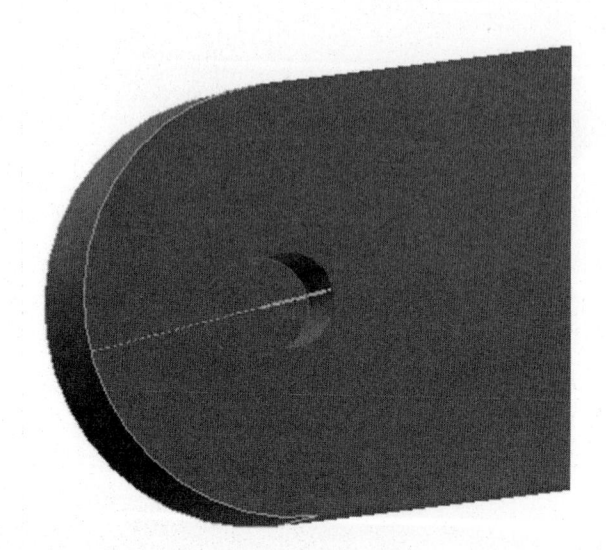

62. Left-click on **OK** as shown in Figure 7-57.

FIGURE 7-57

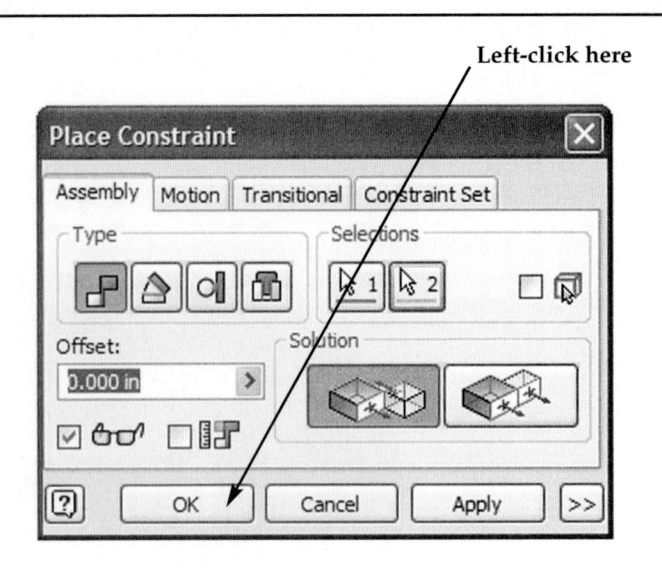

63. Your screen should look similar to Figure 7-58.

FIGURE 7-58

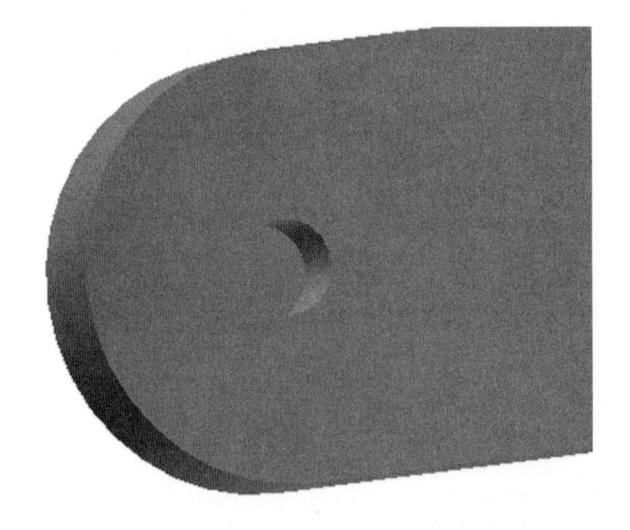

64. Right-click anywhere around the drawing. A pop-up menu will appear. Left-click on **Home View** as shown in Figure 7-59.

FIGURE 7-59

Left-click here

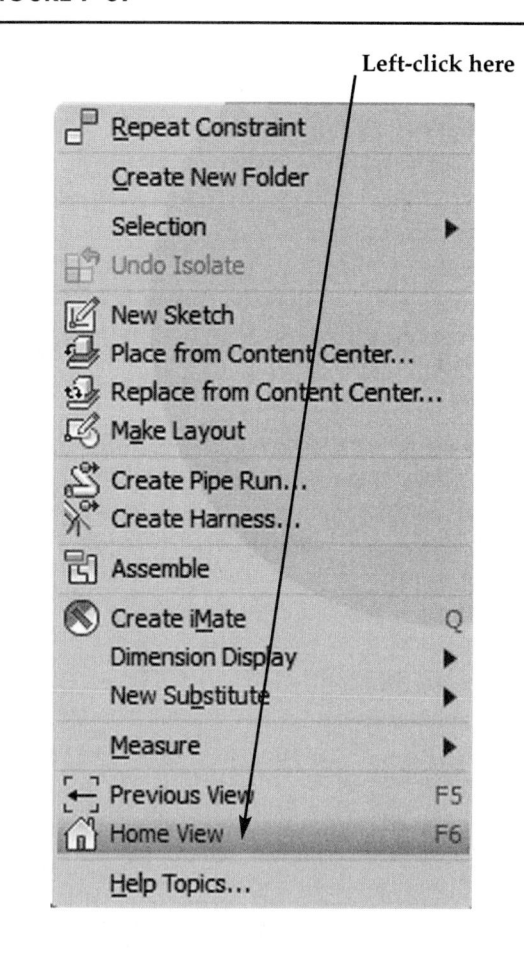

65. Your screen should look similar to Figure 7-60. If the crankshaft is not visible, it is embedded into the pistoncase. Simply left-click on it (holding the left mouse button down) and move the crank out of the pistoncase.

FIGURE 7-60

Crankshaft visible

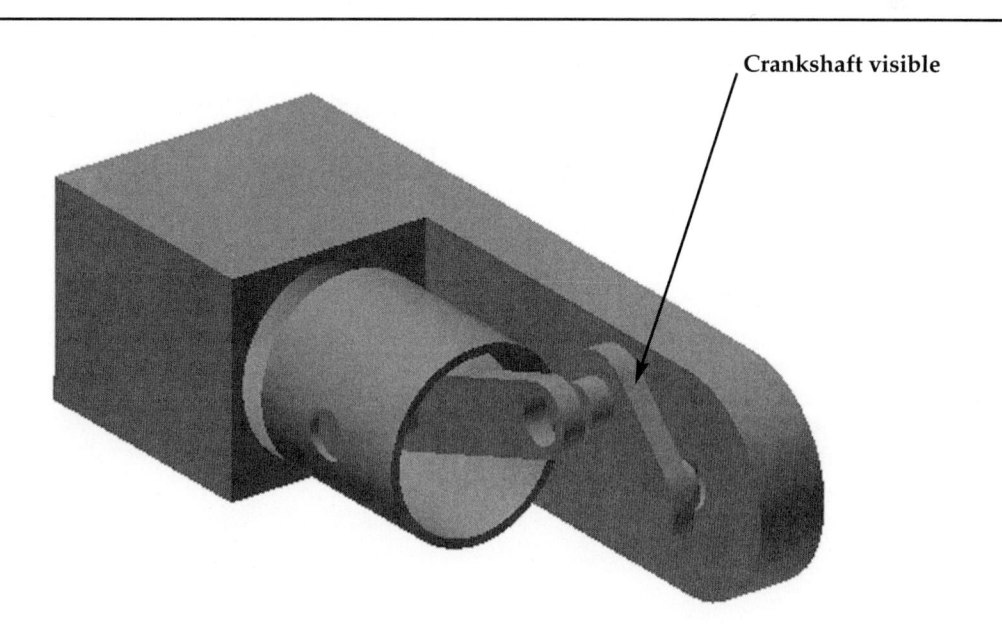

66.   Move the cursor to the upper middle portion of the screen and left-click on **Constrain**. The Place Constraint dialog box will appear as shown in Figure 7-61.

FIGURE 7-61

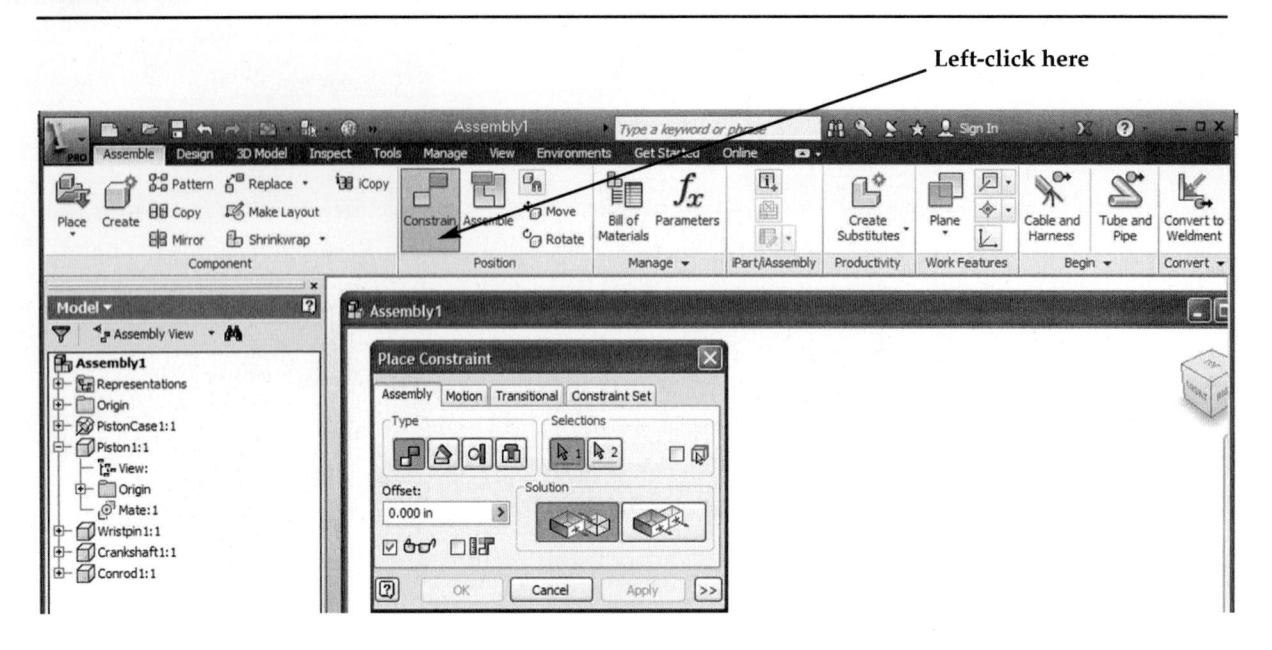

67.   Left-click on the "Flush" icon as shown in Figure 7-62.

FIGURE 7-62

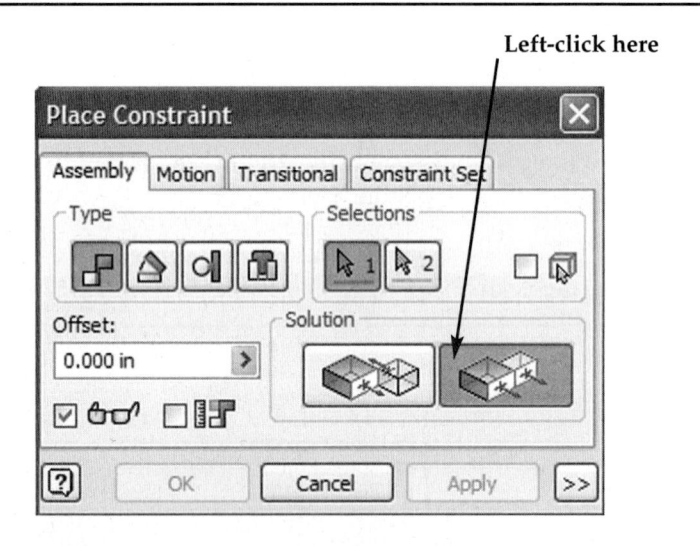

68. Move the cursor to the left side of the connecting rod, causing a red arrow to appear. After a red arrow appears, left-click once as shown in Figure 7-63.

FIGURE 7-63

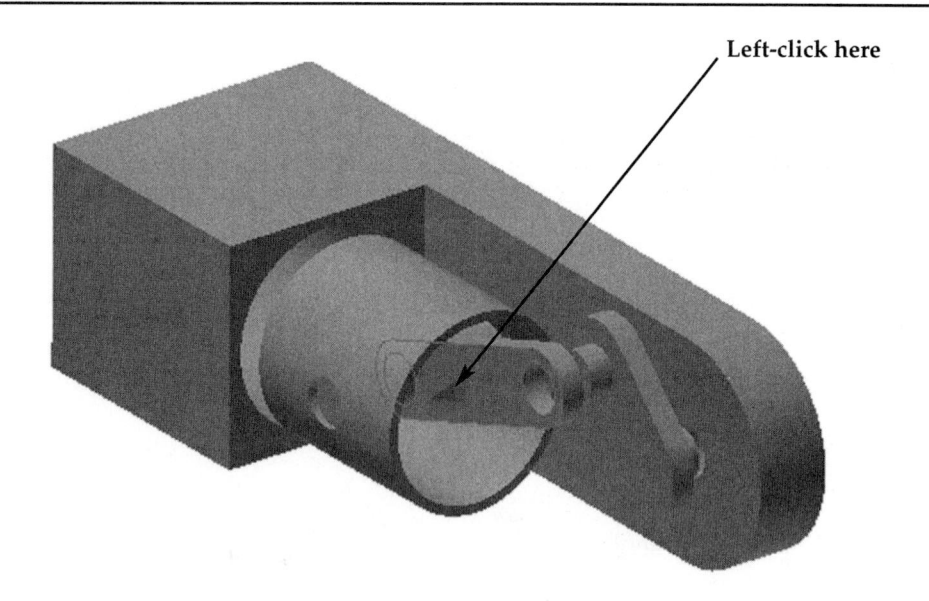

Left-click here

69. Move the cursor to the crankshaft connecting rod pin, causing the red arrow to appear. After a red arrow appears, left-click once as shown in Figure 7-64.

FIGURE 7-64

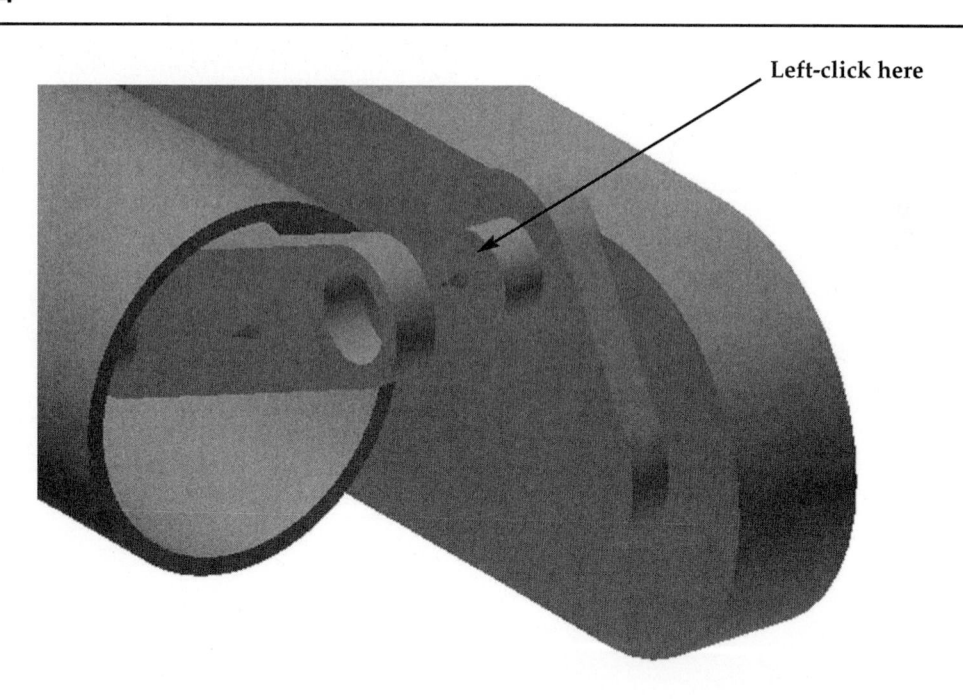

Left-click here

70. Inventor will place the connecting rod flush with the crankshaft connecting rod pin as shown in Figure 7-65.

FIGURE 7-65

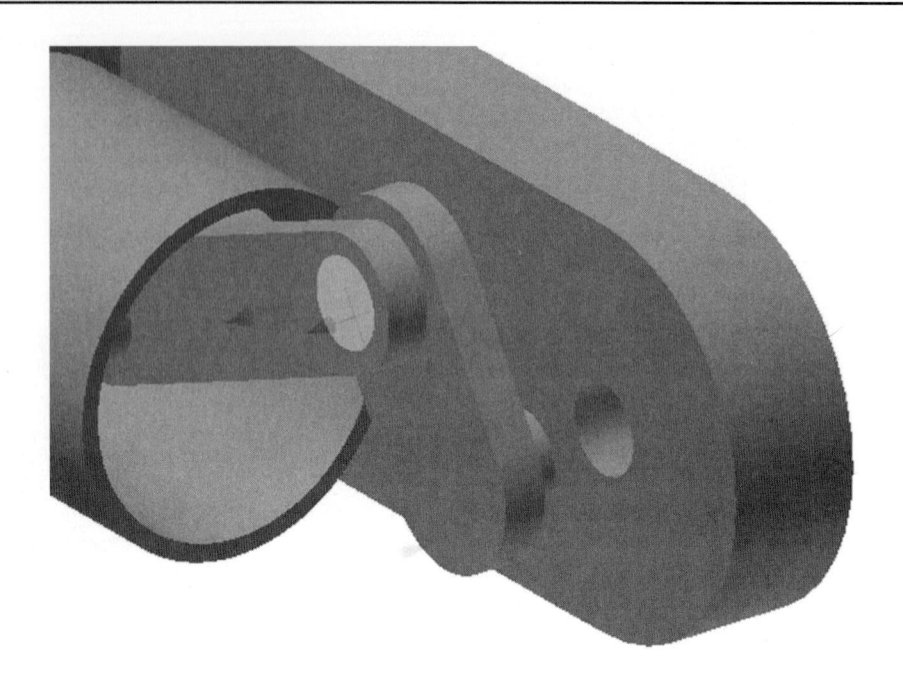

71. Left-click on **OK** as shown in Figure 7-66.

FIGURE 7-66

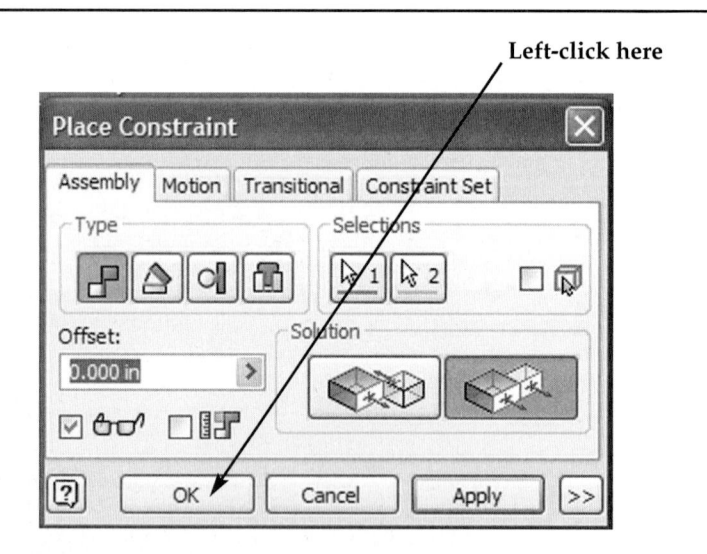

EDIT/MODIFY PARTS WHILE IN THE ASSEMBLY PANEL

72.   Your screen should look similar to Figure 7-67.

FIGURE 7-67

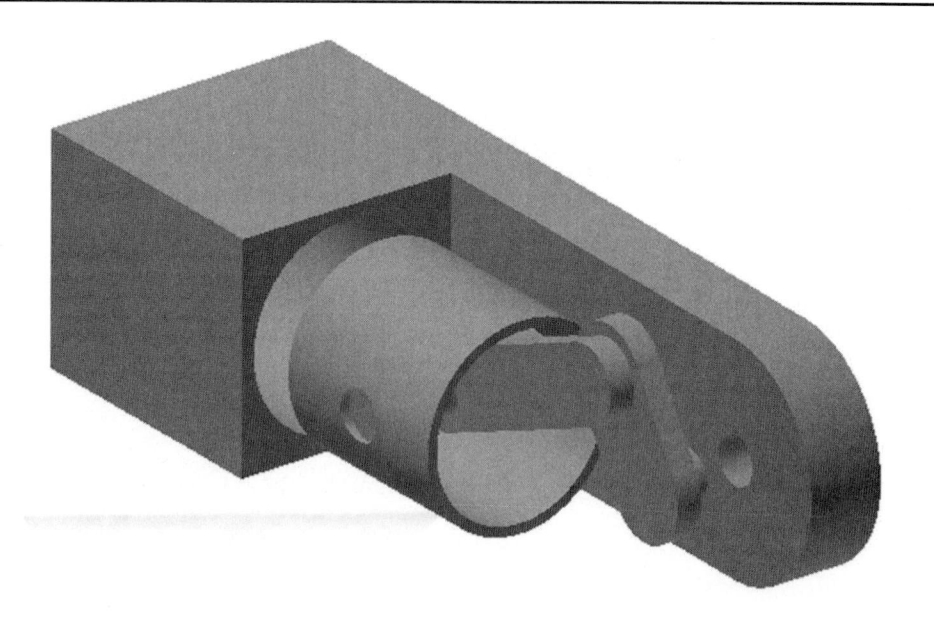

73.   The length of the connecting rod must be modified. Move the cursor over the connecting rod, causing the edges to turn red as shown in Figure 7-68.

FIGURE 7-68

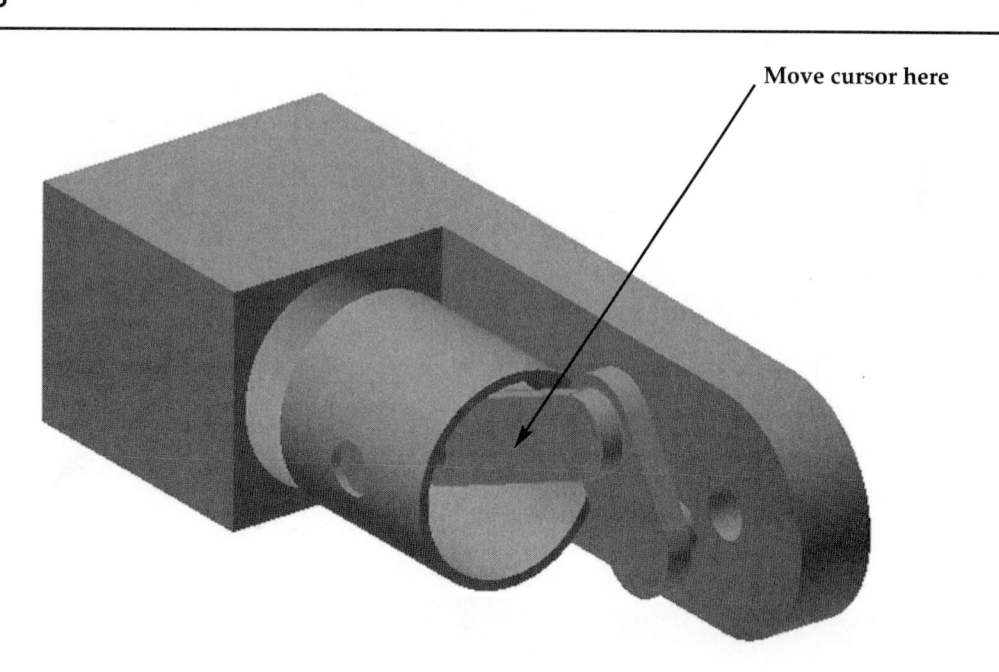

Move cursor here

74. Double-click (left-click) on the connecting rod. All other parts will become grayed as shown in Figure 7-69.

FIGURE 7-69

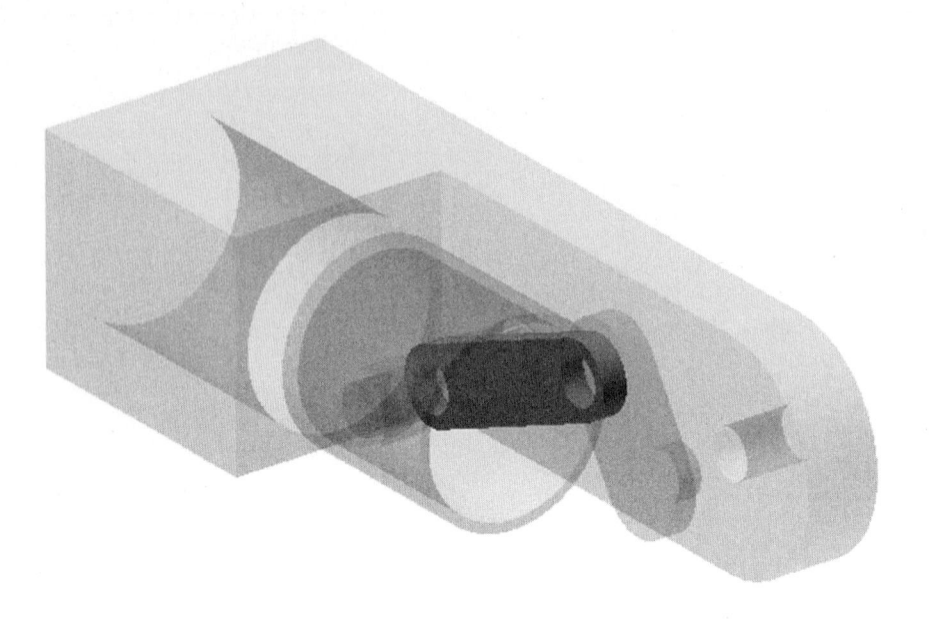

75. Notice that the part tree at the lower left of the screen has changed. All of the branches related to all other parts are grayed (inactive). The branches that illustrate the connecting rod are white (active) as shown in Figure 7-70.

FIGURE 7-70

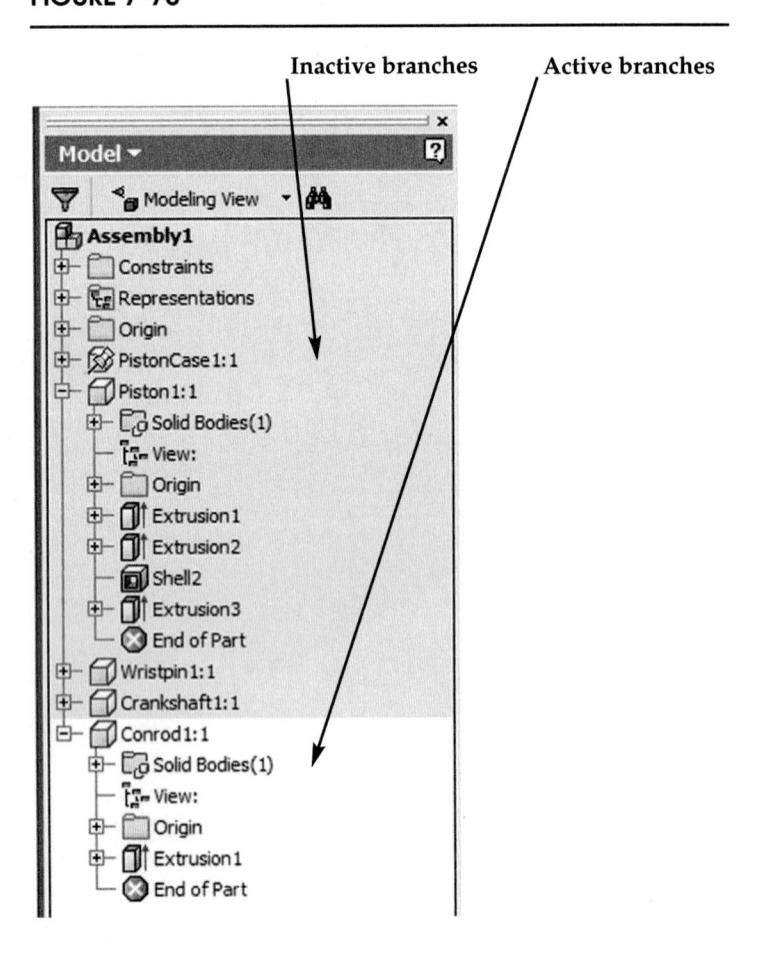

76. Left-click on the "plus" sign next to the text, "Extrusion1" as shown in Figure 7-71.

FIGURE 7-71

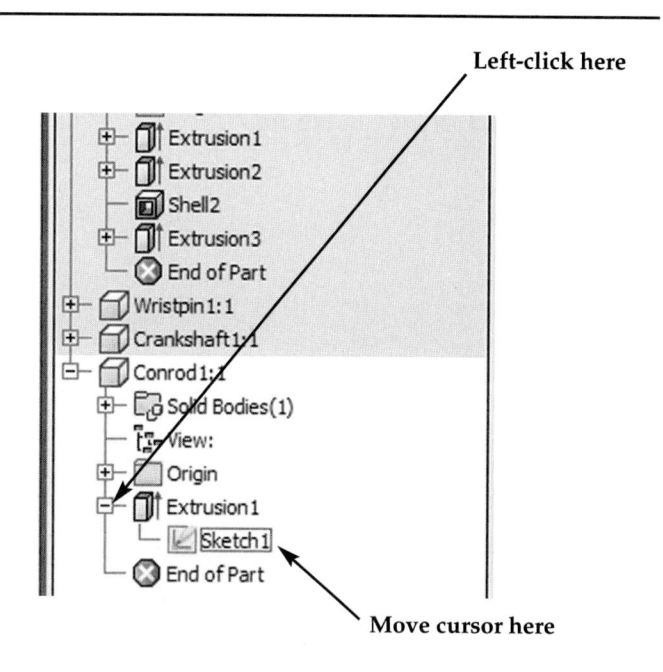

77. Move the cursor over the text, "Sketch1," causing a red box to appear around the text. Notice at the same time the sketch will appear in red on the connecting rod as shown in Figure 7-72.

FIGURE 7-72

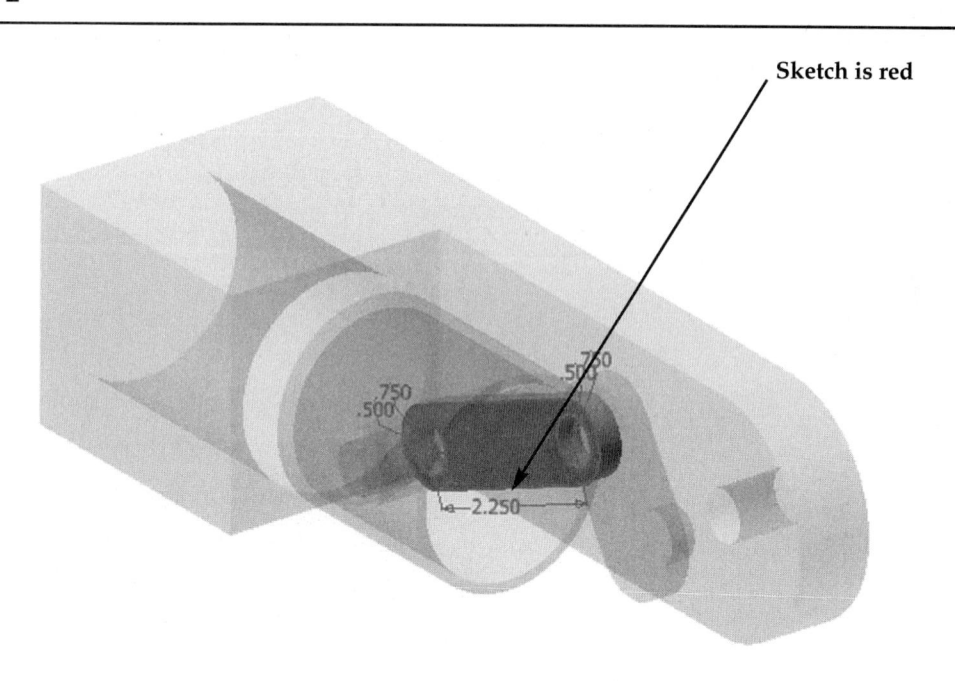

78. Right-click on **Sketch1** while the red box is visible around the text. A pop-up menu will appear. Left-click on **Edit Sketch** as shown in Figure 7-73.

FIGURE 7-73

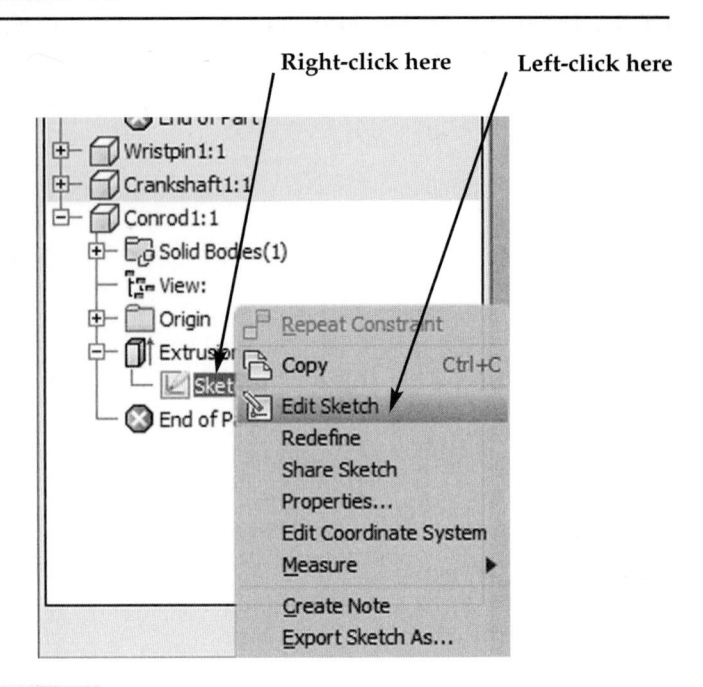

79. Your screen should look similar to Figure 7-74.

FIGURE 7-74

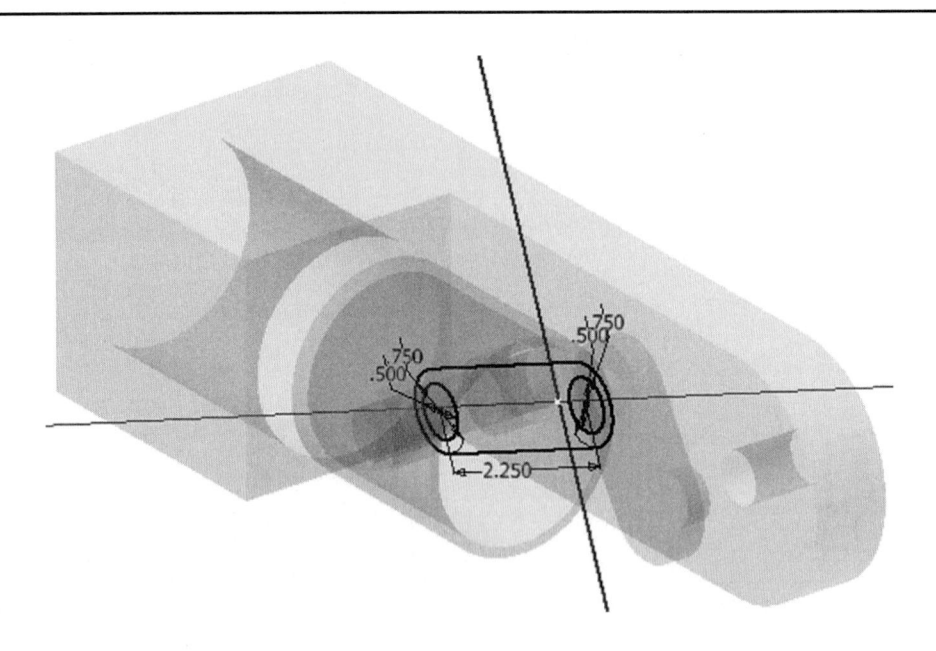

80. Move the cursor over the 2.25 dimension. After it turns red, double-click the left mouse button. The Edit Dimension dialog box will appear as shown in Figure 7-75.

FIGURE 7-75

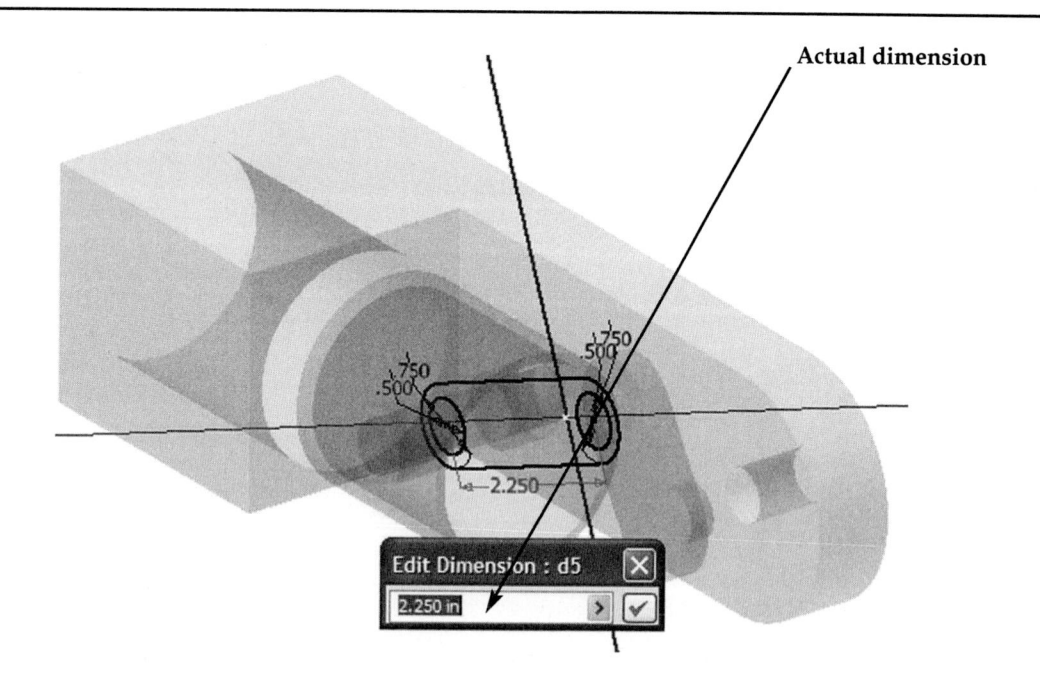

81. While the text is still highlighted, enter **4.75** as shown in Figure 7-76 and press **Enter** on the keyboard.

FIGURE 7-76

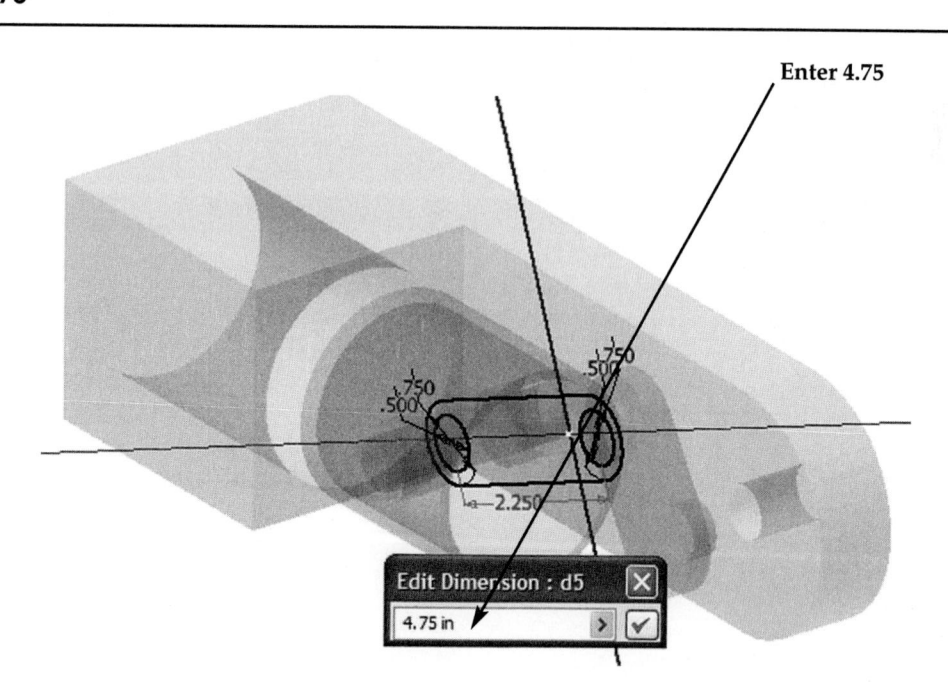

82.   The length of the connecting rod will become 4.75 inches as shown in Figure 7-77.

FIGURE 7-77

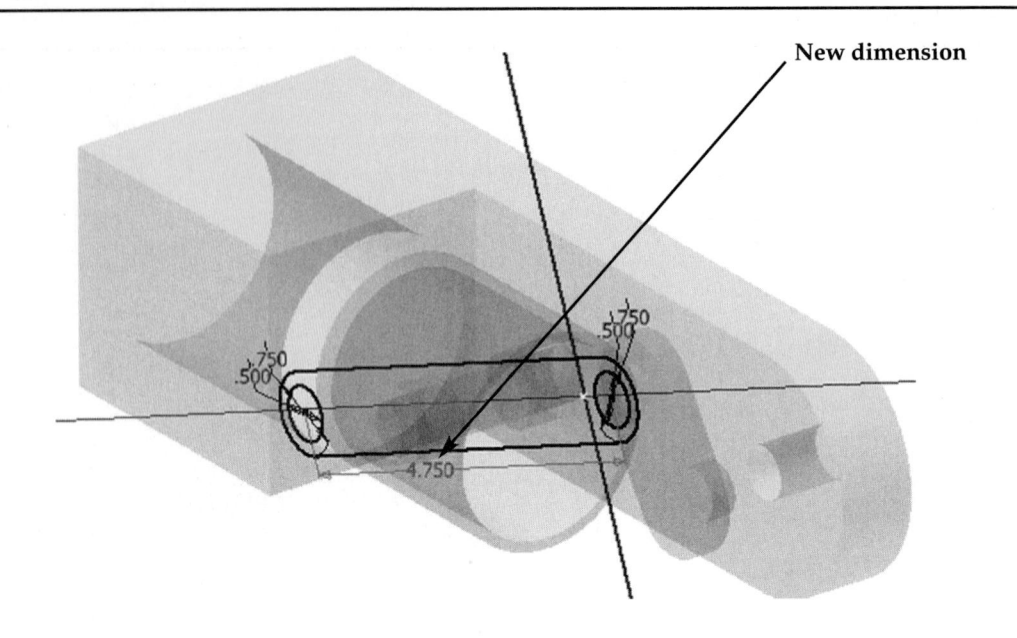

83.   Move the cursor to the upper middle portion of the screen and left-click the **Manage** tab. Left-click on the drop-down arrow below **Update**. Left-click on **Update** as shown in Figure 7-78.

FIGURE 7-78

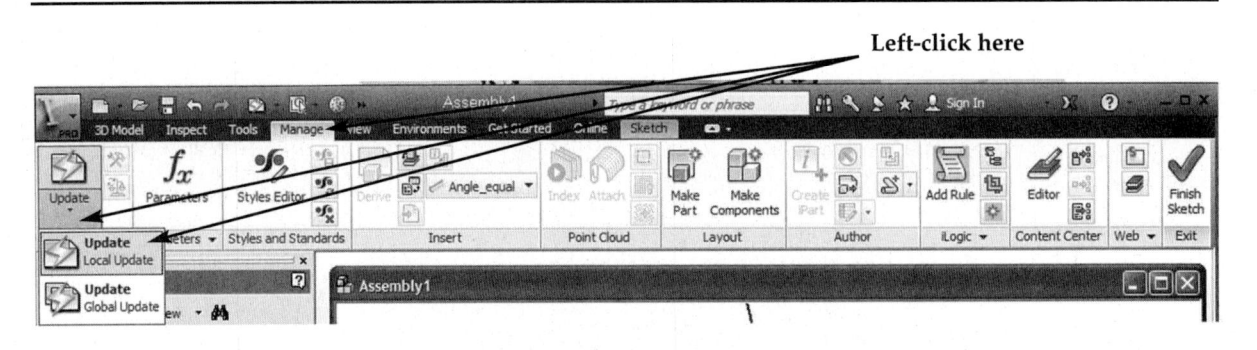

84. Inventor will update the change made to the sketch in the Part Features Panel as shown in Figure 7-79.

FIGURE 7-79

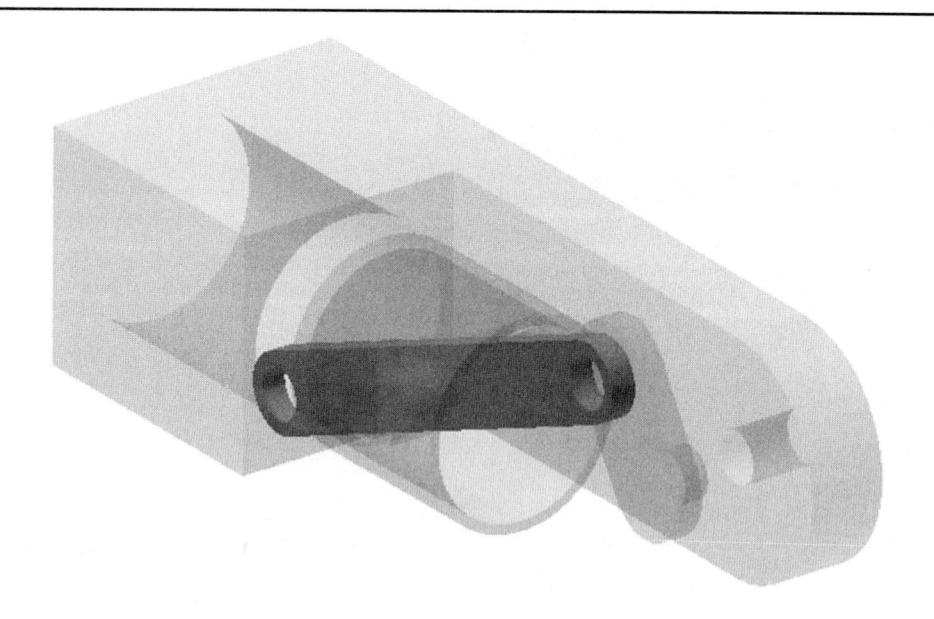

85. Move the cursor to the upper right portion of the screen and left-click on the drop-down arrow below **Return**. Left-click on **Return** as shown in Figure 7-80.

FIGURE 7-80

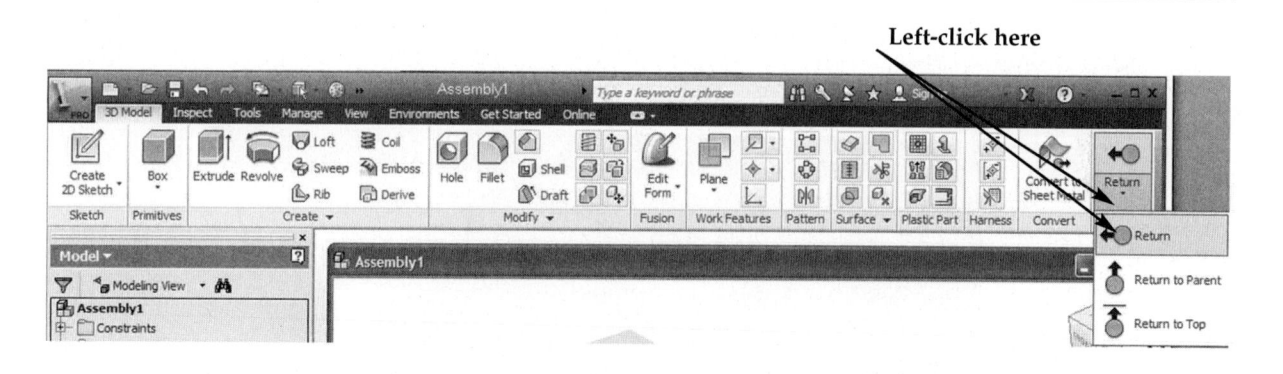

86. Inventor will return to the Assembly Panel displaying the changes made to the connecting rod. Your screen should look similar to Figure 7-81.

FIGURE 7-81

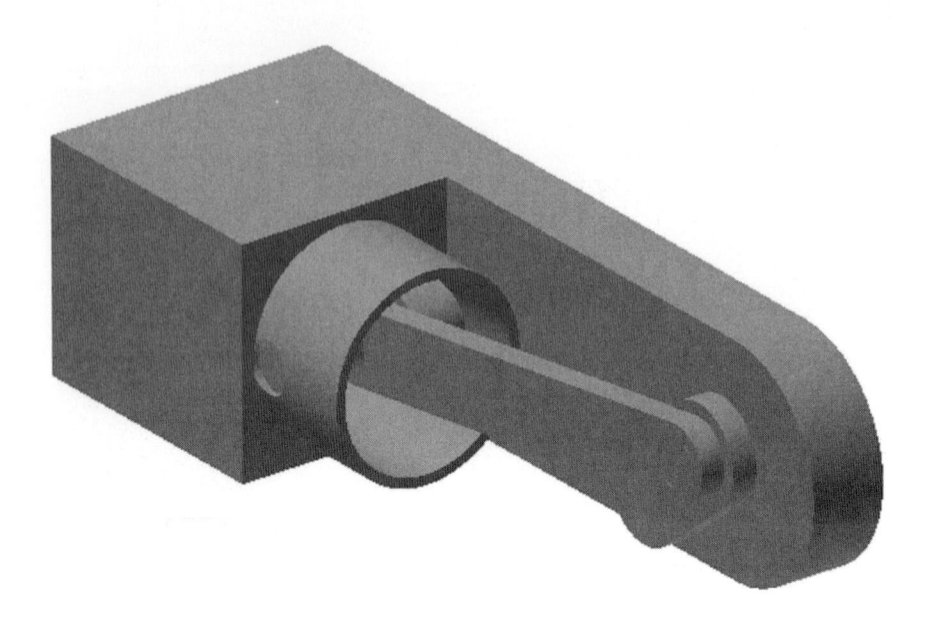

87. The length of the crankshaft pin also must be modified. Move the cursor over the crankshaft as shown in Figure 7-82. The edges will turn red.

FIGURE 7-82

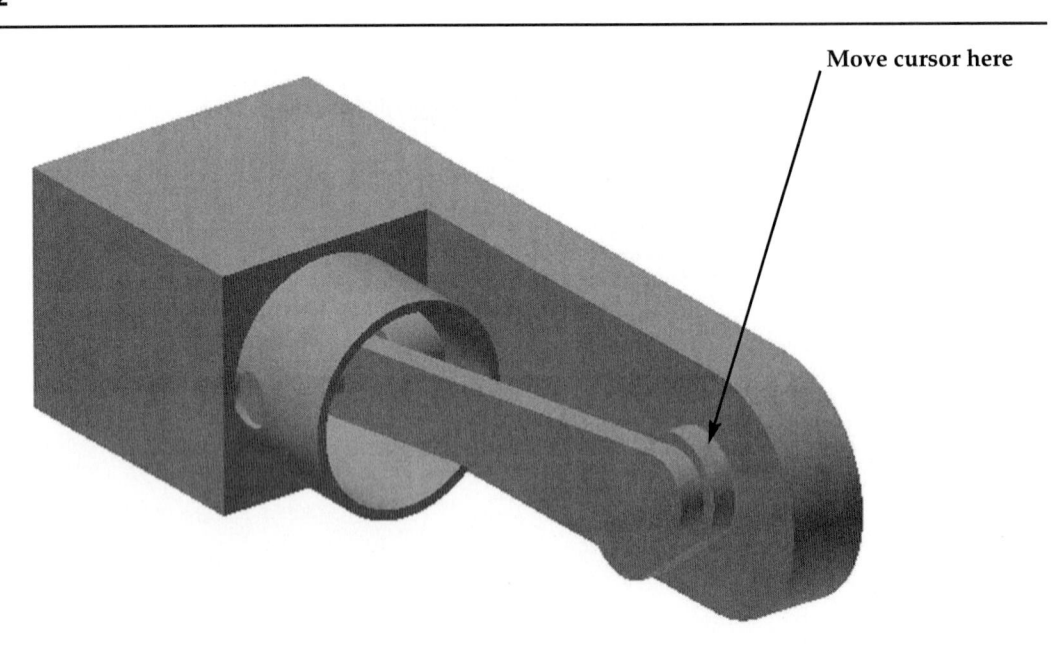

Move cursor here

88. Double-click (left-click) on the crankshaft. All other parts will become grayed as shown in Figure 7-83.

FIGURE 7-83

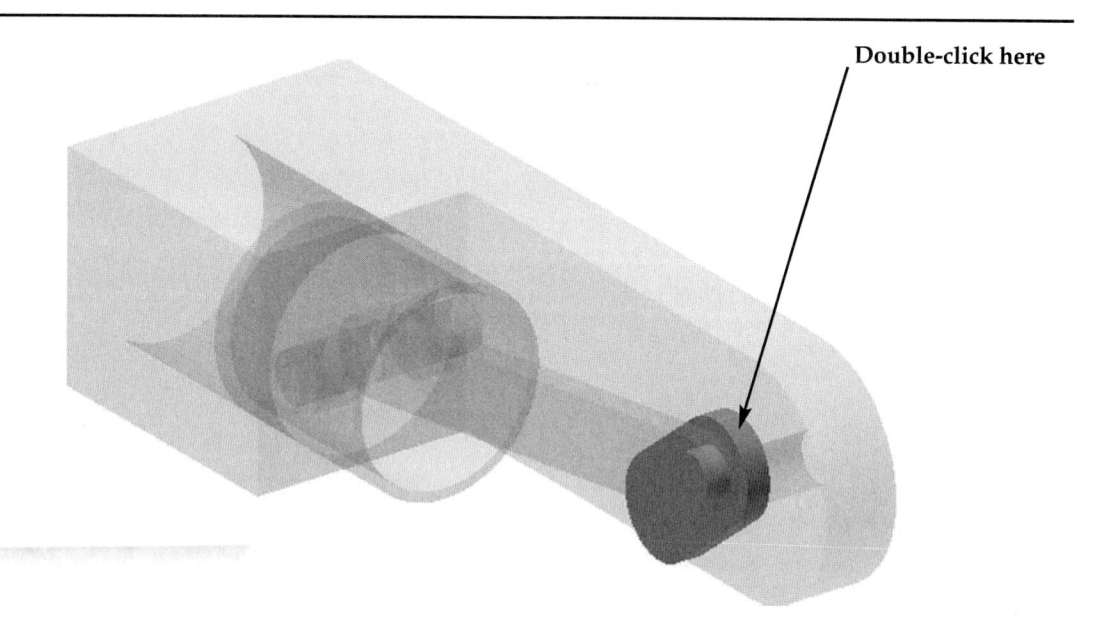

Double-click here

89. Notice that the part tree at the lower left of the screen has changed. All of the branches related to all other parts are grayed (inactive). The branches that illustrate the crankshaft are white (active) as shown in Figure 7-84.

FIGURE 7-84

Inactive branches   Active branches

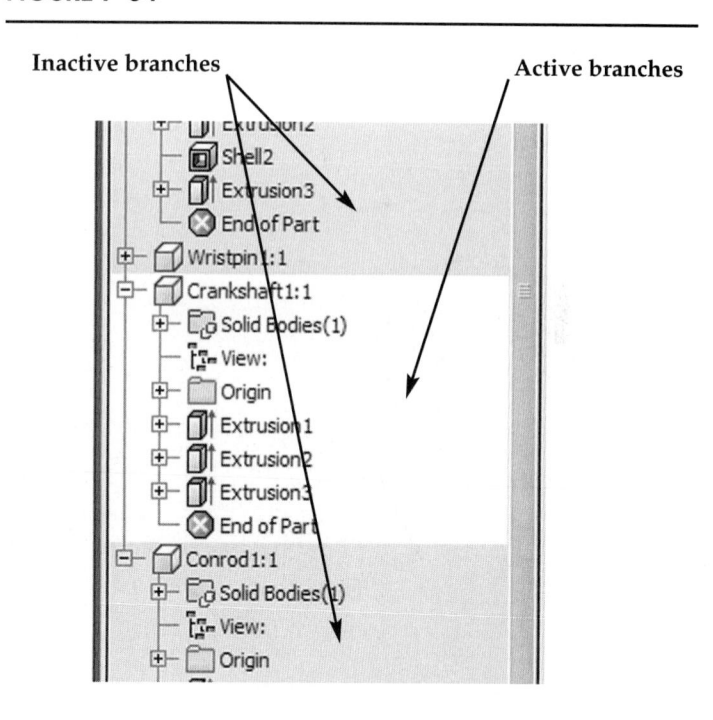

90.   Right-click on **Extrusion3**. A pop-up menu will appear. Left-click on **Edit Feature** as shown in Figure 7-85.

FIGURE 7-85

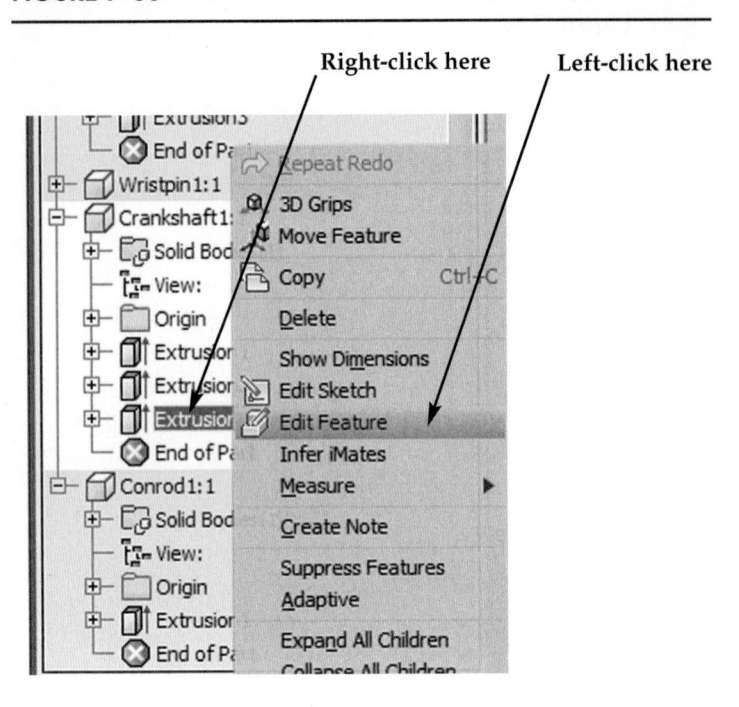

91.   The Extrude dialog box will appear. Enter **2.00** for the extrusion distance and left-click on **OK** as shown in Figure 7-86.

FIGURE 7-86

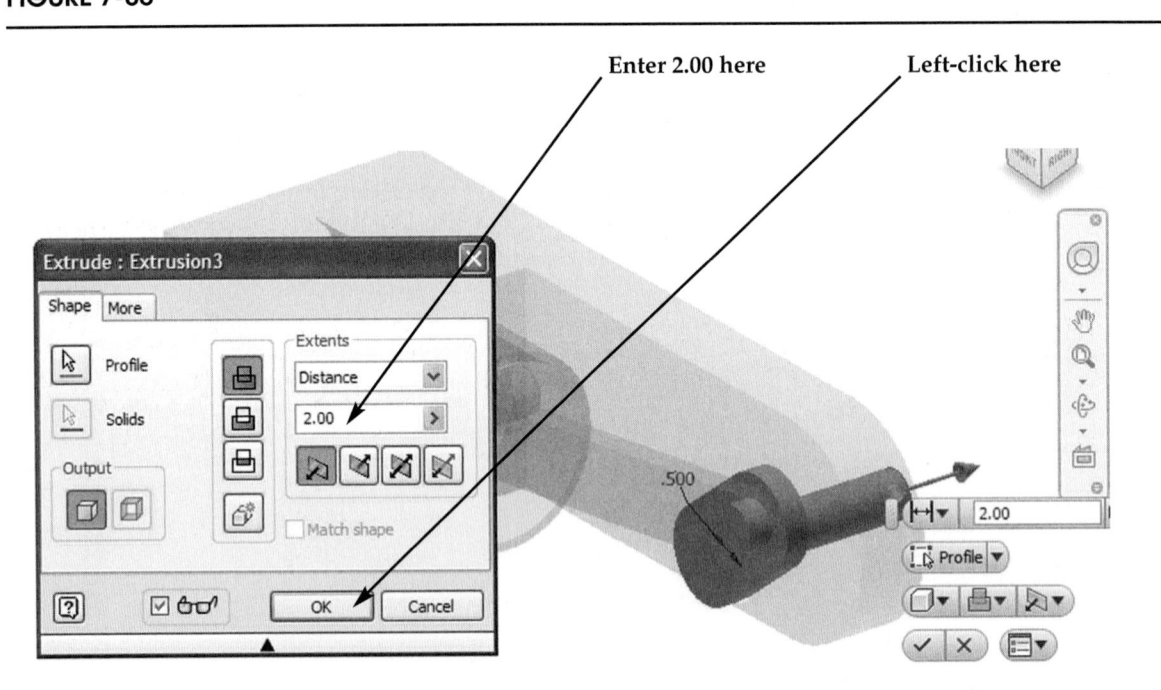

92. Inventor will update the change made to the sketch in the Part Features Panel as shown in Figure 7-87.

FIGURE 7-87

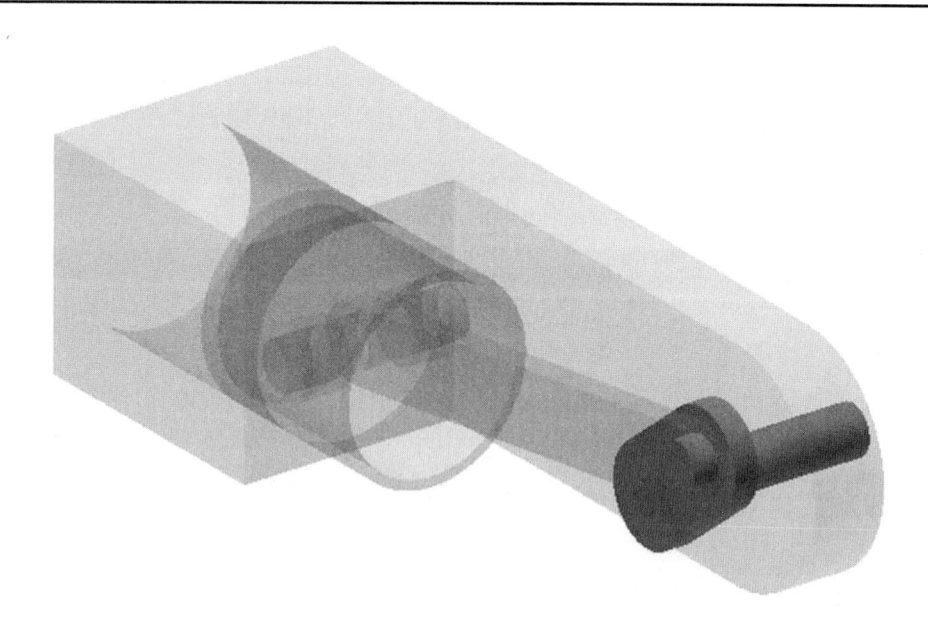

93. Move the cursor to the upper left portion of the screen and left-click on the **Manage** tab. Left-click on the drop-down arrow under **Update**. Left-click on **Update** as shown in Figure 7-88.

FIGURE 7-88

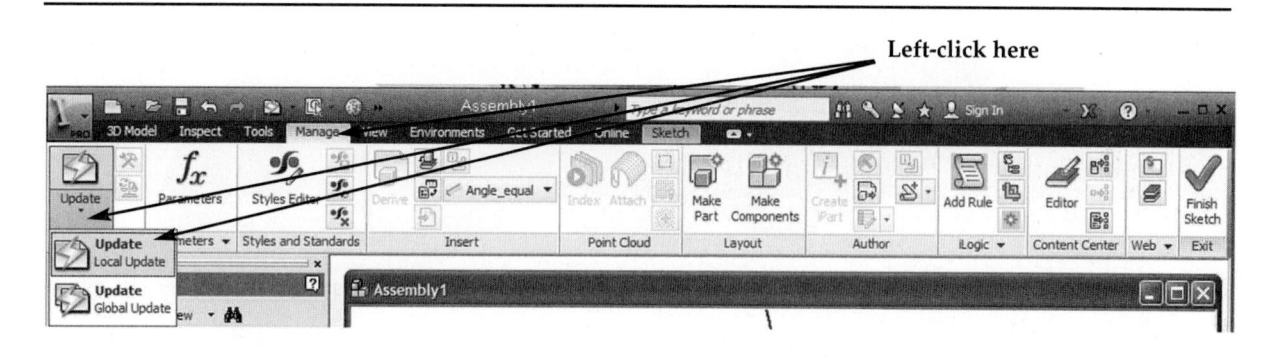

94. Move the cursor to the upper right portion of the screen and left-click on the drop-down arrow under **Return**. A drop-down menu will appear. Left-click on **Return** as shown in Figure 7-89.

FIGURE 7-89

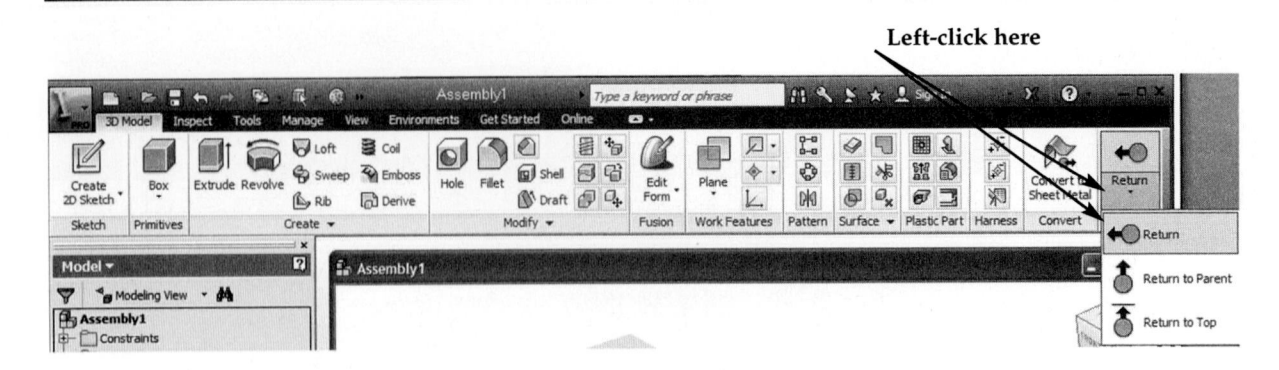

95. Inventor will return to the Assembly Panel, displaying the changes made to the crankshaft. Your screen should look similar to Figure 7-90.

FIGURE 7-90

ASSIGN COLORS TO DIFFERENT PARTS IN THE ASSEMBLY PANEL

96.   Move the cursor to the upper middle portion of the screen and left-click on the double arrows. Left-click on the drop-down arrow at the right. A drop-down menu will appear. Left-click on **Materials** as shown in Figure 7-91.

FIGURE 7-91

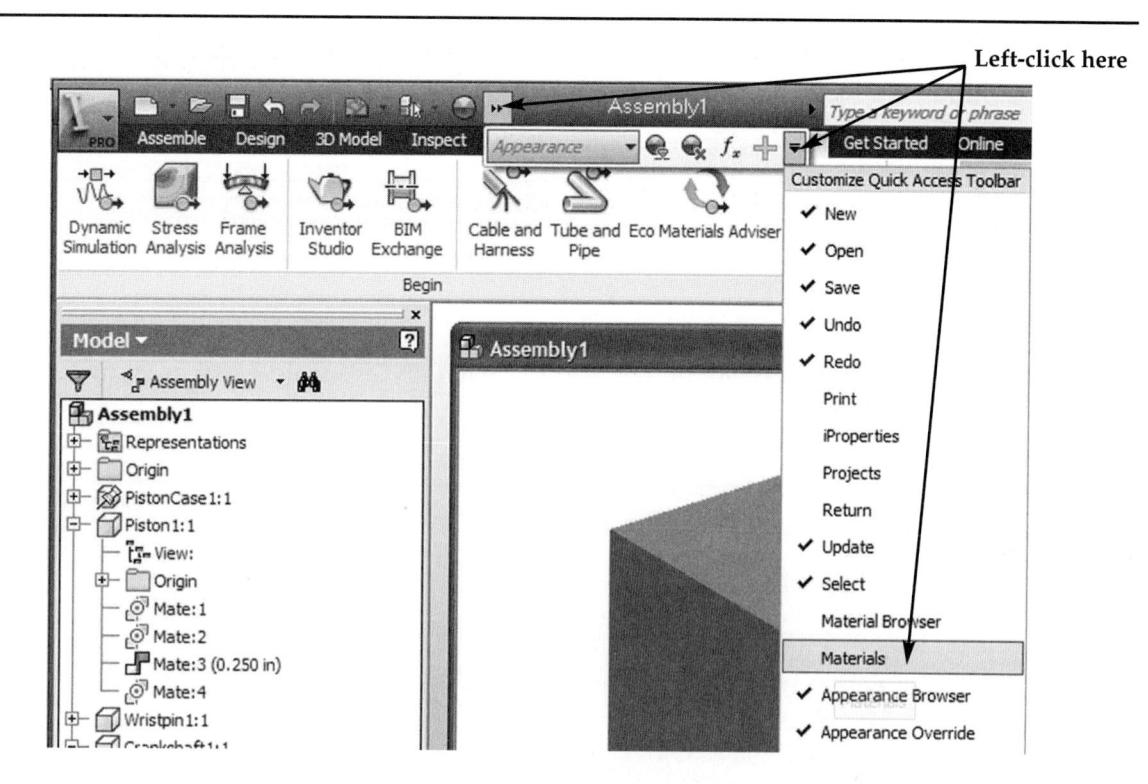

97.   Move the cursor to the left portion of the Piston Case and left-click once as shown in Figure 7-92.

FIGURE 7-92

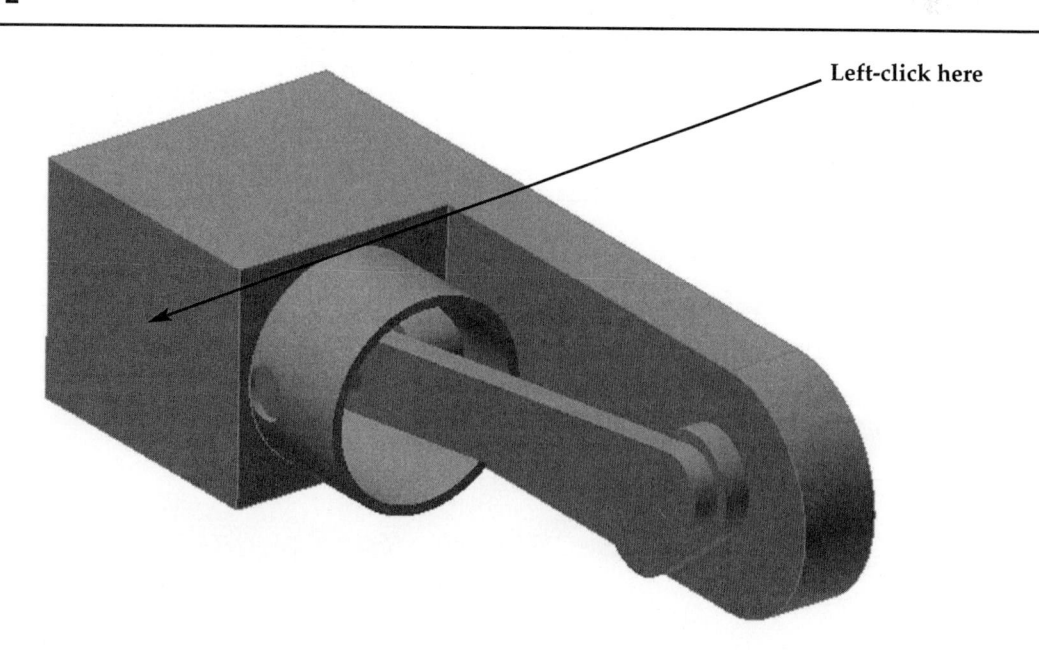

98. Move the cursor to the upper right portion of the screen and left-click on the drop-down arrow next to the text, "PAEK Plastic." A drop-down menu will appear. Scroll down to **Polycarbonate, Clear** and left-click once as shown in Figure 7-93.

FIGURE 7-93

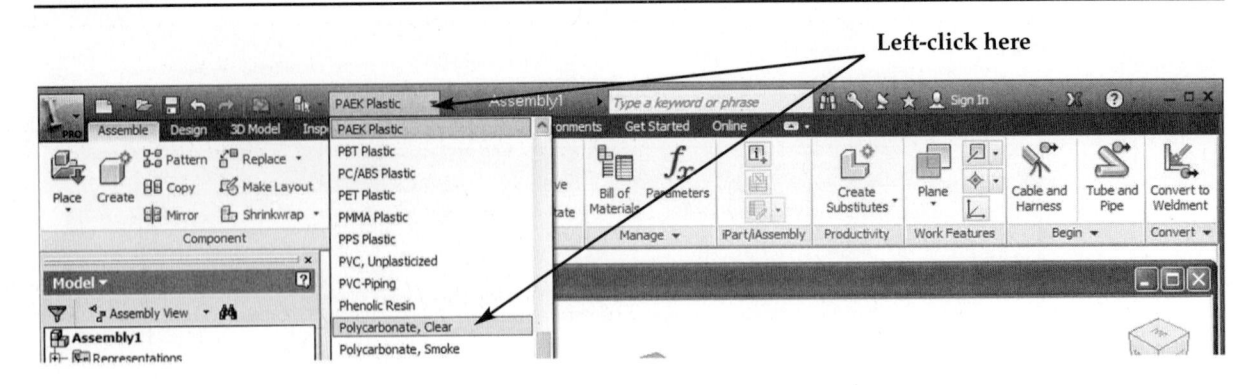

99. Inventor will change the color of the piston case to Polycarbonate, Clear as shown in Figure 7-94.

FIGURE 7-94

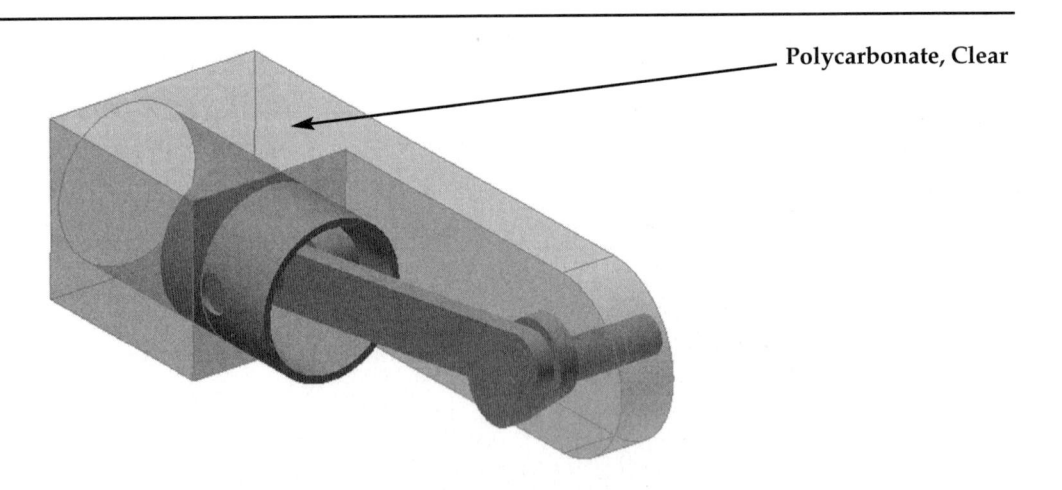

100. Move the cursor to any portion of the piston, causing the edges to turn red and left-click as shown in Figure 7-95.

FIGURE 7-95

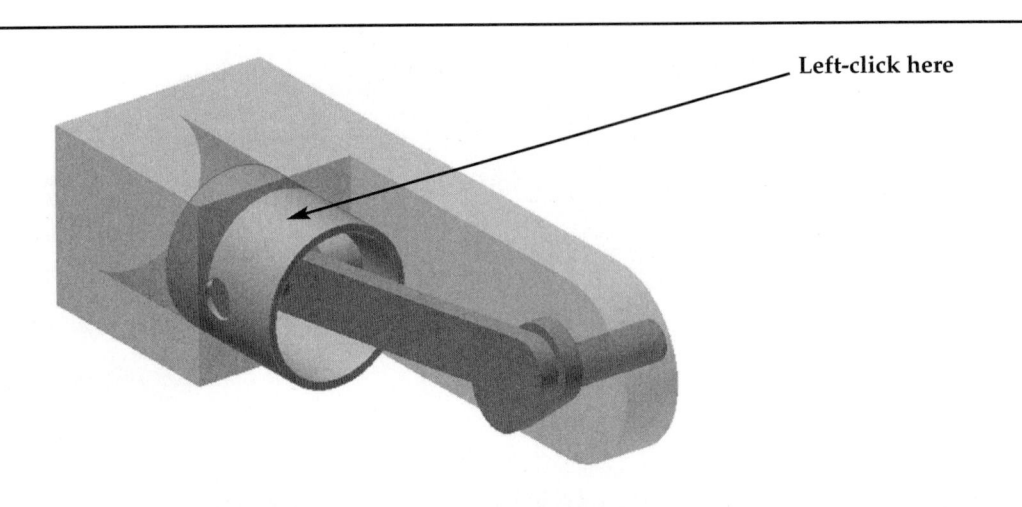

101. Move the cursor to the upper right portion of the screen and left-click on the drop-down arrow next to the text, "Polycarbonate." A drop-down menu will appear. Scroll down to **Stainless Steel** and left-click as shown in Figure 7-96.

FIGURE 7-96

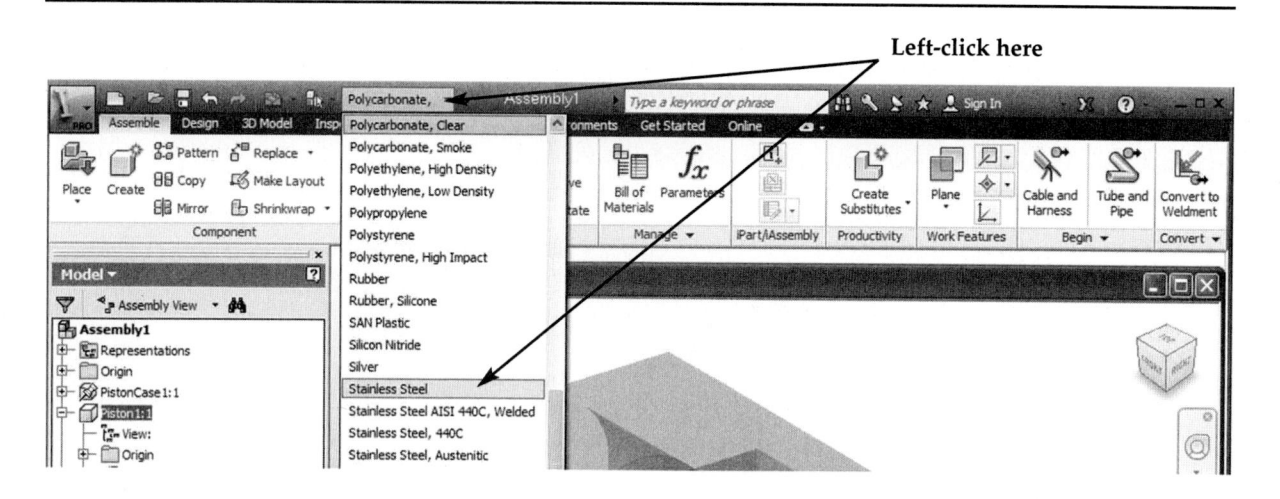

102. Inventor will change the color of the piston to clear polished blue as shown in Figure 7-97.

FIGURE 7-97

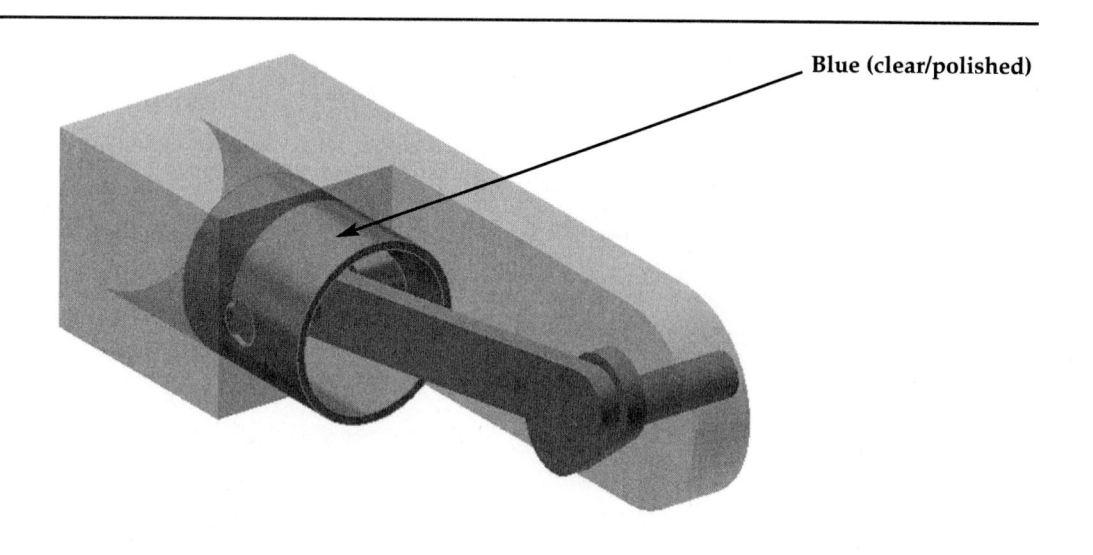

103. Using the same procedure, change the connecting rod color to **SAN Plastic** as shown in Figure 7-98.

FIGURE 7-98

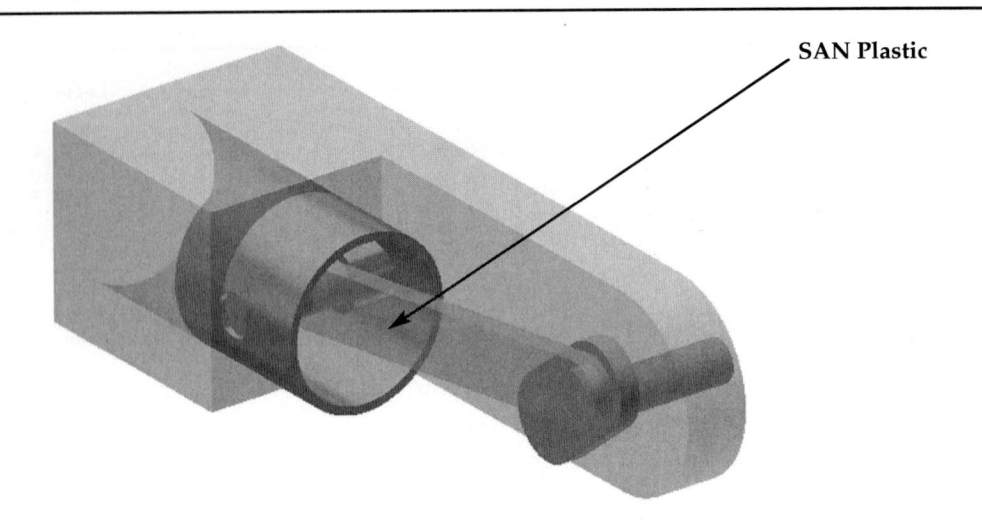

SAN Plastic

104. Move the cursor to the face of the connecting rod, causing the edges to turn red. After the edges turn red, left-click (holding the left mouse button down) and drag the cursor in a circle, causing the crankshaft to turn. Rotate the crankshaft upward to the position shown in Figure 7-99.

FIGURE 7-99

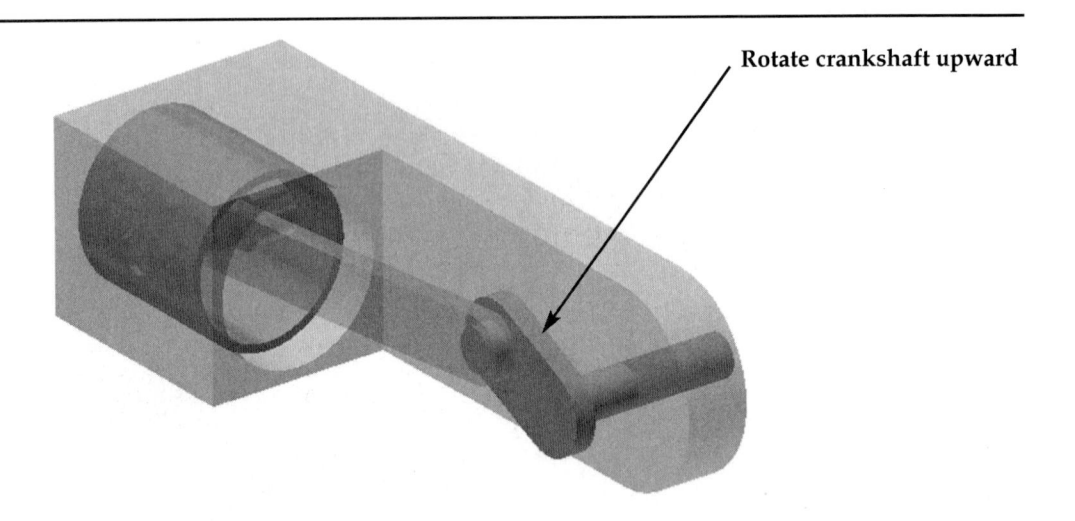

Rotate crankshaft upward

DRIVE CONSTRAINTS TO SIMULATE MOTION

105. Move the cursor to the upper left portion of the screen and left-click on the **Assemble** tab. Left-click on **Constrain**. The Place Constraint dialog box will appear. Left-click on the "Angle Constraint" icon as shown in Figure 7-100.

FIGURE 7-100

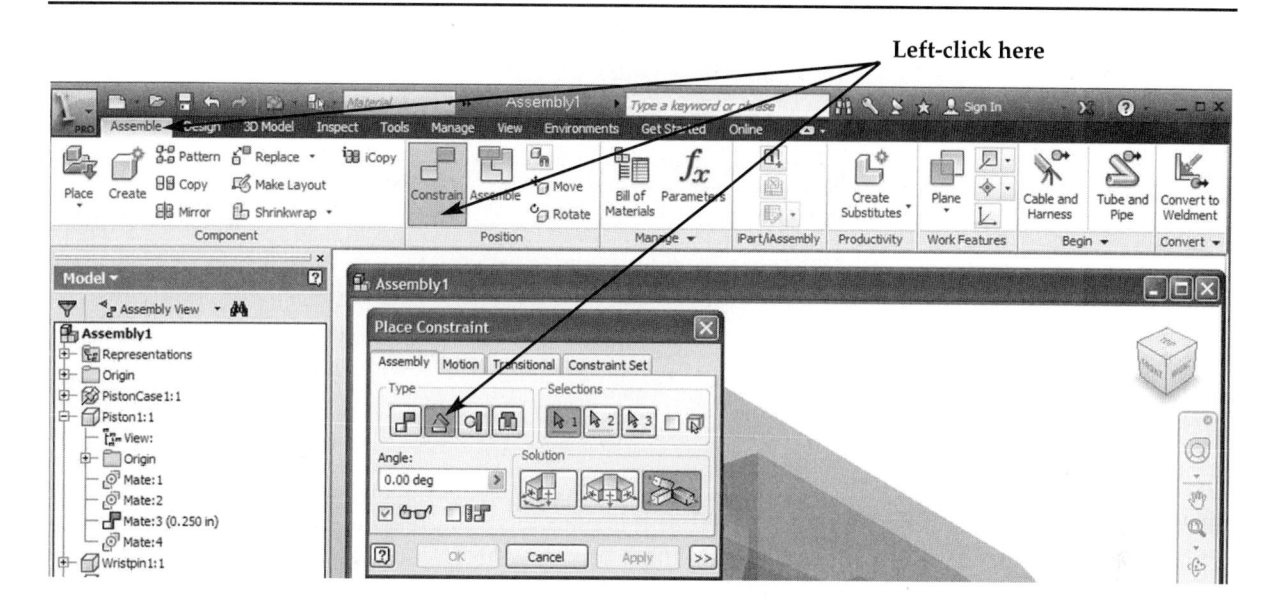

106. Move the cursor to the top portion of the crankshaft, causing a red arrow to appear. Left-click as shown in Figure 7-101. You may have to zoom in to select the surface.

FIGURE 7-101

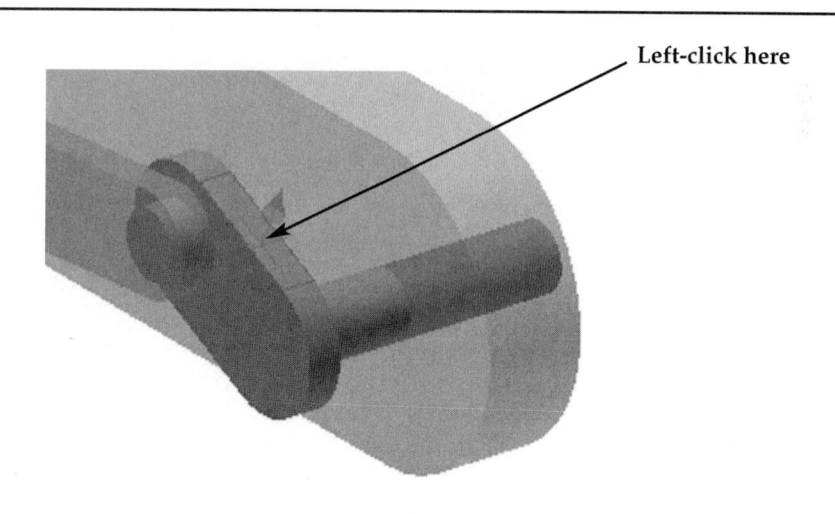

107. Move the cursor to the side of the piston case, causing a red arrow to appear. Left-click once as shown in Figure 7-102.

FIGURE 7-102

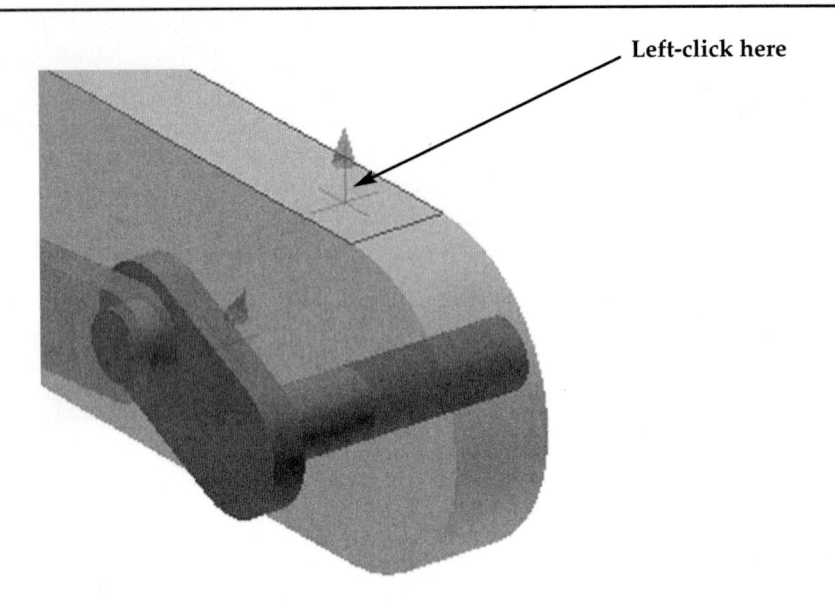

Left-click here

108. Inventor will rotate the crankshaft so that it is parallel (0 degrees) to the side of the piston case. If 20 or 30 degrees were entered in the Angle box, Inventor would rotate the crankshaft to a position 20 or 30 degrees from the side of the piston case. When 0 is entered into the Angle box, Inventor will rotate the crankshaft parallel to the piston case side as shown in Figure 7-103.

FIGURE 7-103

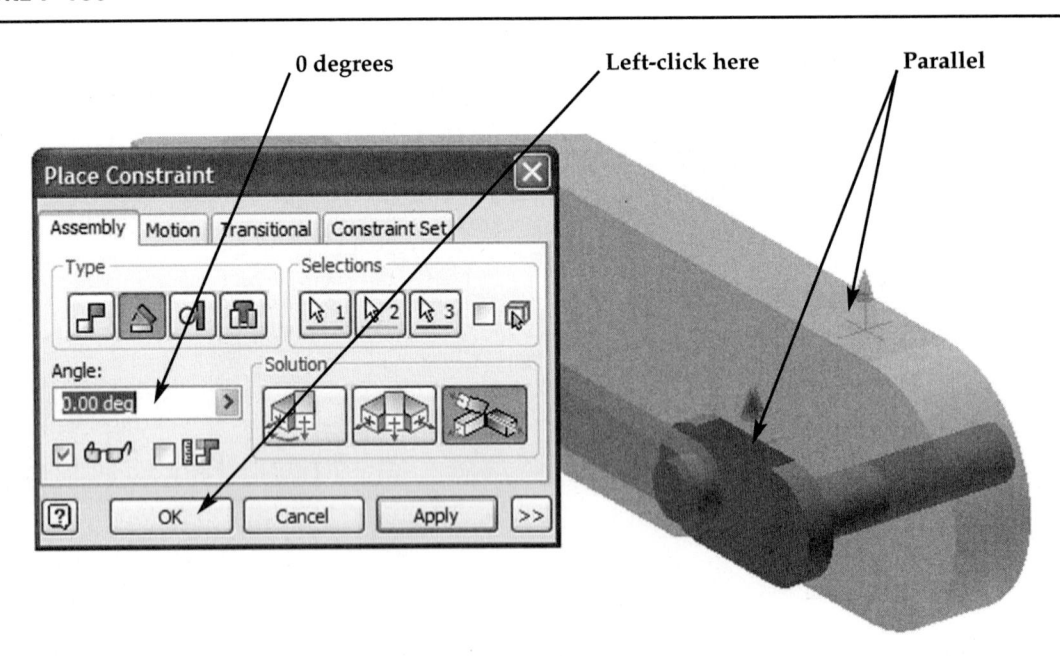

0 degrees   Left-click here   Parallel

109. Left-click on **OK** as shown in Figure 7-103.

110. Move the cursor to the lower left portion of the screen to the part tree. Scroll down to **Angle:1** and right-click once. A pop-up menu will appear. Left-click on **Drive Constraint** as shown in Figure 7-104.

FIGURE 7-104

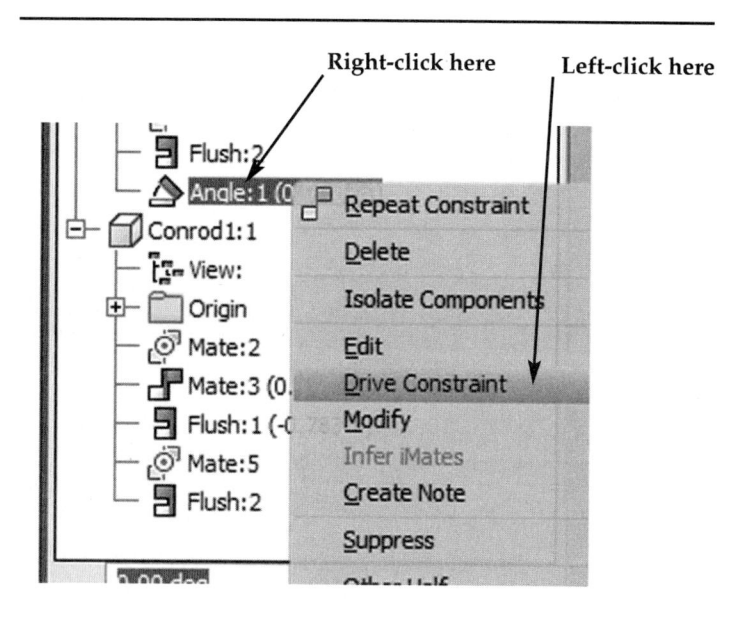

111. The Drive Constraint dialog box will appear. Enter **0** degrees under "Start." Enter **360000** degrees under "End." Left-click on the double arrows at the far right lower corner of the dialog box as shown in Figure 7-105.

FIGURE 7-105

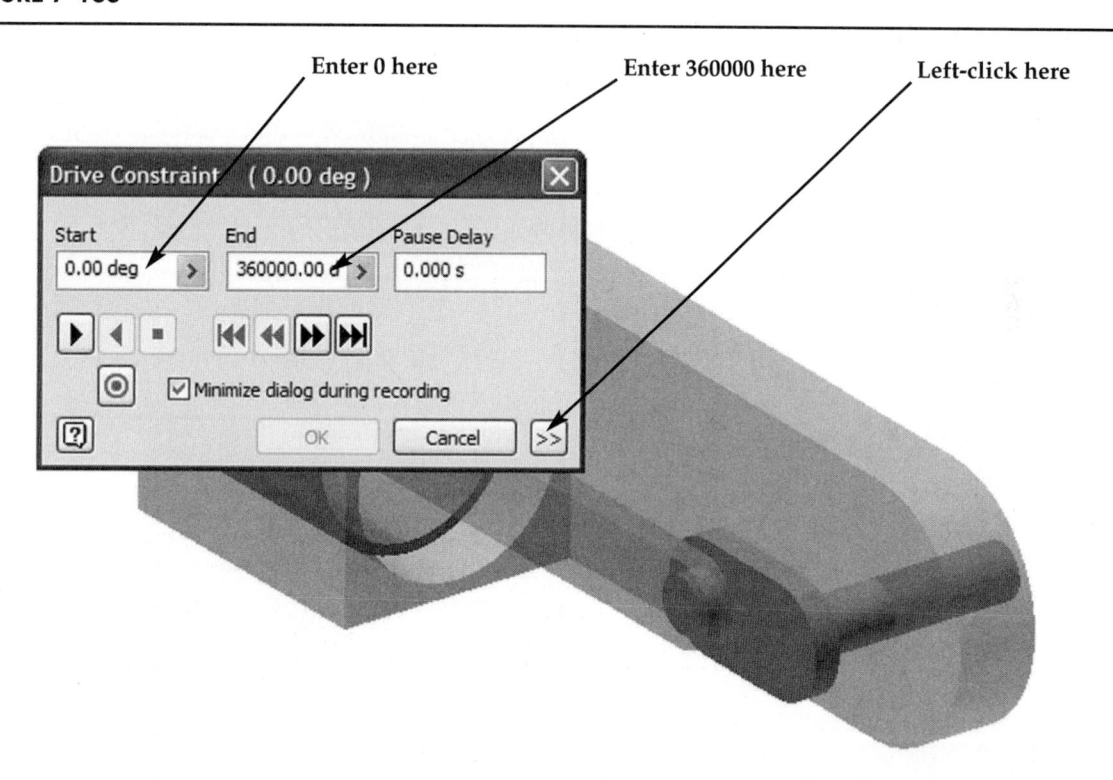

112. The Drive Constraint dialog box will expand, providing more options. Enter **10** for number of degrees as shown in Figure 7-106.

FIGURE 7-106

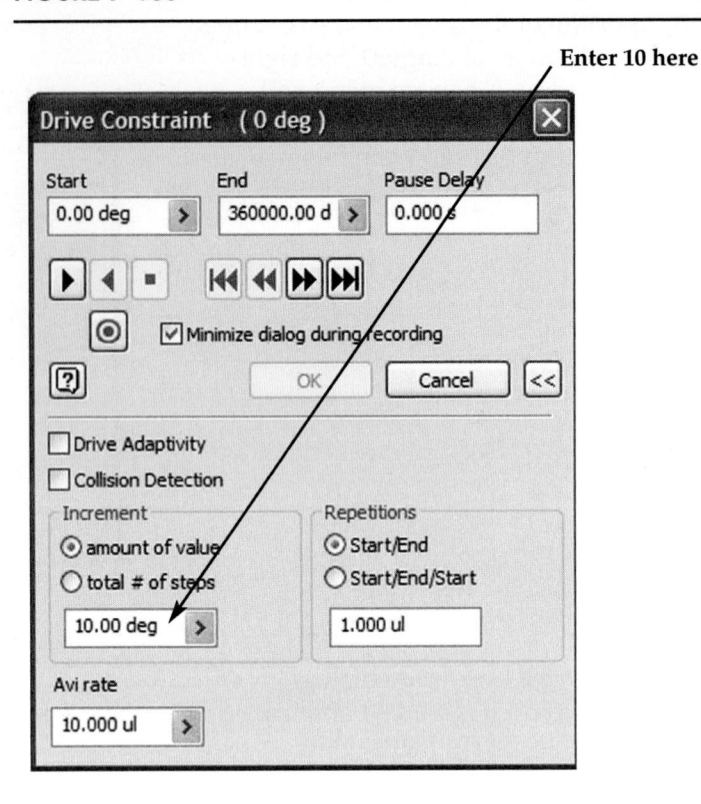

113. Use the "Zoom" option to zoom out. Use the "Pan" option to move the assembly off to the side. Left-click on the "Play" icon as shown in Figure 7-107.

FIGURE 7-107

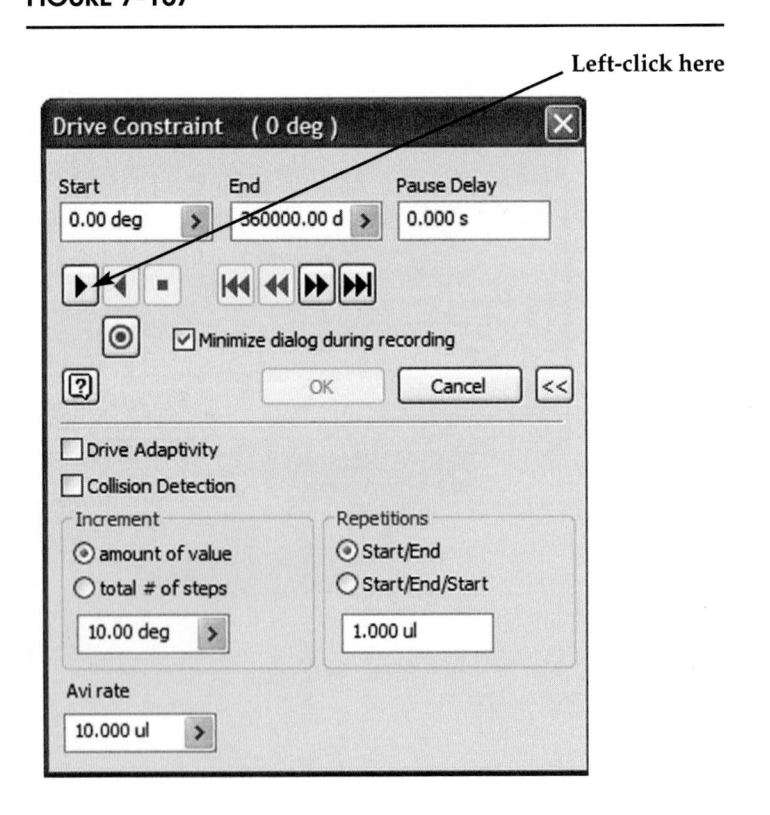

114. Inventor will animate the part, causing the crankshaft to rotate.

115. Left-click on the "Stop" icon. The animation will stop. Left-click on the "Minimize" icon. The Drive Constraint dialog box will get smaller. Left-click on the "Rewind" icon. This will rewind the animation back to 0 degrees as shown in Figure 7-108.

FIGURE 7-108

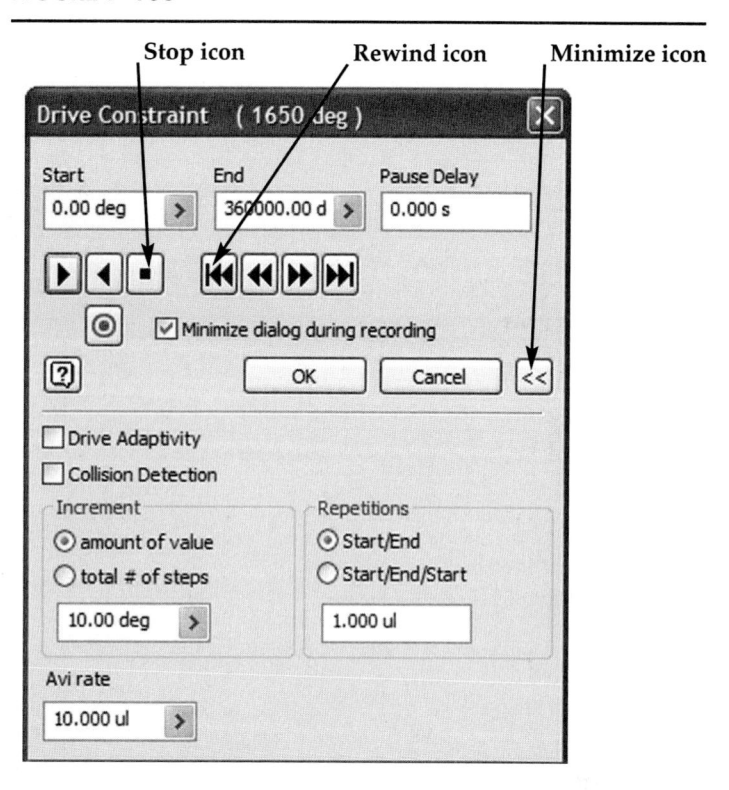

CREATE AN .AVI OR .WMV FILE WHILE IN THE ASSEMBLY PANEL

116. Left-click on the "Play" icon and immediately left-click on the "Record" icon as shown in Figure 7-109.

FIGURE 7-109

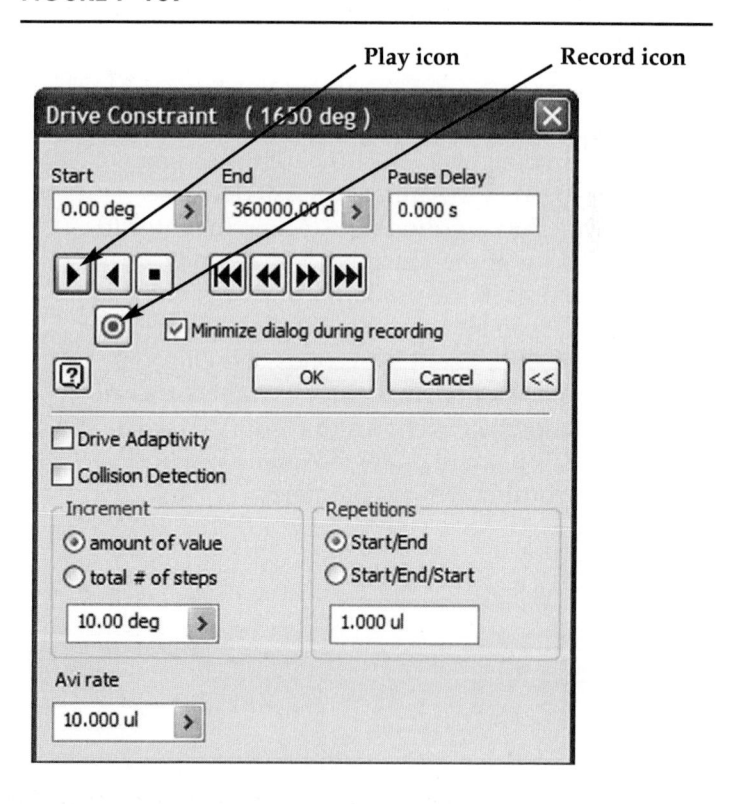

117. The Save As dialog box will appear. Save the file where it can be easily retrieved later and left-click on **OK**.

118.   The ASF Export Properties dialog box will appear. Left-click on **OK** as shown in Figure 7-110.

FIGURE 7-110

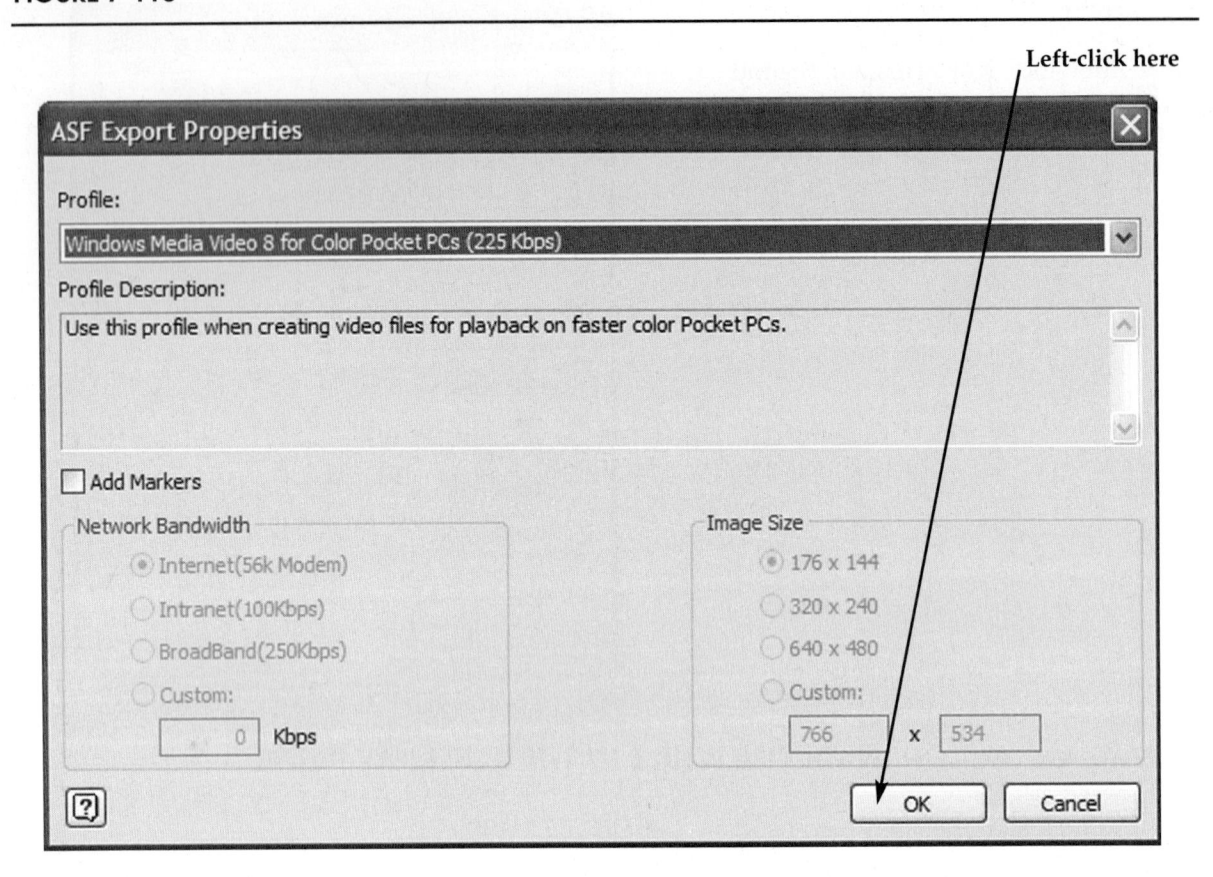

119.   While Inventor is recording the simulation, the Drive Constraint dialog box will minimize in the lower left corner of the screen. The speed of the animation will decrease during the recording time. Allow Inventor to record for approximately 15-30 seconds. Inventor is in the process of creating a .wmv file that can be viewed in Windows Media Player. After about 30 seconds, left-click on the Close symbol in the upper right corner of the dialog box as shown in Figure 7-111. The Drive Constraint dialog box will close and the recording will be complete.

FIGURE 7-111

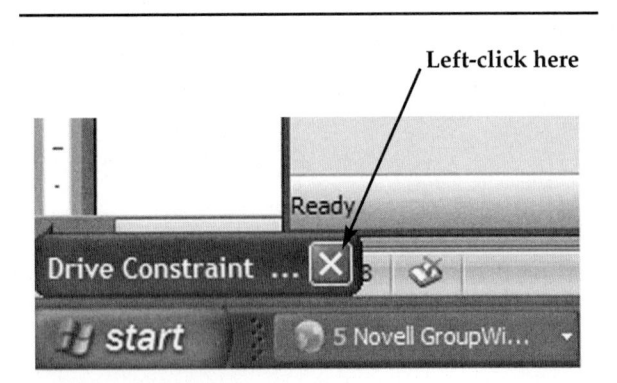

120. Go to the location where the file was saved and double-click on it.

121. Windows Media Player or Real Player will play the file. The file can also be opened in either Windows Media Player or Real Player.

122. Save the Inventor file (.iam) as Chapter 7 Assembly1.iam where it can be easily retrieved at a later time. The Save dialog box will appear. The dialog box will ask if you want to save the assembly itself along with any changes that were made to individual parts that make up the assembly. You can elect to save or not save changes made to individual parts. Left-click on **OK** as shown in Figure 7-112.

FIGURE 7-112

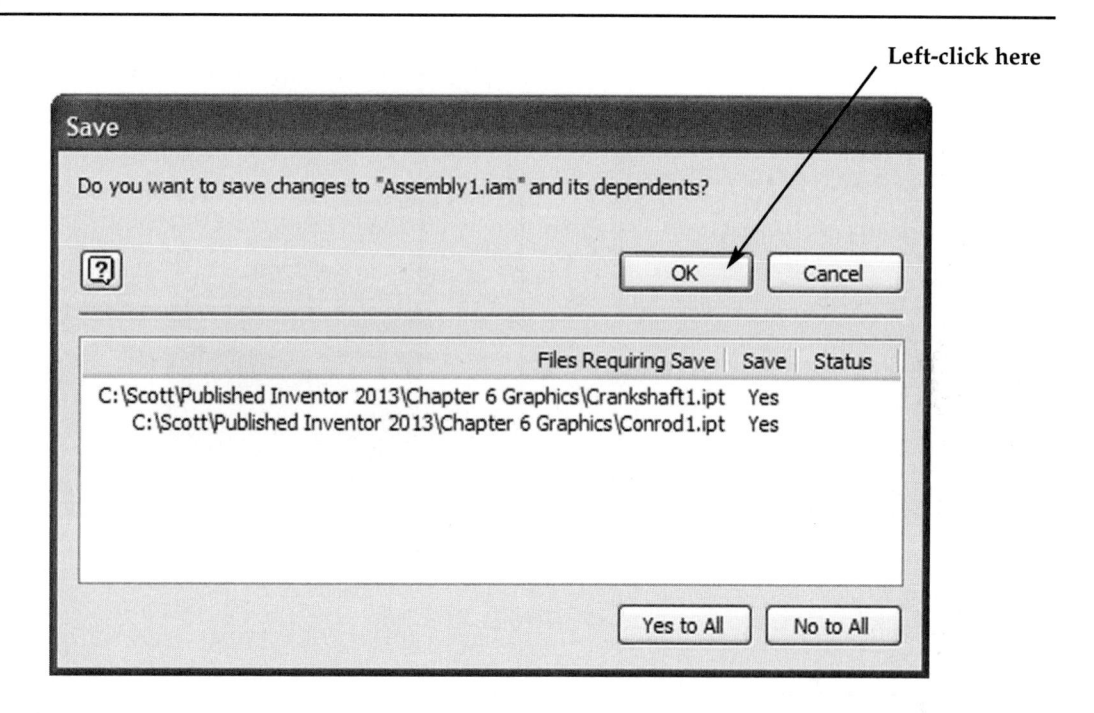

CHAPTER PROBLEMS

Using the dimensions from Chapter 6, modify the PistonCase housing to accept 2 pistons. The length of the shorter crankshaft pin will need to be increased by .25 inches to .050 inches in order to accommodate the additional connection rod as shown.

Start by opening the original Pistoncase1 part file and editing the housing. Then open the original Crankshaft1 part file and increase the extrusion distance of the shorter crankshaft pin to 0.50 inches.

PROBLEM 7-1

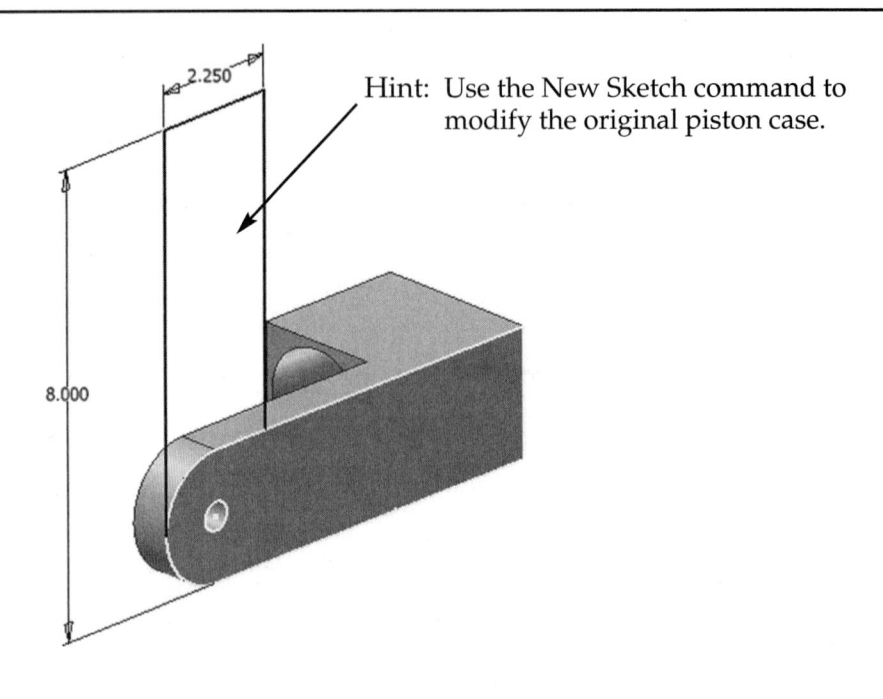

Hint: Use the New Sketch command to modify the original piston case.

PROBLEM 7-2

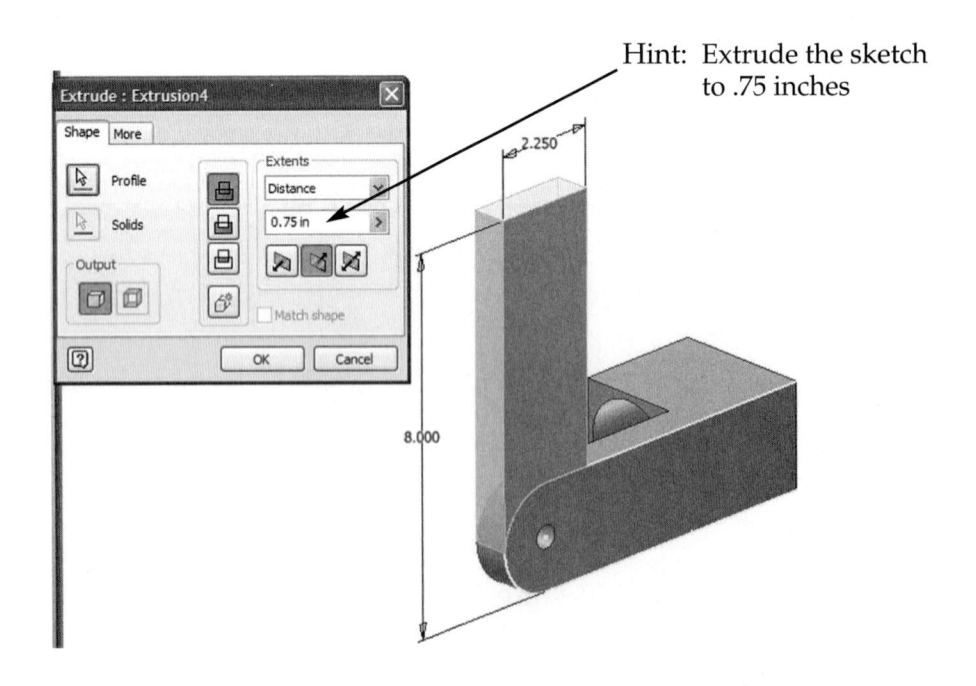

Hint: Extrude the sketch to .75 inches

PROBLEM 7-3

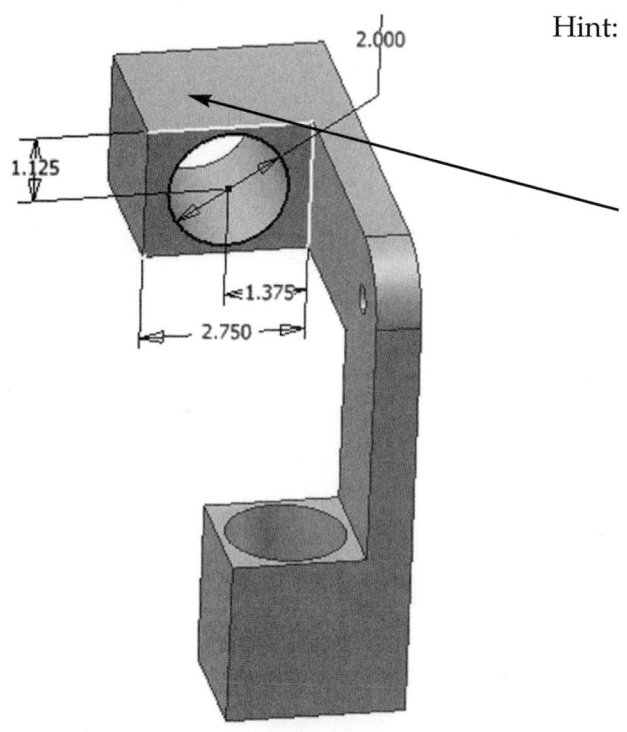

2.000

1.125

1.375

2.750

Hint: Use the Rectangle command to create a new sketch and extrude it to 2.750 inches. Then complete the sketch shown below.

Extrude the rectangle sketch to 2.75 inches

PROBLEM 7-4

Hint: Increase the Extrusion distance on one crankshaft pin from .25 to 2.00 inches (if this was not already completed from Chapter 7). Increase the extrusion distance from .25 inches to .50 inches on the other pin as shown.

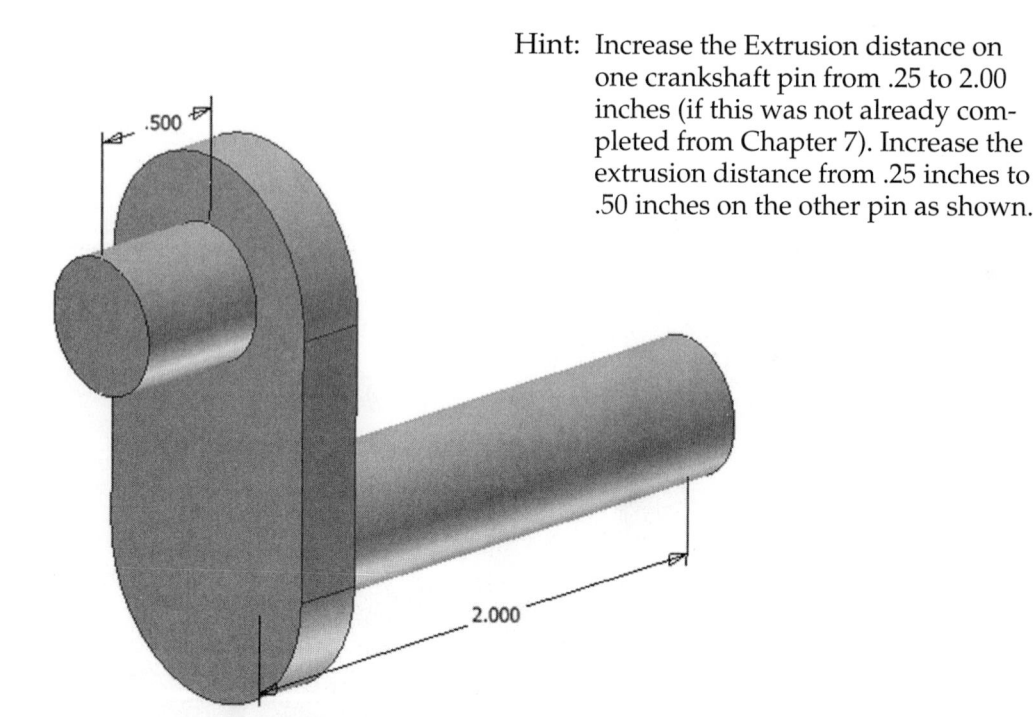

.500

2.000

PROBLEM 7-5

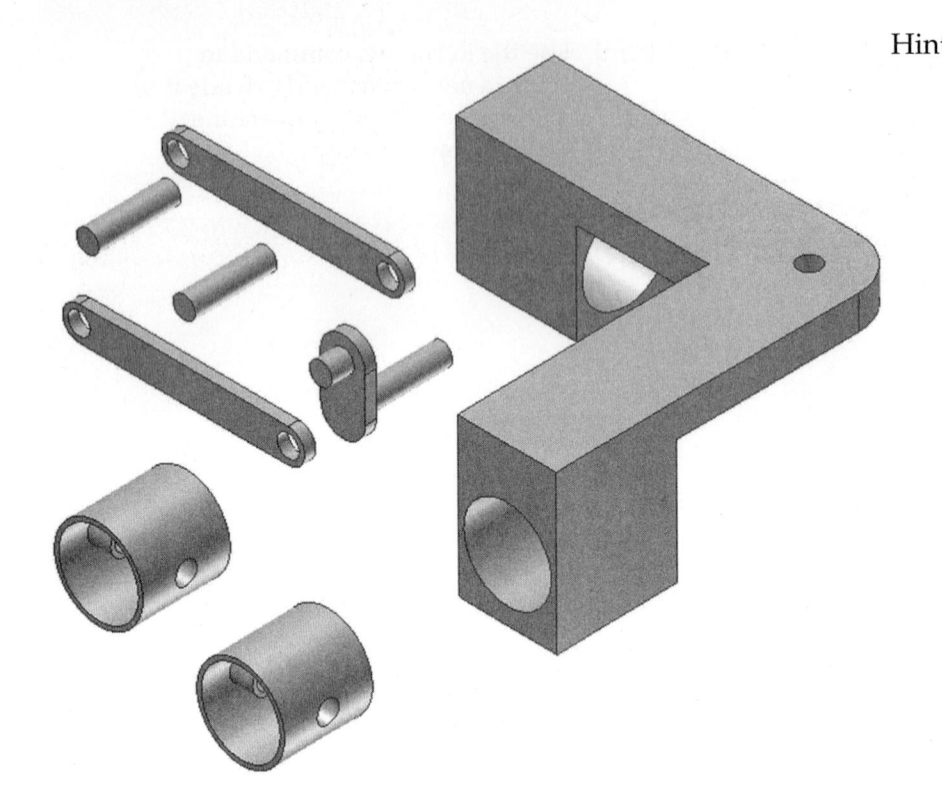

Hint: Using the Place command, import the parts shown below into the Assembly Panel. You will need 2 pistons, 2 connecting rods and 2 wrist pins, 1 piston case and 1 crankshaft.

PROBLEM 7-6

Hint: Using the dimensions from Chapter 6, assemble all the parts in the same manner as Chapter 6 into a new assembly. When finished your screen should look similar to what is shown below.

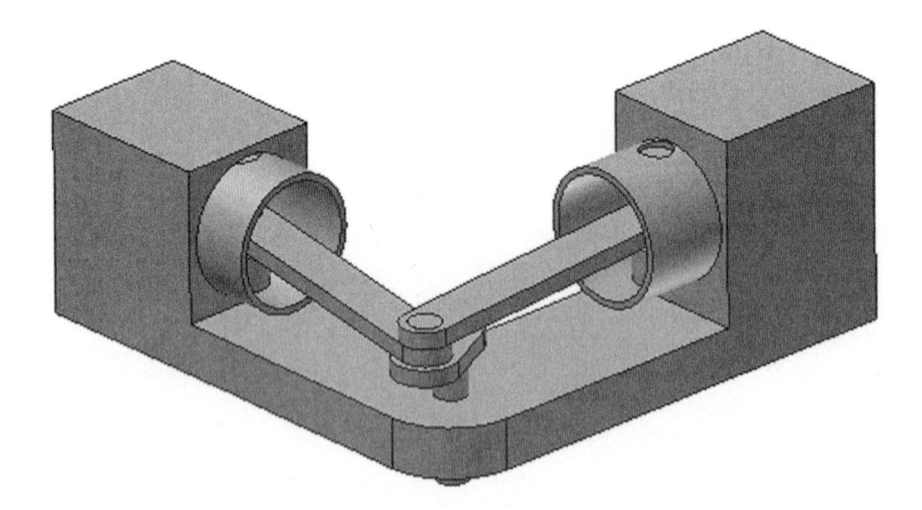

INTRODUCTION TO THE PRESENTATION PANEL

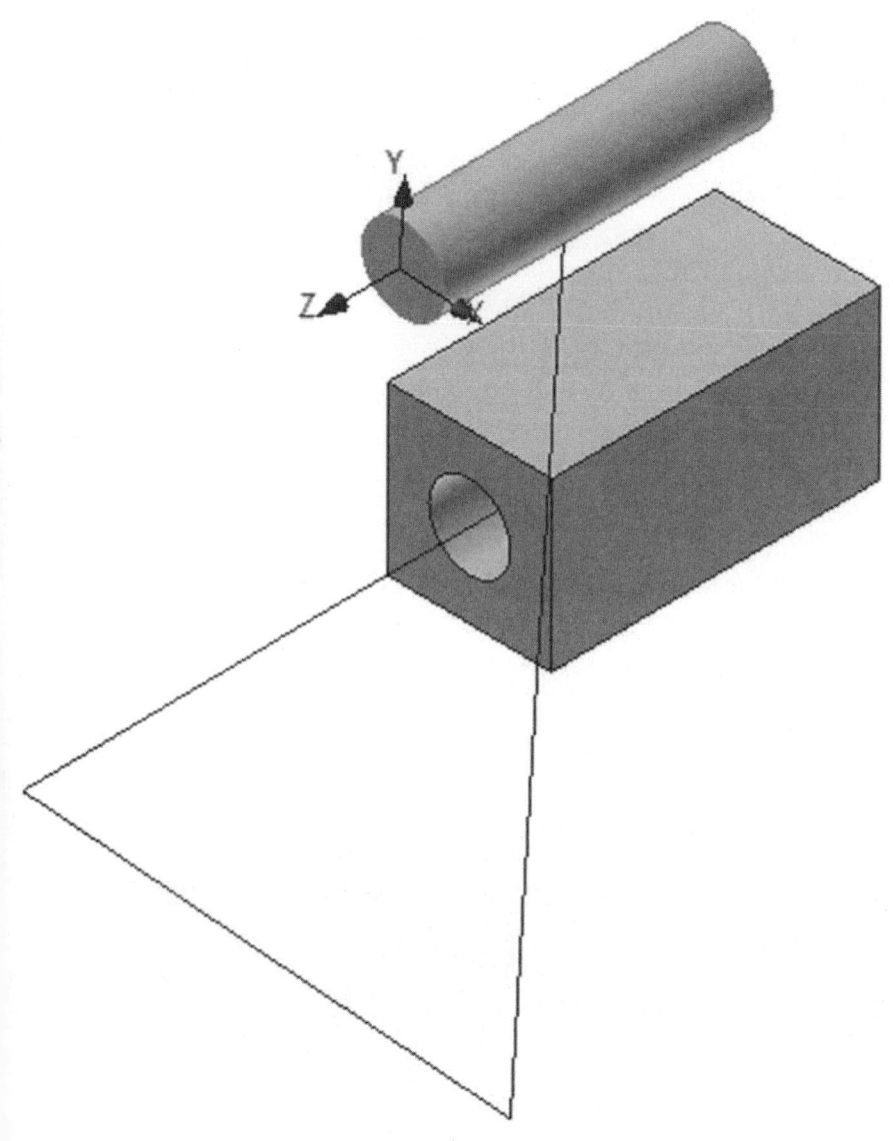

OBJECTIVES

1. Import existing solid models into the Presentation Panel

2. Design parts trails in the Presentation Panel

Chapter 8 includes instruction on how to design the presentation shown above.

1.   Start Inventor 2013 by referring to Chapter 1.

2.   After Inventor is running, begin by creating the parts shown in Figure 8-1.

FIGURE 8-1

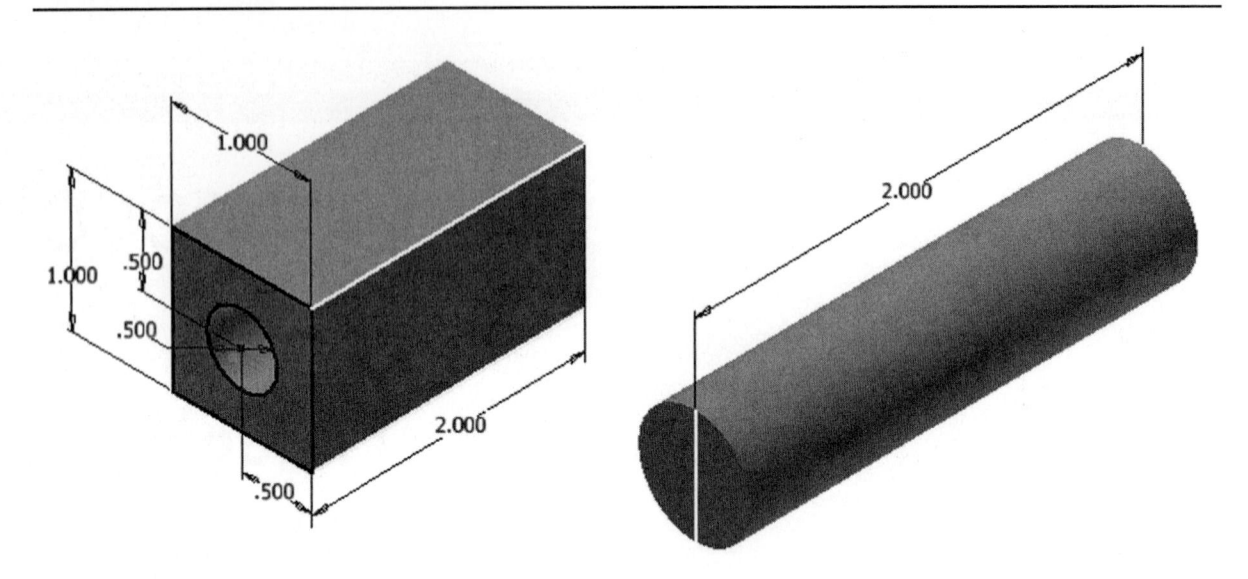

3.   Save the block as Part 1.ipt. and save the pin as Part 2.ipt where they can be easily retrieved at a later time. Close both files.

4.   Move the cursor to the upper left portion of the screen and left-click on the **New** icon as shown in Figure 8-2.

FIGURE 8-2

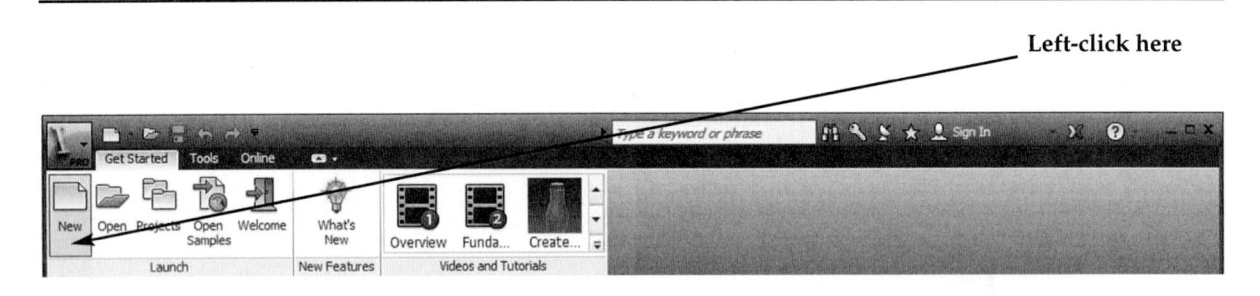

5.   The Create New File dialog box will appear. Select the **English** folder and **Standard (in).iam** and left-click on **Create** as shown in Figure 8-3.

FIGURE 8-3

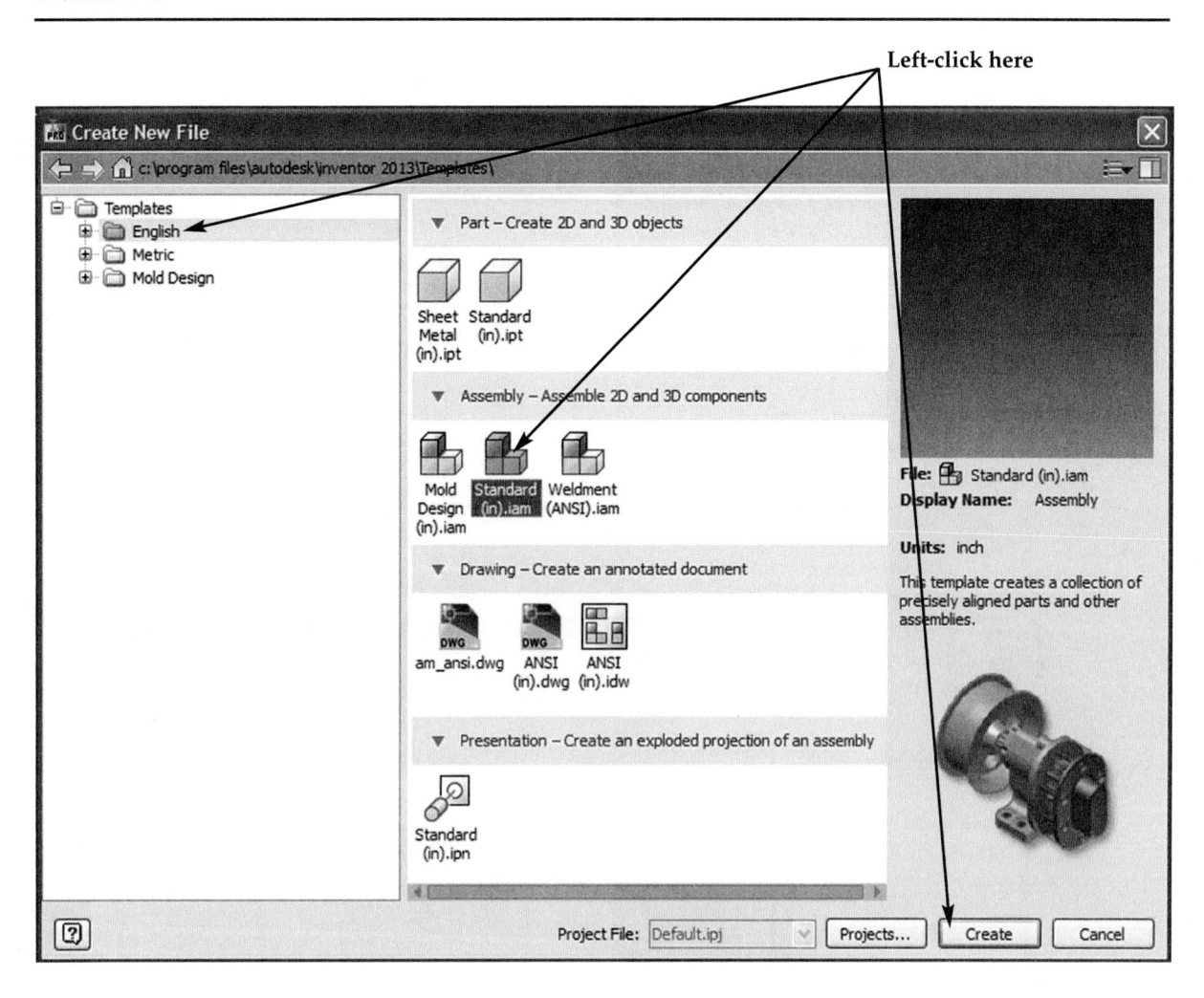

6.   The Assembly Panel will open.

7. Your screen should look similar to Figure 8-4. If the Assembly Panel tools are not visible, left-click on the **Assemble** tab as shown in Figure 8-4.

FIGURE 8-4

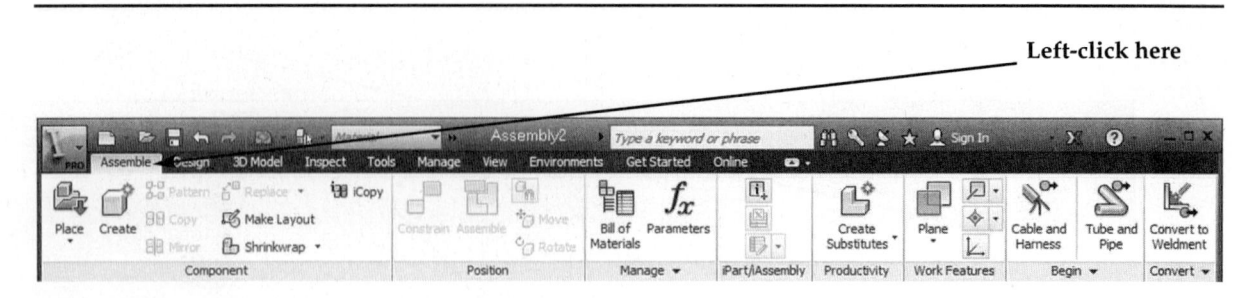

8. Move the cursor to the upper left portion of the screen and left-click on **Place** as shown in Figure 8-5.

FIGURE 8-5

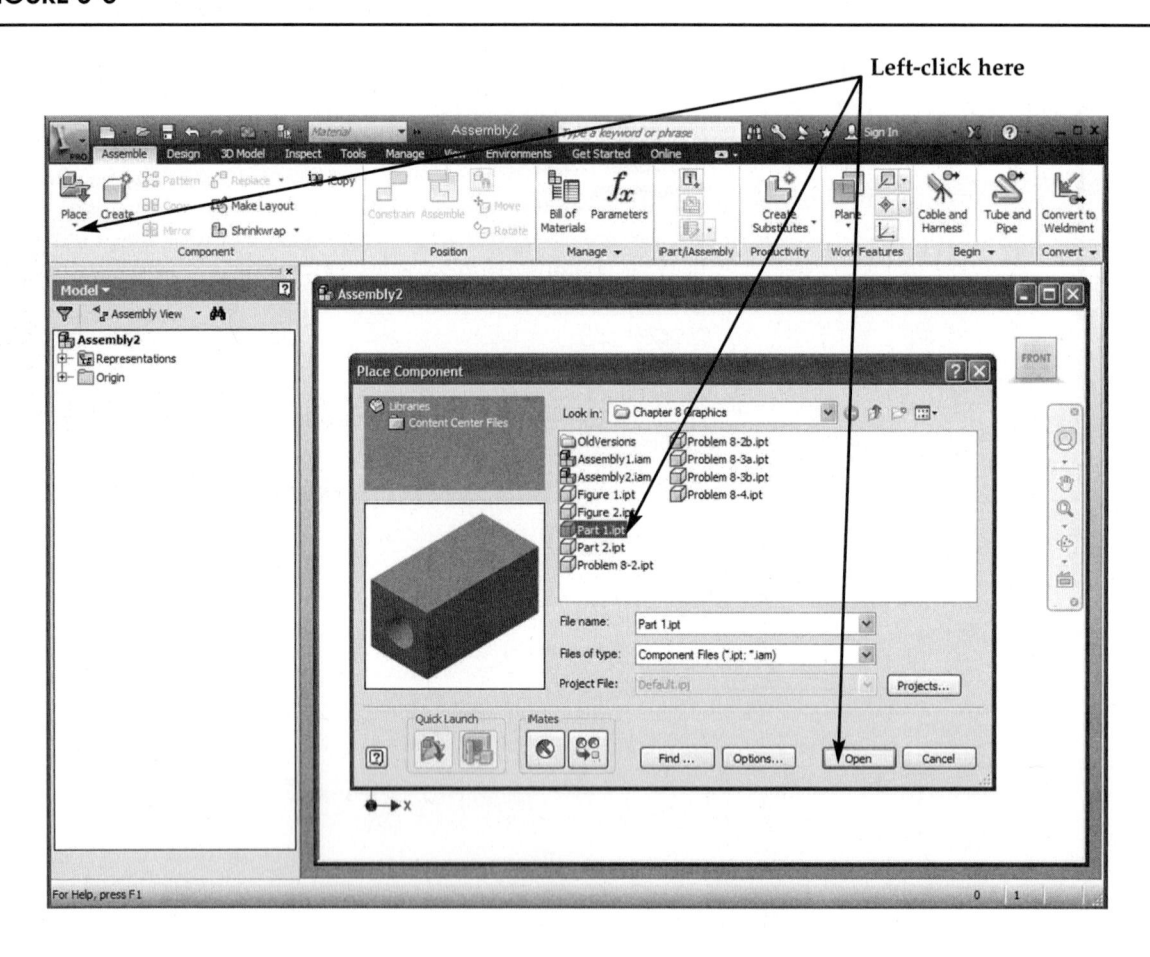

9. The Open dialog box will appear. Left-click on **Part 1.ipt**. Left-click on **Open** as shown in Figure 8-5.

10. The block will appear attached to the cursor. Do <u>not</u> left-click. Press **Esc** on the keyboard. Your screen should look similar to Figure 8-6.

FIGURE 8-6

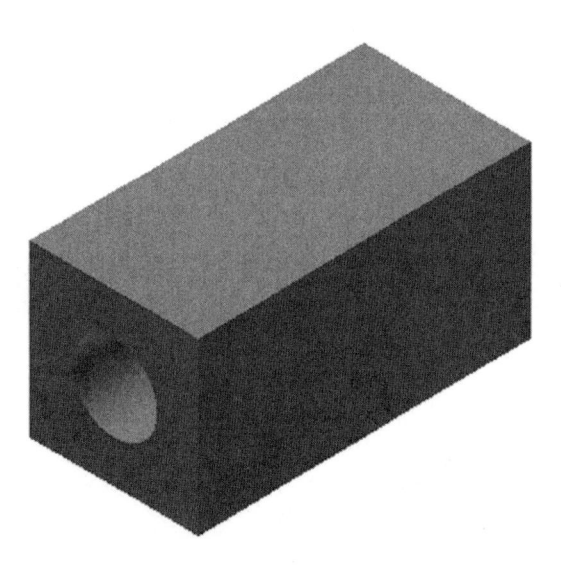

11. Move the cursor to the upper left portion of the screen and left-click on **Place/Place Component** as shown in Figure 8-7.

FIGURE 8-7

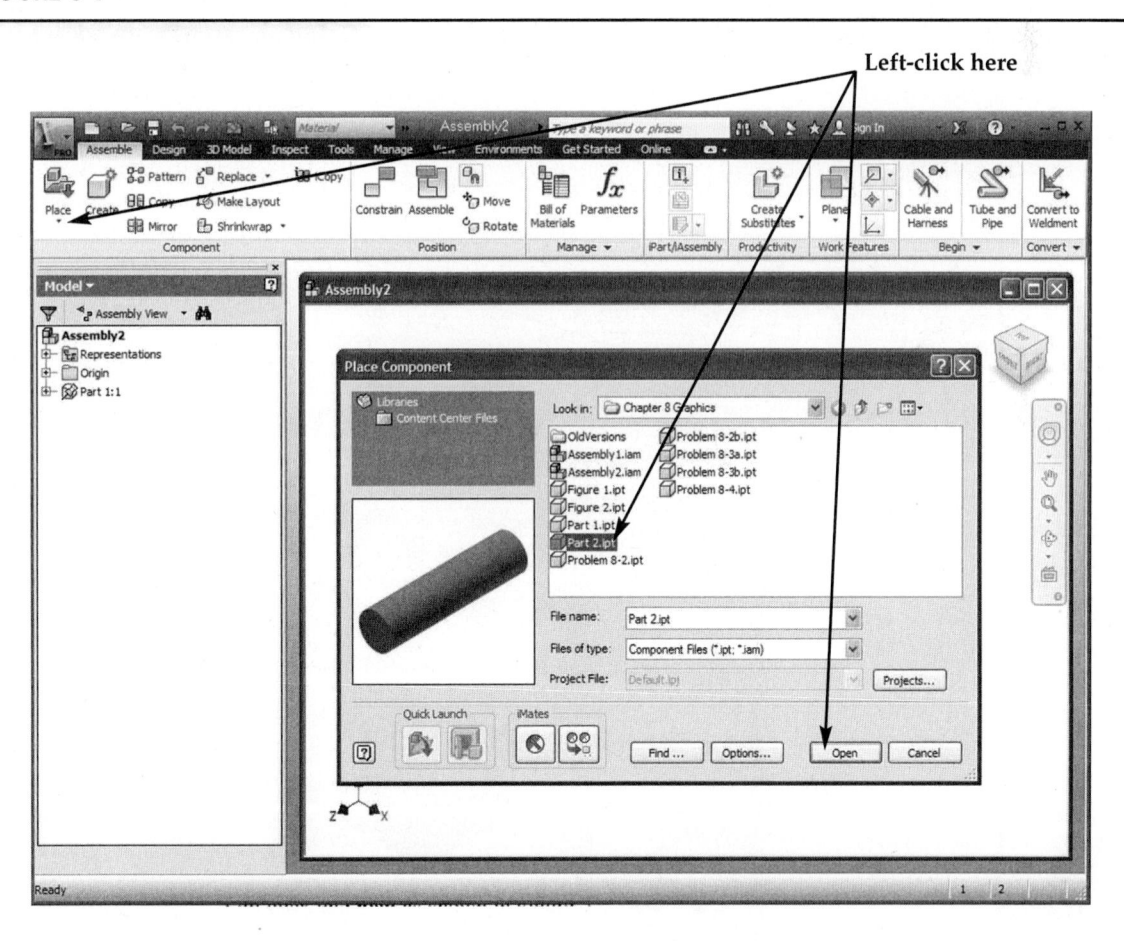

12. The Place Component dialog box will appear. Left-click on **Part 2.ipt**. Left-click on **Open** as shown in Figure 8-7.

13. The pin will appear attached to the cursor. Place the pin near the block and left-click once. Press **Esc** on the keyboard. Your screen should look similar to Figure 8-8.

FIGURE 8-8

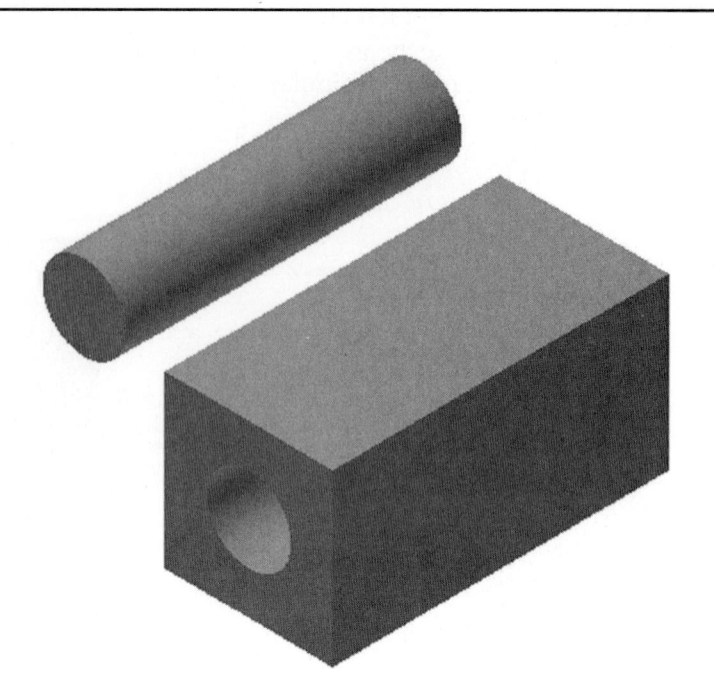

14. Move the cursor to the middle left portion of the screen and left-click on **Constrain** as shown in Figure 8-9.

FIGURE 8-9

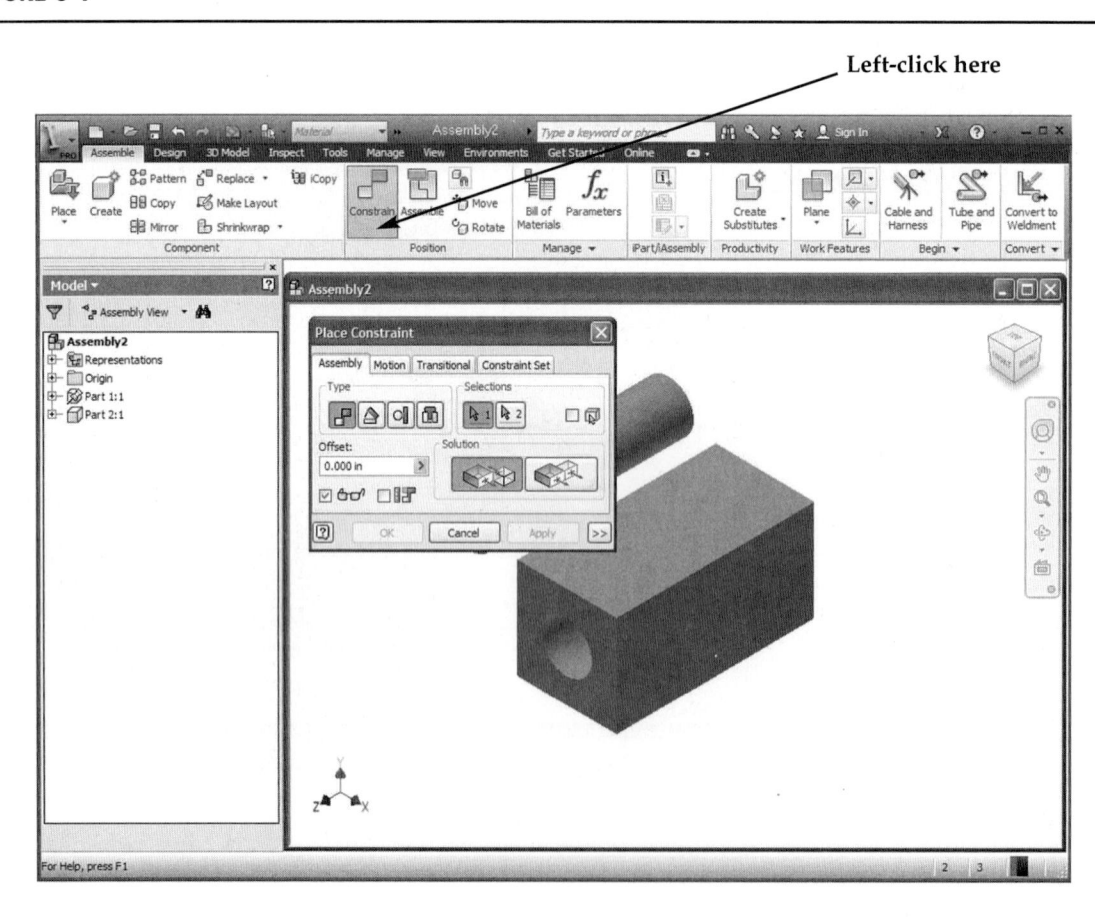

15.   The Place Constraint dialog box will appear as shown in Figure 8-9.

16.   Move the cursor over the hole in the block. A red dashed center line will appear. Left-click once as shown in Figure 8-10.

FIGURE 8-10

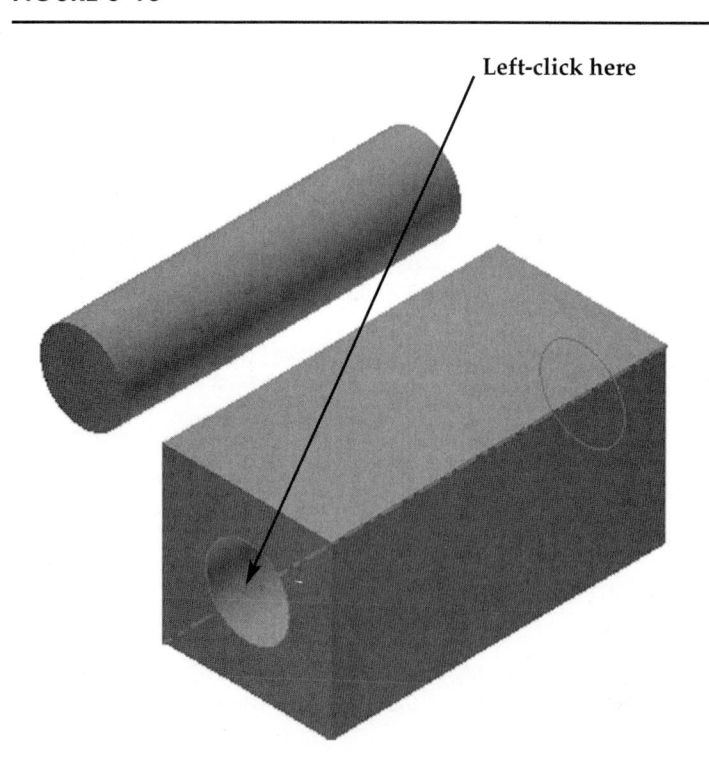

Left-click here

17.   Move the cursor over the pin. A red dashed center line will appear. Left-click once as shown in Figure 8-11.

FIGURE 8-11

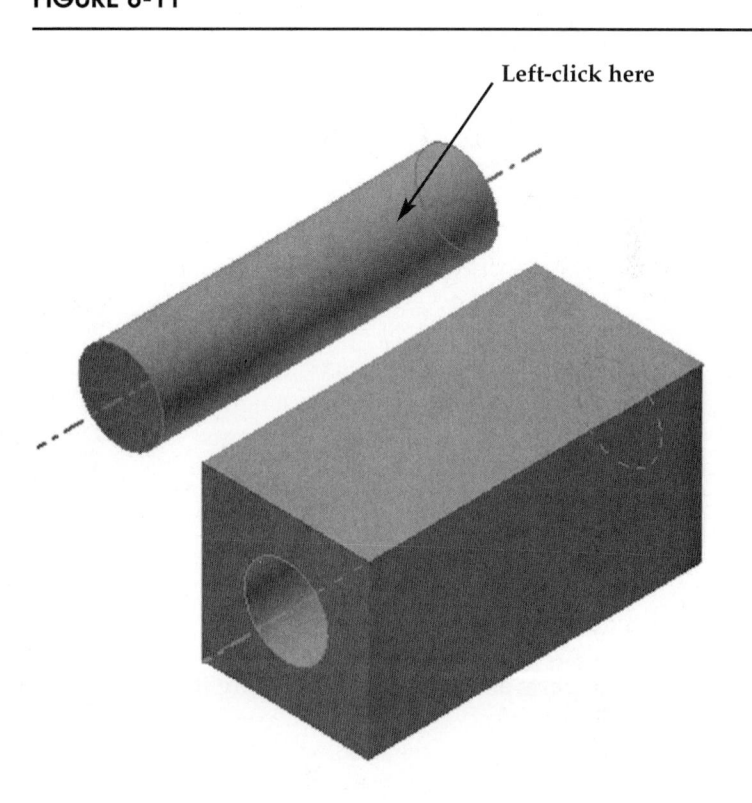

Left-click here

18.   Inventor will insert the pin into the block. Your screen should look similar to Figure 8-12.

FIGURE 8-12

19.   Typically, a surface constraint would be added to prevent the pin from sliding back and forth in the block. However, this Assembly will be used in the Presentation Panel. A surface constraint will not be added because the pin must slide in and out of the block.

20.   Move the cursor to the center of the pin. Left-click (holding down the left mouse button) and slide the pin flush with the outside of the block as shown in Figure 8-13.

FIGURE 8-13

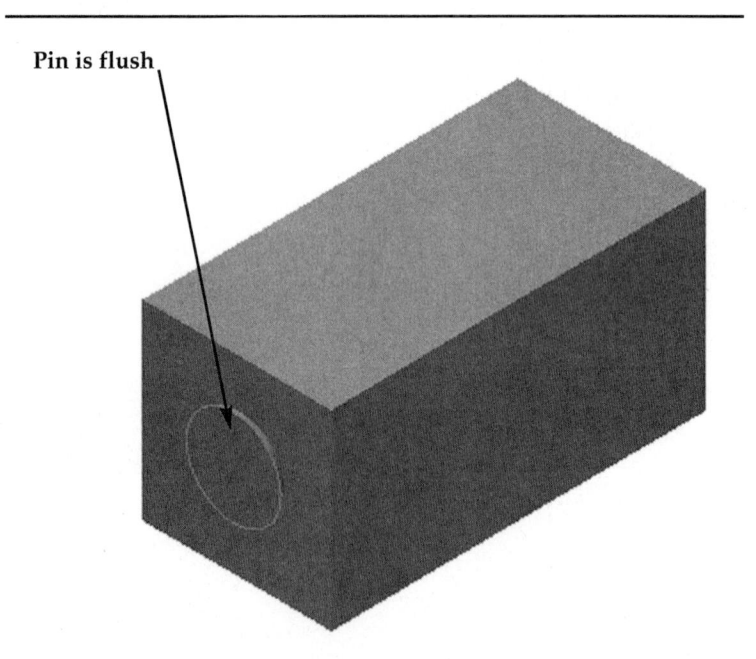

Pin is flush

21.   Save the parts as Assembly1.iam where it can be easily retrieved later. Leave the file open at this time.

IMPORT EXISTING ASSEMBLY MODELS INTO THE PRESENTATION PANEL

22. Move the cursor to the upper left portion of the screen and left-click on the "New" icon as shown in Figure 8-14.

FIGURE 8-14

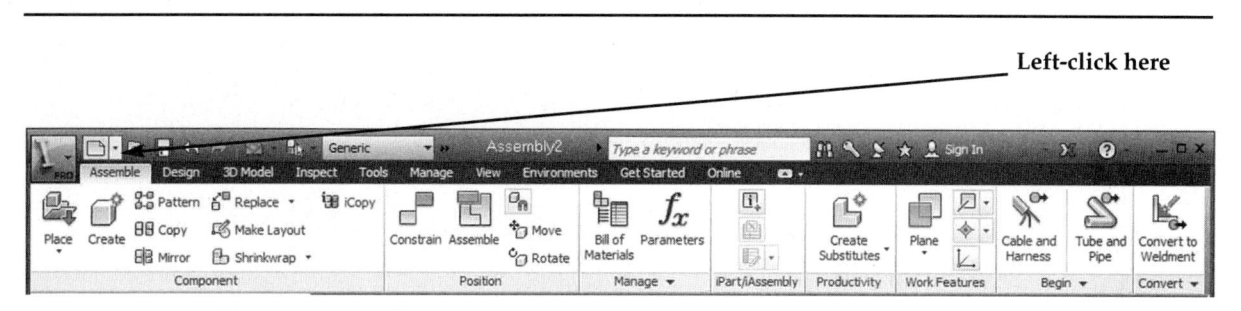

23. The New File dialog box will appear. Select the **English** tab and **Standard (in).ipn**. Left-click on **Create** as shown in Figure 8-15.

FIGURE 8-15

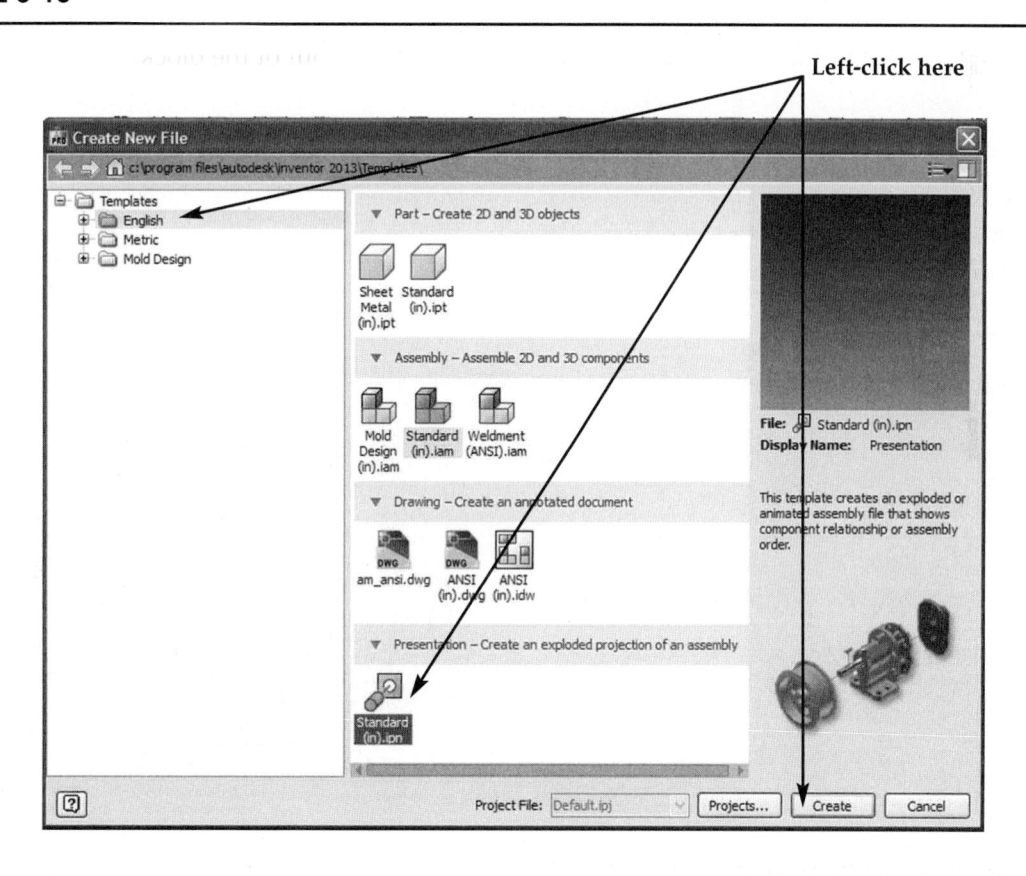

24.   The Presentation Panel will open as shown in Figure 8-16.

FIGURE 8-16

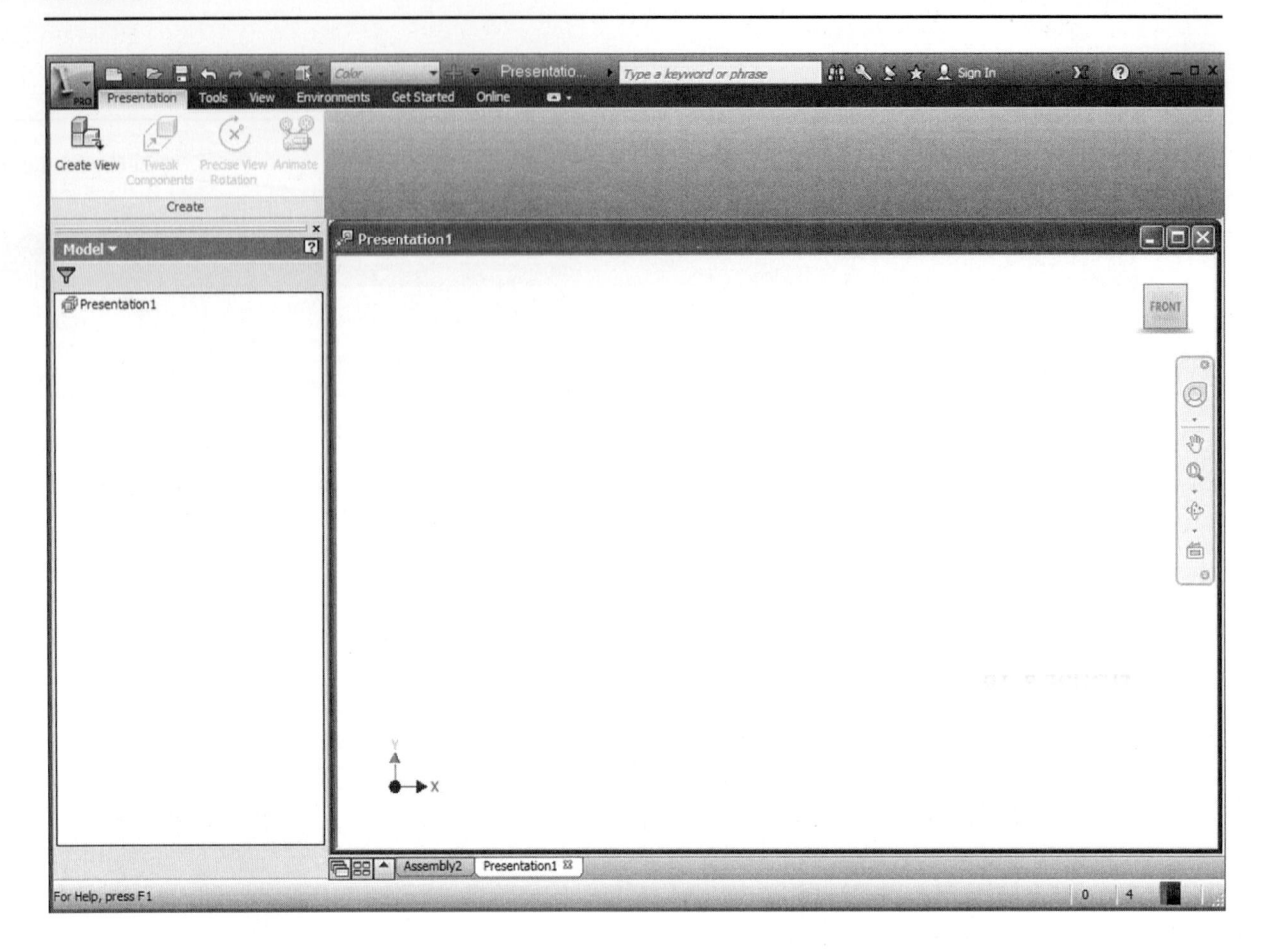

25.   Move the cursor to the upper left portion of the screen and left-click on **Create View** as shown in Figure 8-17.

FIGURE 8-17

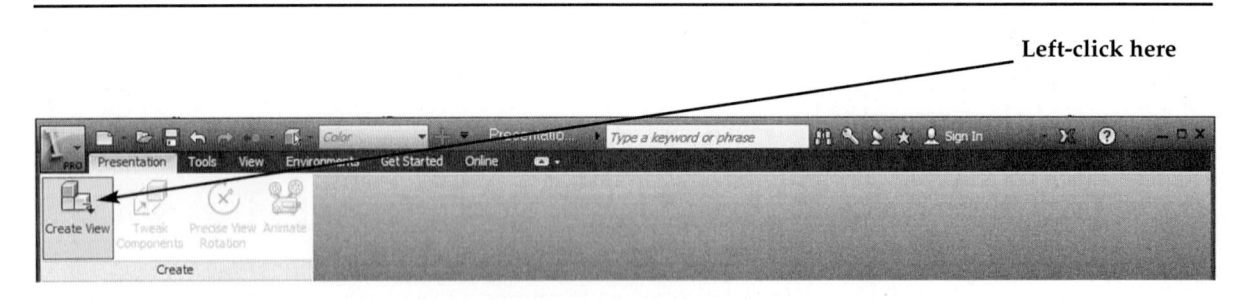

26. The Select Assembly dialog box will appear. Left-click on the "Explore" icon located at the upper right portion of the dialog box if the assembly does not appear. Left-click on **OK** as shown in Figure 8-18. If an error message appears, left-click on **OK**.

FIGURE 8-18

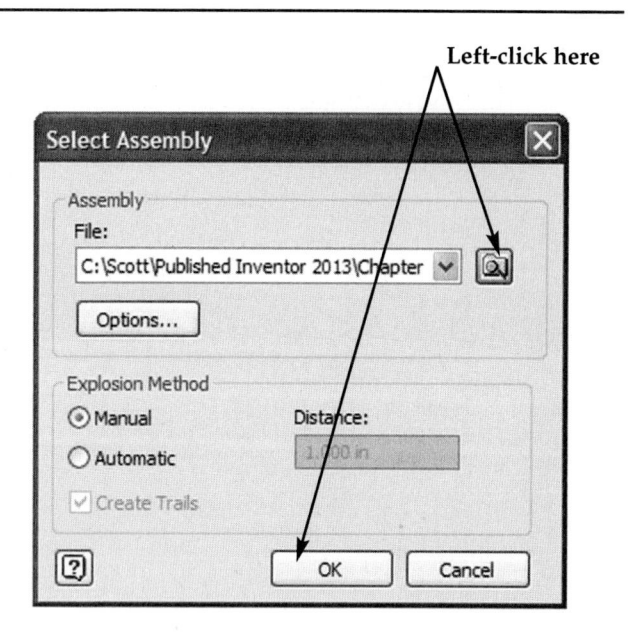

27. Left-click on **Assembly1.iam**. Left-click on **Open** as shown in Figure 8-19.

FIGURE 8-19

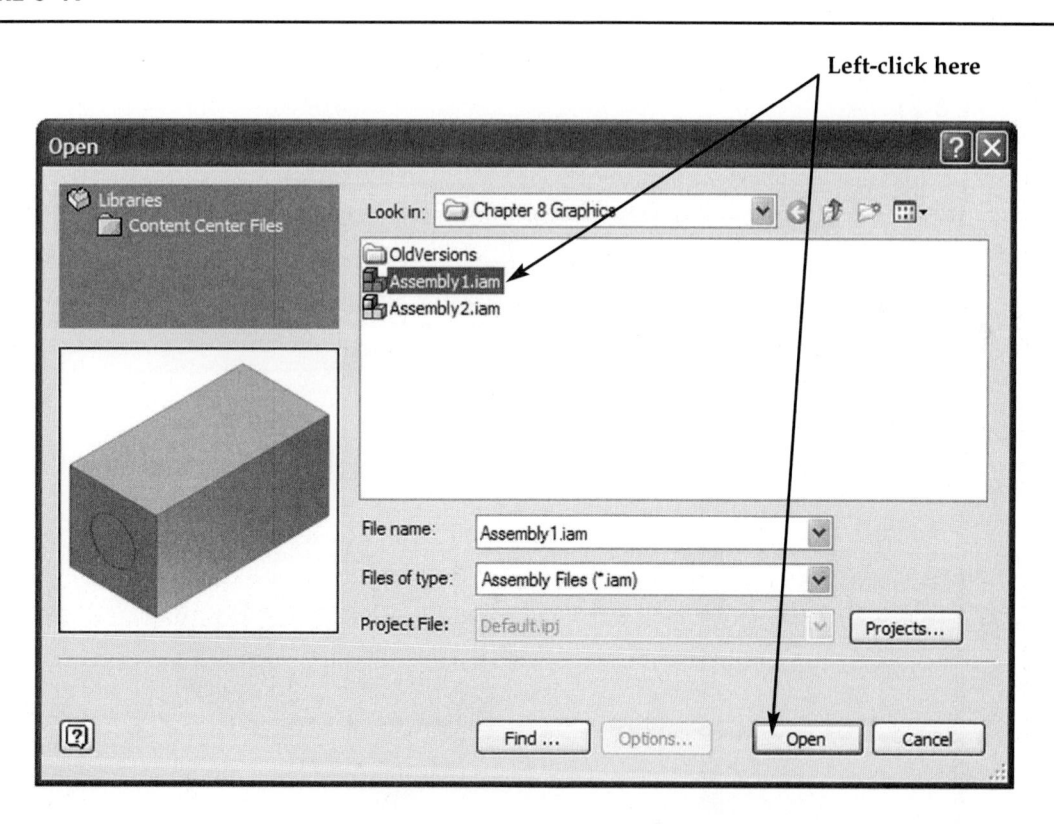

28. The Presentation Panel will only read assembly drawings. Assembly drawings are imported into the Presentation Panel in order to create an .ipn file (Inventor Presentation).

29. The Select Assembly dialog box will open. Left-click on **OK** as shown in Figure 8-20.

FIGURE 8-20

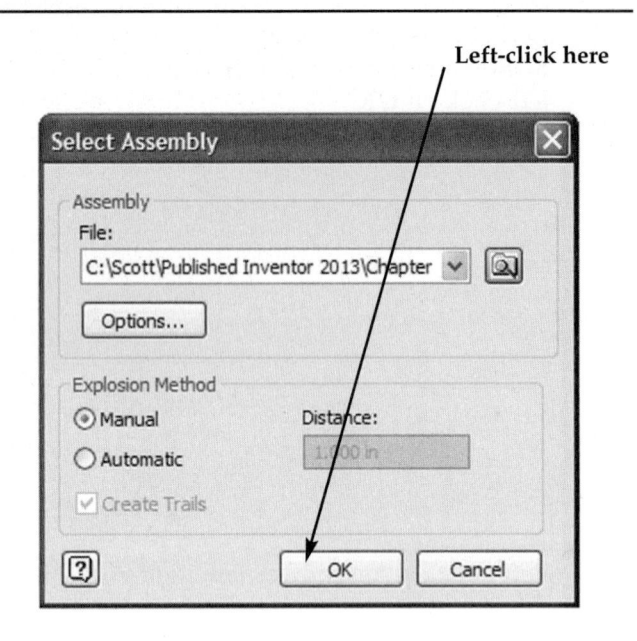

30. Your screen should look similar to Figure 8-21.

FIGURE 8-21

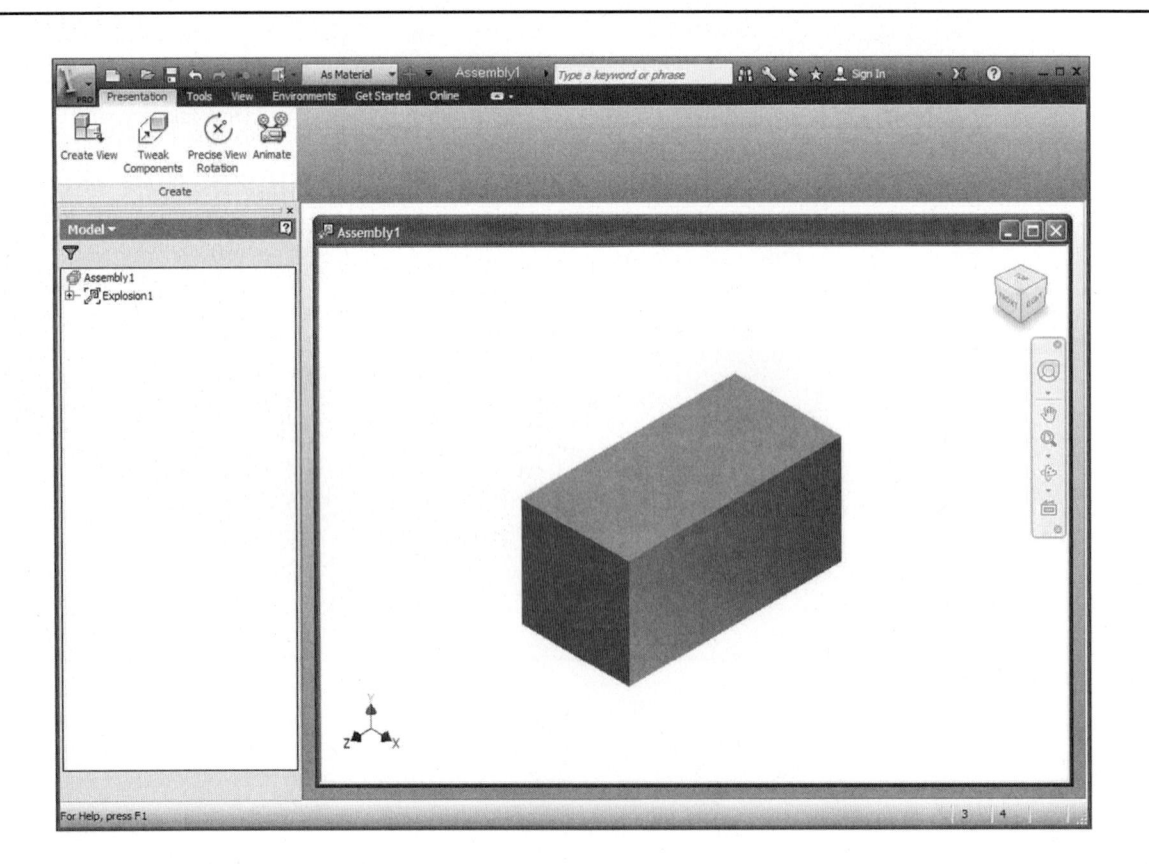

DESIGN PARTS TRAILS IN THE PRESENTATION PANEL

31.   Move the cursor to the upper left portion of the screen and left-click on **Tweak Components**. Left-click on the **Z** icon as shown in Figure 8-22.

FIGURE 8-22

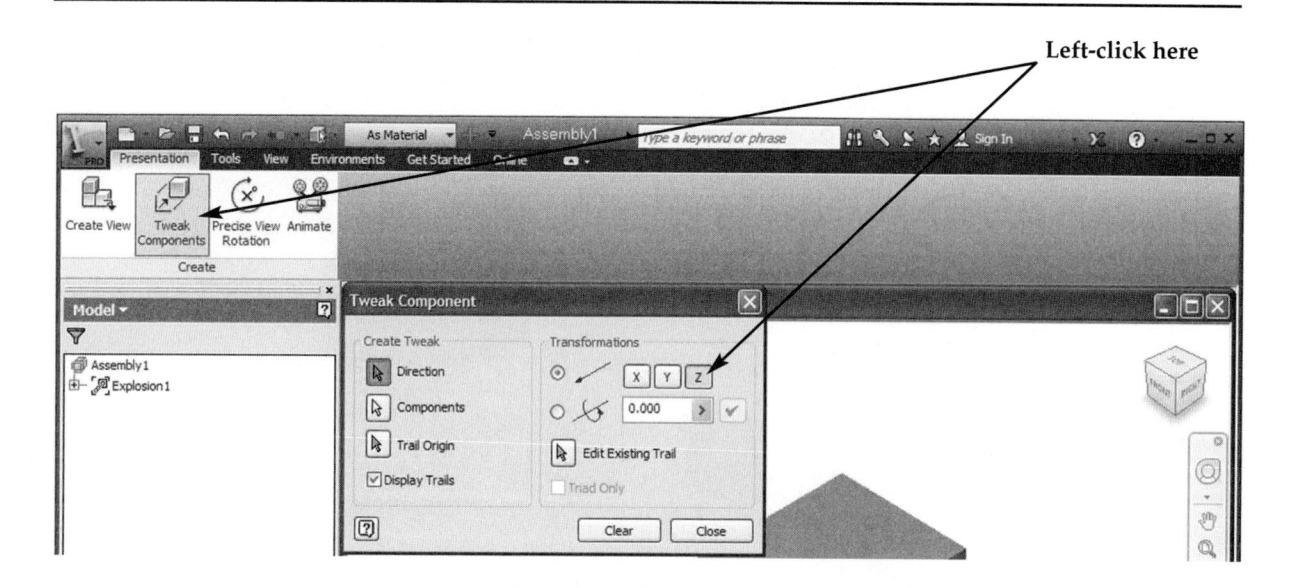

Left-click here

32.   Move the cursor to the face of the pin. An origin symbol will be attached to the cursor. After the origin symbol appears, left-click once as shown in Figure 8-23.

FIGURE 8-23

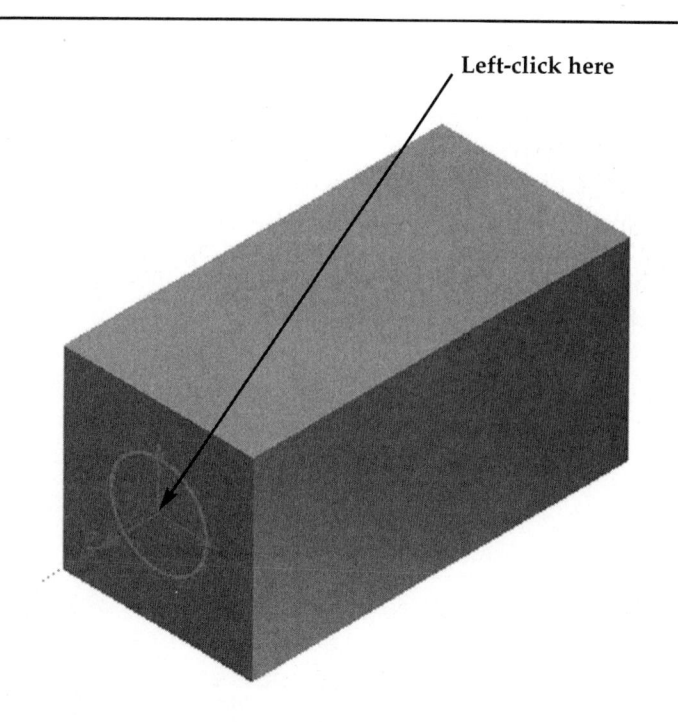

Left-click here

33.   Move the cursor to the face of the pin. Left-click (holding the left mouse button down) and drag the pin out of the block towards the lower left portion of the screen as shown in Figure 8-24. Notice the blue line coming out of the hole in the block. This is the "trail" that the pin will follow. This is the "Z" axis.

FIGURE 8-24

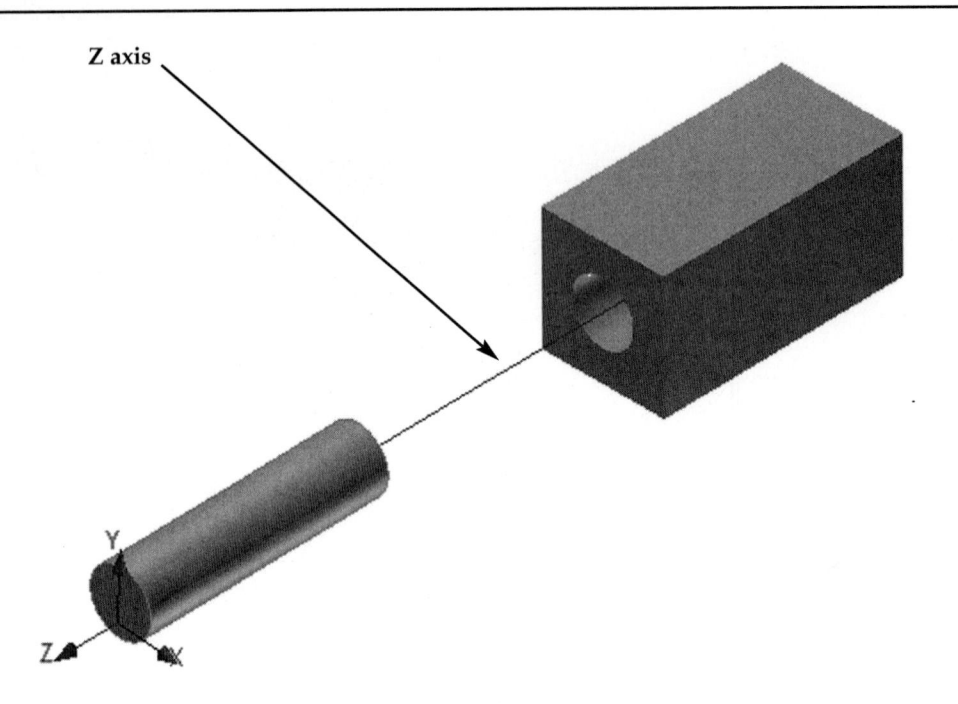

34.   Enter **3.000** for the distance that appears below X, Y, Z. This is the distance the pin has traveled on the Z axis. The text can be highlighted so users can enter any numerical distance. Move the cursor to the Tweak Component dialog box and left-click on **X** as shown in Figure 8-25.

FIGURE 8-25

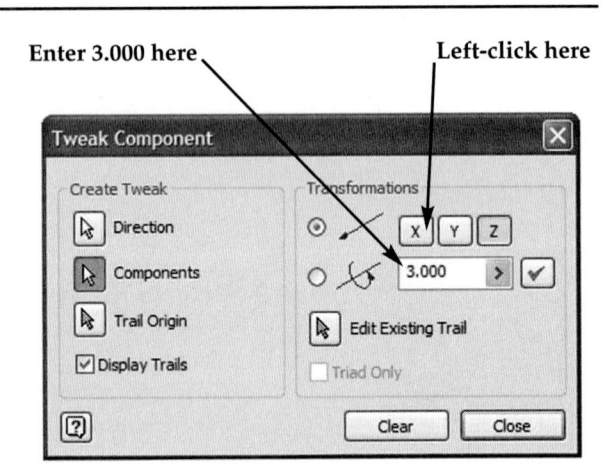

35. Move the cursor to the center of the pin. Left-click (holding the left mouse button down) and drag the pin from the end of the "Z" trail, towards the lower right portion of the screen as shown in Figure 8-26. Notice the direction change of the blue line. This is the "trail" that the pin will follow. This is the "X" axis.

FIGURE 8-26

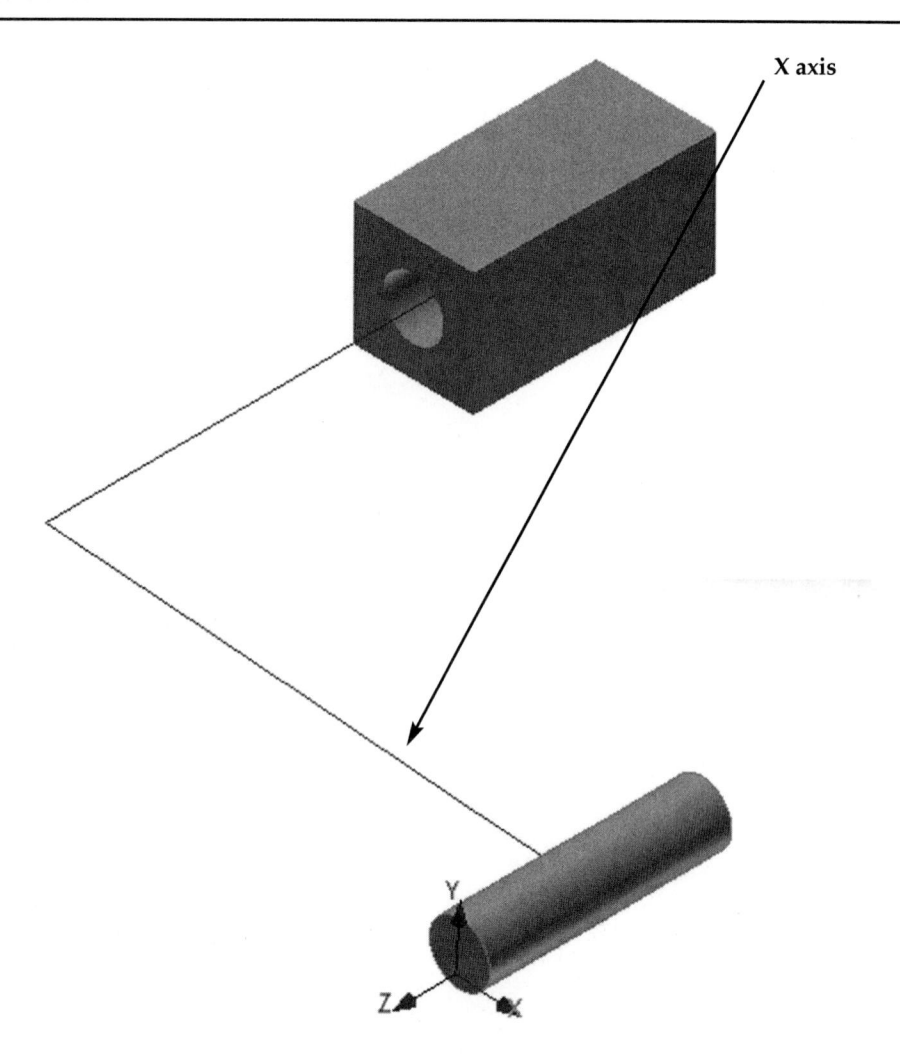

X axis

36. Enter **-3.000** for the distance. Move the cursor to the Tweak Component dialog box and left-click on **Y** as shown in Figure 8-27.

FIGURE 8-27

Enter -3.000 here

Left-click here

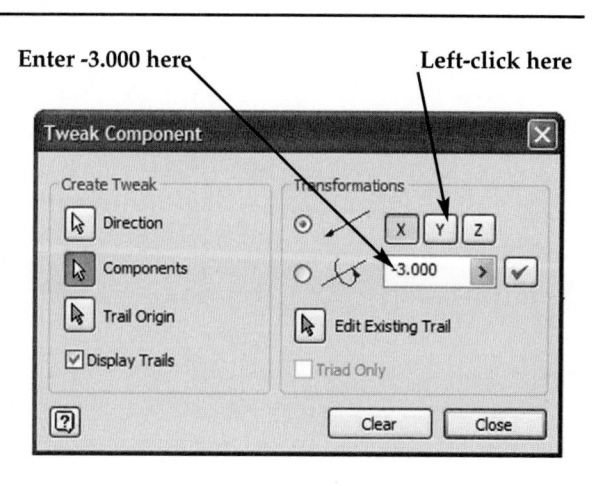

37. Move the cursor to the center of the pin. Left-click (holding the left mouse button down) and drag the pin from the end of the "Y" trail upward towards the middle portion of the screen as shown in Figure 8-28. Notice the direction change of the blue line. This is the "trail" that the pin will follow. This is the "Y" axis.

FIGURE 8-28

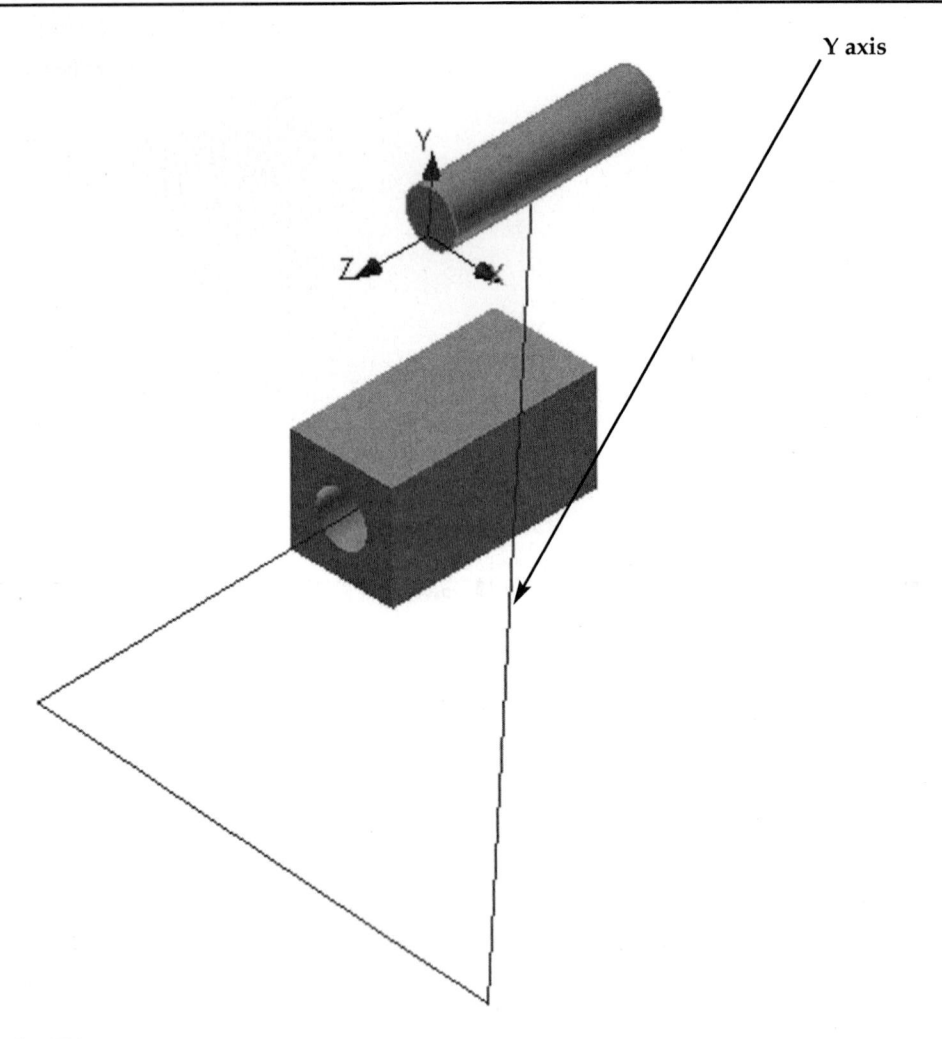

38. Enter **5.000** for the Distance. Left-click on **Clear** and then **Close** as shown in Figure 8-29.

FIGURE 8-29

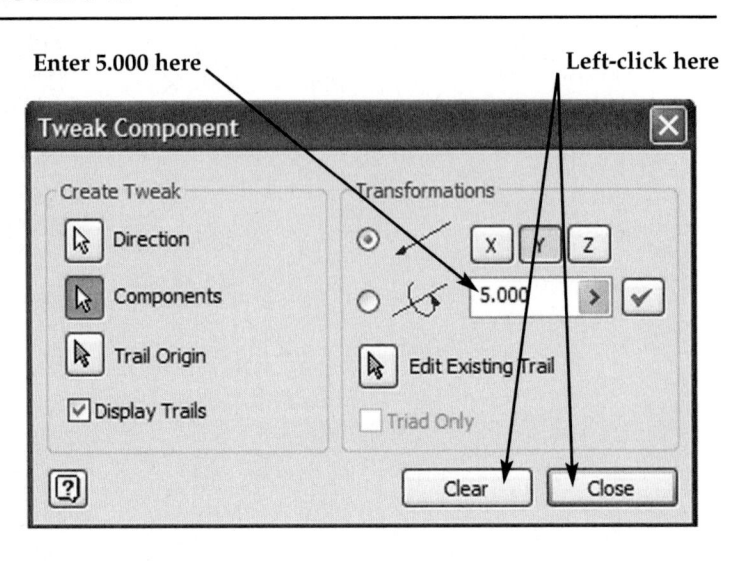

39.   The Tweak Component dialog box will close.

40.   Move the cursor to the upper left portion of the screen and left-click on **Animate**. The Animation dialog box will appear. Left-click on "Play" as shown in Figure 8-30.

FIGURE 8-30

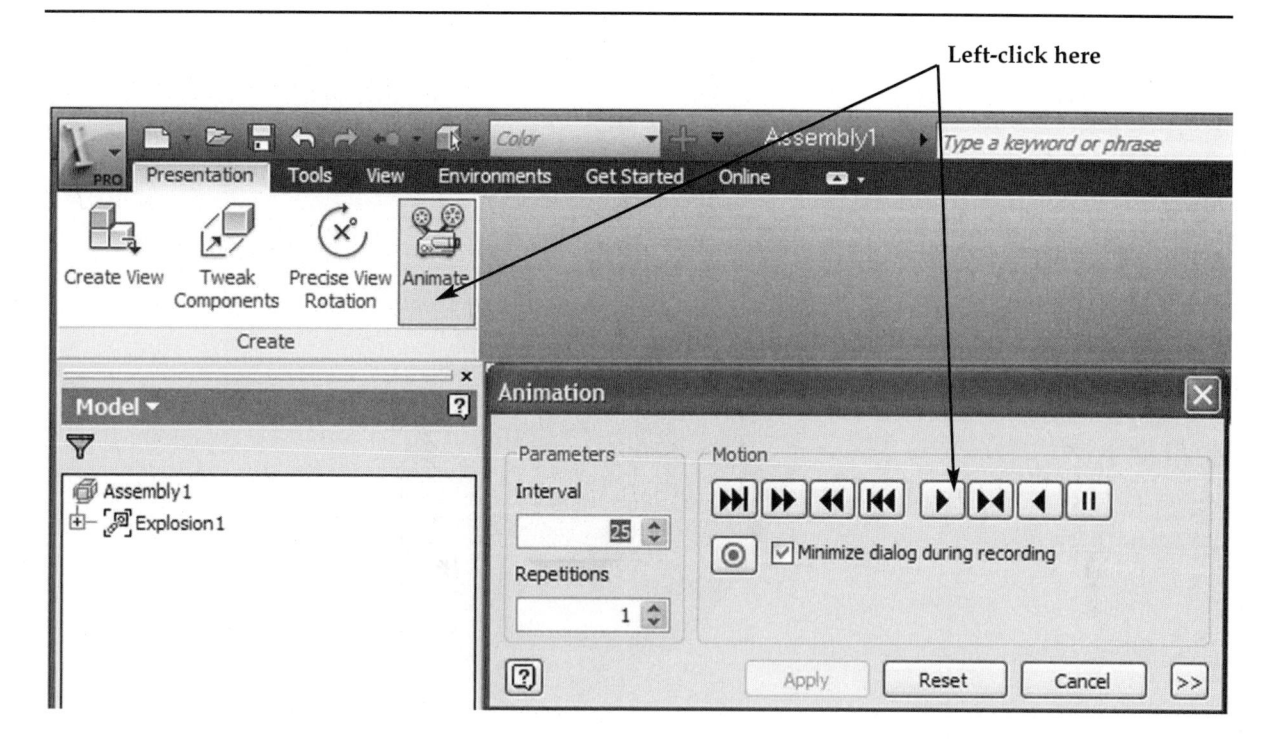

41.   Inventor will animate the parts. The pin should follow the part trail back to the hole in the block. Your screen should look similar to Figure 8-31.

FIGURE 8-31

CHAPTER PROBLEMS

Create the following parts and use them to design Inventor Presentations.

PROBLEM 8-1

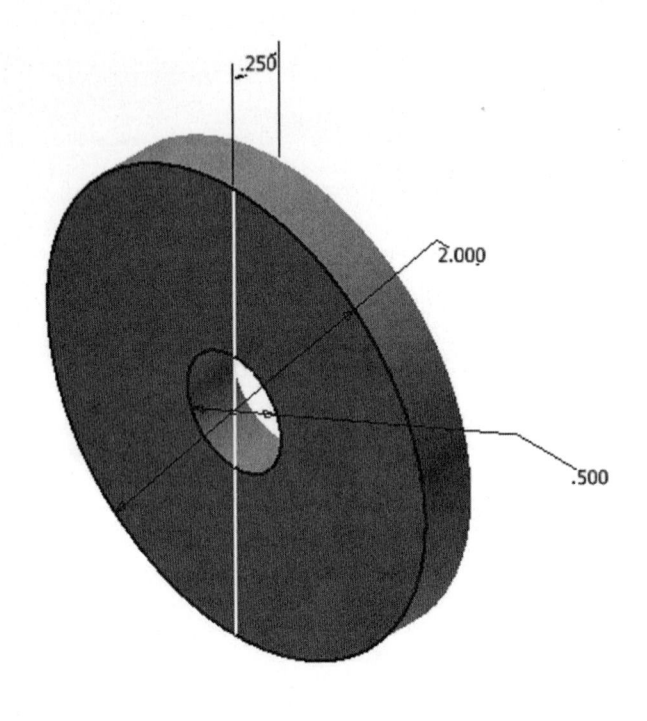

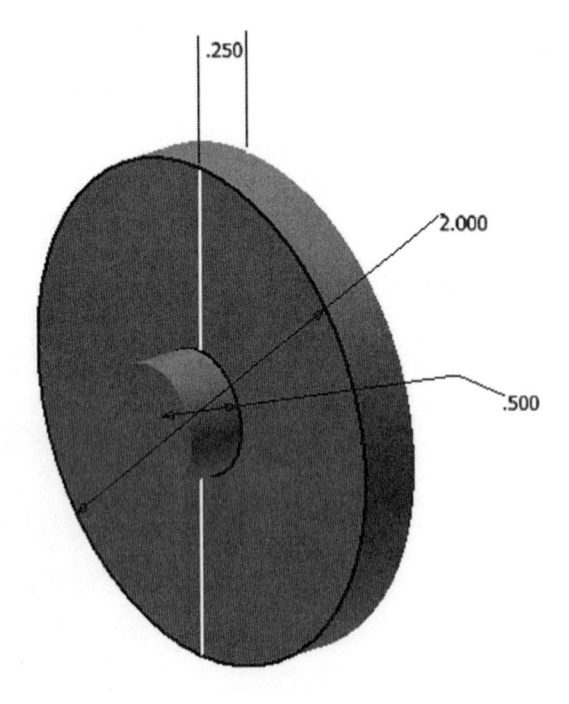

PROBLEM 8-2

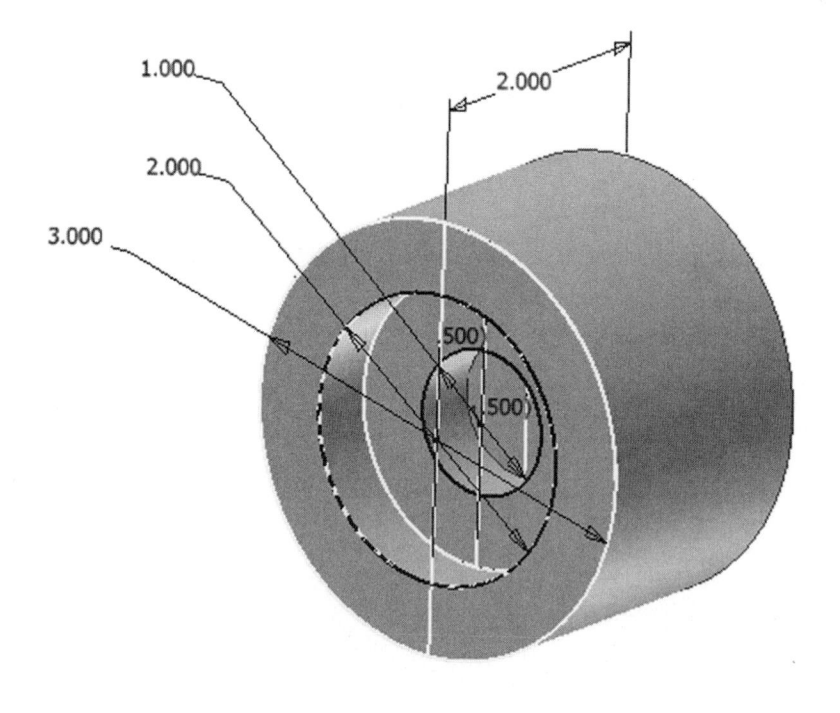

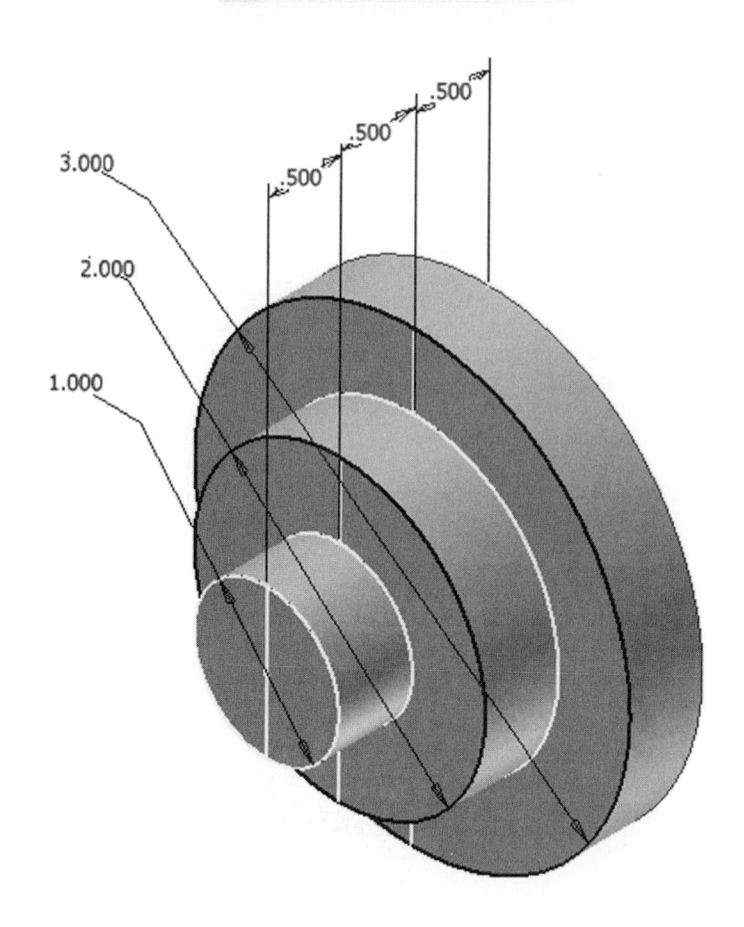

PROBLEM 8-3

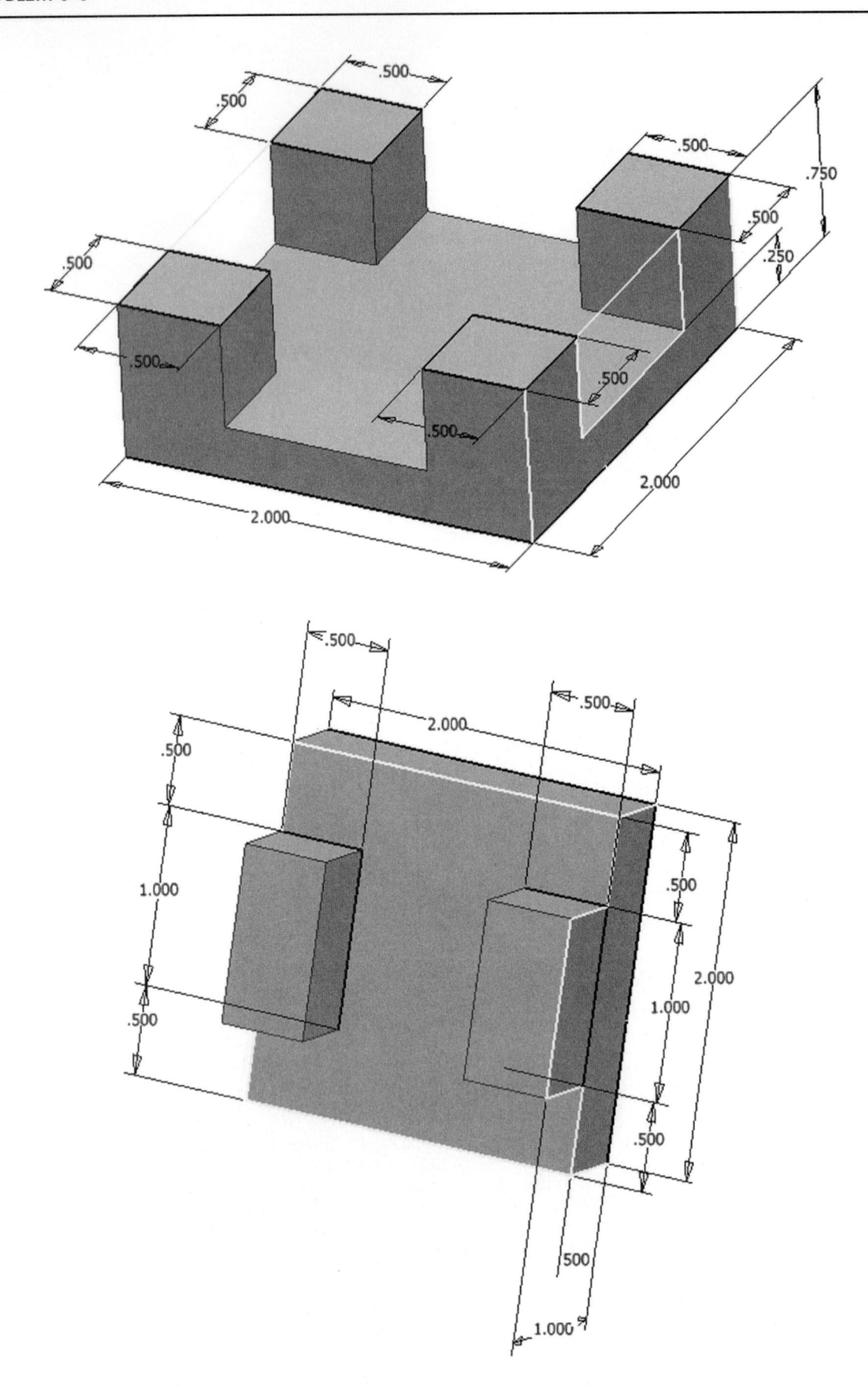

PROBLEM 8-4

Import the following part twice into the Assembly Panel and then into the Presentation Panel.

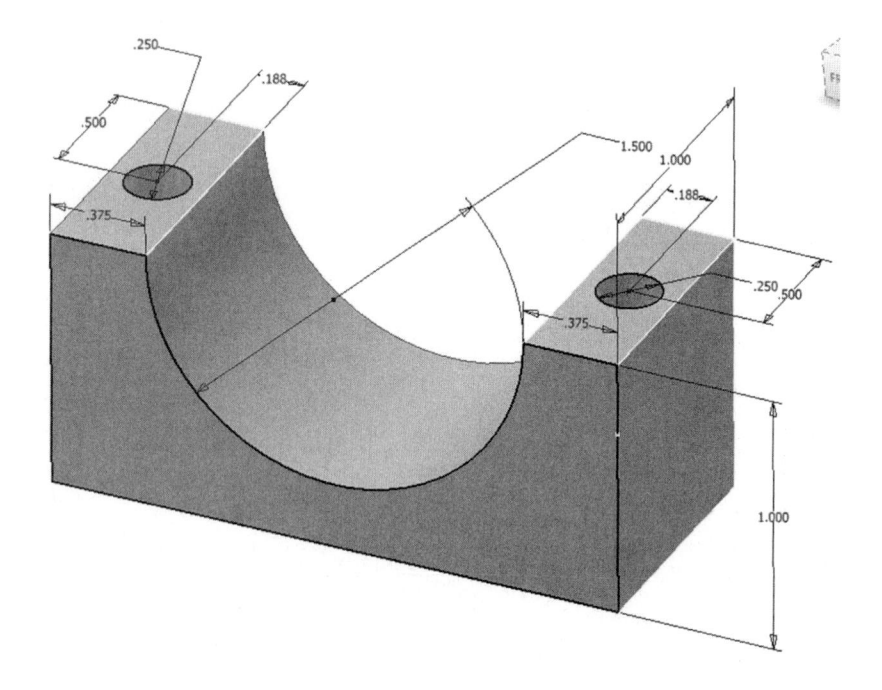

Your screen should look similar to what is shown below.

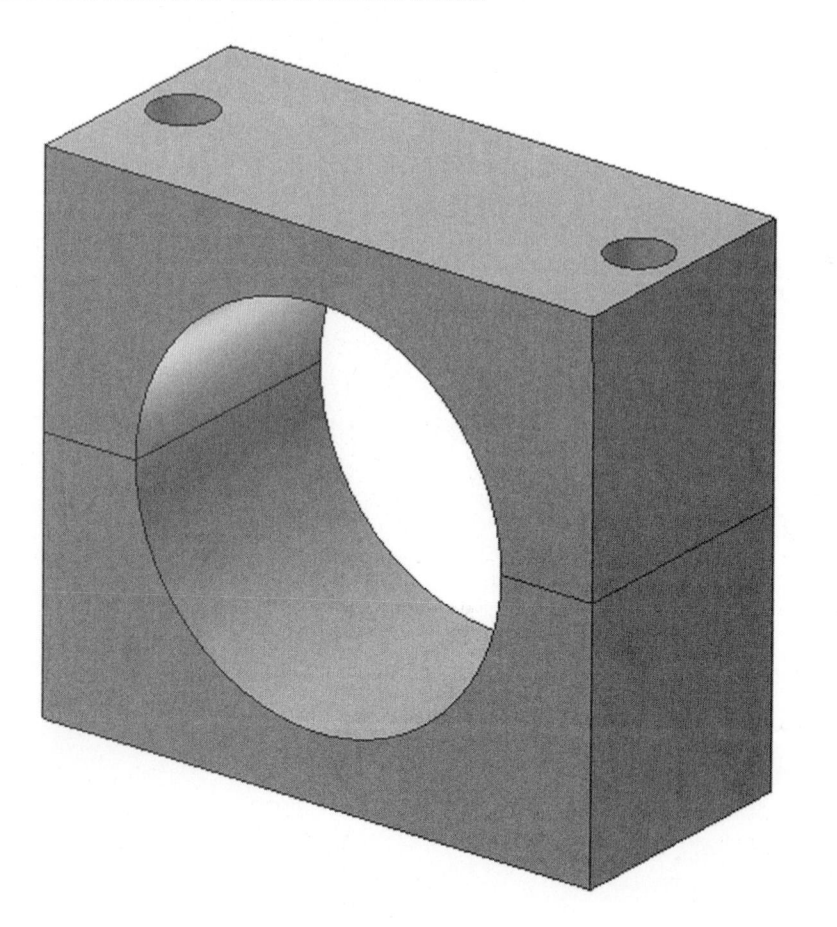

9

INTRODUCTION TO ADVANCED COMMANDS

OBJECTIVES

1. Use the Sweep command
2. Use the Rectangular Pattern command
3. Use the Loft command
4. Use the Work Plane command
5. Use the Coil command

Chapter 9 includes instruction on how to design the parts shown above.

CREATE A SWEEP USING THE SWEEP COMMAND

1. Start Inventor by referring to Chapter 1.

2. After Inventor is running, begin a New Sketch.

3. Move the cursor to the upper left portion of the screen and left-click on **Rectangle** as shown in Figure 9-1.

FIGURE 9-1

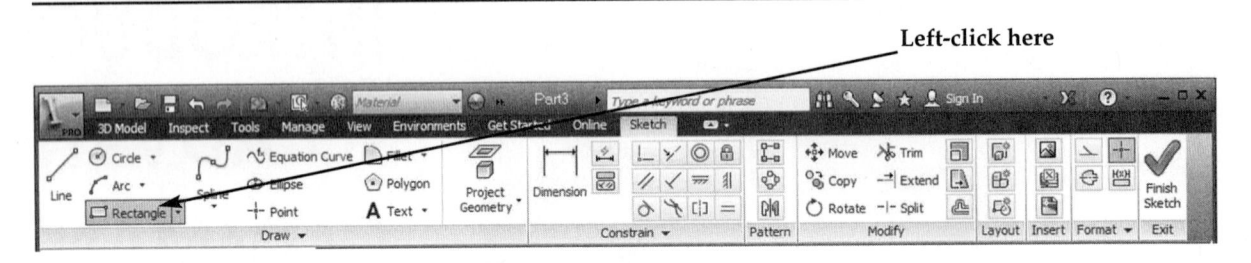

4. Complete the sketch shown below. Once complete, right-click anywhere on the screen. A pop-up menu will appear. Left-click **Finish 2D Sketch** as shown in Figure 9-2.

FIGURE 9-2

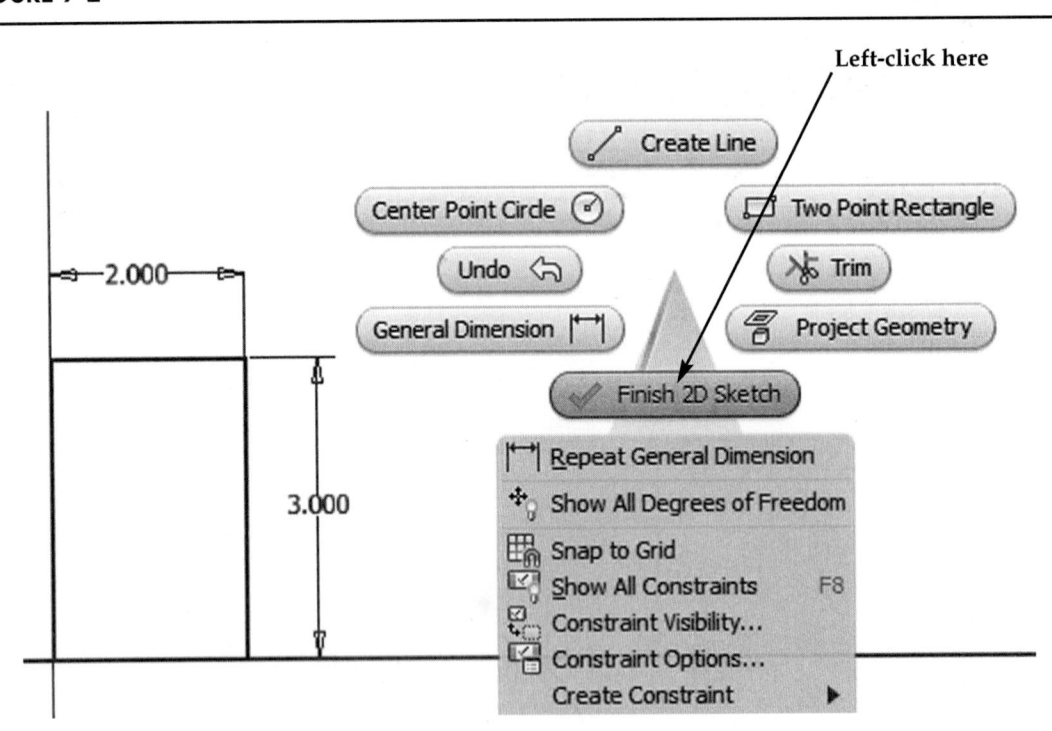

5.   Your screen should look similar to Figure 9-3.

FIGURE 9-3

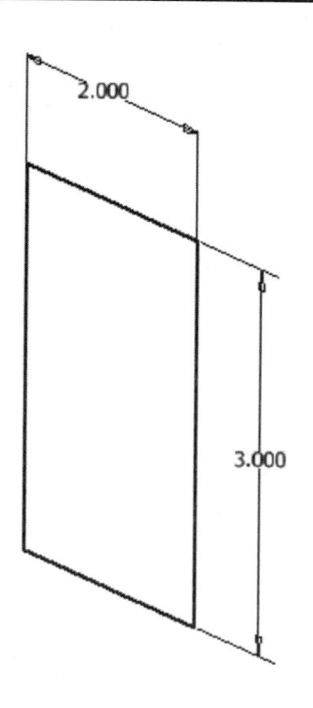

6.   Move the cursor over the text, "YZ Plane," causing a red box to appear. Right-click once. A pop-up menu will appear. Left-click on **New Sketch** as shown in Figure 9-4.

FIGURE 9-4

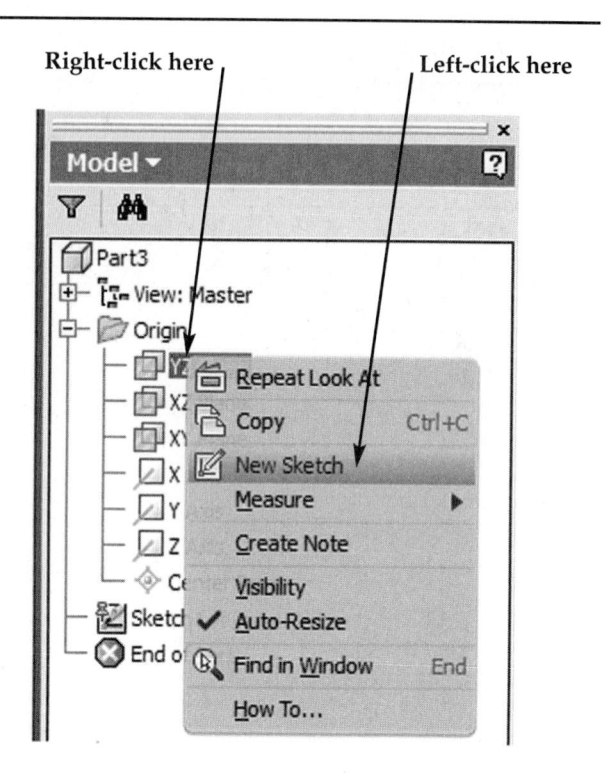

7. Your screen should look similar to Figure 9-5.

FIGURE 9-5

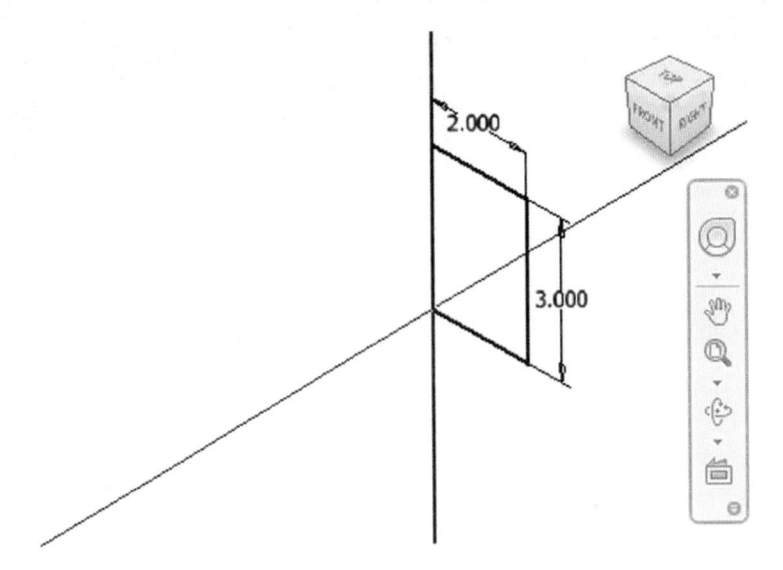

8. Complete the following sketch. The angle of the lines can be estimated. The sketch lines must intersect with the corner of the 2-inch by 3-inch box as shown in Figure 9-6. Remember to use the **Aligned** dimension function (while the dimension is attached to the cursor, right-click, causing a pop-up menu to appear then left-click on **Aligned**). Exit out of the Sketch Panel.

FIGURE 9-6

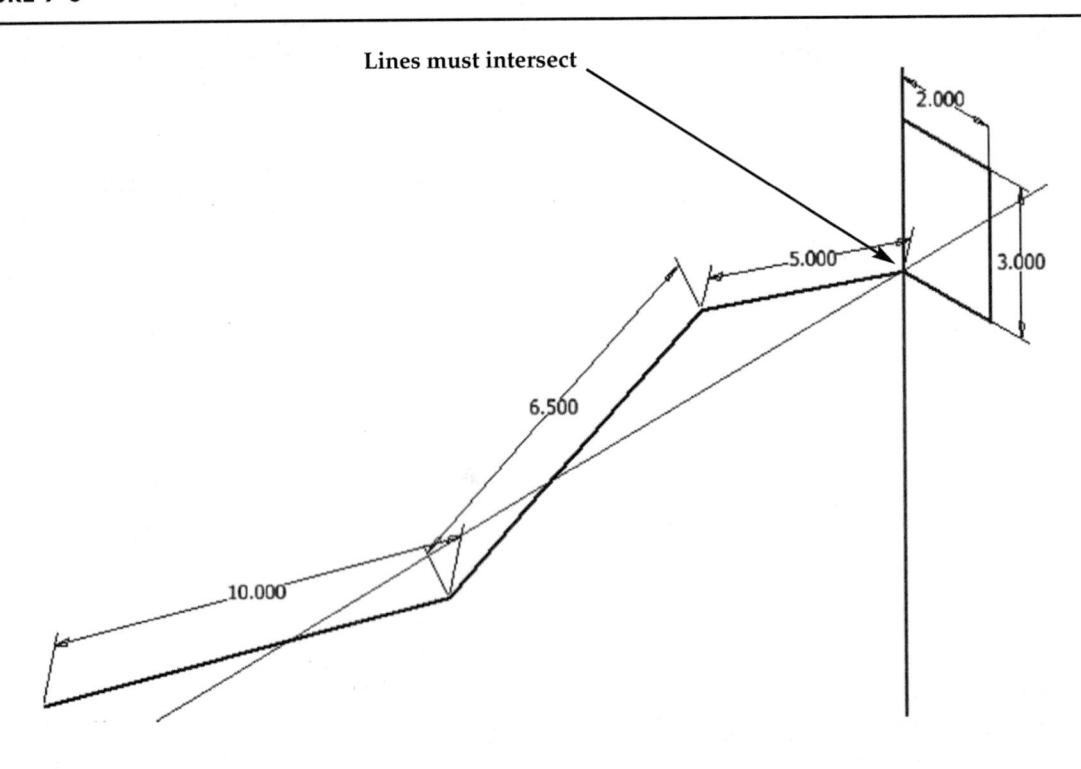

9. Your screen should look similar to Figure 9-7.

FIGURE 9-7

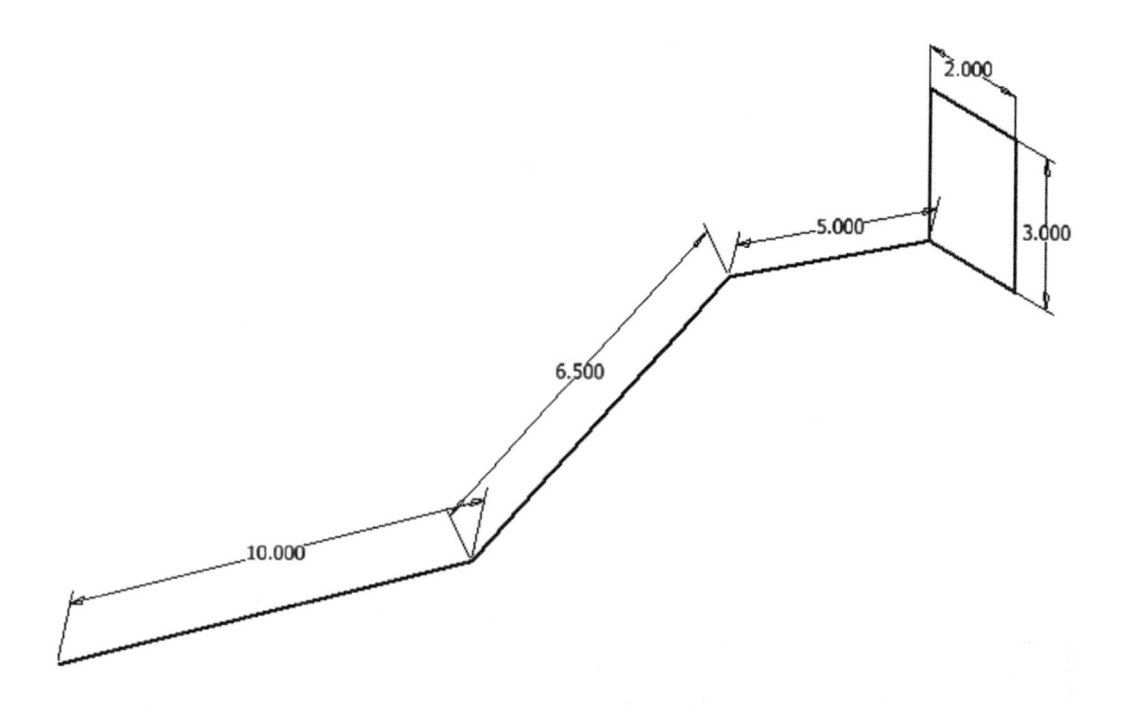

10. Move the cursor to the upper left portion of the screen and left-click on **Sweep**. Move the cursor over the sweep line, causing it to turn red and left-click once as shown in Figure 9-8.

FIGURE 9-8

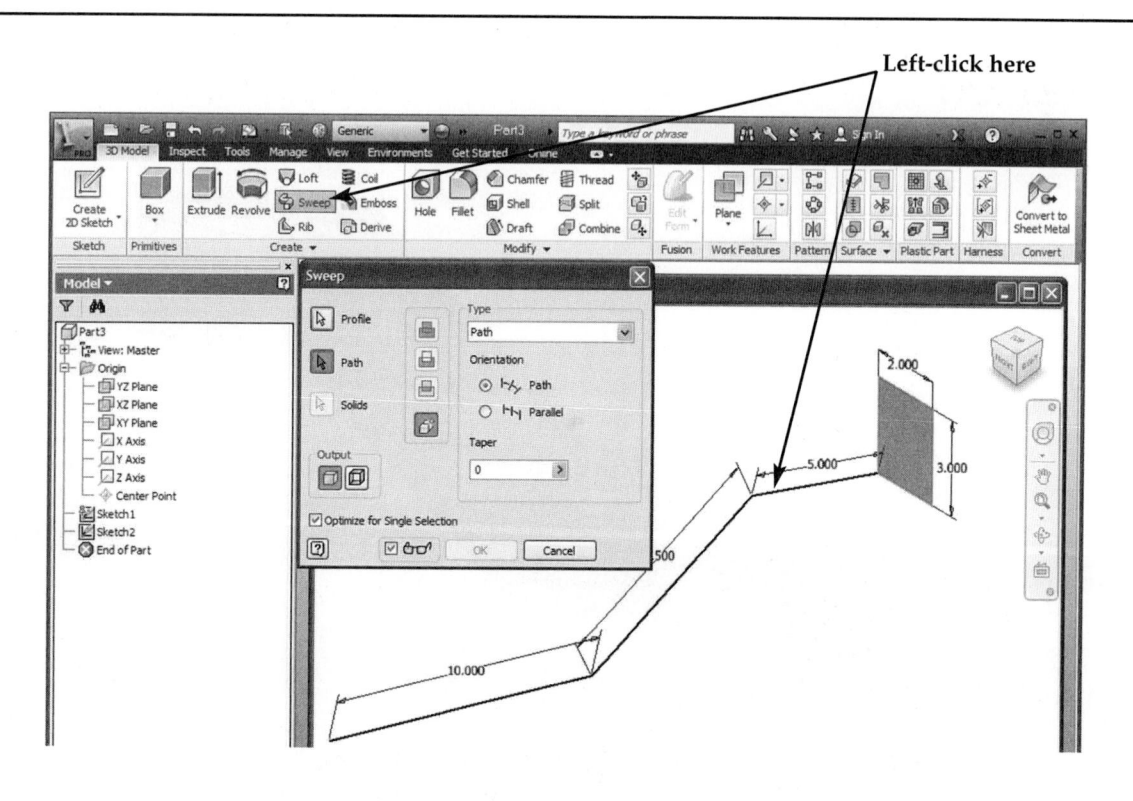

11.   A preview of the sweep will appear as shown in Figure 9-9.

FIGURE 9-9

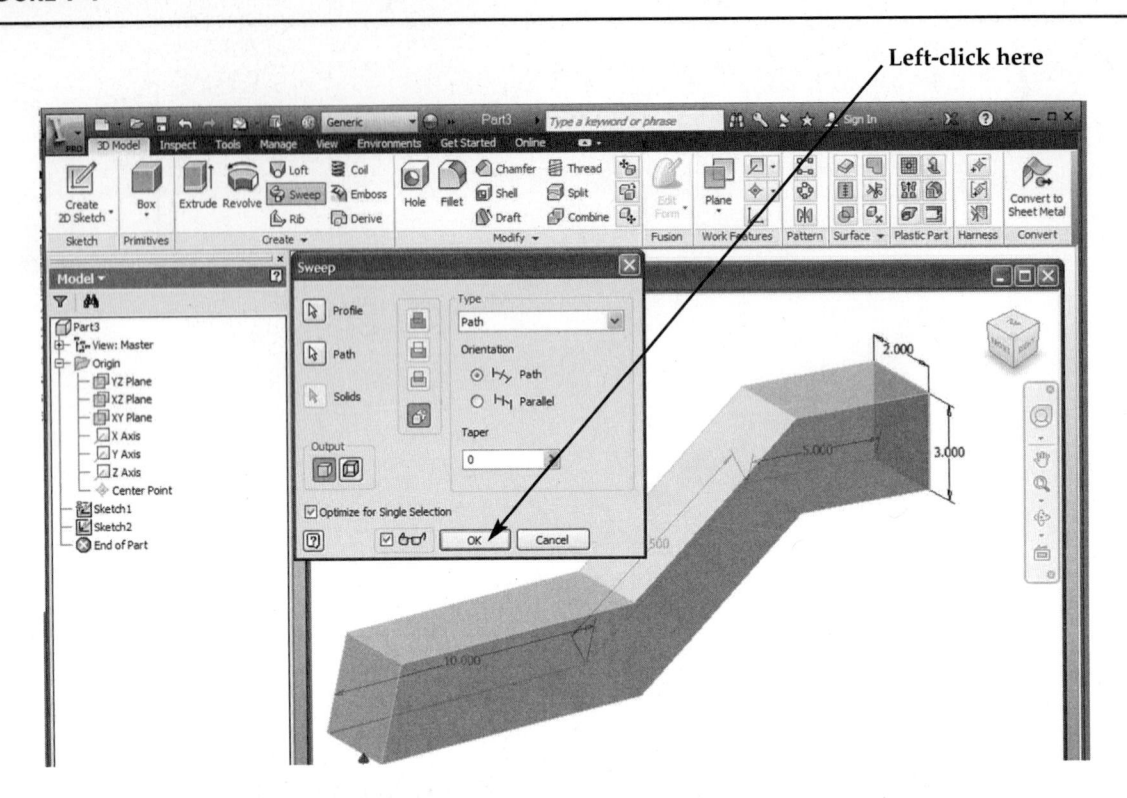

12.   Left-click on **OK** as shown in Figure 9-9.

13.   Your screen should look similar to Figure 9-10.

FIGURE 9-10

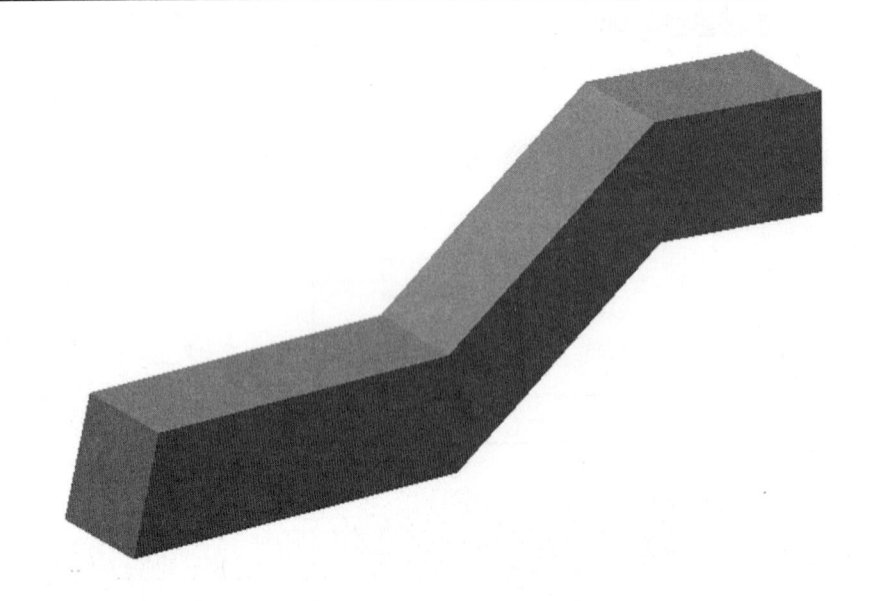

14. Use the **Shell** command to shell the tubing as shown in Figure 9-11.

FIGURE 9-11

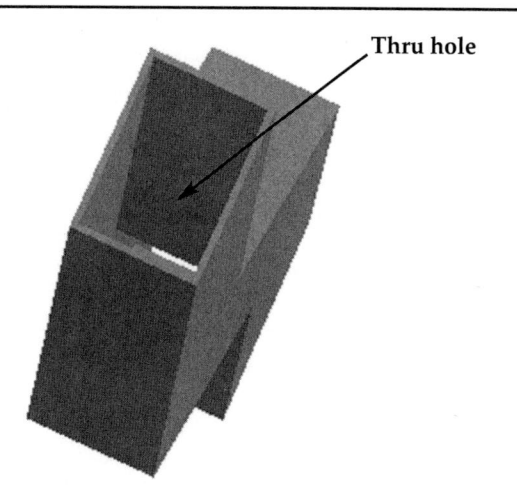

Thru hole

15. Begin a new sketch as shown in Figure 9-12. Use the "View Face/Look At" command to gain a perpendicular view of the surface.

FIGURE 9-12

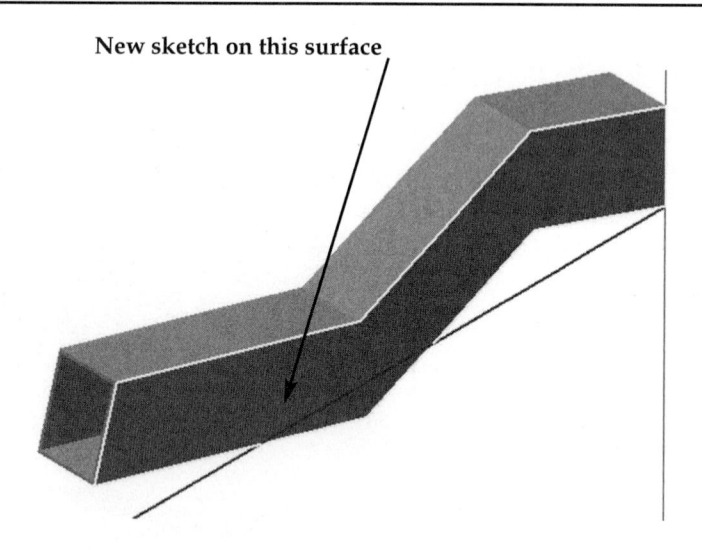

New sketch on this surface

16. Your screen should look similar to Figure 9-13.

FIGURE 9-13

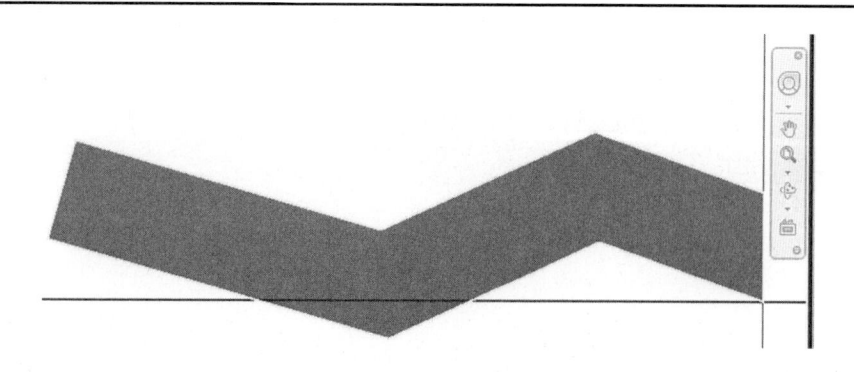

17.   Complete the following sketch. You may have to zoom in as shown in Figure 9-14.

FIGURE 9-14

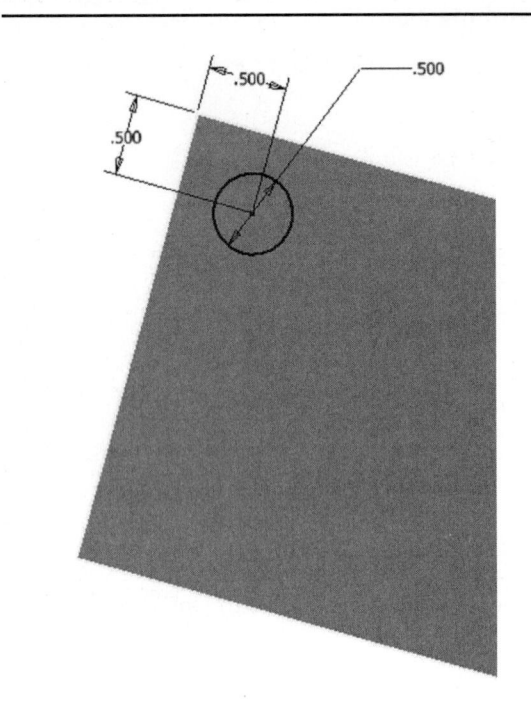

18.   Exit out of the Sketch area and change the view to Home View as shown in Figure 9-15.

FIGURE 9-15

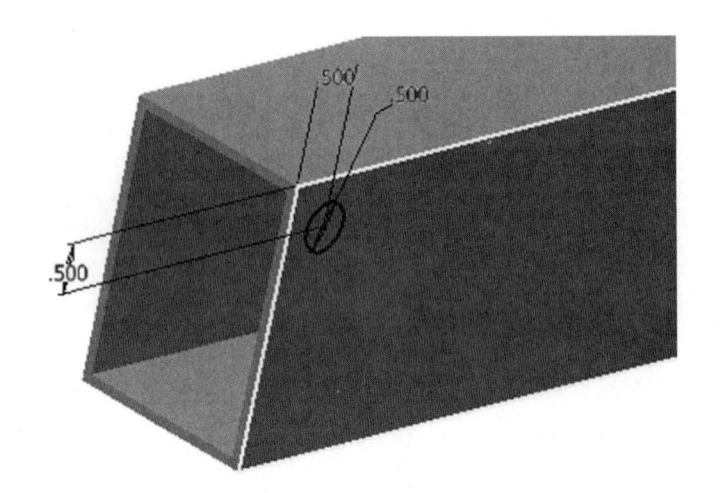

USE THE RECTANGULAR PATTERN COMMAND

19.   Use the **Extrude** command to cut the hole out as shown in Figure 9-16.

FIGURE 9-16

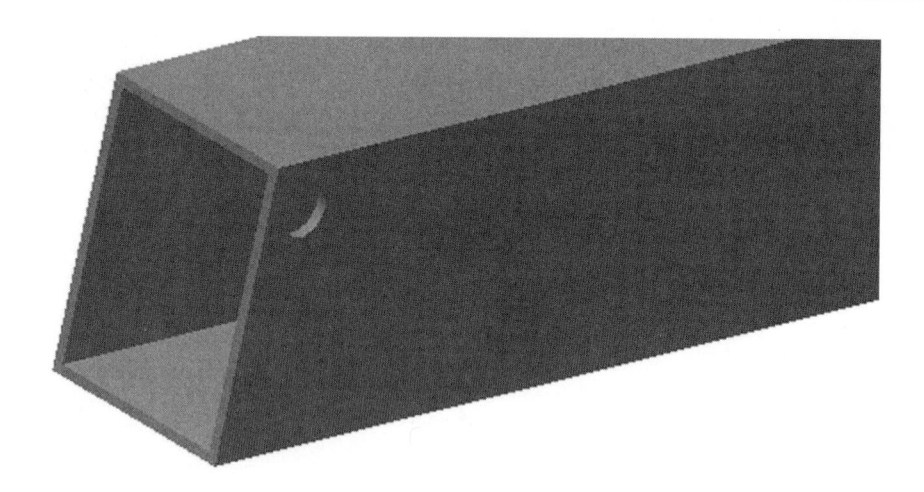

20.   Move the cursor to the upper middle portion of the screen and left-click on the "Rectangular Pattern" icon. The Rectangular Pattern dialog box will appear as shown in Figure 9-17.

FIGURE 9-17

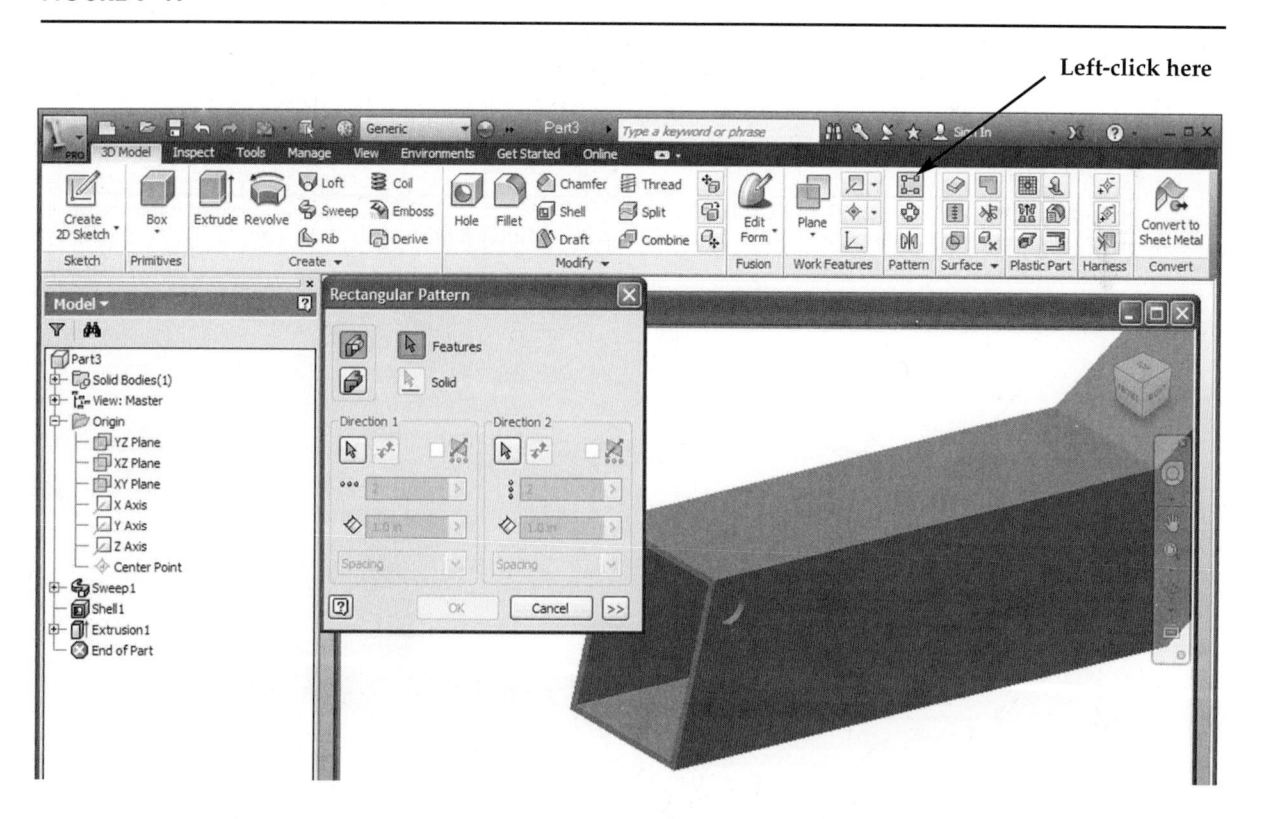

21. Move the cursor inside the hole and left-click once as shown in Figure 9-18.

FIGURE 9-18

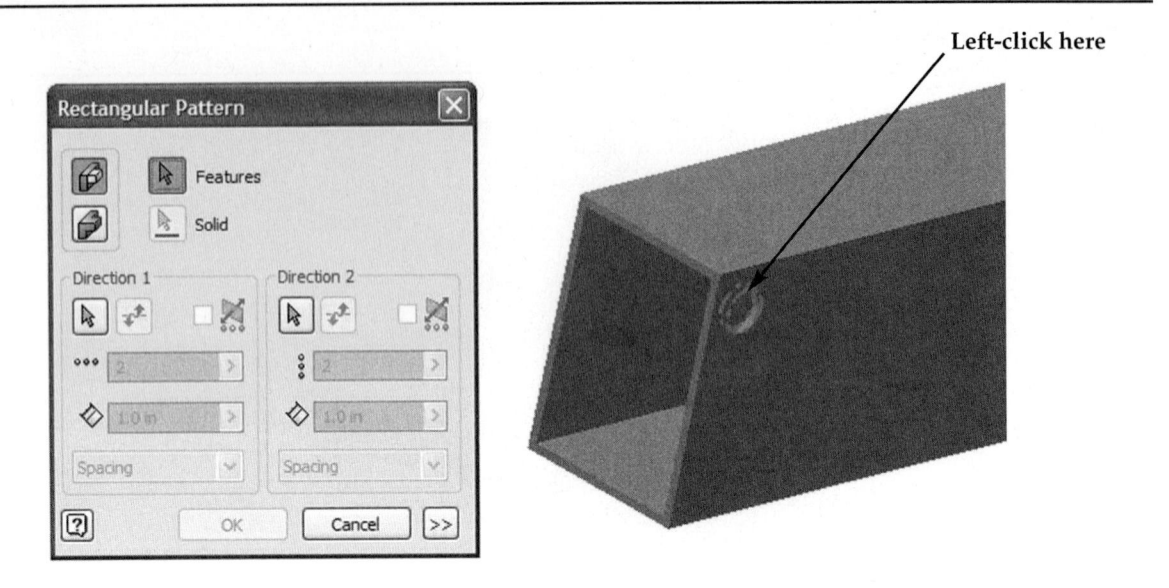

22. Left-click on the arrow under **Direction 1**. Now left-click on the top edge of the part. If there is a need to reverse the direction of the pattern (so the pattern does not travel off the part), left-click on the Flip icon (icon with arrows pointing up and down). Inventor will provide a preview of the holes it will pattern in the X direction. Now left-click on the arrow under **Direction 2**. Now left-click on the side edge of the part. Enter **8** for the number of occurrences in Direction **1** and **3** for the number of occurrences in Direction **2**. Enter **1.0** and **.875** for the distance between the circles respectively. Left-click on **OK** as shown in Figure 9-19.

FIGURE 9-19

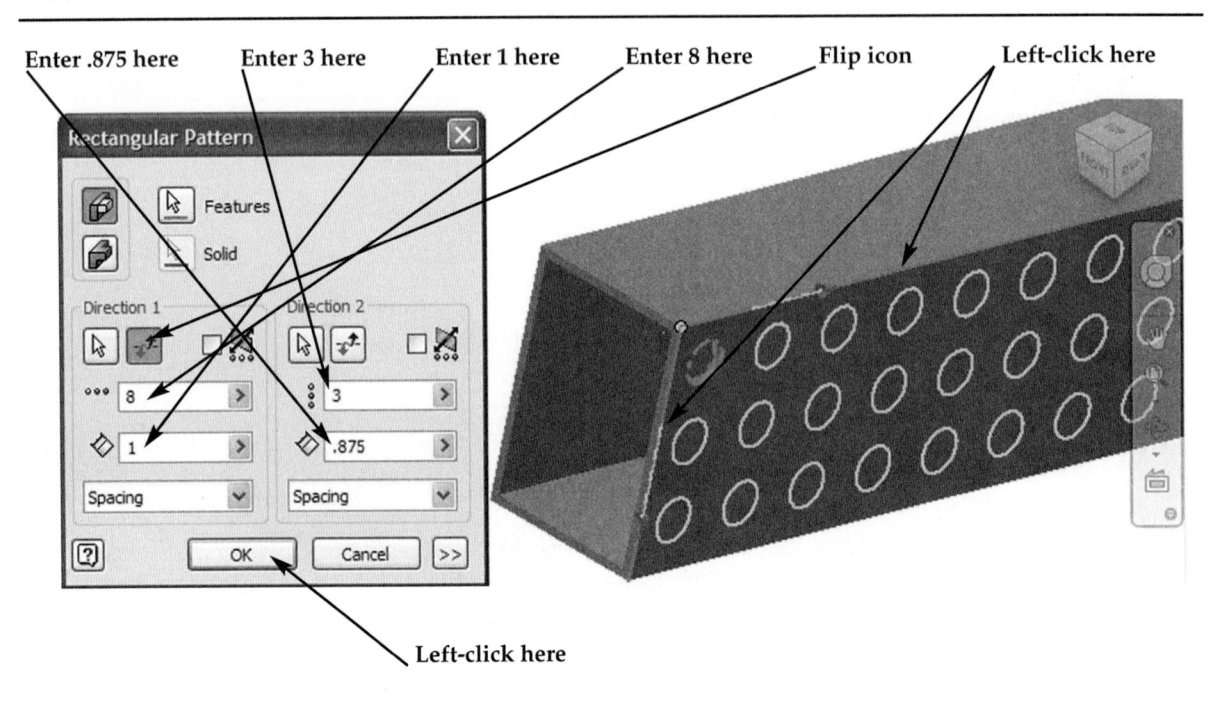

CREATE A LOFT USING THE LOFT COMMAND

23.   Your screen should look similar to Figure 9-20.

FIGURE 9-20

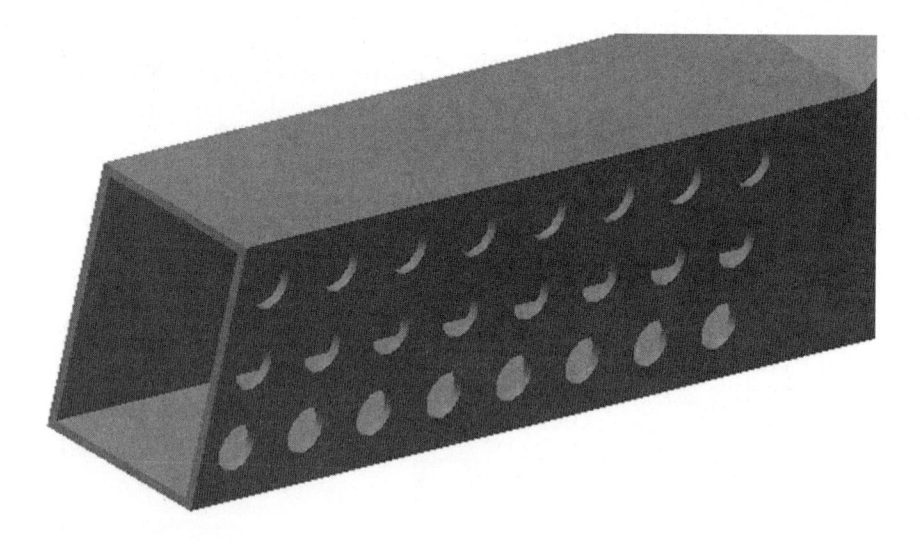

24.   Begin a new drawing. Complete the sketch shown below. Exit out of the Sketch Panel as shown in Figure 9-21.

FIGURE 9-21

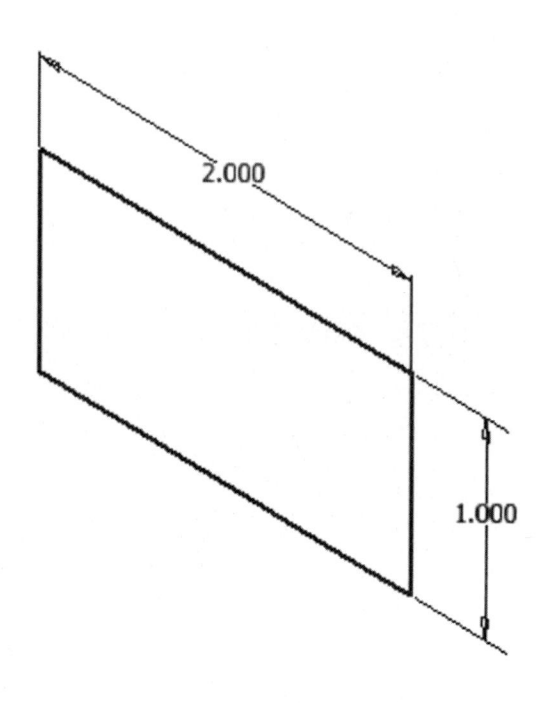

25. Left-click on the "plus" sign to the left of the text, "Origin." The part tree will expand as shown in Figure 9-22.

FIGURE 9-22

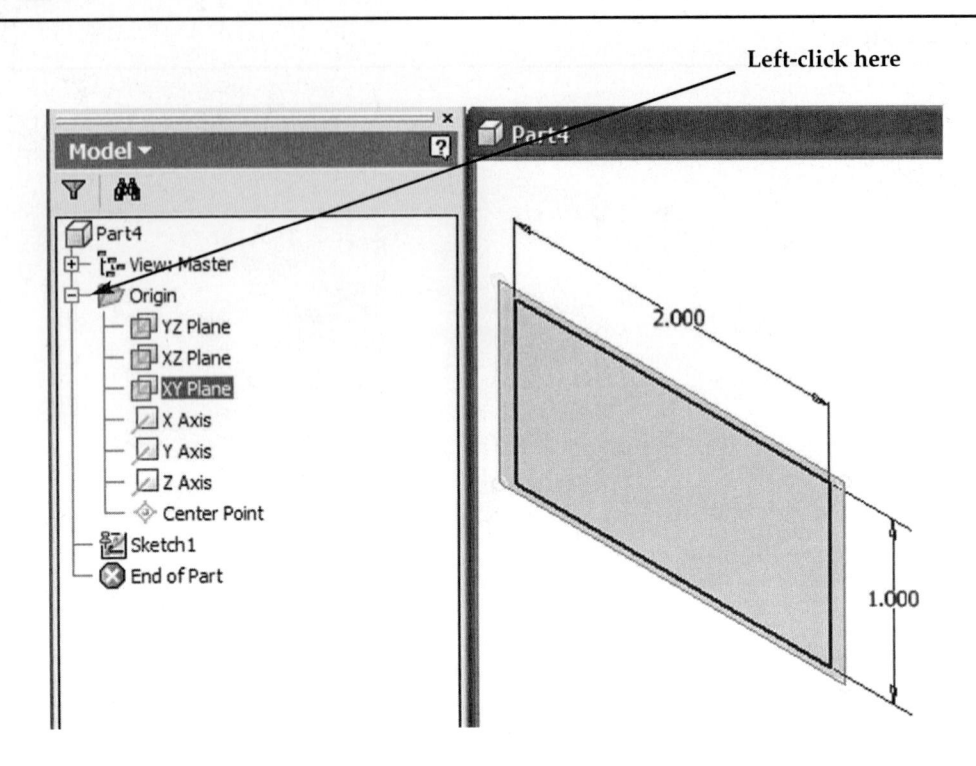

26 Move the cursor to the upper right portion of the screen and left-click on **Plane** as shown in Figure 9-23.

FIGURE 9-23

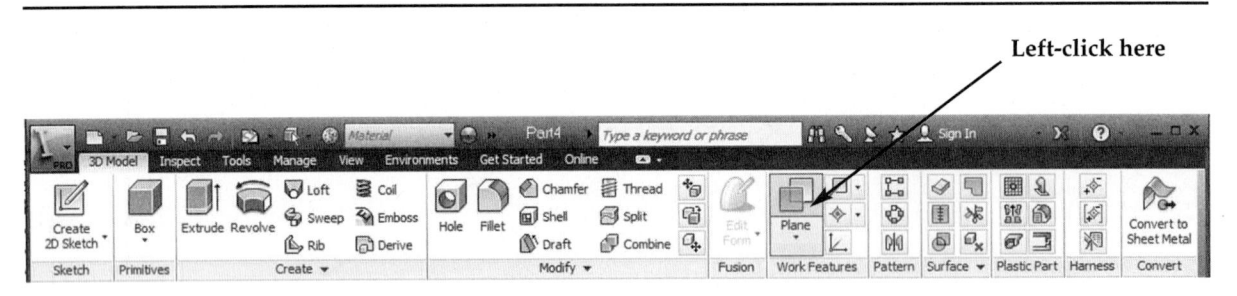

27. Left-click on **XY Plane** in the part tree. The "XY Plane" text will become highlighted as shown in Figure 9-24.

FIGURE 9-24

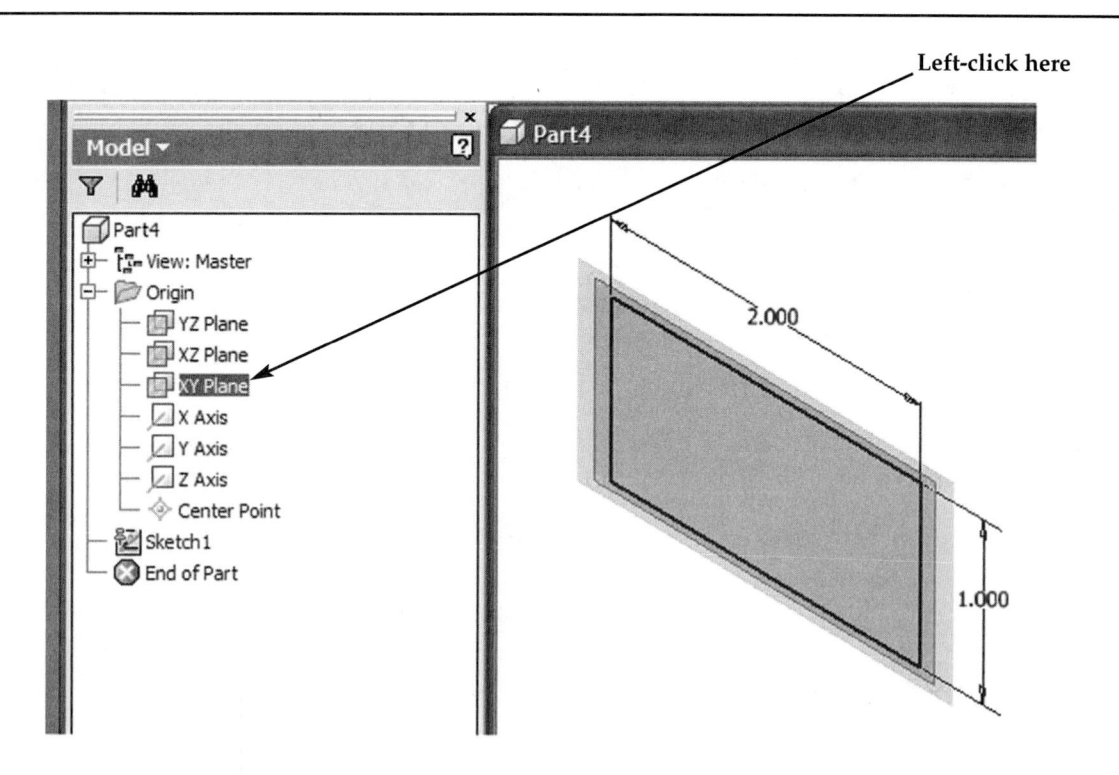

28. Move the cursor to the center of the sketch and left-click (holding the left mouse button down), dragging the cursor to the lower left portion of the screen. Enter **.500** as shown in Figure 9-25 and press the **Enter** key on the keyboard.

FIGURE 9-25

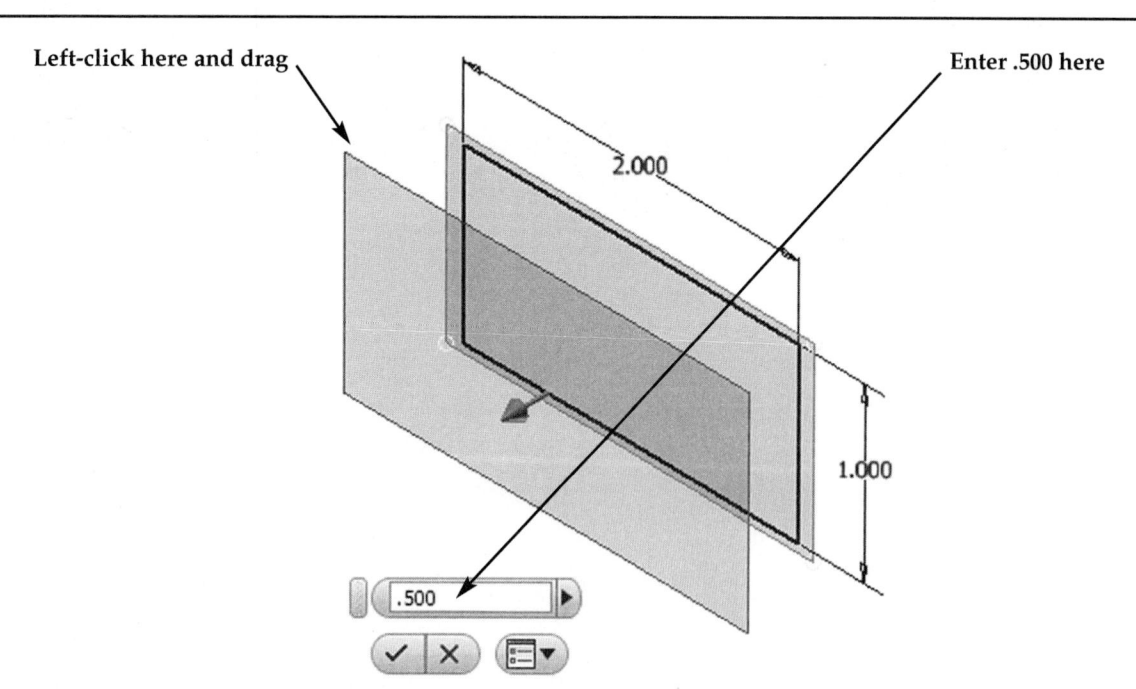

29. Your screen should look similar to Figure 9-26.

FIGURE 9-26

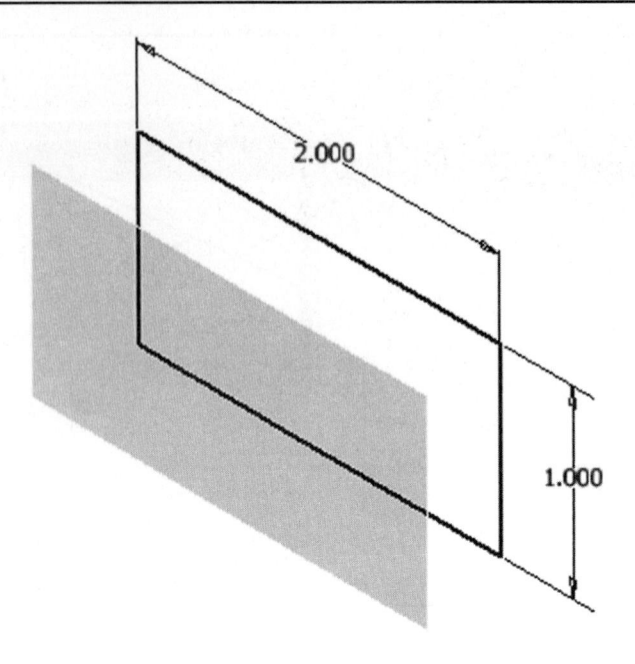

30. Begin a New Sketch on the newly created Plane as shown in Figure 9-27.

FIGURE 9-27

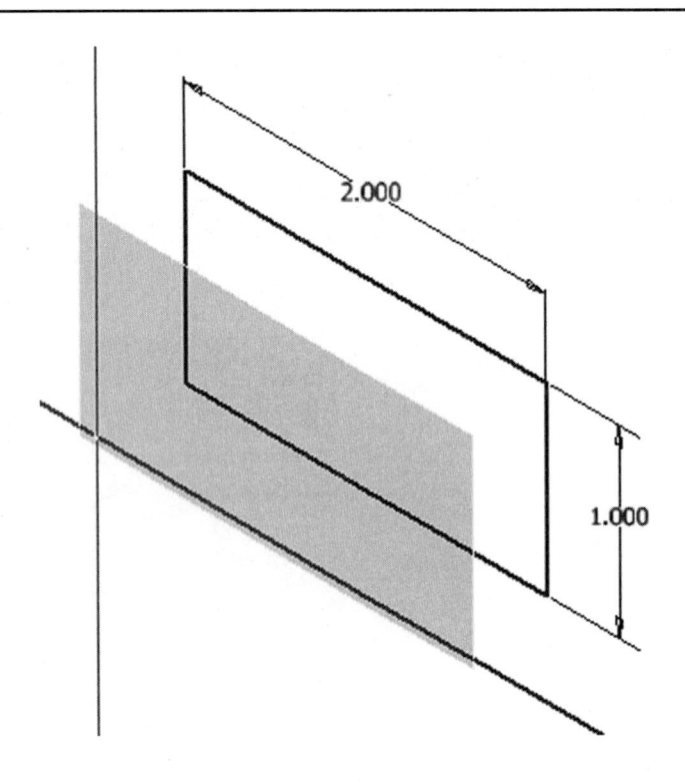

31. Complete the sketch shown in Figure 9-28. Estimate the location and size of the circle and exit the Sketch Panel.

FIGURE 9-28

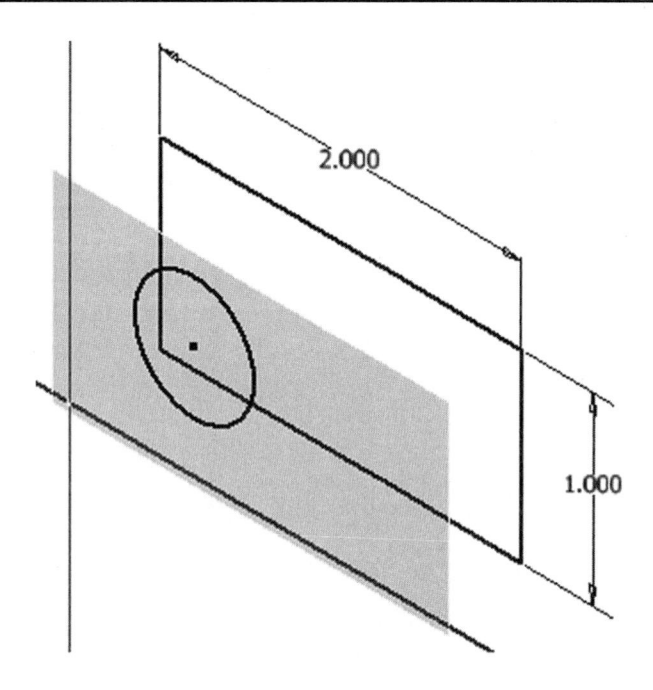

32. Move the cursor to the upper right portion of the screen and left-click **Plane** as shown in Figure 9-29.

FIGURE 9-29

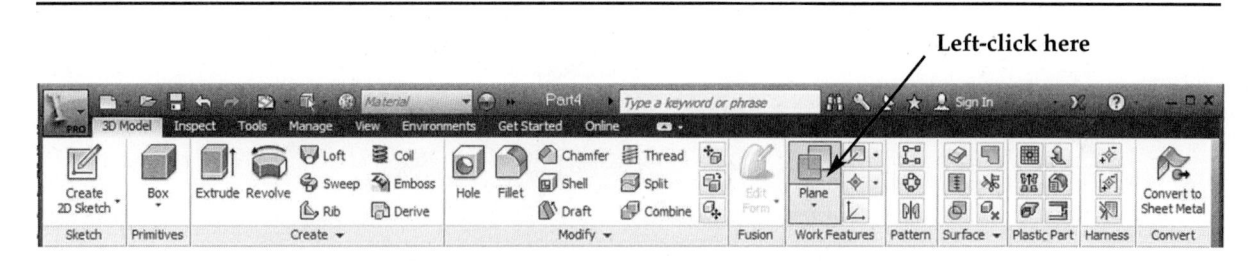

33. Left-click on **XY Plane** in the part tree. The "XY Plane" text will become highlighted as shown in Figure 9-30.

FIGURE 9-30

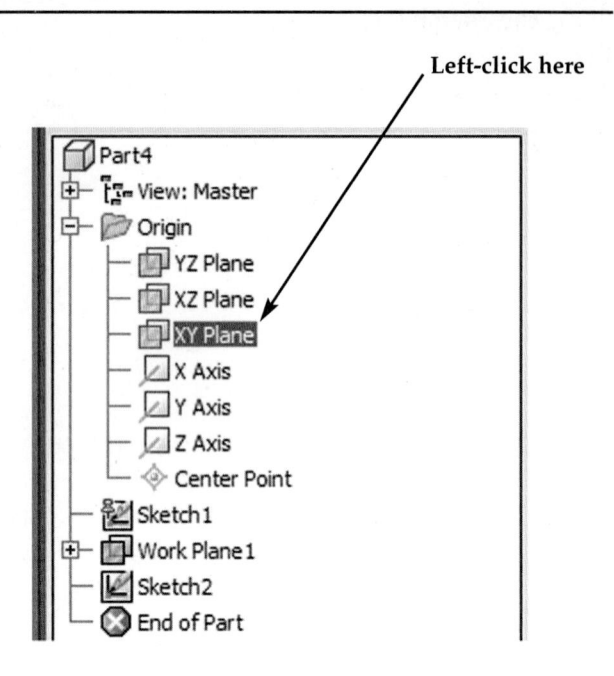

34. Move the cursor to the center of the sketch and left-click (holding the left mouse button down), dragging the cursor to the lower left portion of the screen. Enter **1.000** as shown in Figure 9-31 and press the **Enter key** on the keyboard.

FIGURE 9-31

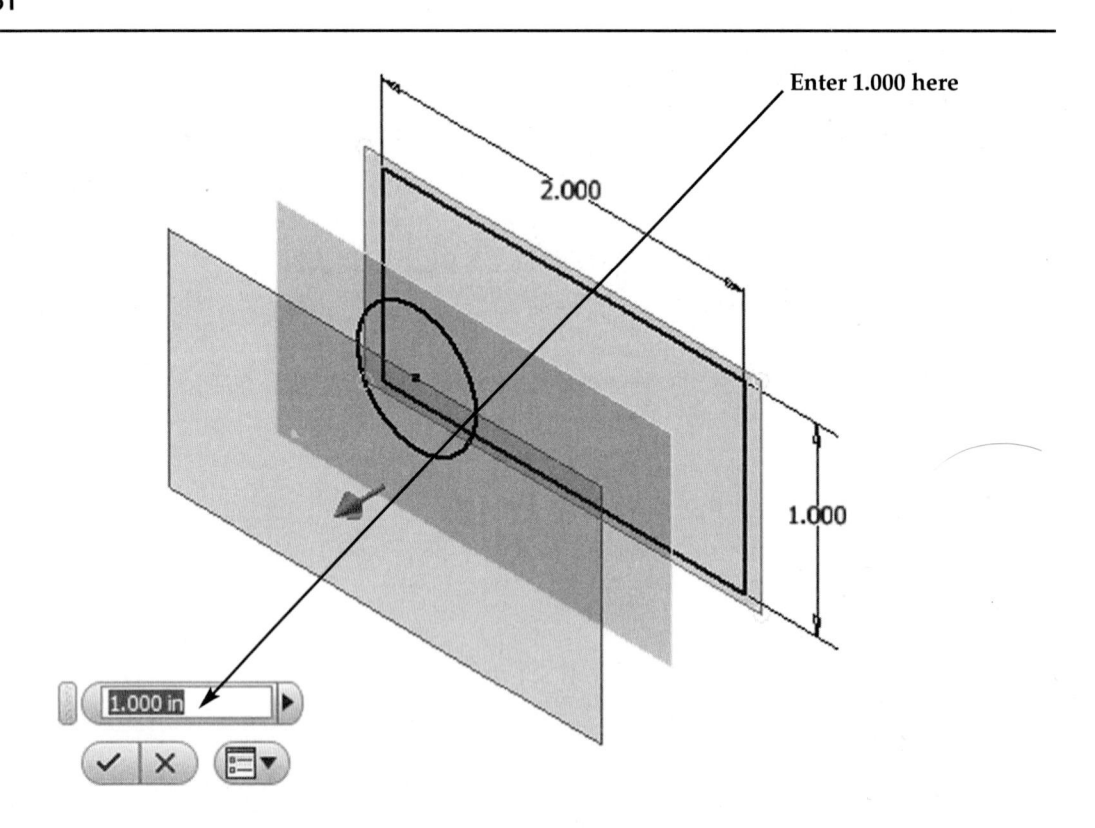

35.   Complete the sketch shown in Figure 9-32. Estimate the location and size of the rectangle and exit the Sketch Panel.

FIGURE 9-32

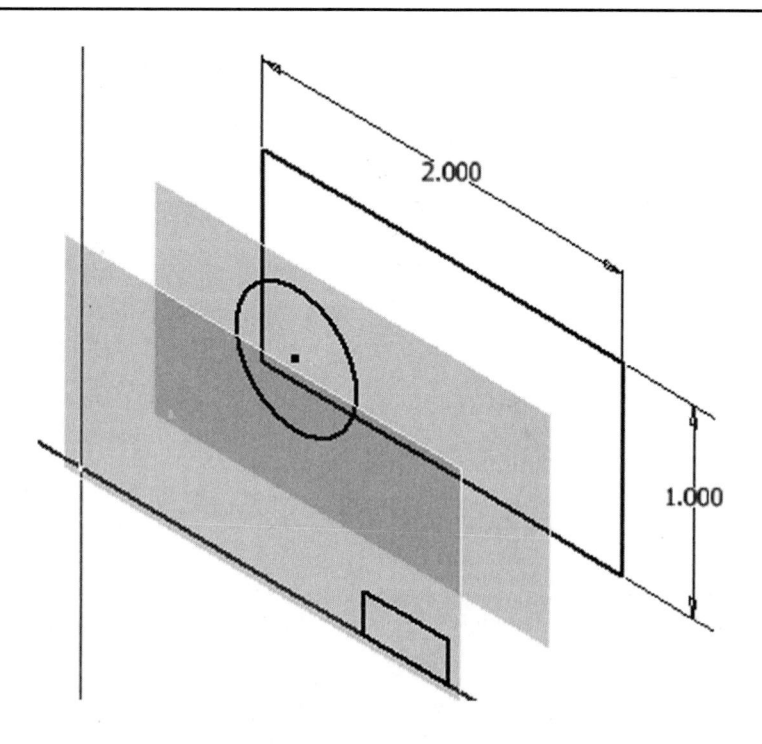

36.   Move the cursor to the upper left portion of the screen and left-click on **Loft** as shown in Figure 9-33.

FIGURE 9-33

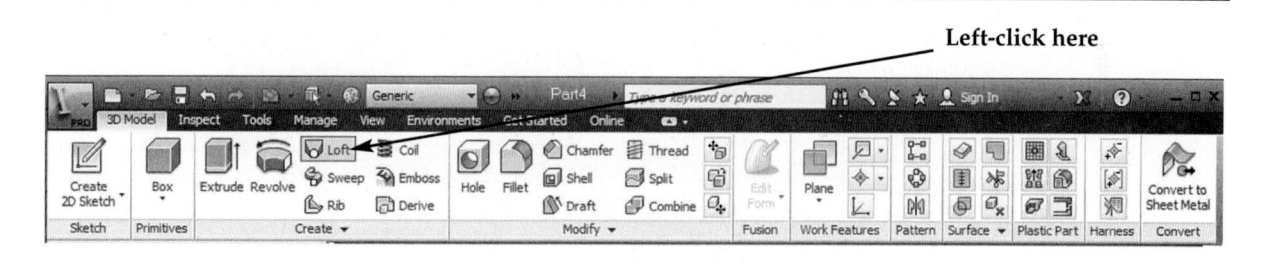

37. The Loft dialog box will appear as shown in Figure 9-34.

FIGURE 9-34

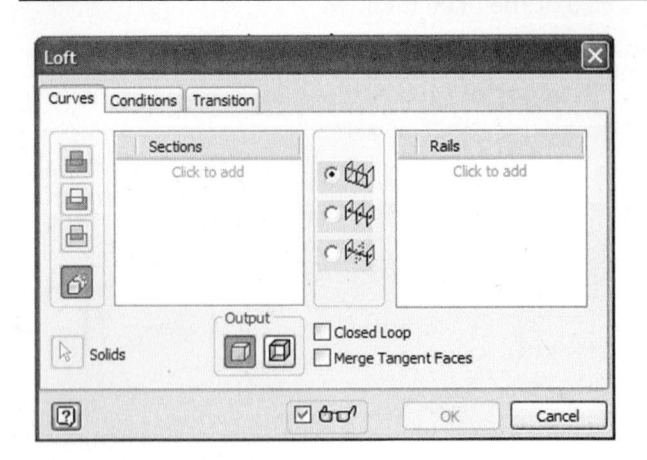

38. Left-click on each of the sketches as shown in Figure 9-35.

FIGURE 9-35

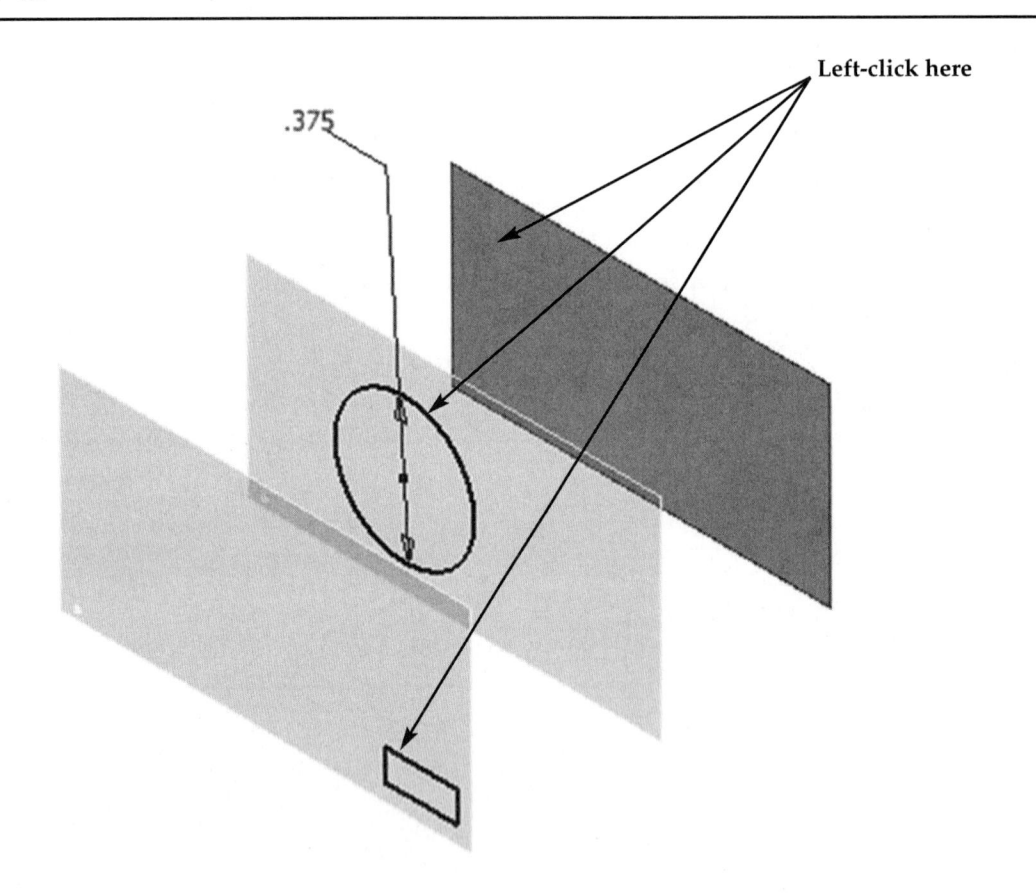

.375

Left-click here

39.   Inventor will provide a preview of the loft as shown in Figure 9-36.

FIGURE 9-36

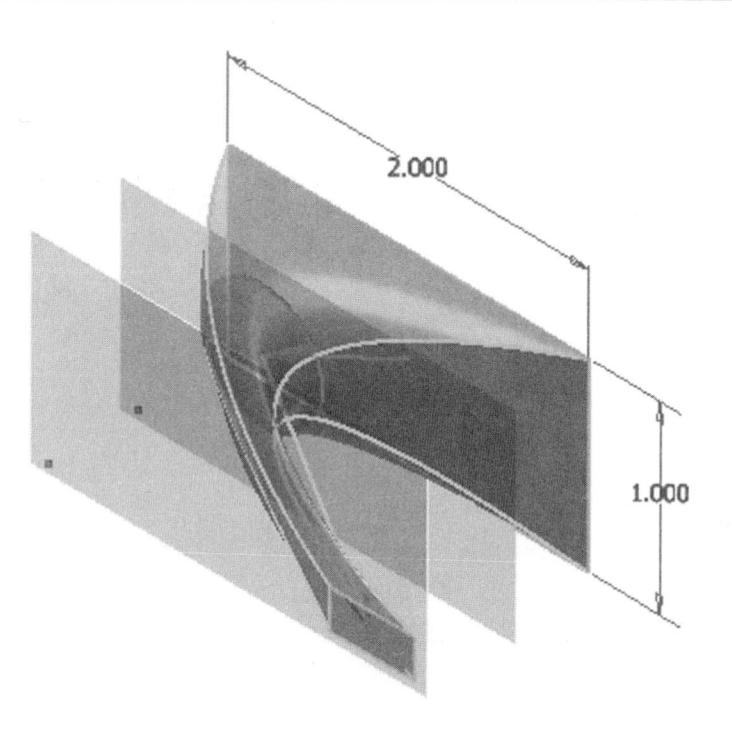

40.   Left-click on **OK**.

41.   Your screen should look similar to Figure 9-37.

FIGURE 9-37

42.　To hide the work planes, move the cursor over the edge of the work plane, causing the edges to turn red. Right-click once. A pop-up menu will appear. Left-click on **Visibility**. Hide both work planes as shown in Figure 9-38.

FIGURE 9-38

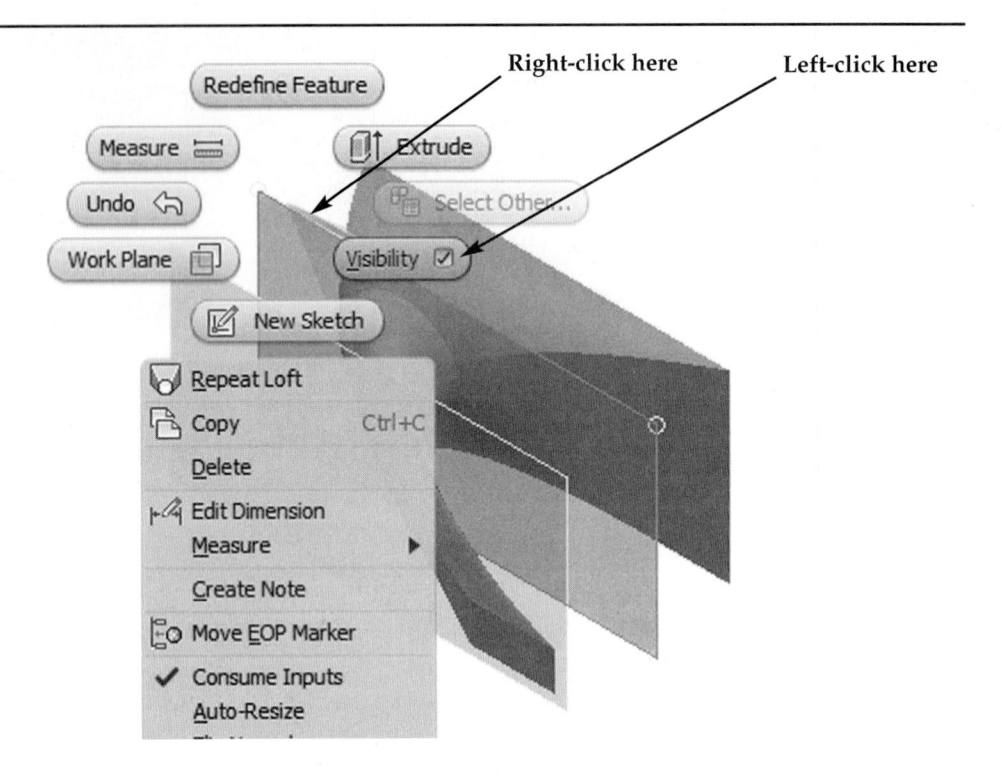

43.　Your screen should look similar to Figure 9-39.

FIGURE 9-39

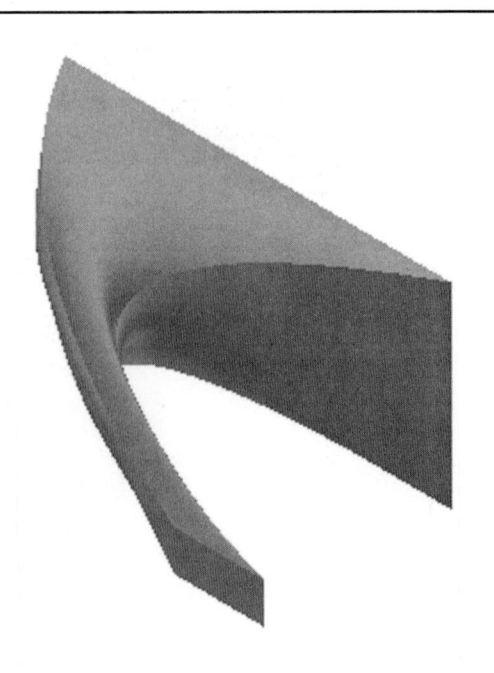

CREATE A COIL USING THE COIL COMMAND

44.   If more accuracy is required for each sketch, geometry can be projected from one work plane to another as previously described in Chapter 6. Geometry can also be sized and located using the Dimension command as discussed in previous chapters.

45.   Begin a new drawing.

46.   Complete the sketch shown in Figure 9-40 and exit the Sketch Panel.

FIGURE 9-40

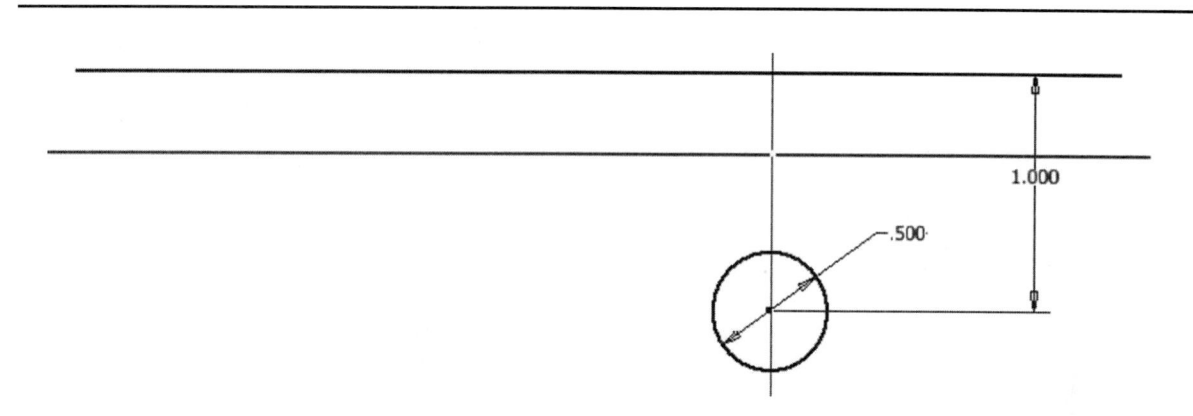

47.   Move the cursor to the middle left portion of the screen and left-click on **Coil**. Left-click on the horizontal line above the circle as shown in Figure 9-41.

FIGURE 9-41

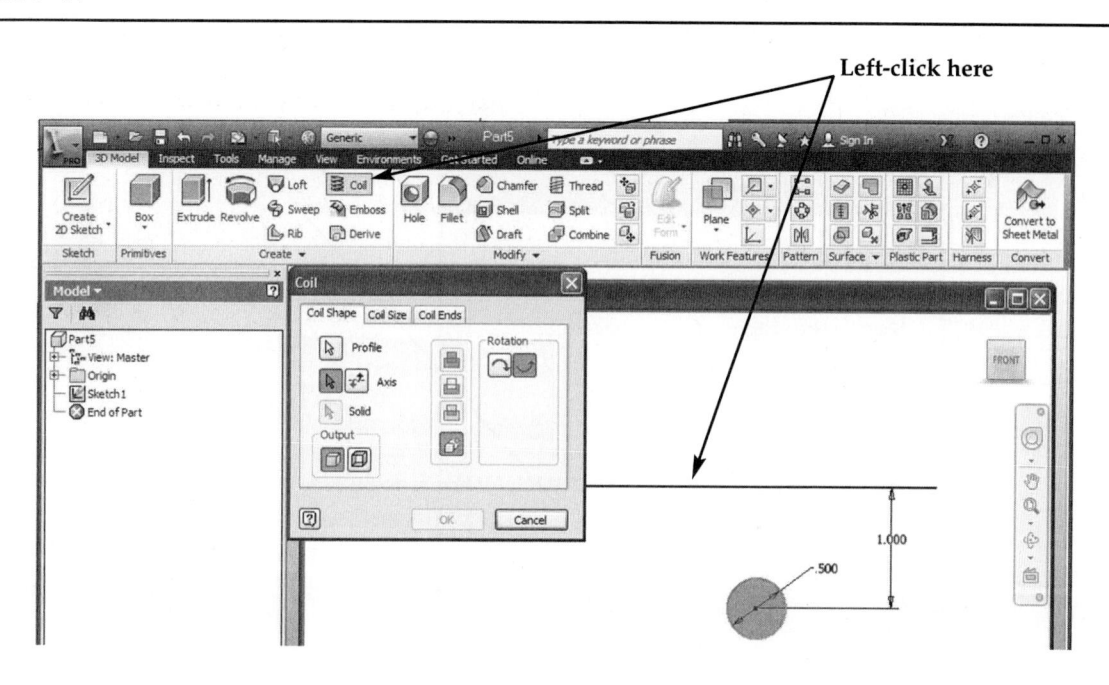

48.   Inventor will provide a preview of the coil. Left-click on the **Axis** icon to reverse the direction of the coil as shown in Figure 9-42.

FIGURE 9-42

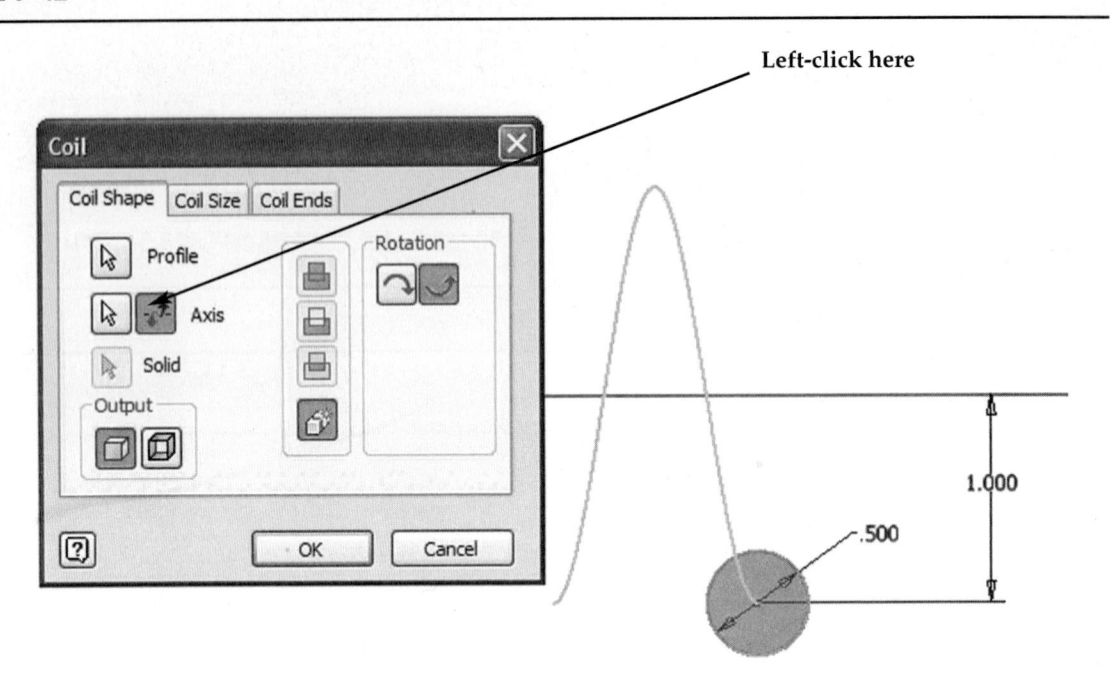

49.   Left-click on the **Coil Size** tab. Under Revolution enter **10**. Left-click on **OK** as shown in Figure 9-43.

FIGURE 9-43

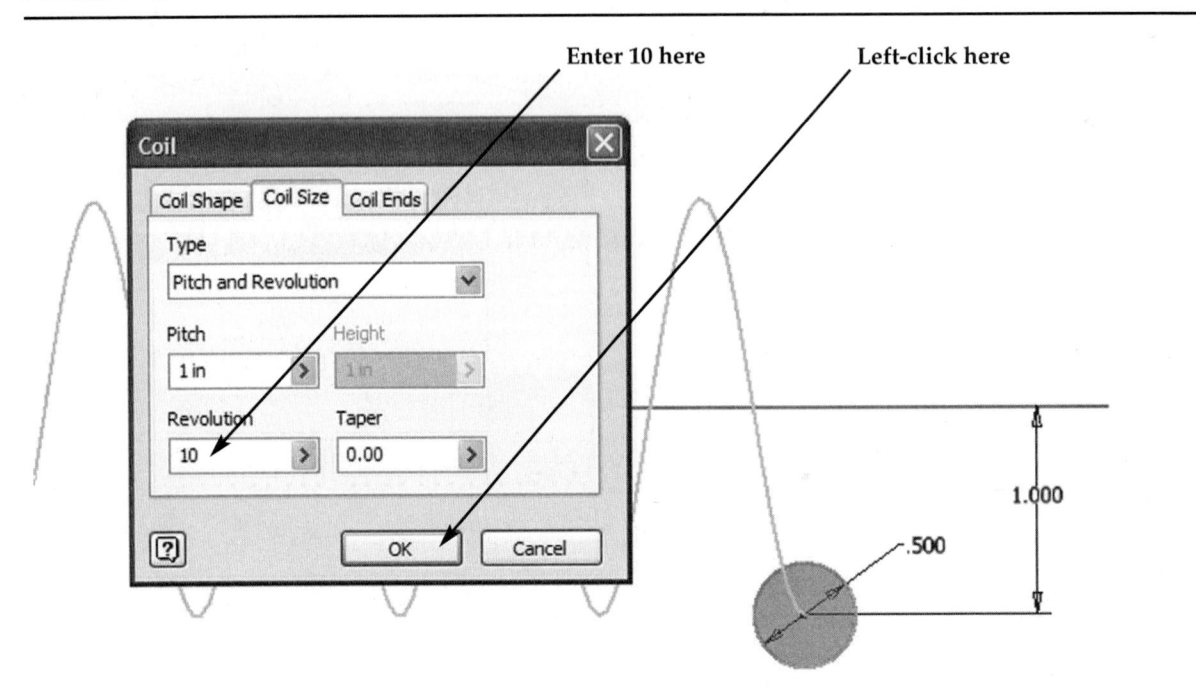

50.   Your screen should look similar to Figure 9-44.

FIGURE 9-44

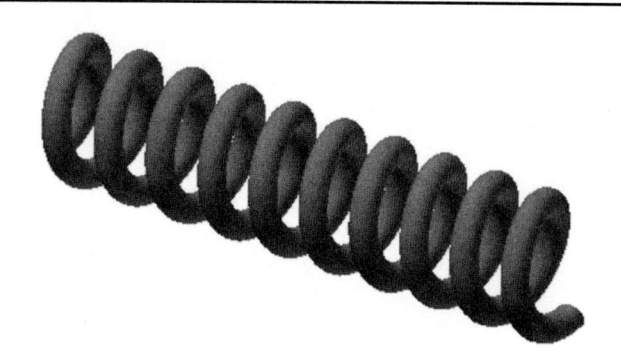

CHAPTER PROBLEMS

Complete the following problem sketches.

PROBLEM 9-1

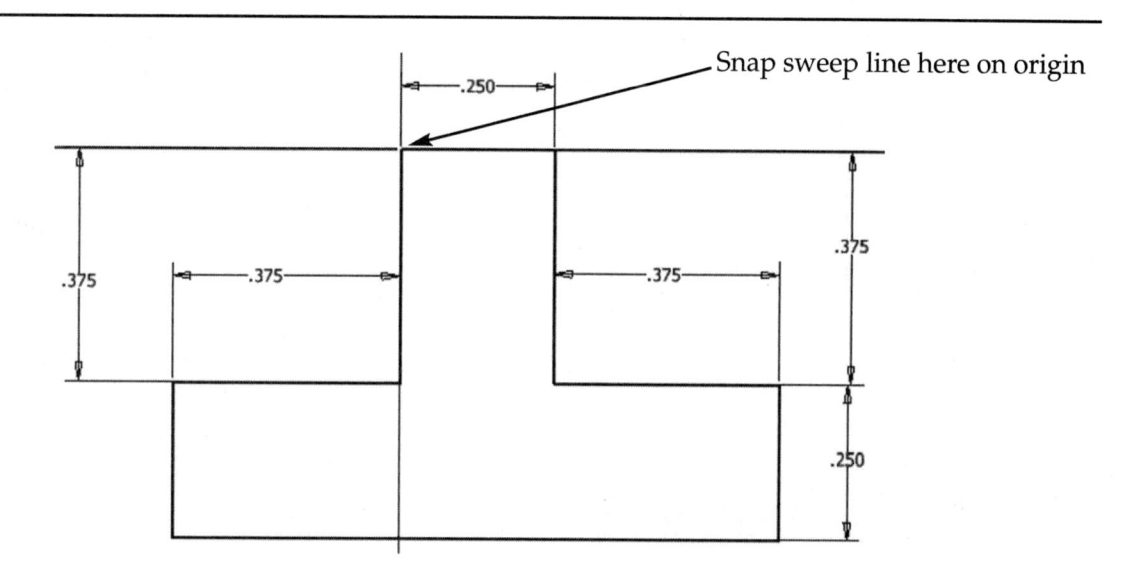

Create the following sweep line. Sweep the sketch along the sweep line shown below.

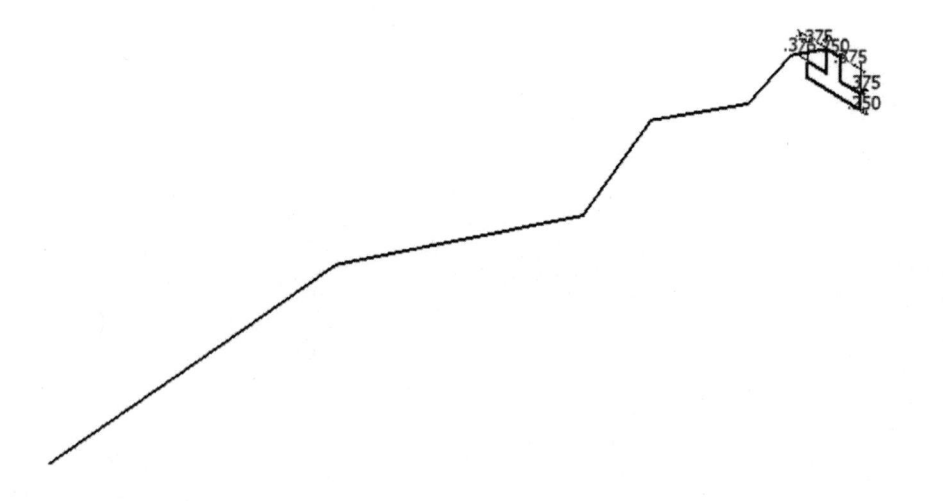

PROBLEM 9-2

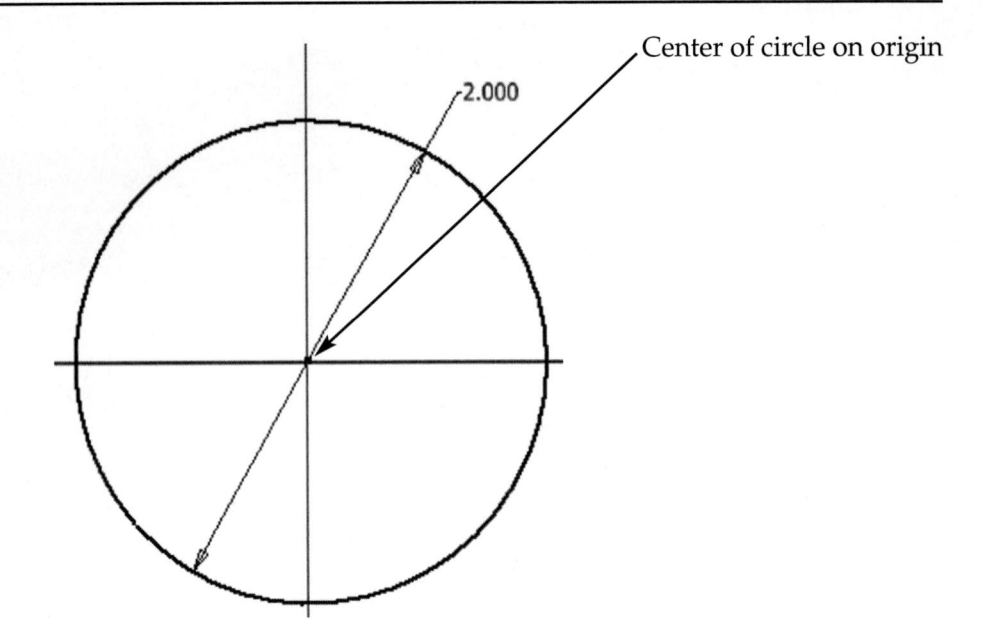

Create the following Sweep line. The center of the circle must intersect the sweep line as shown.

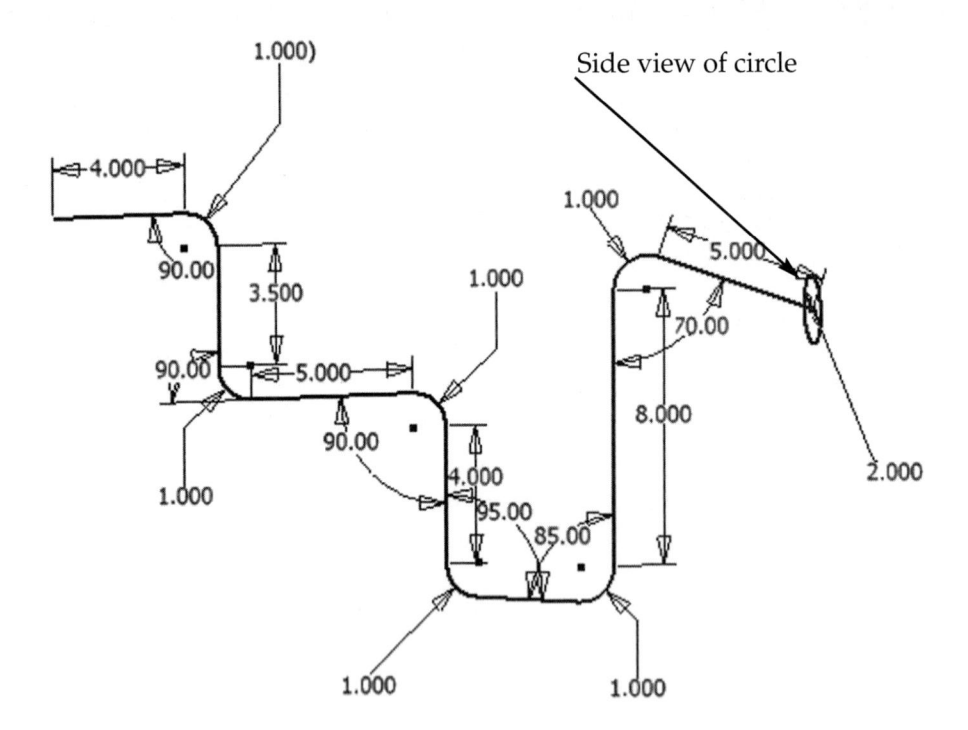

CHAPTER 10

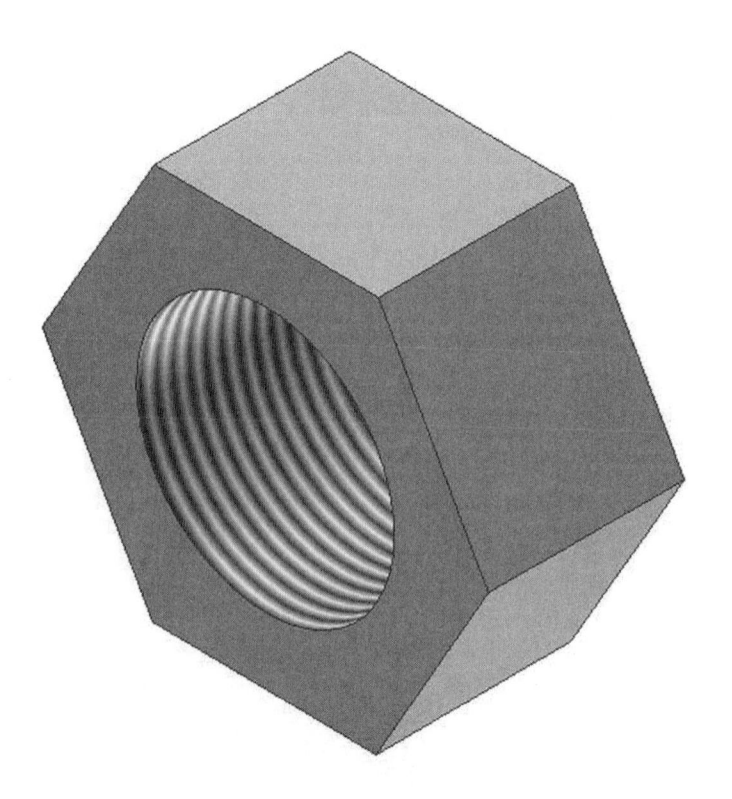

INTRODUCTION TO CREATING THREADS

OBJECTIVES

1. Use the Polygon command
2. Create threads in a solid model
3. Interpret threads specs

Chapter 10 includes instruction on how to create threads in a small part and how to design the part shown above.

CREATE A POLYGON

1. Start Inventor by referring to Chapter 1. After Inventor is running, begin a New Sketch.

2. Move the cursor to the middle left portion of the screen and left-click on **Polygon** as shown in Figure 10-1.

FIGURE 10-1

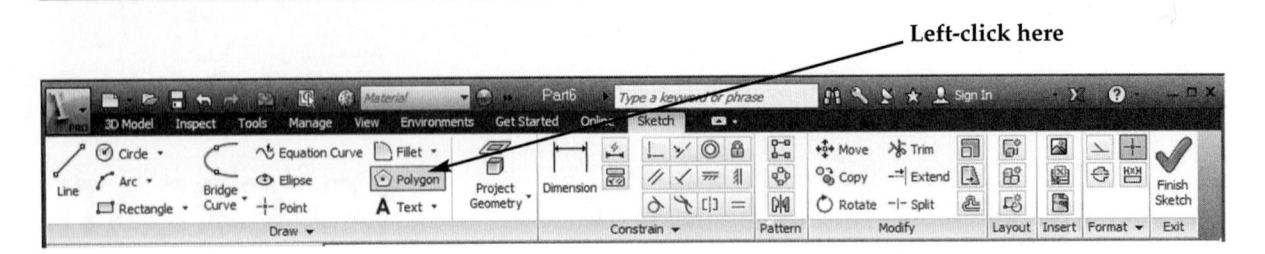

3. The Polygon dialog box will appear. Left-click on the "Circumscribed" icon. This will be used to define the distance across the flats (basically what size wrench or socket would be used to loosen or tighten the nut). Enter **6** for the number of sides as shown in Figure 10-2.

FIGURE 10-2

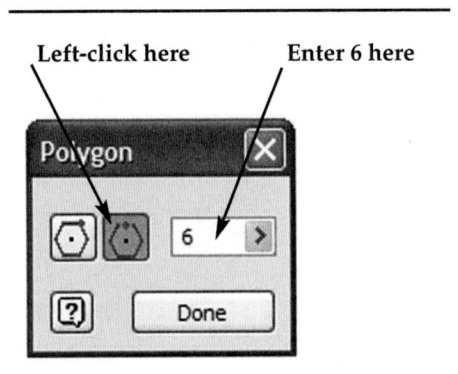

4. Left-click on the origin and drag the cursor out to the left, similar to creating a circle as shown in Figure 10-3.

FIGURE 10-3

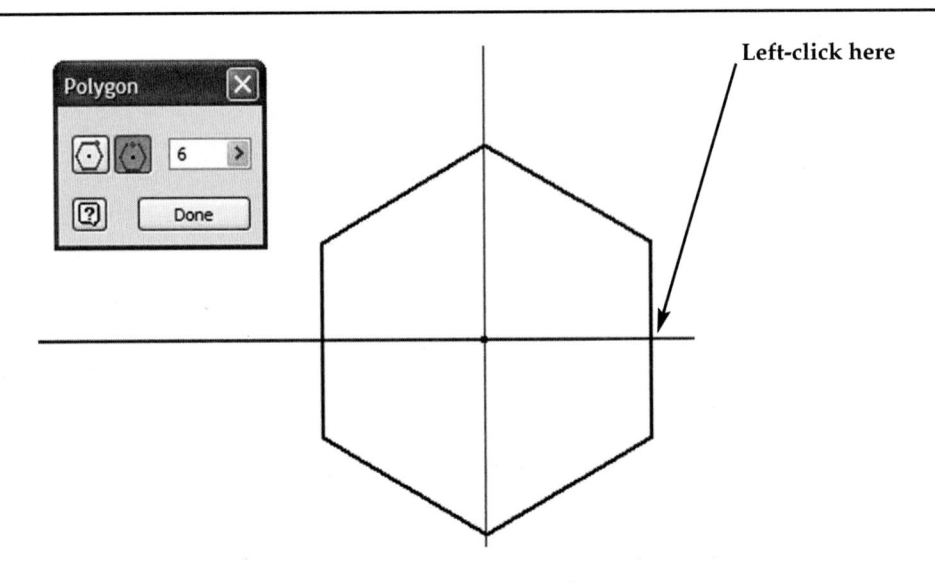

5.   Left-click on **Done** as shown in Figure 10-4.

FIGURE 10-4

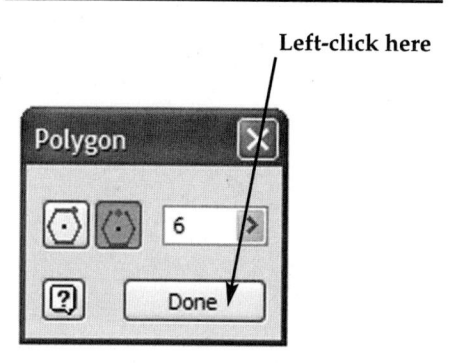

6.   Complete and dimension the sketch as shown in Figure 10-5.

FIGURE 10-5

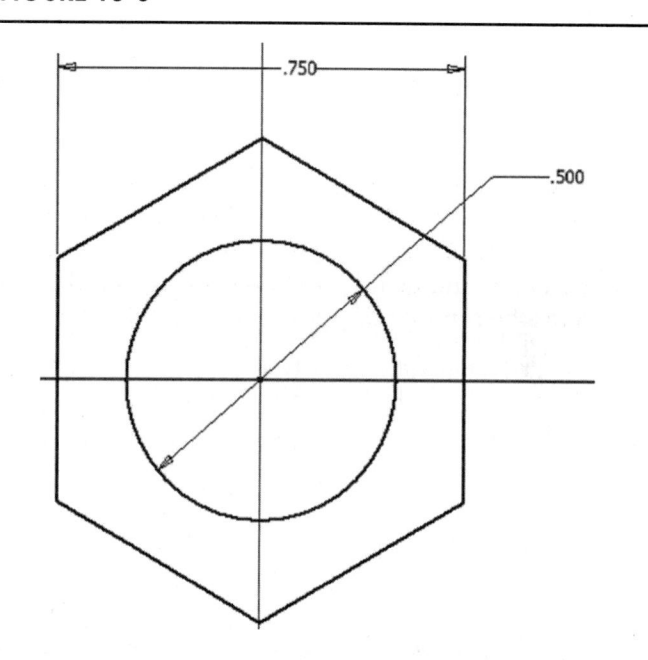

7.   Exit out of the Sketch Panel as shown in Figure 10-6.

FIGURE 10-6

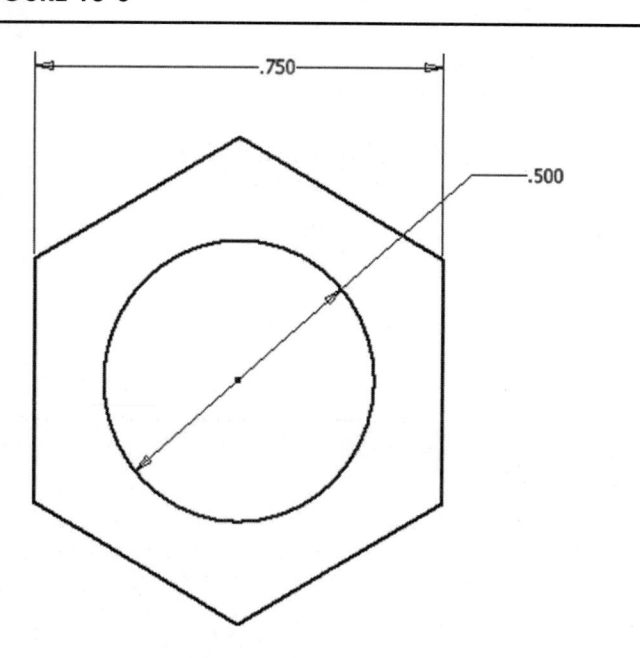

8.   Rotate the part around to gain an isometric view of the part as shown in Figure 10-7.

FIGURE 10-7

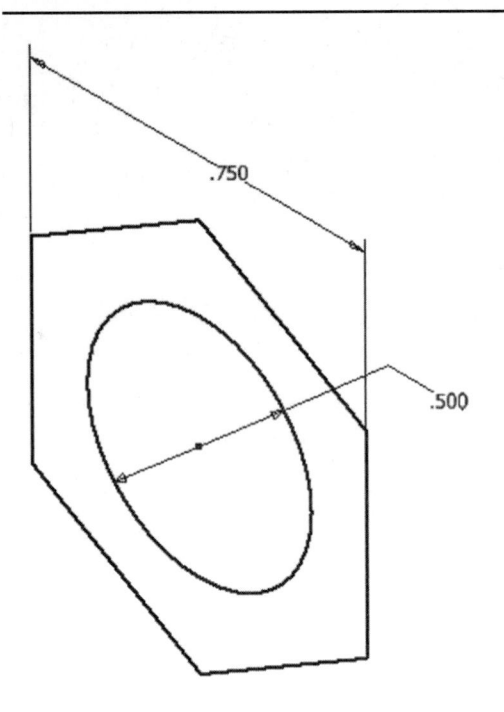

9.   Extrude the part a distance of **.375** inches as shown in Figure 10-8.

FIGURE 10-8

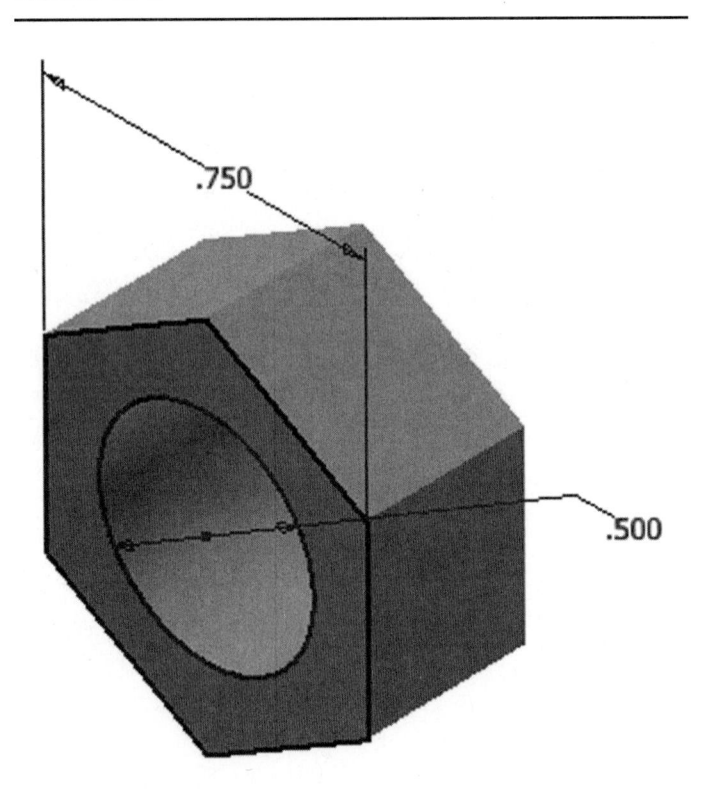

CREATE THREADS

10. Move the cursor to the upper middle portion of the screen and left-click on **Thread**. The Thread dialog box will appear as shown in Figure 10-9.

FIGURE 10-9

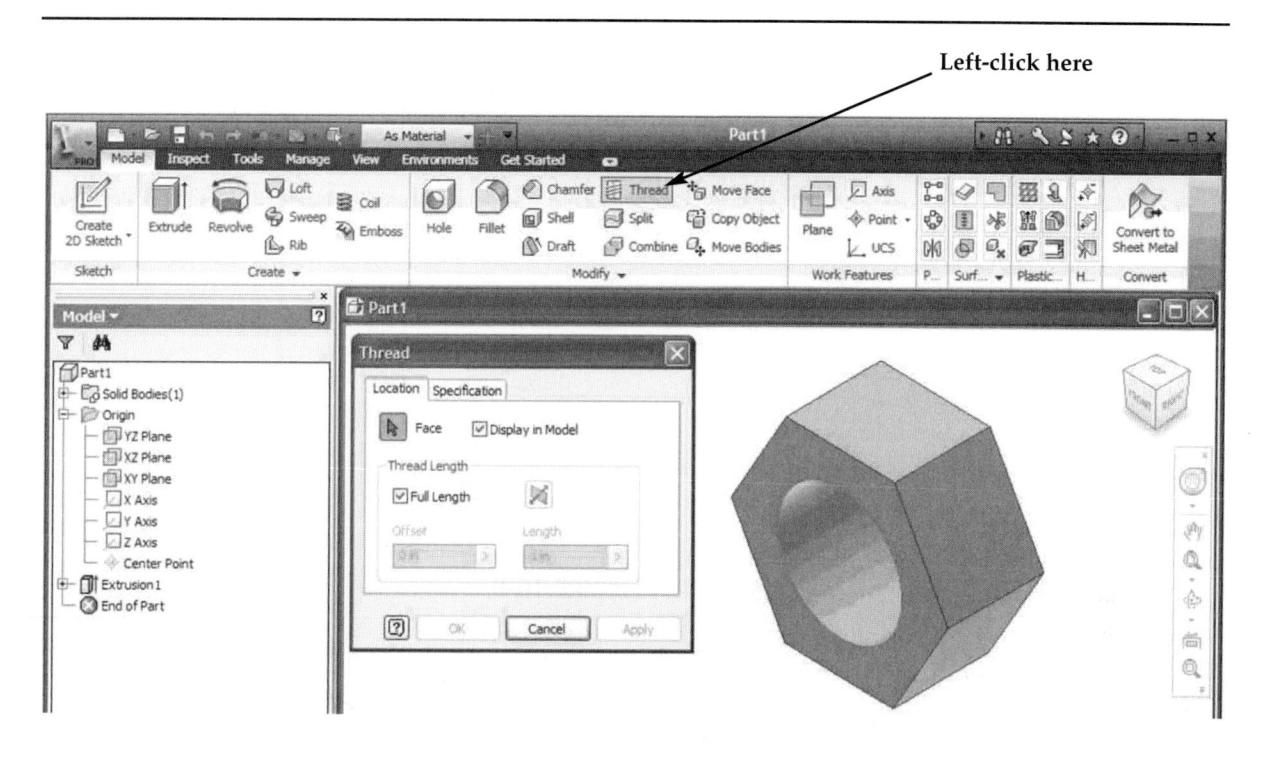

11. Move the cursor inside the hole and left-click once as shown in Figure 10-10.

FIGURE 10-10

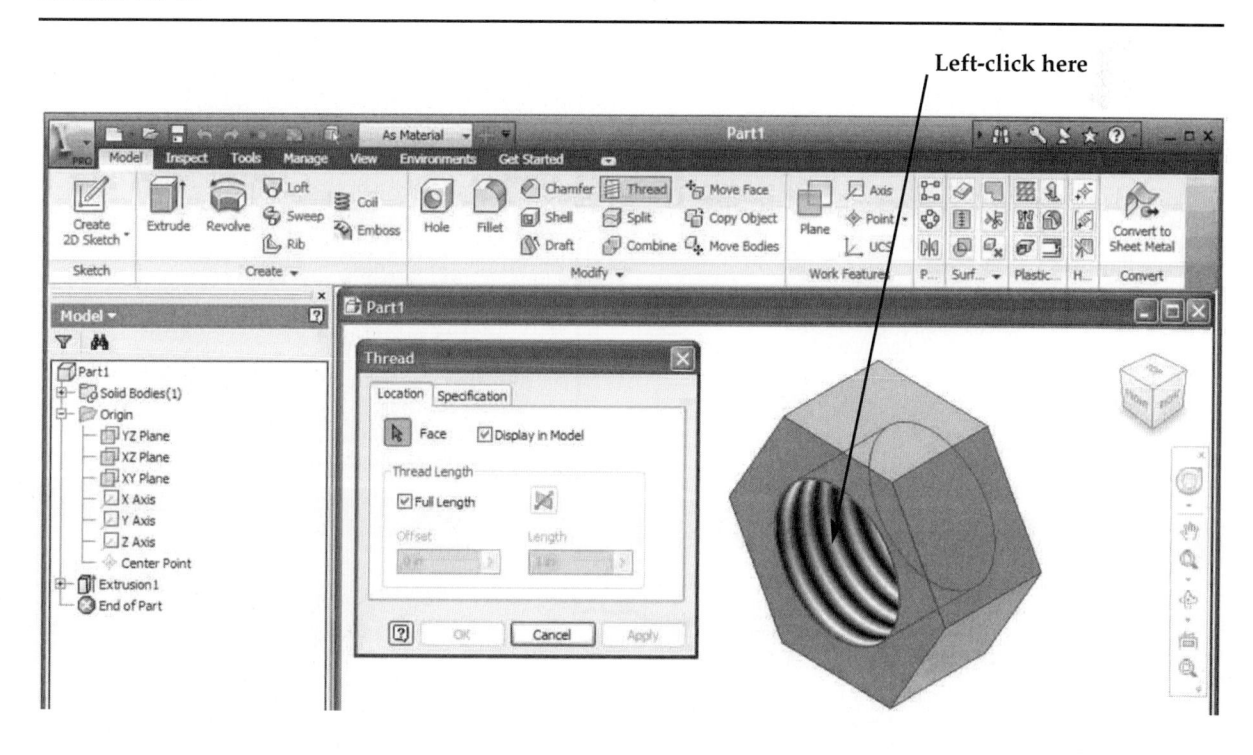

12. Left-click on the **Specification** tab. Left-click on the drop-down arrow located underneath **Designation**. A drop-down menu will appear. A list of compatible threads will appear. The first number refers to the diameter of the thread (bolt or hole). The second number refers to the pitch (in the case, the number of threads per inch). The letters indicate what type of thread and whether the threads are coarse or fine as shown in Figure 10-11.

FIGURE 10-11

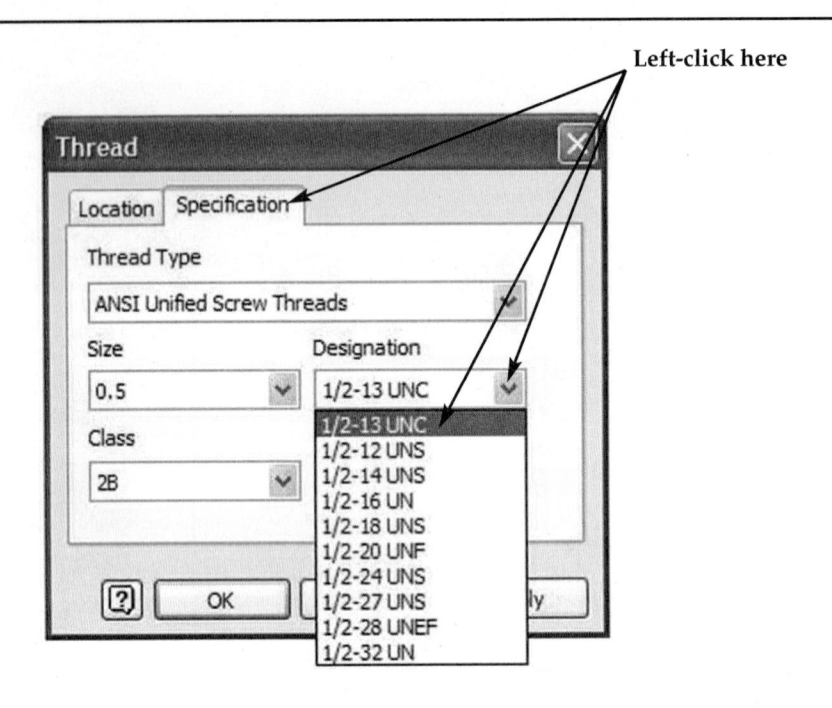

13. Inventor will provide a preview of the different thread types selected as shown in Figure 10-12.

FIGURE 10-12

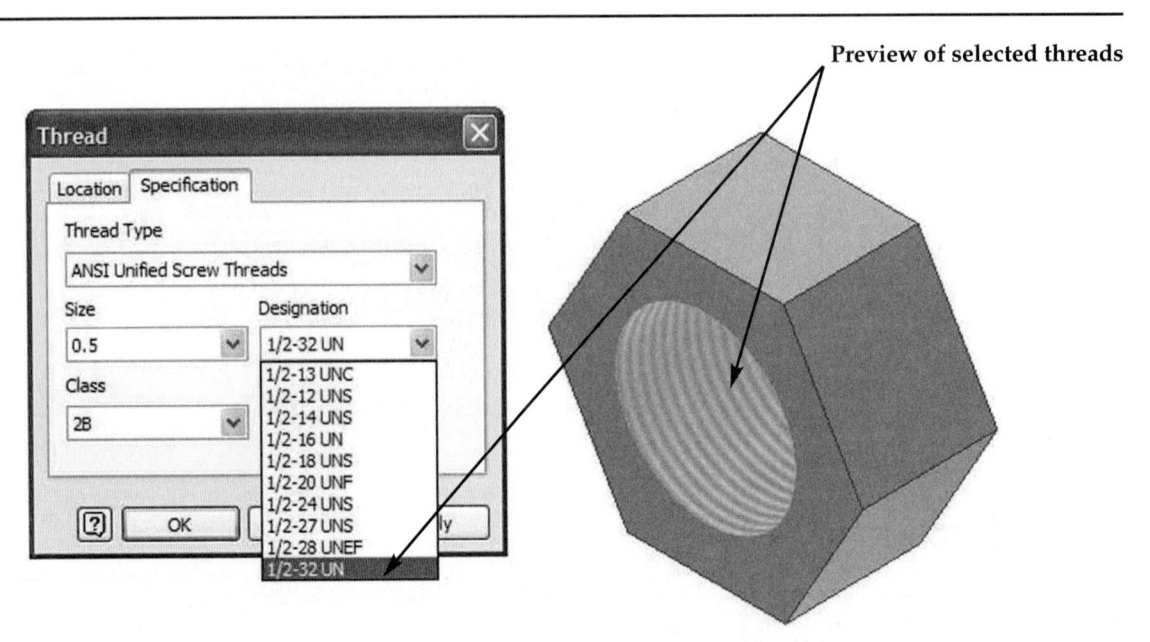

14.   Select a thread of your choice and left-click on **OK** as shown in Figure 10-13.

FIGURE 10-13

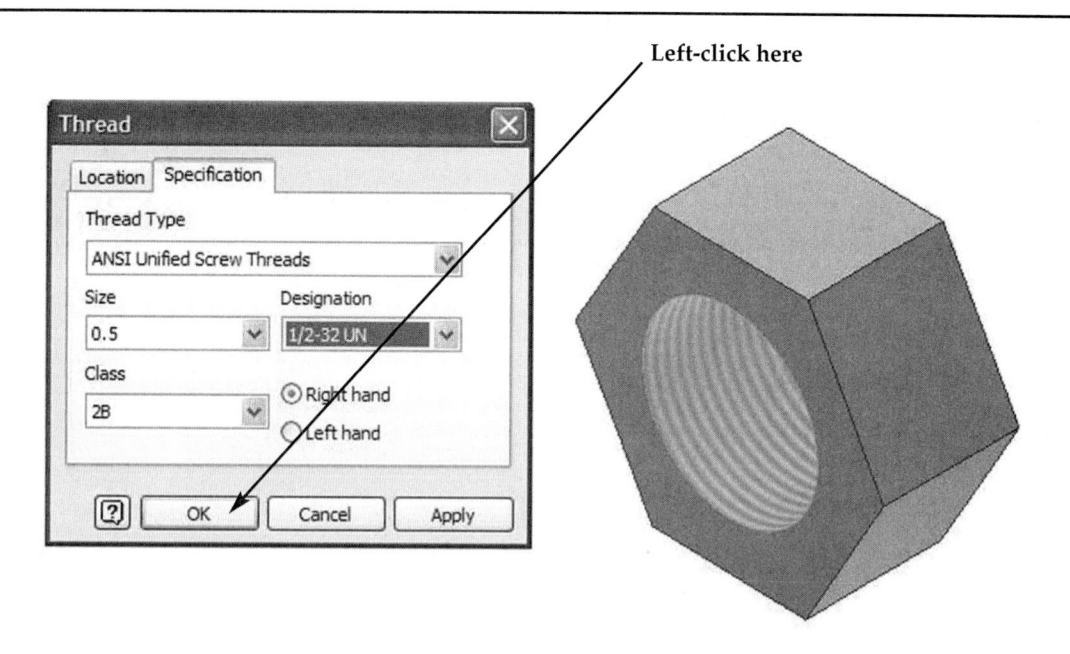

15.   Your screen should look similar to Figure 10-14.

FIGURE 10-14

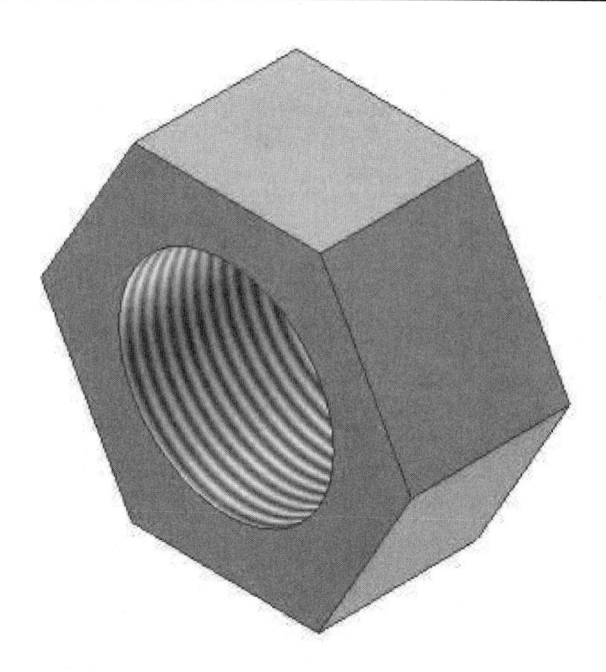

CHAPTER

ADVANCED WORK PLANE PROCEDURES

Chapter 11 includes instruction on how to create multiple points on a solid model and use these points to create an offset work plane like the part shown above.

OBJECTIVES

1. Create points on a solid model
2. Use the Split command
3. Create an offset/oblique Work Plane

1.   Start Inventor by referring to Chapter 1. After Inventor is running, begin a New Sketch.

2.   Complete the sketch shown in Figure 11-1.

FIGURE 11-1

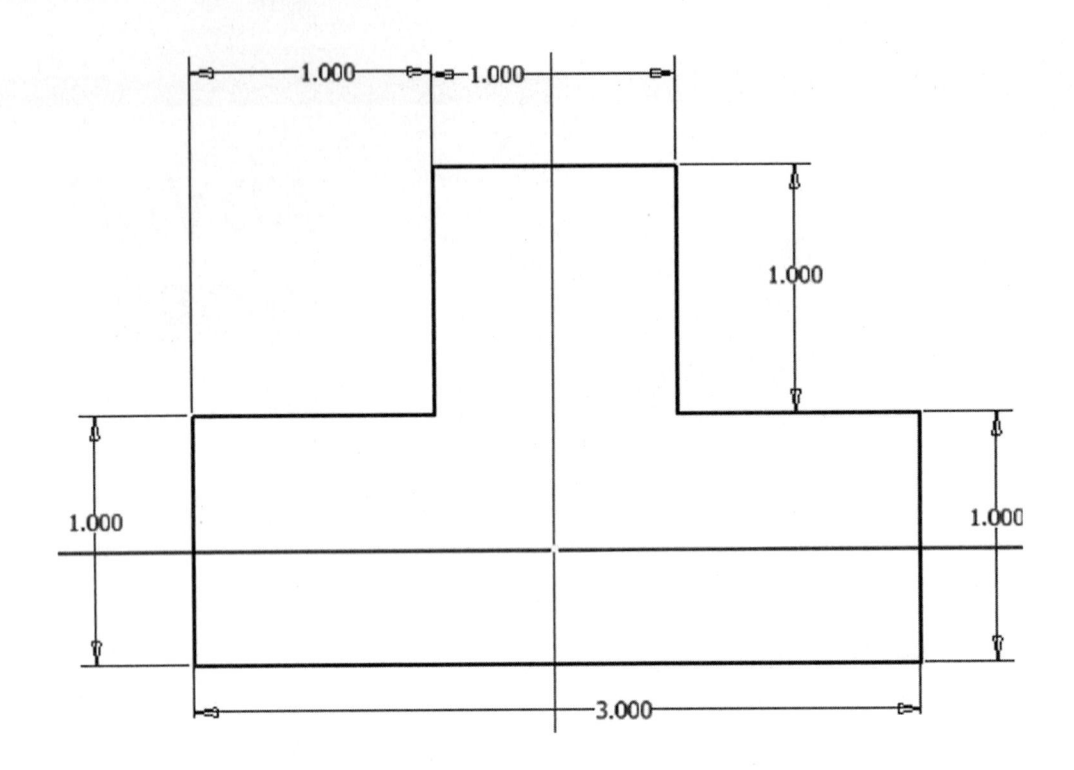

3.   Exit the Sketch Panel. Extrude the sketch to a thickness of **4.00** inches as shown in Figure 11-2.

FIGURE 11-2

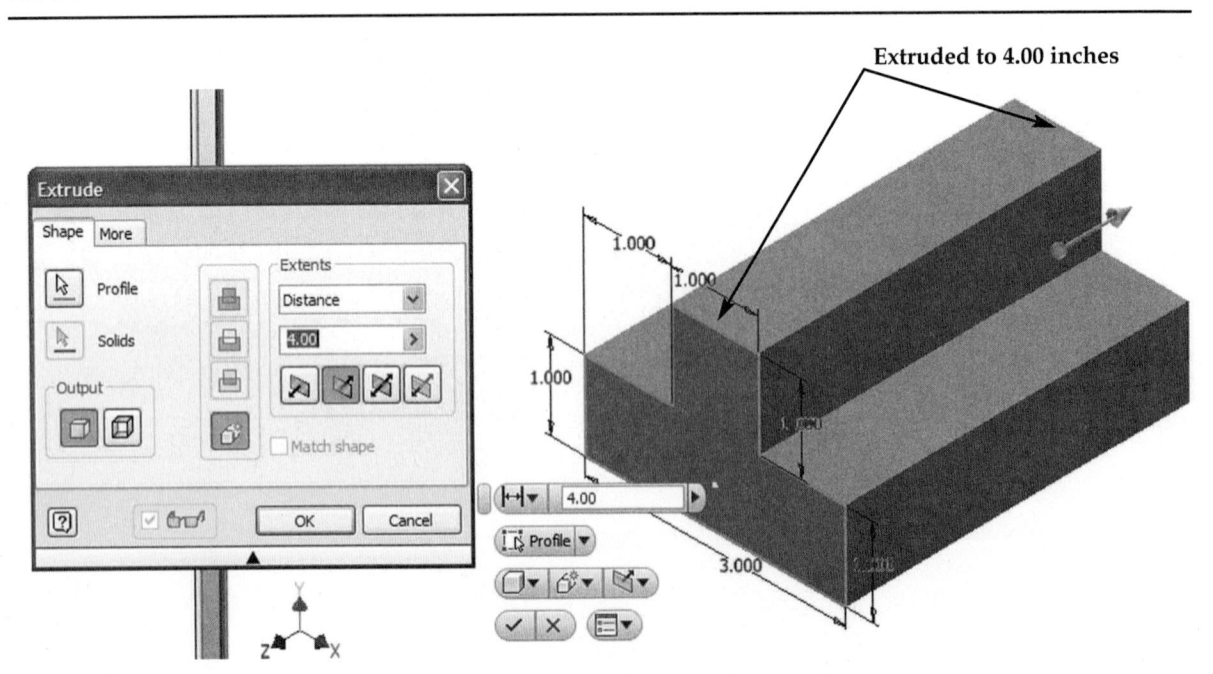

4.   Your screen should look similar to Figure 11-3.

FIGURE 11-3

5.   Use the **Rotate** command to rotate the part to gain a perpendicular view of the front surface as shown in Figure 11-4.

FIGURE 11-4

6. Use the **Rectangle** command to complete the sketch as shown in Figure 11-5. All dimensions are .188.

FIGURE 11-5

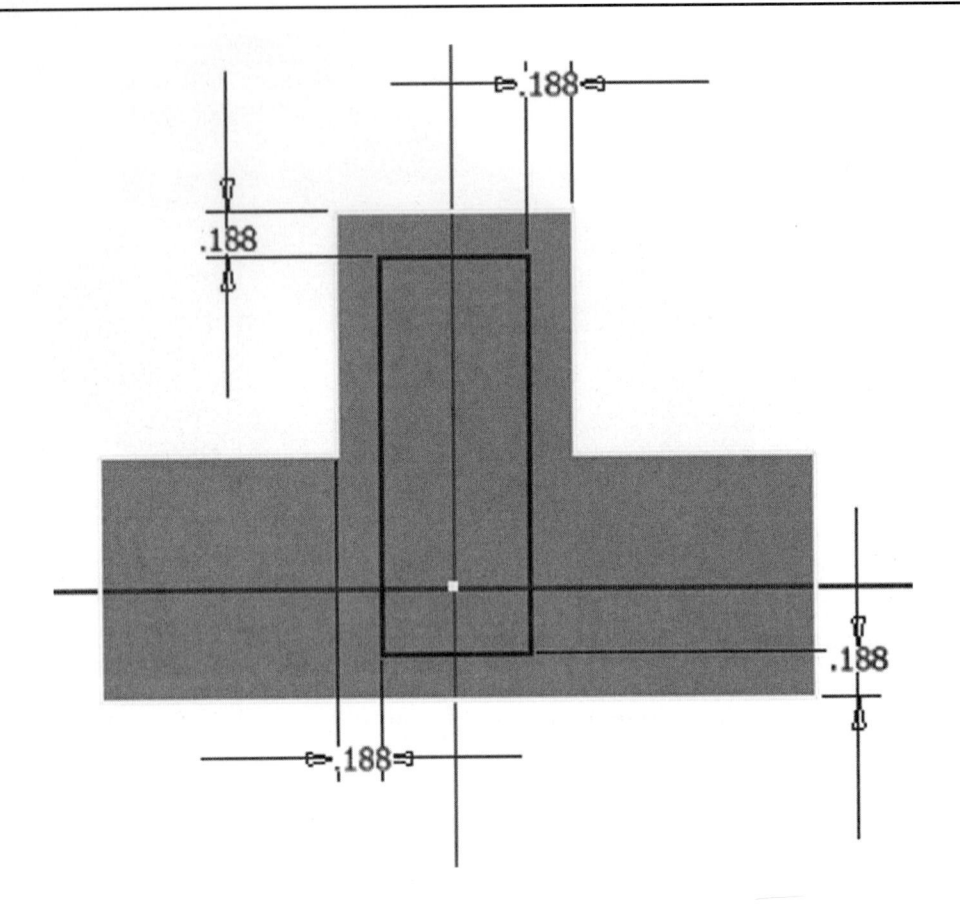

7. Extrude (cut) the rectangle a distance of **4.00** inches, creating a square hole as shown in Figure 11-6.

FIGURE 11-6

8.   Rotate the part around to gain an Isometric (Home View) of the part as shown in Figure 11-7.

FIGURE 11-7

9.   Rotate the part around to gain an Isometric View of the side of the part as shown in Figure 11-8.

FIGURE 11-8

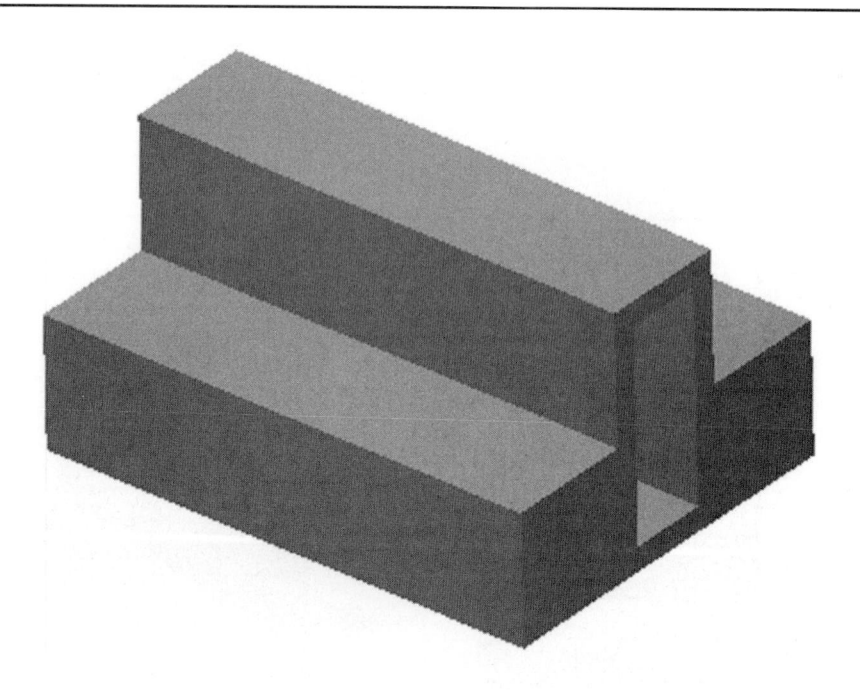

CREATE POINTS ON MULTIPLE SKETCHES

10.   Begin a **New Sketch** on the front surface of the part as shown in Figure 11-9.

FIGURE 11-9

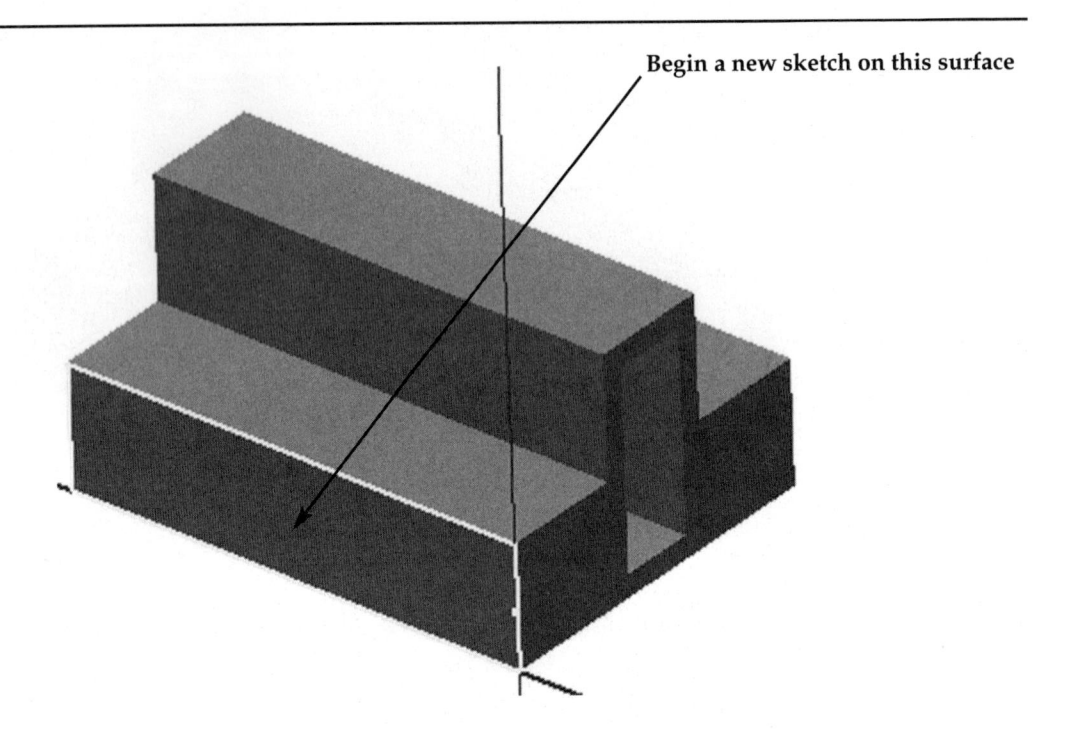

Begin a new sketch on this surface

11.   Complete the sketch (2 points) as shown in Figure 11-10.

FIGURE 11-10

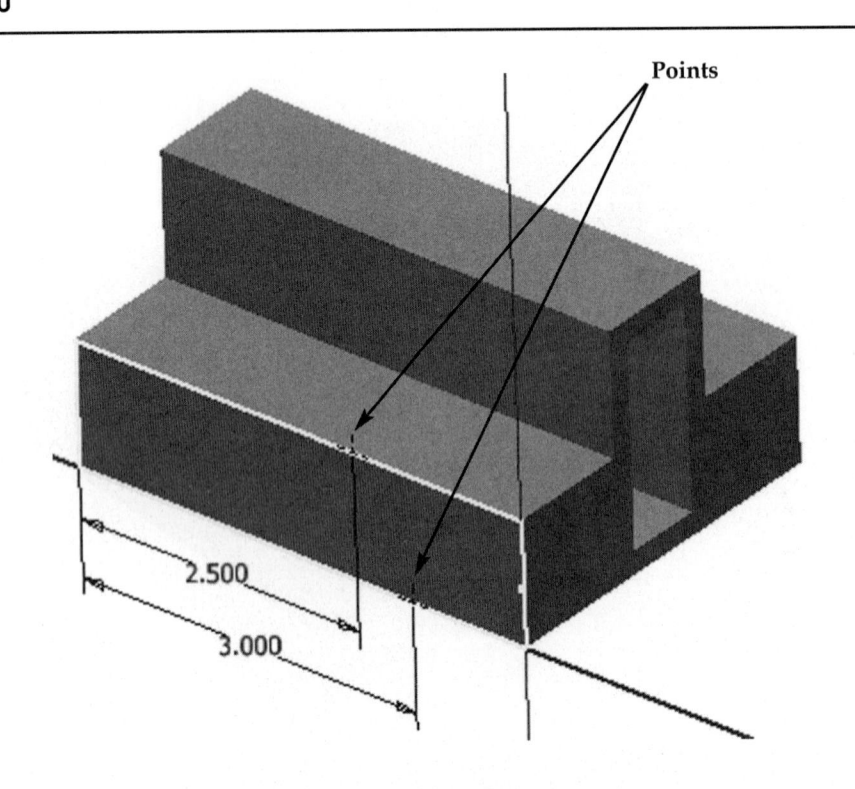

Points

2.500

3.000

12.   Exit out of the Sketch as shown in Figure 11-11.

FIGURE 11-11

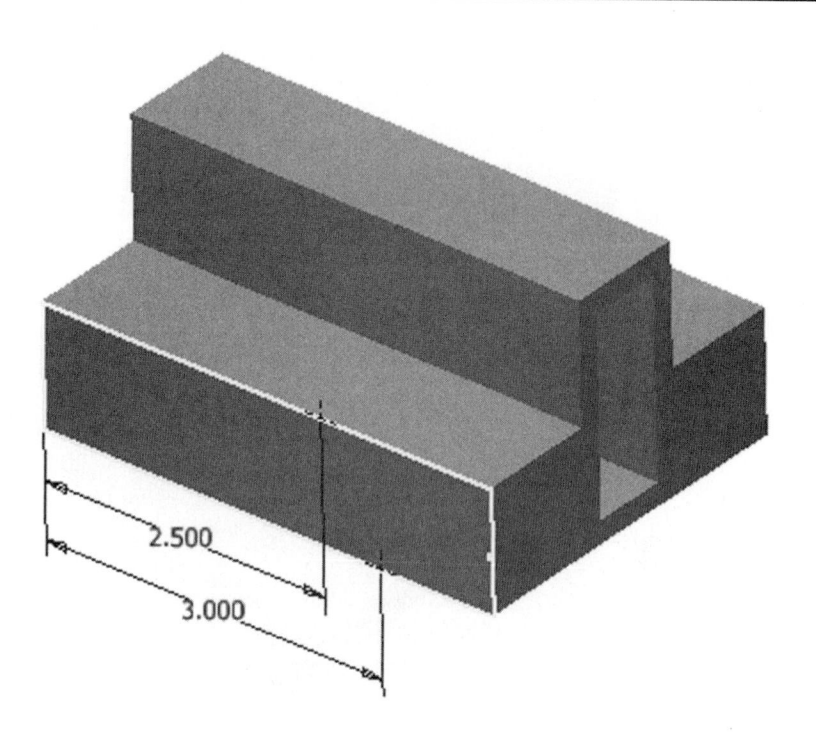

13.   Rotate the part upward as shown in Figure 11-12.

FIGURE 11-12

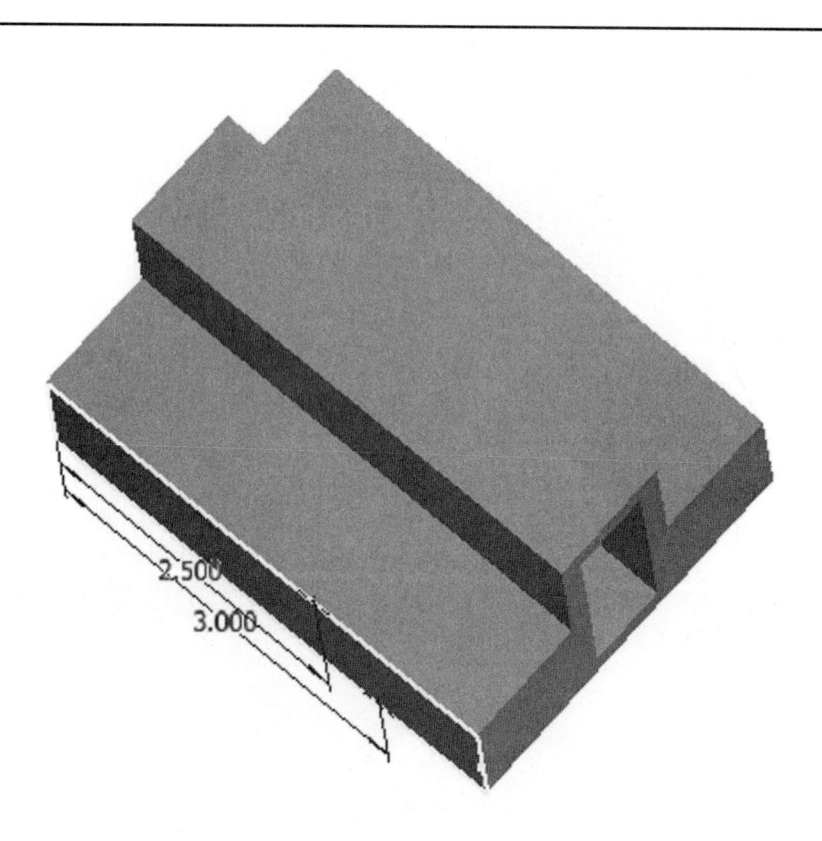

14. Begin a **New Sketch** on the surface shown in Figure 11-13.

FIGURE 11-13

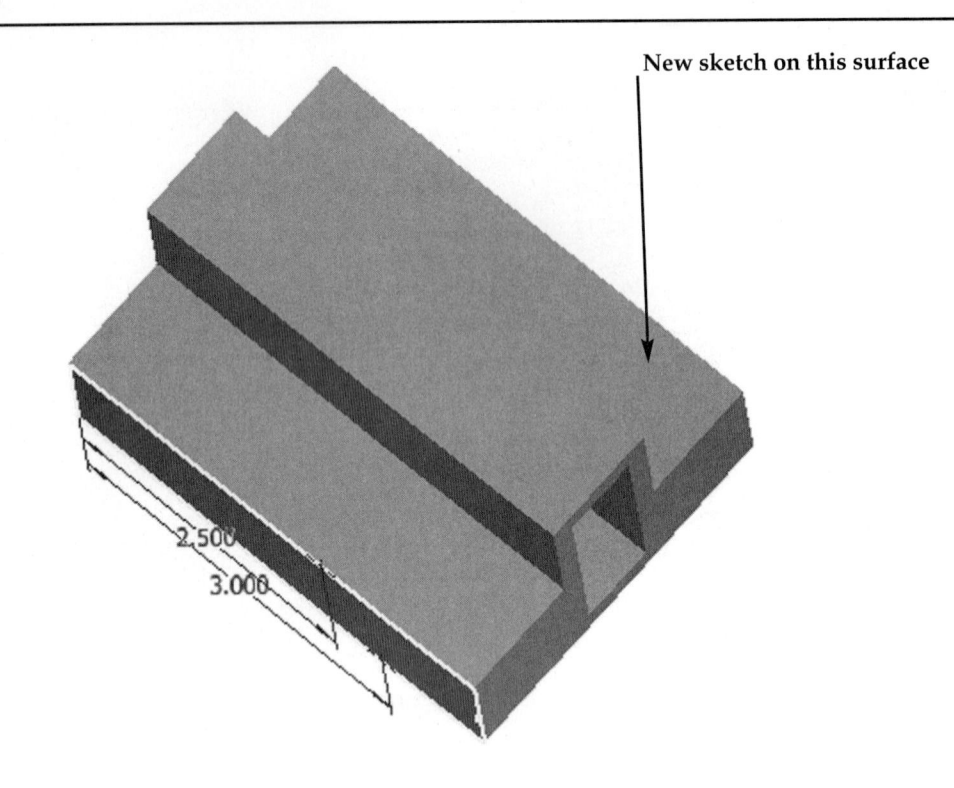

15. Create a **Point** on the surface as shown in Figure 11-14.

FIGURE 11-14

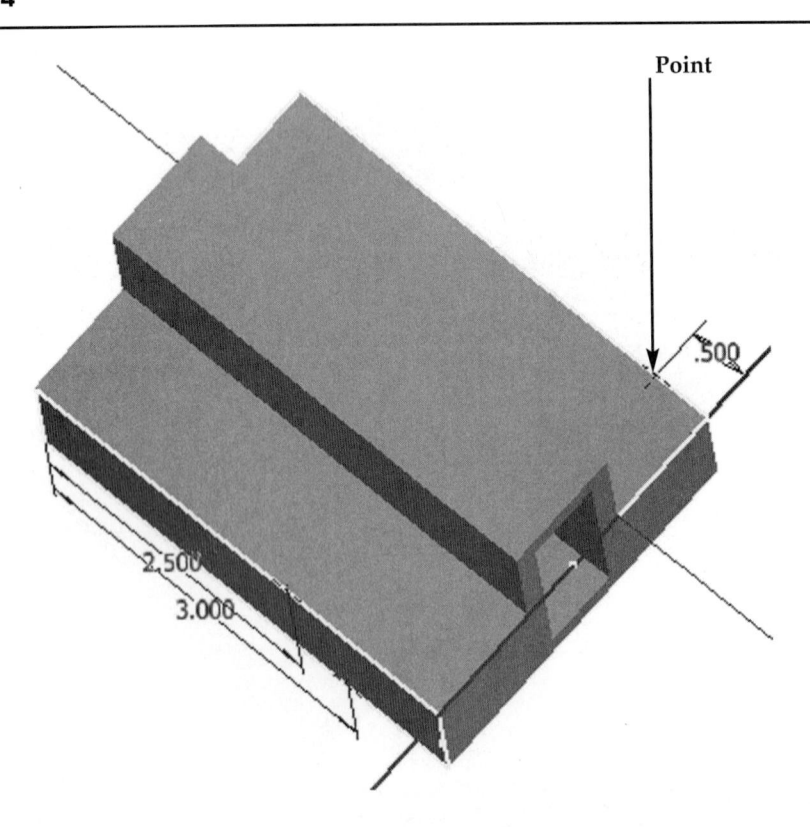

USE THESE POINTS TO CREATE AN OFFSET WORK PLANE

16.   Exit the Sketch as shown in Figure 11-15.

FIGURE 11-15

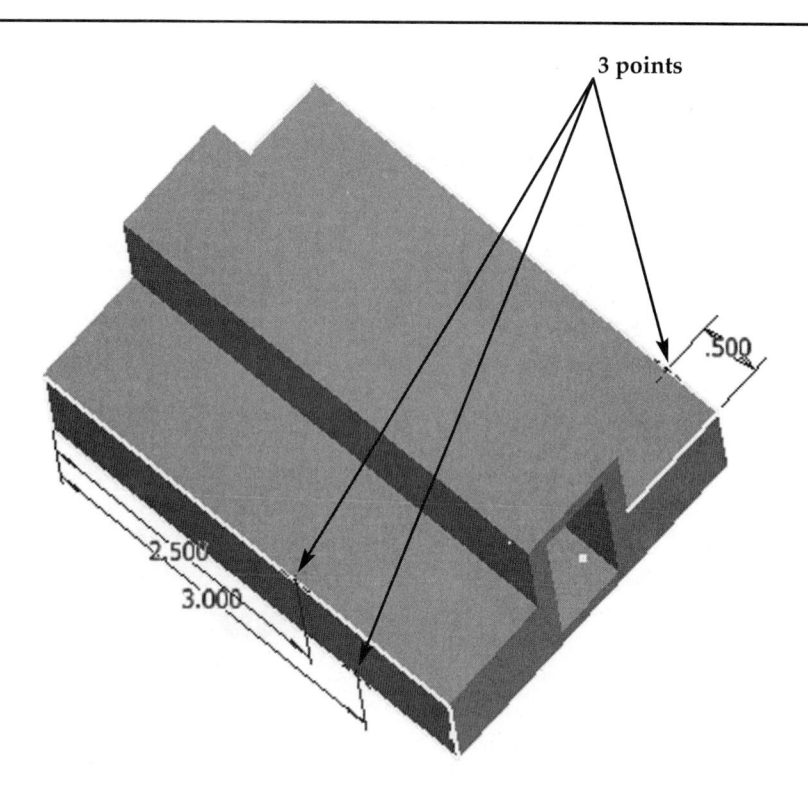

17.   Move the cursor to the upper middle portion of the screen and left-click on **Plane/Work Plane** as shown in Figure 11-16.

FIGURE 11-16

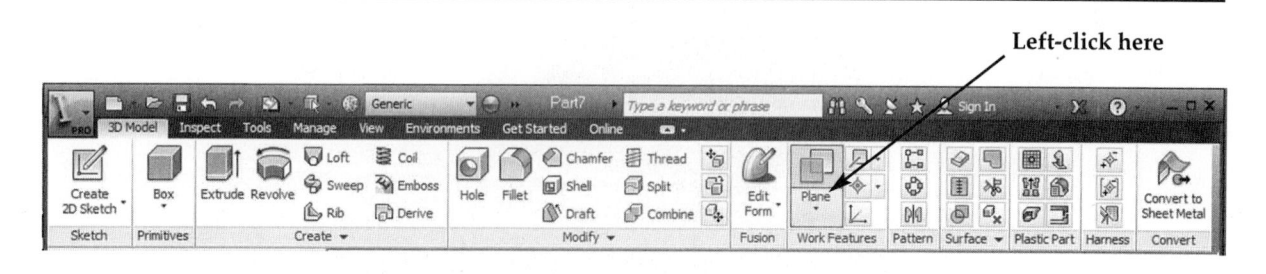

18.   Left-click on each of the 3 points as shown in Figure 11-17.

FIGURE 11-17

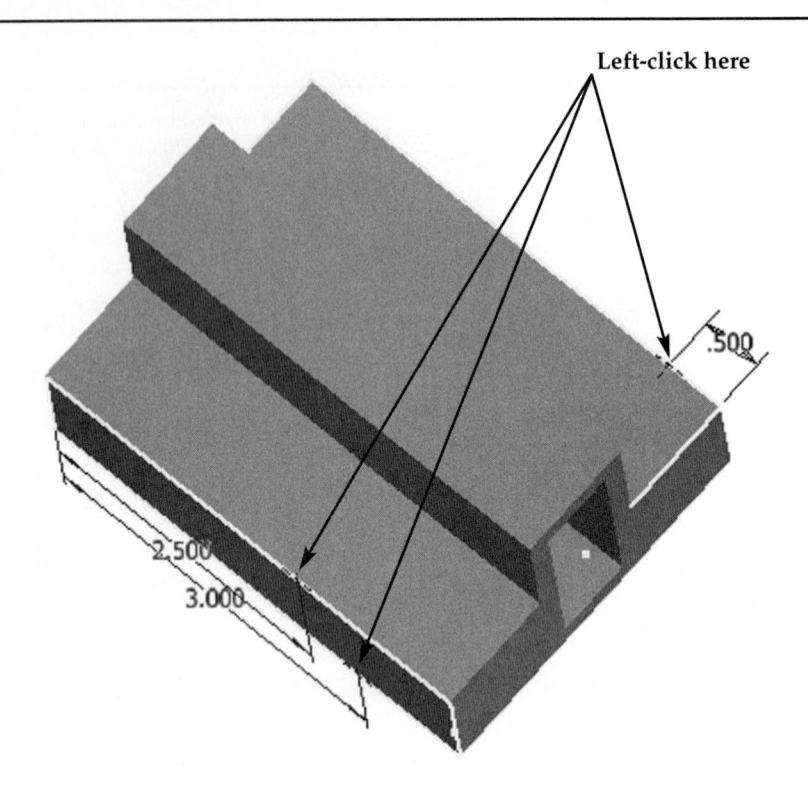

19.   Inventor will create a Work Plane from the 3 points as shown in Figure 11-18.

FIGURE 11-18

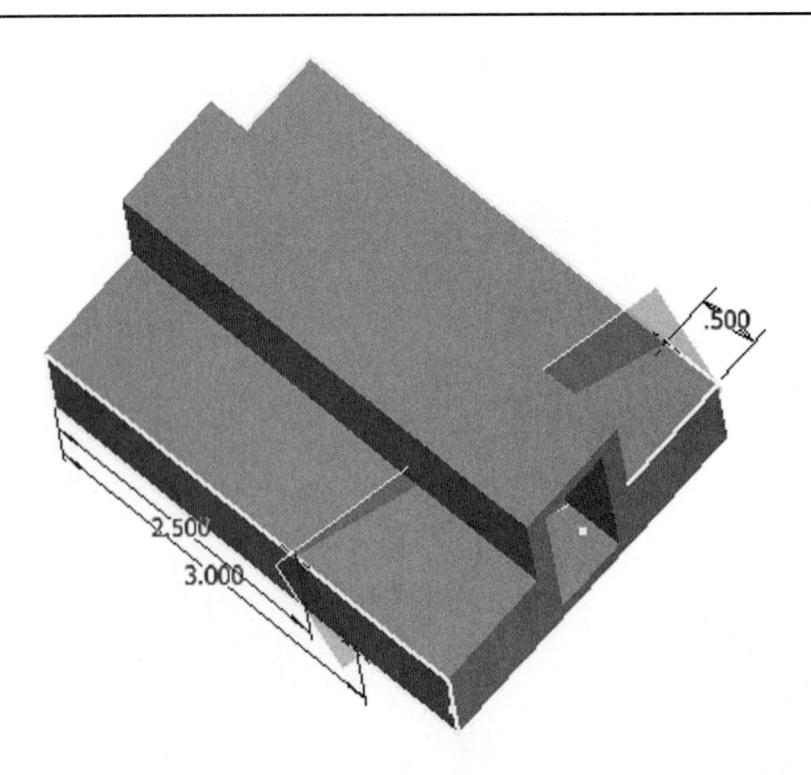

20. Move the cursor to the upper middle portion of the screen and left-click on **Split** as shown in Figure 11-19.

FIGURE 11-19

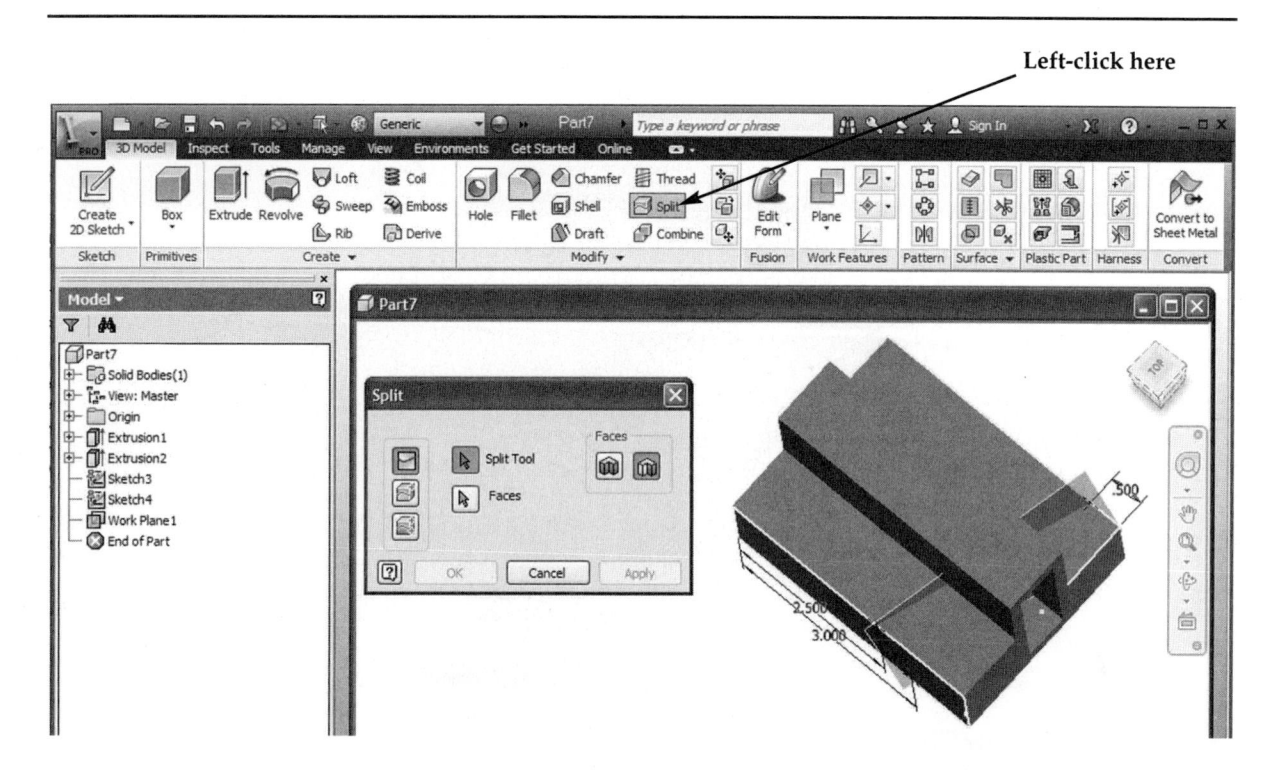

21. Move the cursor over the edge of the newly created plane, causing the edges to turn red and left-click as shown in Figure 11-20.

FIGURE 11-20

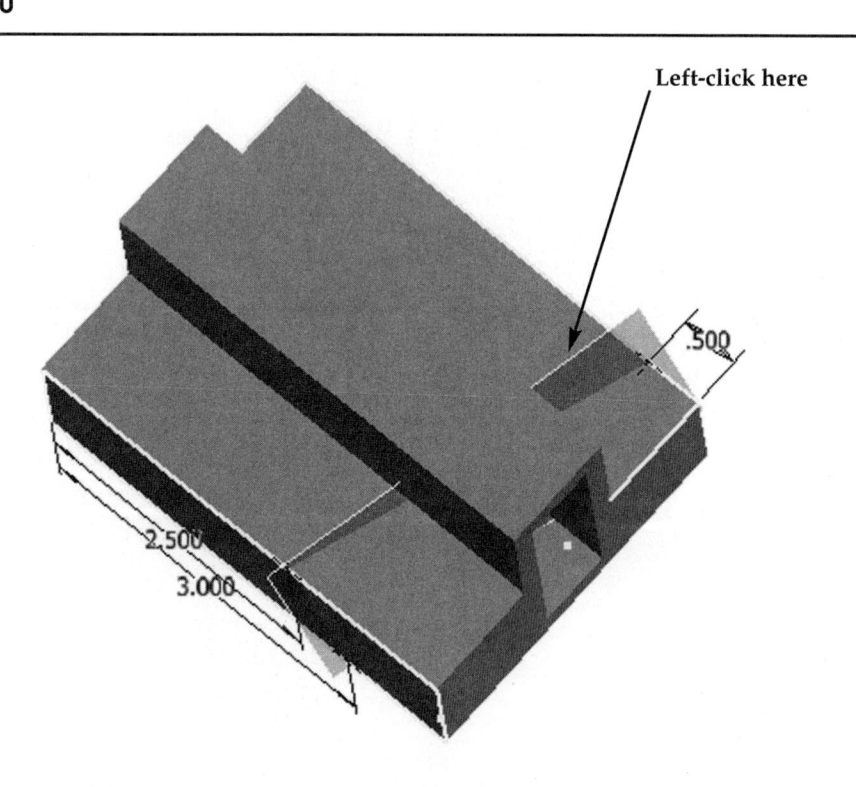

22.   Inventor will create a Split Plane from the work plane as shown in Figure 11-21.

FIGURE 11-21

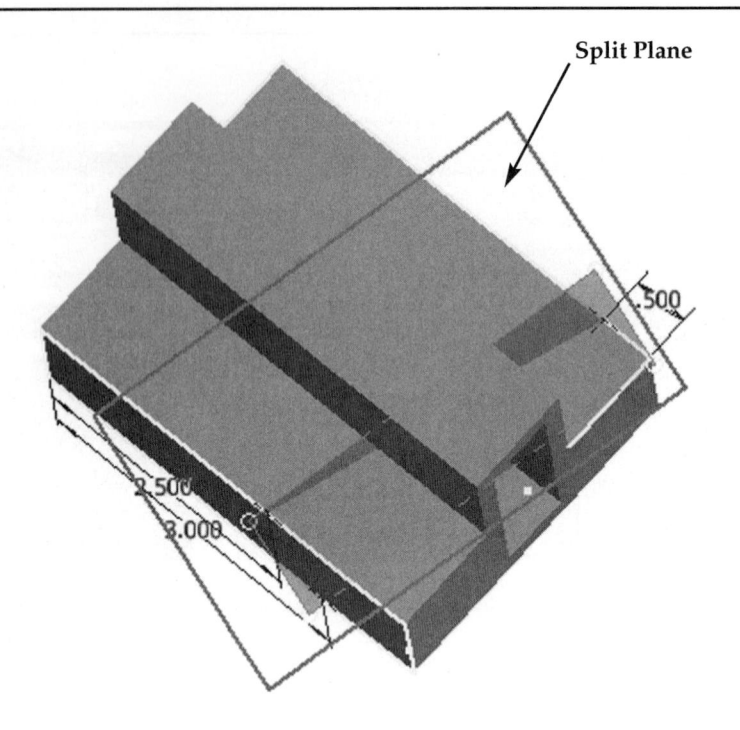

23.   Left-click on the "Trim Solid" icon. Left-click on **OK** as shown in Figure 11-22.

FIGURE 11-22

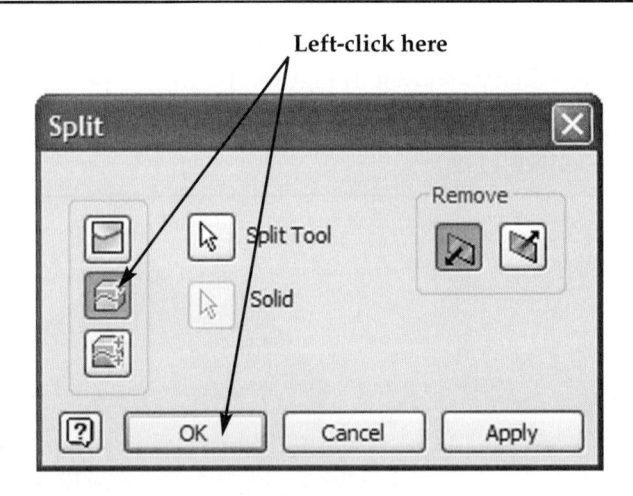

24.   Your screen should look similar to Figure 11-23.

FIGURE 11-23

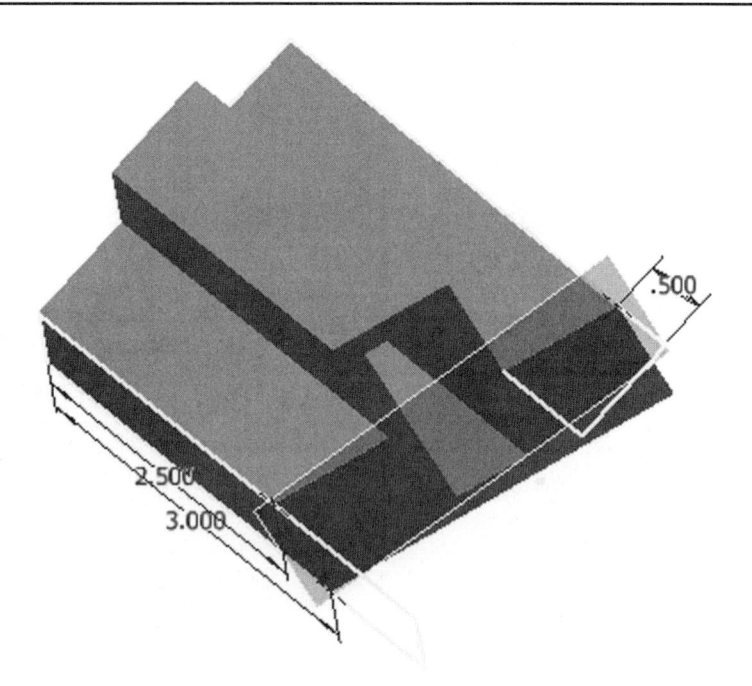

25.   To clean up the appearance of the part, move the cursor over the left portion of the screen and left-click, then right-click on each of the sketches, and the work plane, and turn off the **Visibility** on each item as shown in Figure 11-24.

FIGURE 11-24

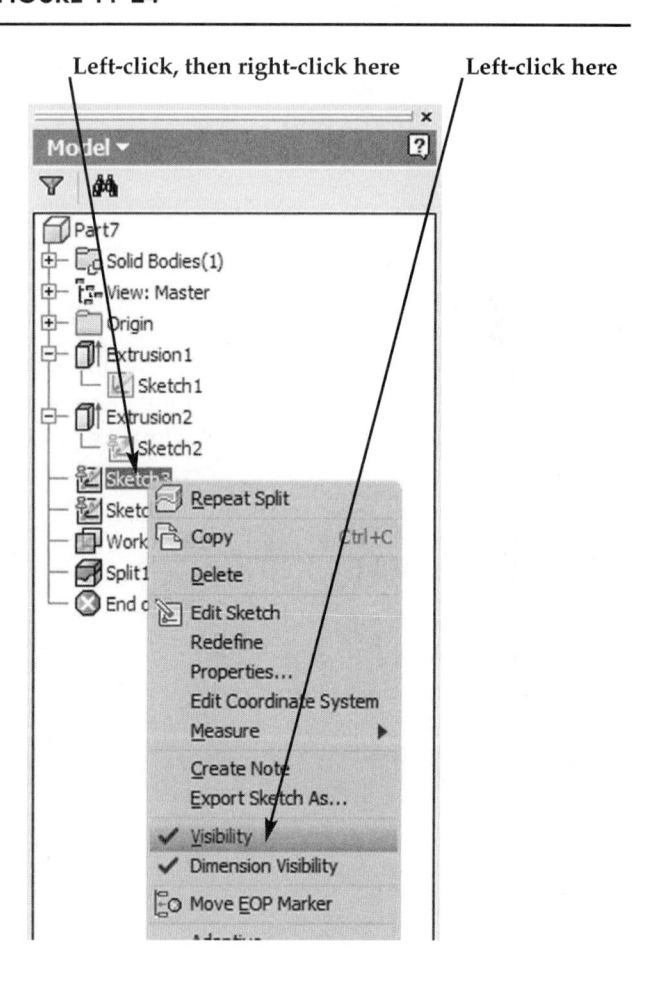

26.   Your screen should look similar to Figure 11-25.

FIGURE 11-25

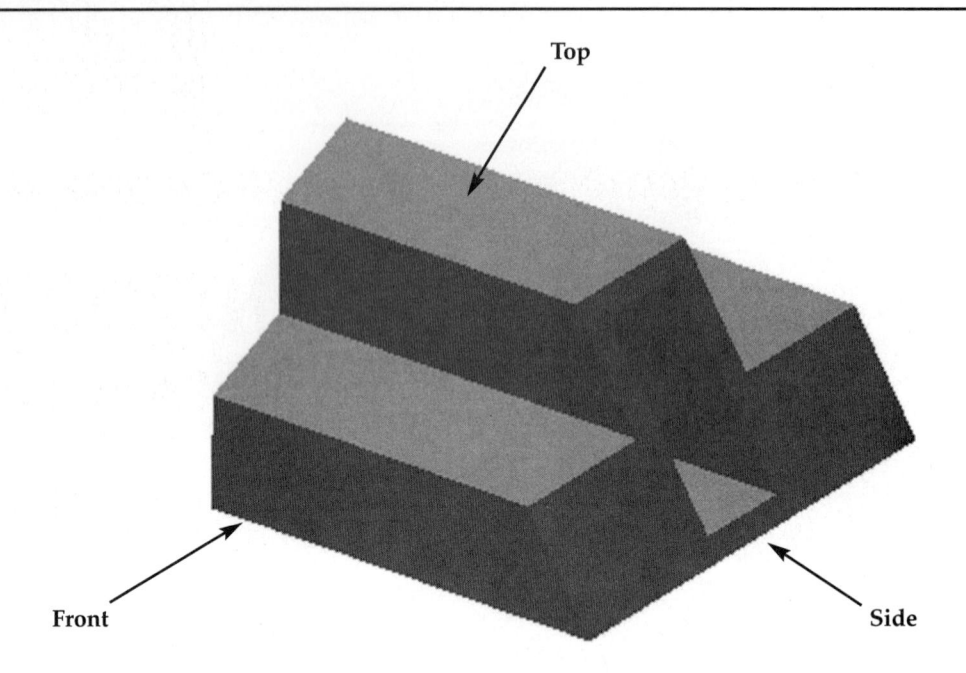

CHAPTER PROBLEM

PROBLEM 11-1

To test your Orthographic skills, create a 3 view Orthographic drawing of the part shown above (Top, Front and Right Hand Side views) in either a 2 Dimensional CAD package (such as Autocad) or by hand on a Drafting board. Show all hidden lines in your 2 D drawing. Once you have completed your 2 Dimensional drawing, refer to Chapter 3 on how to create a 3 View drawing using Inventor (showing all hidden lines), and then compare your "hand drawing" to the Inventor 3 view drawing.

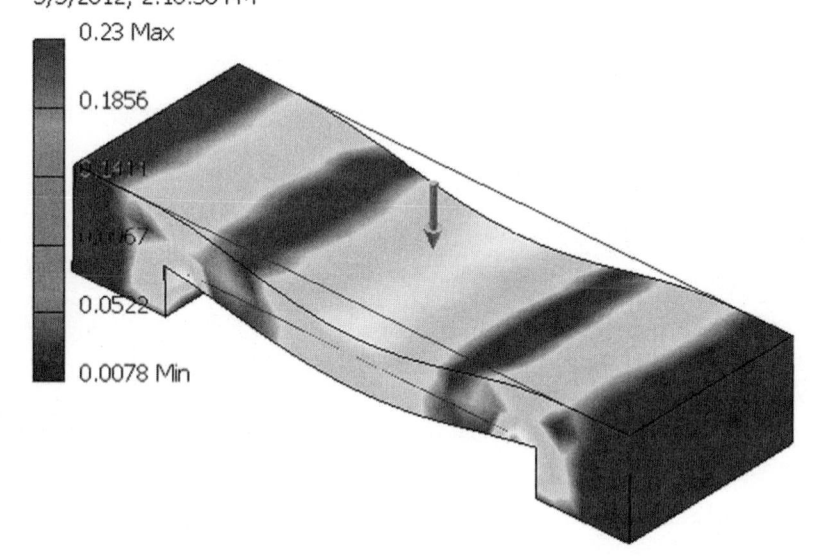

Type: Von Mises Stress
Unit: ksi
3/3/2012, 2:10:50 PM

0.23 Max
0.1856
0.1411
0.1067
0.0522
0.0078 Min

INTRODUCTION TO STRESS ANALYSIS

OBJECTIVES

1. Create a simple part
2. Run Stress Analysis
3. Interpret Stress Analysis results

Chapter 12 includes instruction on how to design the parts for stress analysis including how to analyze the part shown above.

CREATE A SIMPLE PART

1. Start Inventor by referring to Chapter 1.

2. After Inventor is running, begin a New Sketch.

3. Complete the sketch shown in Figure 12-1.

FIGURE 12-1

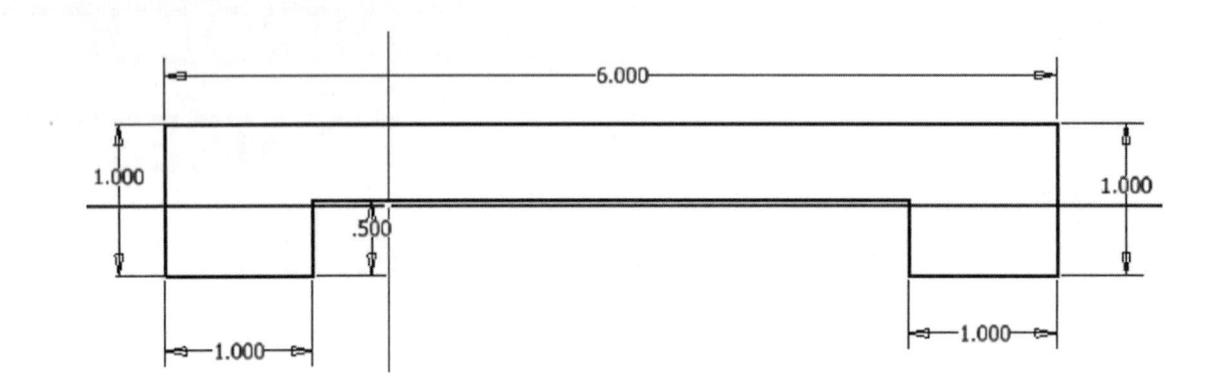

4. Exit the Sketch Panel. Extrude the sketch to a thickness of **2.00** inches as shown in Figure 12-2. Save the part where it can be easily retrieved later.

FIGURE 12-2

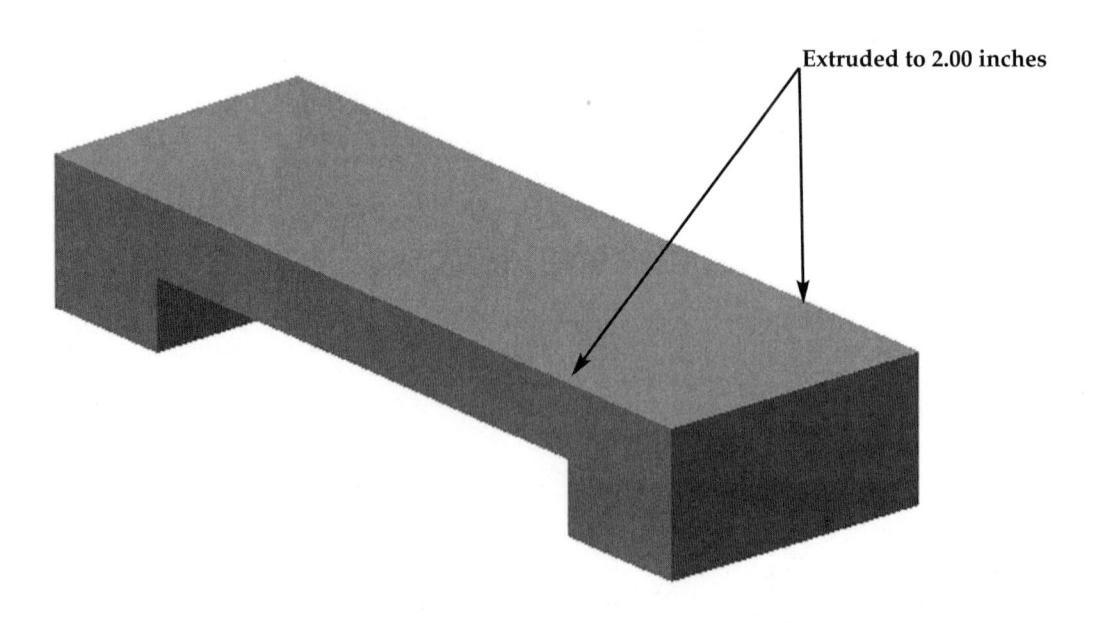

Extruded to 2.00 inches

APPLY MATERIAL TO A SIMPLE PART

5. Move the cursor to the upper middle portion of the screen and left-click on the **Environments** tab. Move the cursor to the upper left portion of the screen and left-click on **Stress Analysis** as shown in Figure 12-3.

FIGURE 12-3

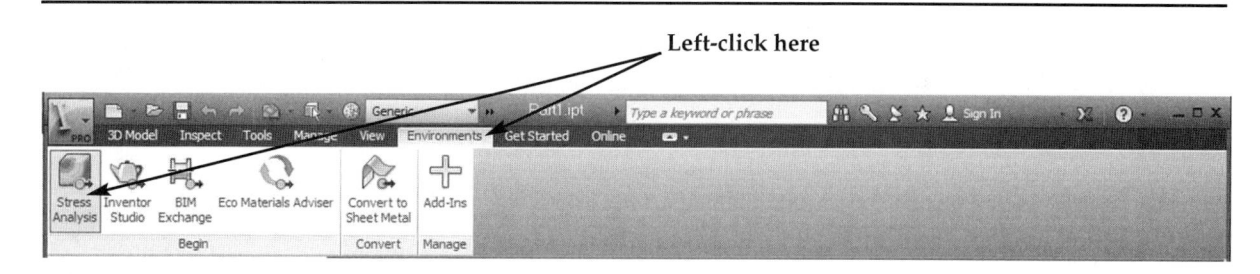

6. Move the cursor to the upper left portion of the screen and left-click on **Create Simulation** as shown in Figure 12-4.

FIGURE 12-4

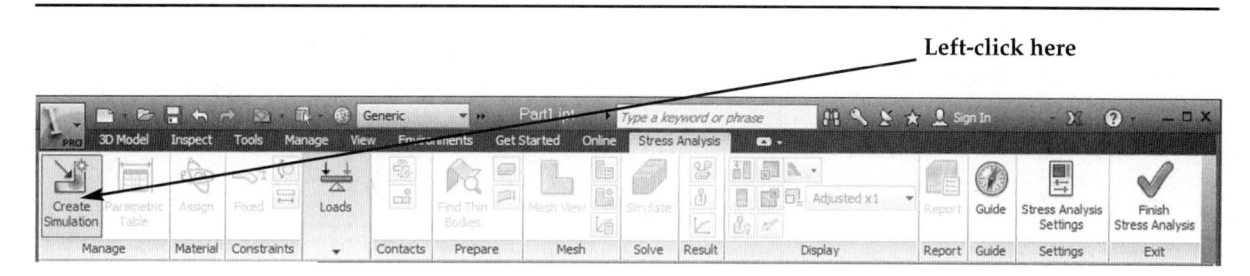

7. The Create New Simulation dialog box will appear. Left-click on **OK** as shown in Figure 12-5.

FIGURE 12-5

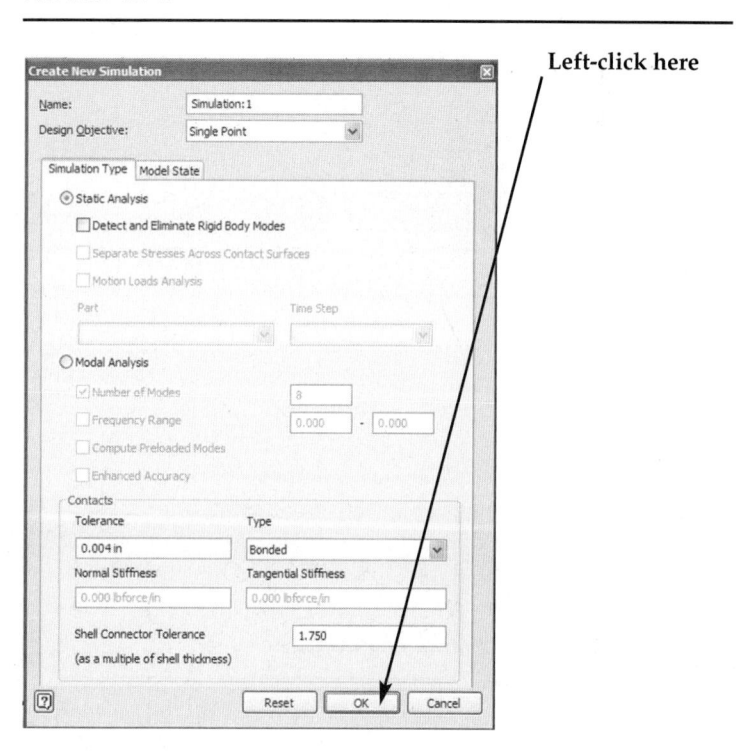

8. Move the cursor to the upper left portion of the screen and left-click on **Assign**. This "assigns" material to the solid model for analysis as shown in Figure 12-6.

FIGURE 12-6

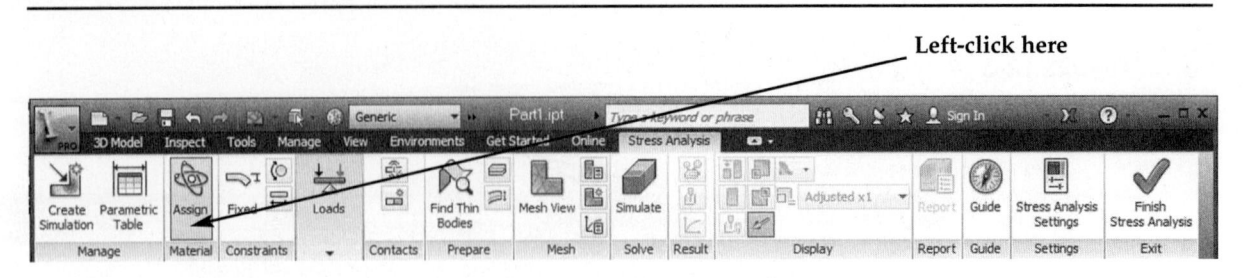

9. The Assign Materials dialog box will appear. Left-click on **Materials** as shown in Figure 12-7.

FIGURE 12-7

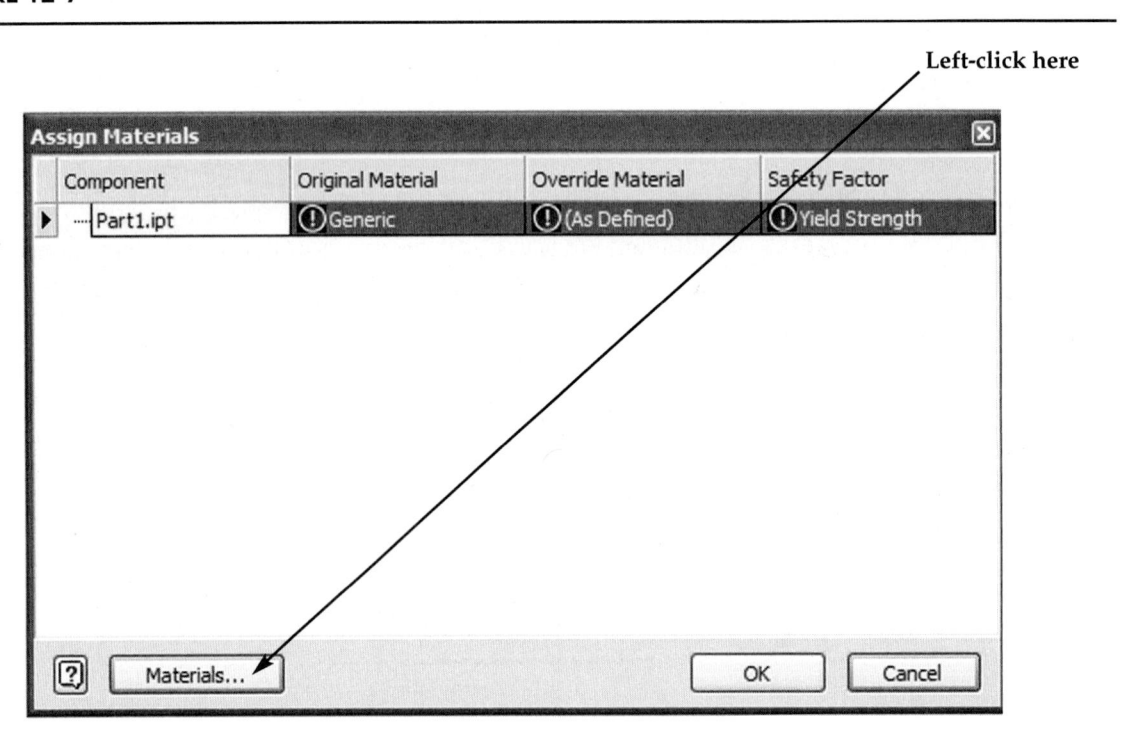

10. The Material Browser dialog box will appear. Left-click on **Autodesk Material Library**. A drop-down menu will appear. Left-click on **Aluminum 6061** as shown in Figure 12-8.

FIGURE 12-8

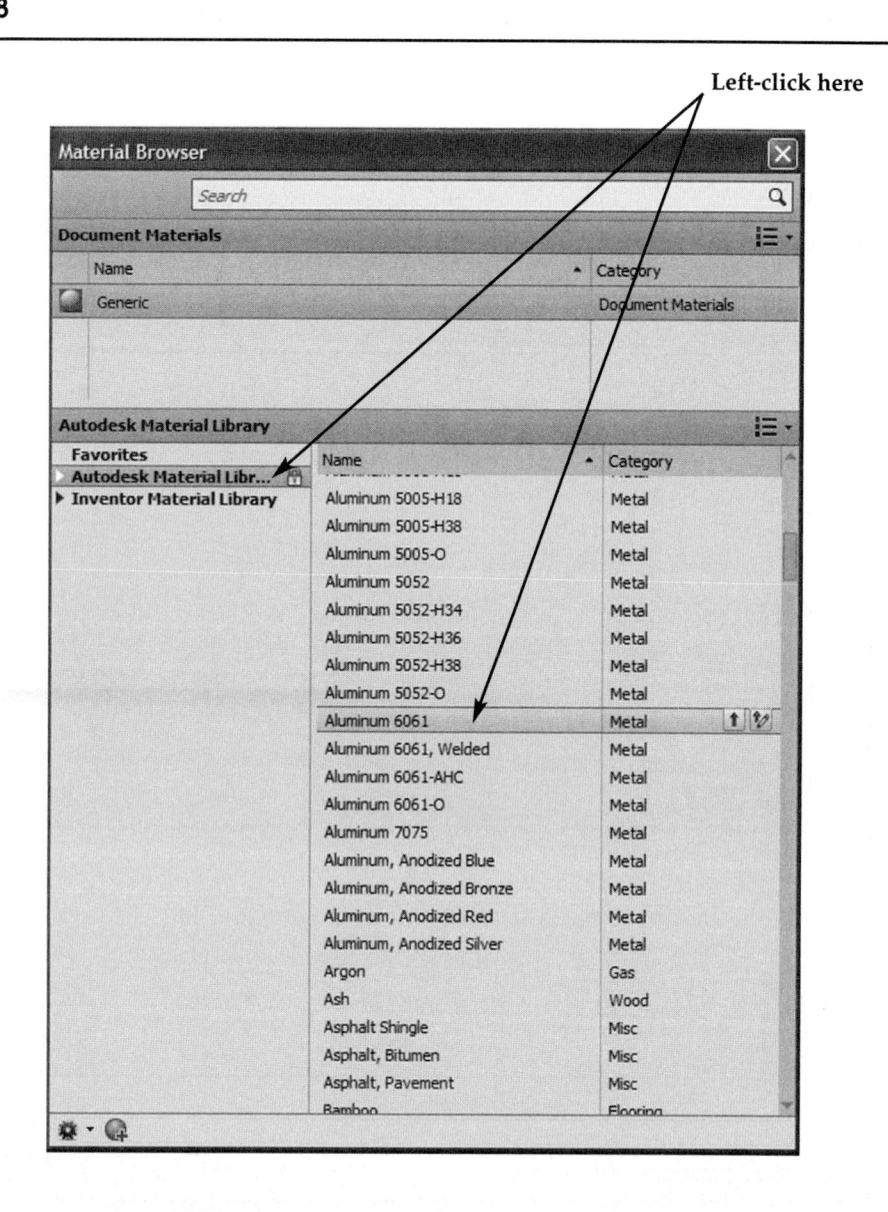

APPLY A FIXTURE TO A SIMPLE PART

11. Close the Material Browser dialog box. The Assign Materials dialog box will re-appear with Aluminum 6061 listed in the Original Material column. Left-click on **OK** as shown in Figure 12-9.

FIGURE 12-9

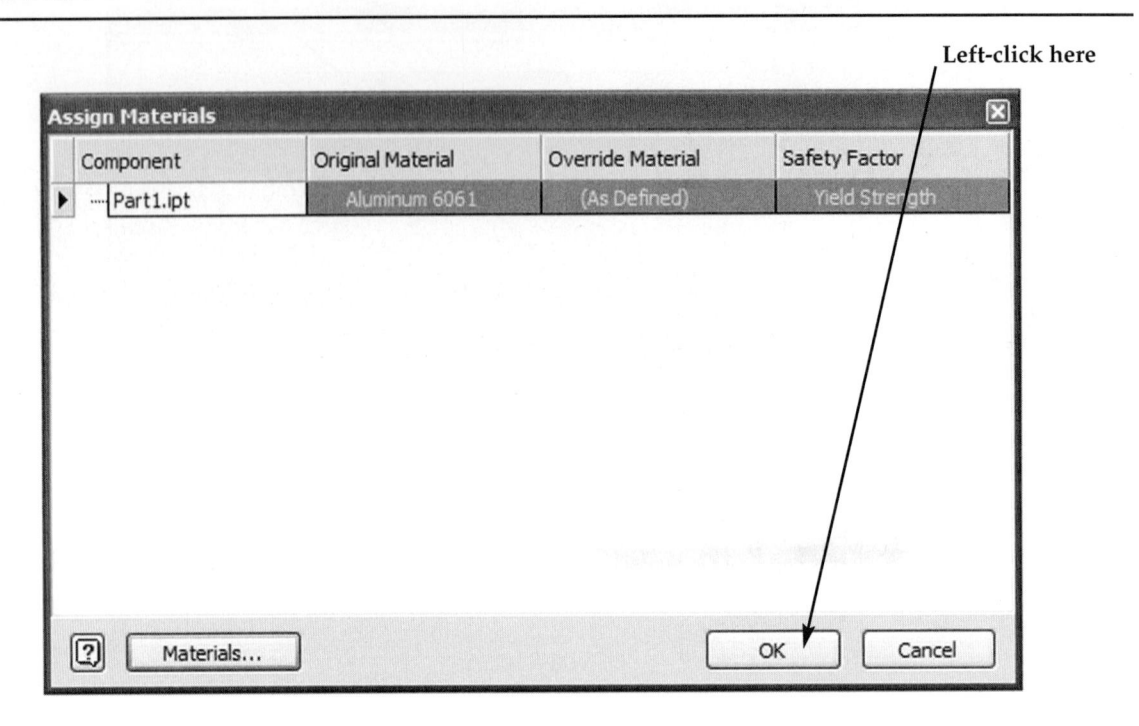

12. Move the cursor to the upper middle portion of the screen and left-click on **Fixed** as shown in Figure 12-10.

FIGURE 12-10

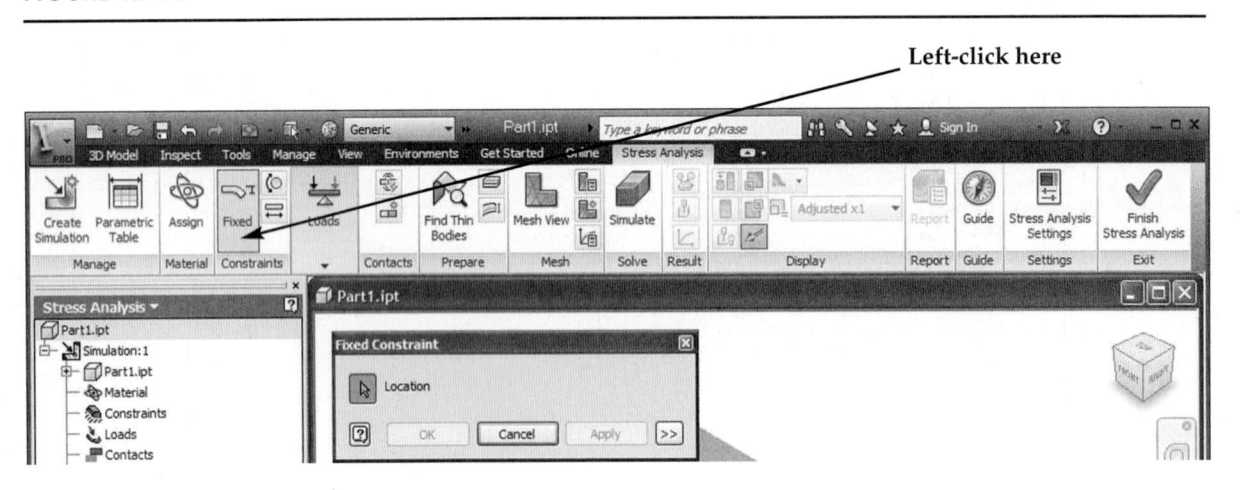

13. The Fixed Constraint dialog box will appear. Rotate the part around to gain access to the bottom side and left-click on the "feet." Then, left-click on **OK** as shown in Figure 12-11.

FIGURE 12-11

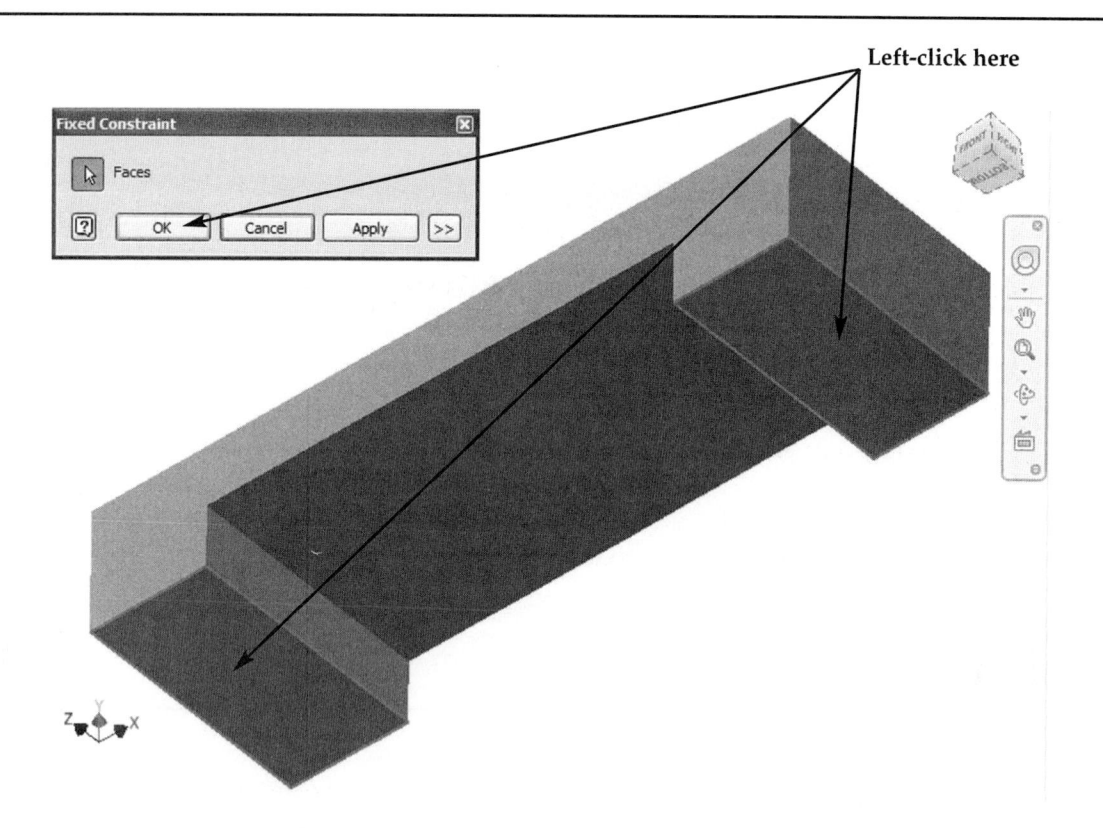

14. Move the cursor to the upper middle portion of the screen and left-click on the drop-down arrow located under **Loads**. A drop-down menu will appear. Left-click on **Force** as shown in Figure 12-12.

FIGURE 12-12

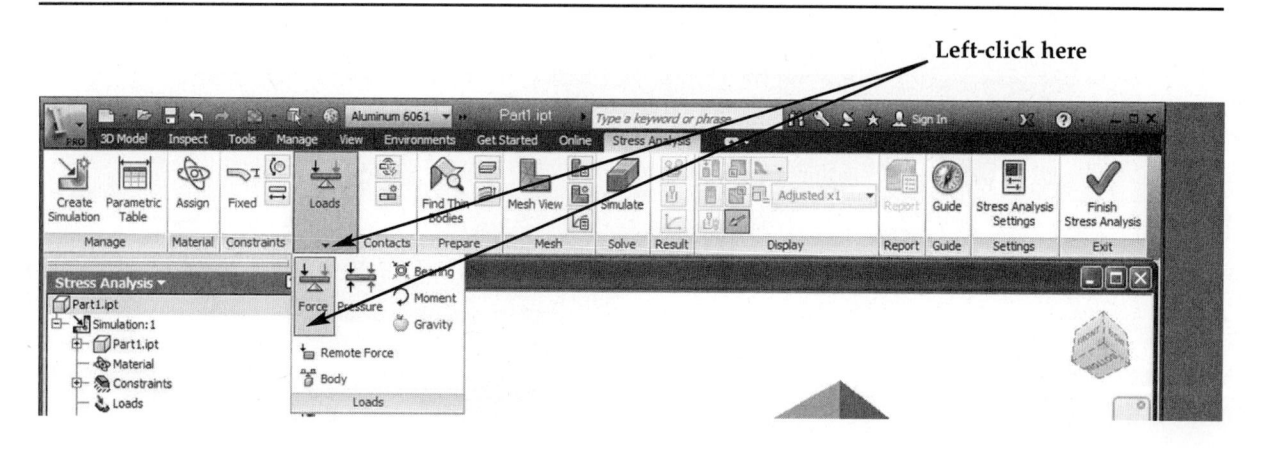

APPLY FORCE TO A SIMPLE PART

15. The Force dialog box will appear. Enter **100 lbs** for the amount of force as shown in Figure 12-13.

FIGURE 12-13

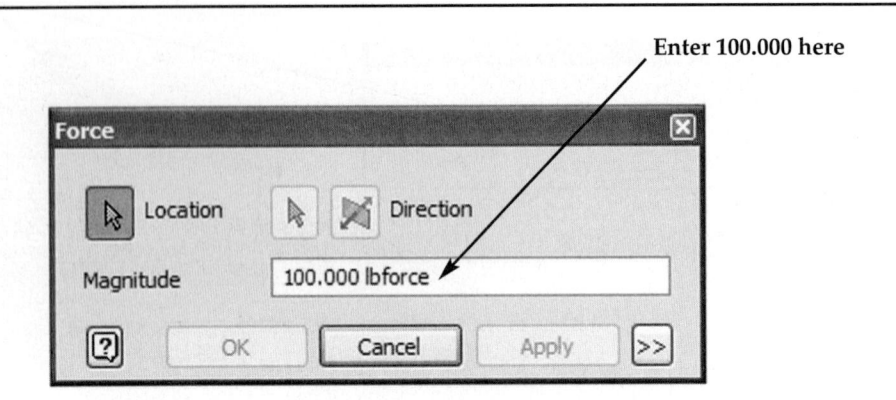

16. Rotate the part around to gain access to the top portion. Left-click on the top portion. Left-click on **OK** as shown in Figure 12-14.

FIGURE 12-14

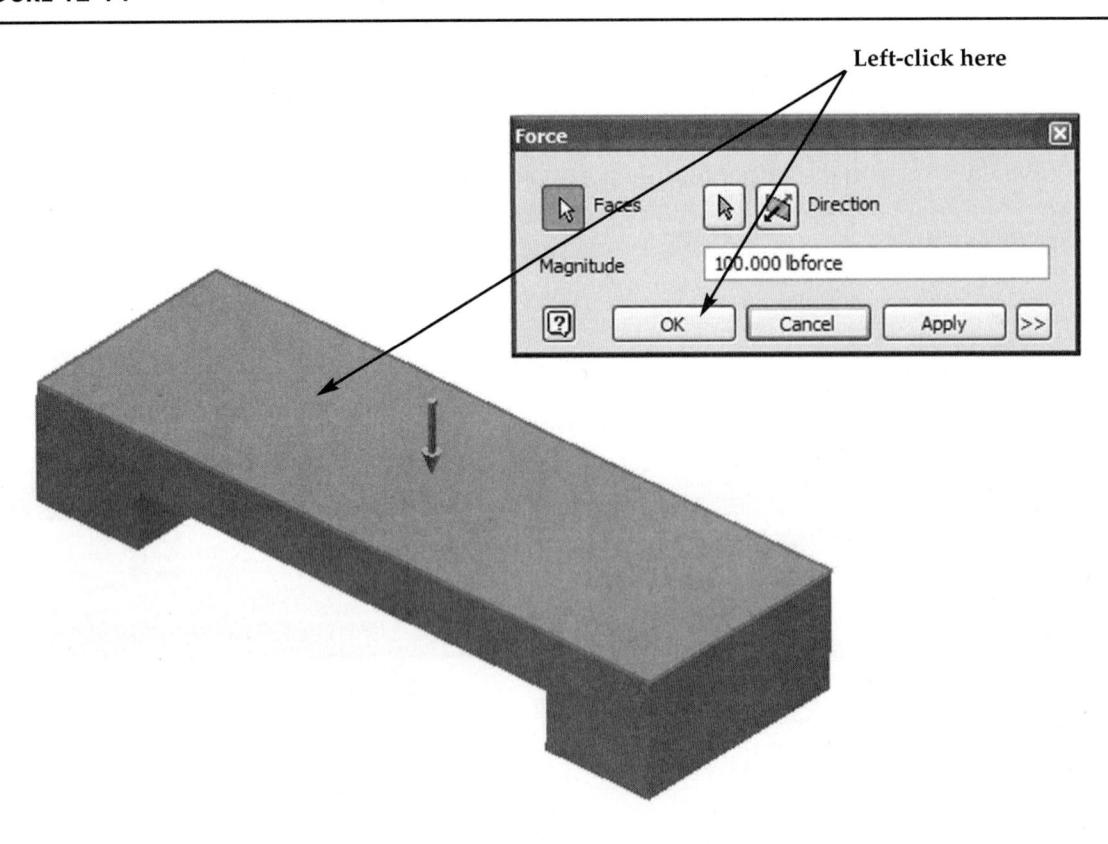

17.   Repeat the same steps to place the Lifter1.ipt file into the assembly. Your screen should look similar to Figure 12-15.

FIGURE 12-15

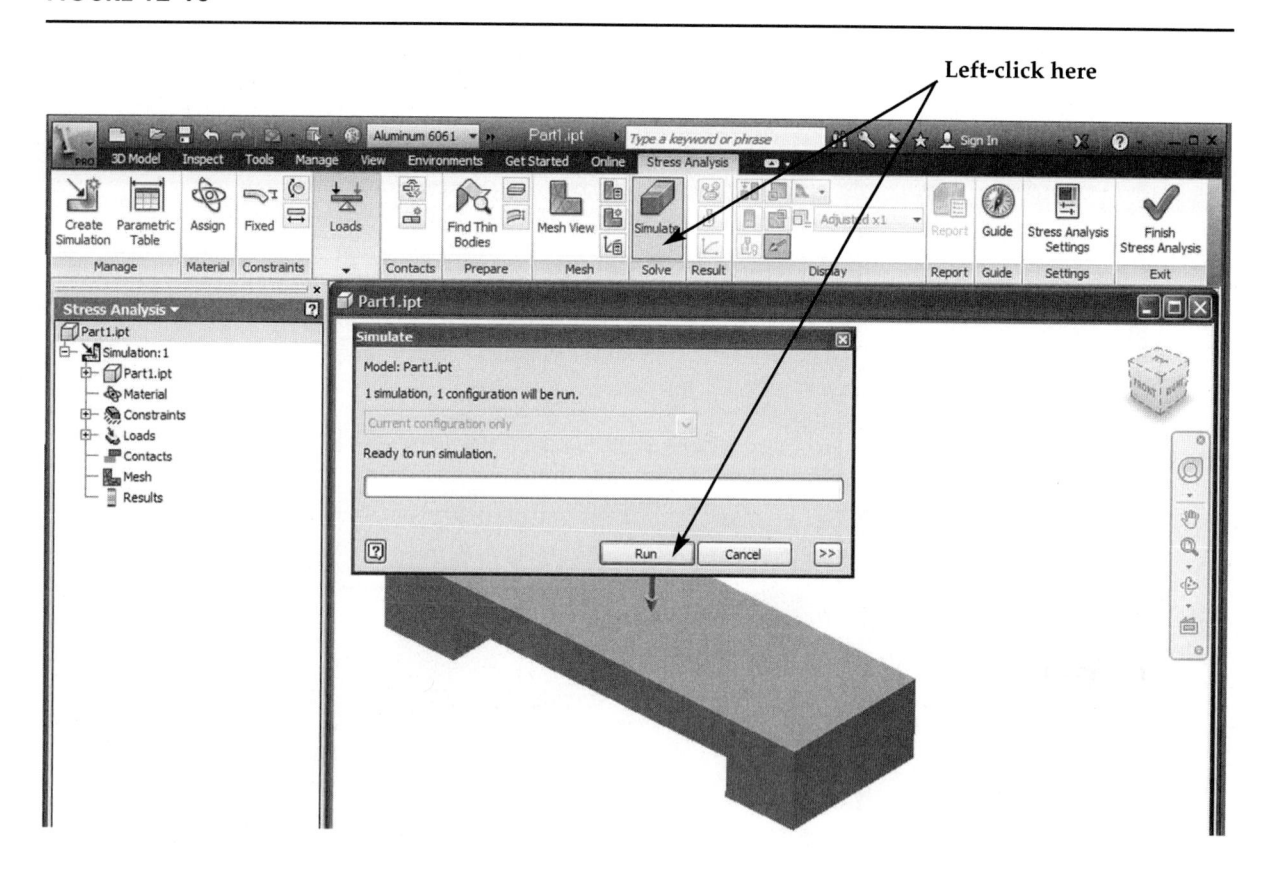

PERFORM A STRESS ANALYSIS ON A SIMPLE PART

18.   Inventor is in the process of analyzing the stress on this part with a force of 100 lbs as shown in Figure 12-16.

FIGURE 12-16

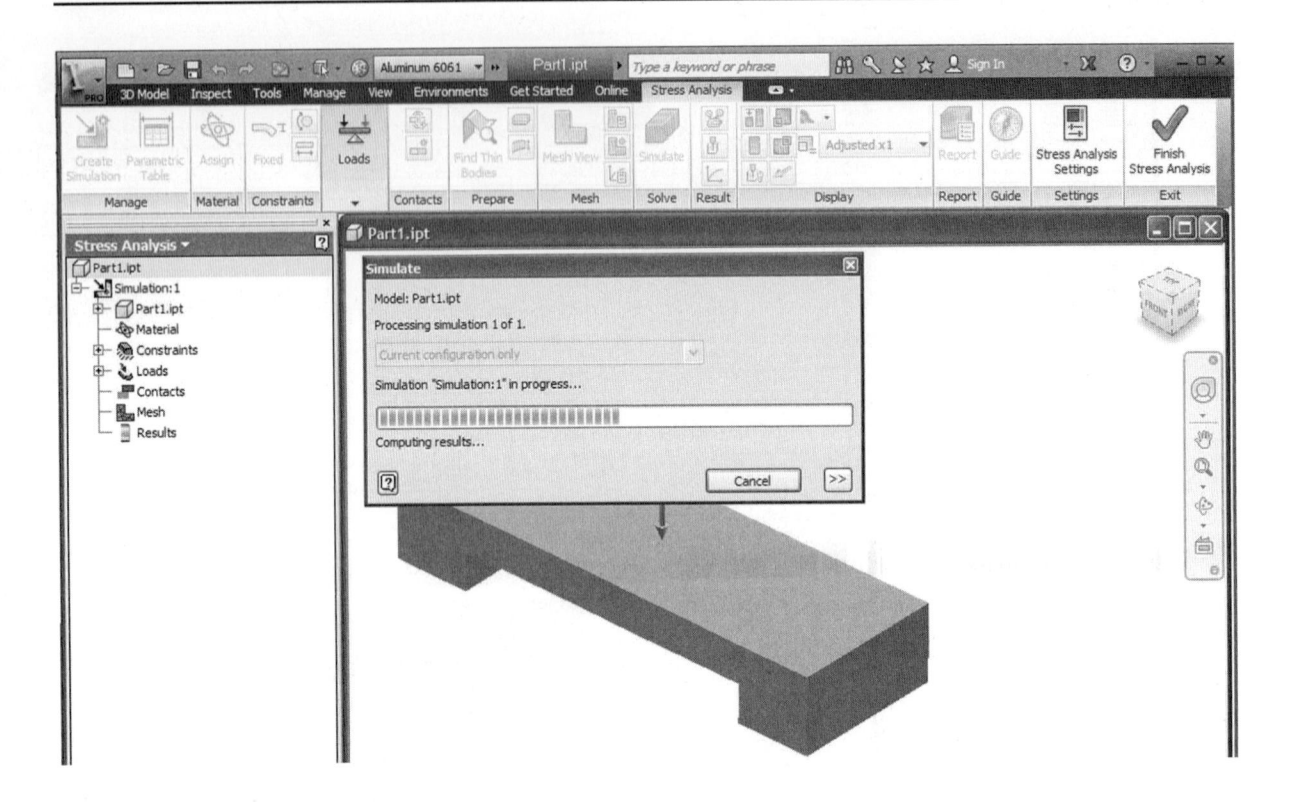

19.   Inventor will display the results of the analysis. High stress areas are identified by any reddish orange areas that appear on the model as shown in Figure 12-17. Rotate the part around to examine any high stress areas on the underside.

FIGURE 12-17

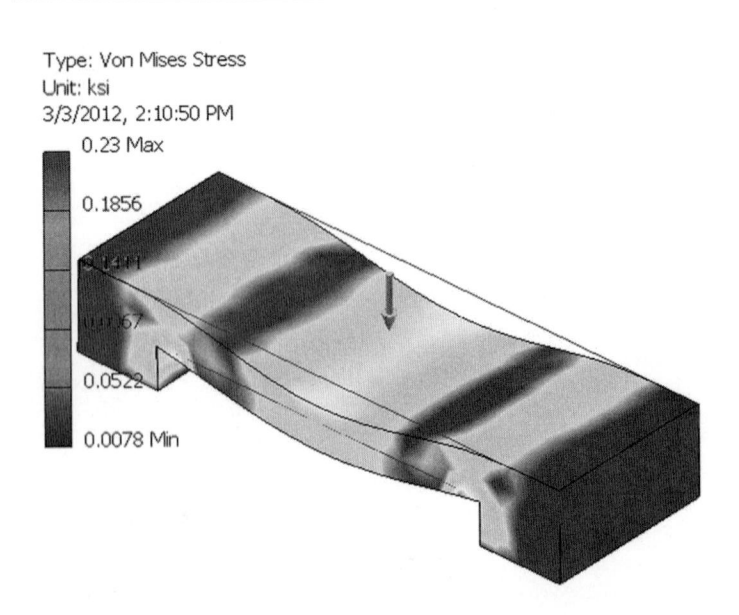

Type: Von Mises Stress
Unit: ksi
3/3/2012, 2:10:50 PM

0.23 Max

0.1856

0.0522

0.0078 Min

INTERPRET RESULTS OF A STRESS ANALYSIS

20. Move the cursor to the upper middle portion of the screen and left-click on the "Animate Results" icon. The Animate Results dialog box will appear. Left-click on "Play." Inventor will animate the stress analysis as shown in Figure 12-18.

FIGURE 12-18

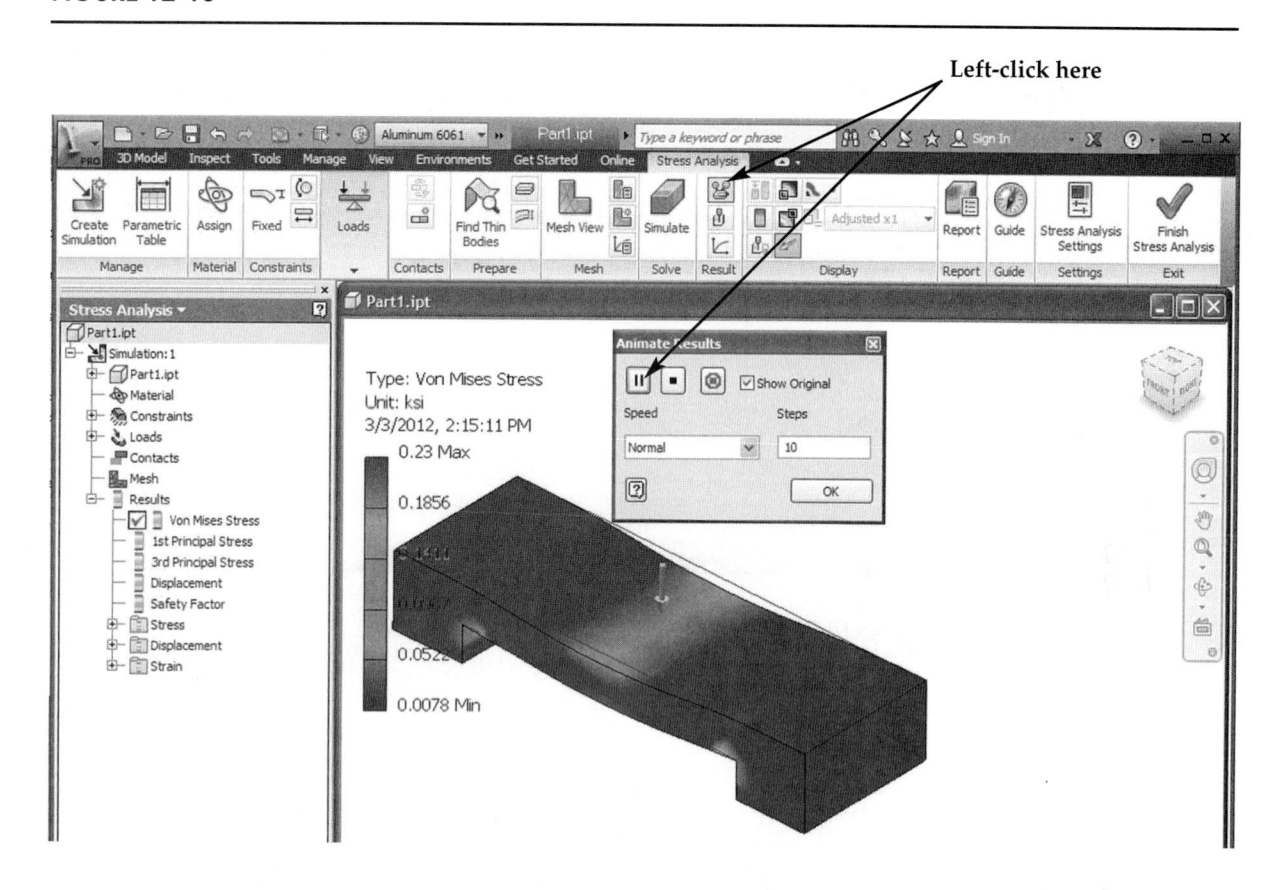

21. Results of the analysis can be viewed by selecting the "Probe" icon or the "Convergence" plot. Results can also be viewed by selecting the **Report** icon. Left-click on **Finish Stress Analysis** as shown in Figure 12-19.

FIGURE 12-19

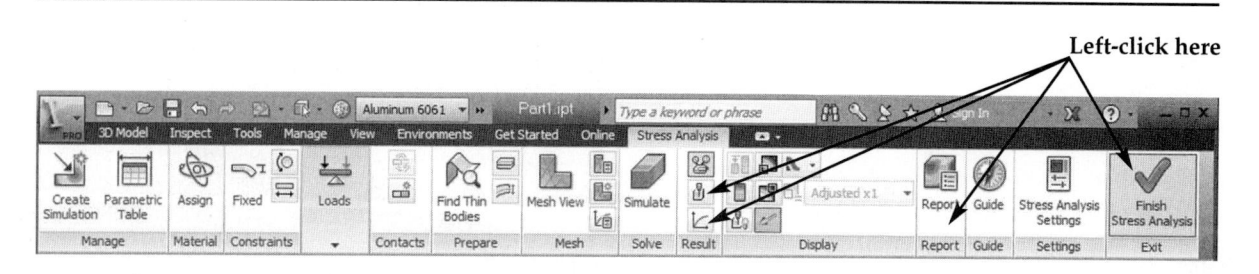

CHAPTER PROBLEMS

PROBLEM 12-1

Complete the following sketch.

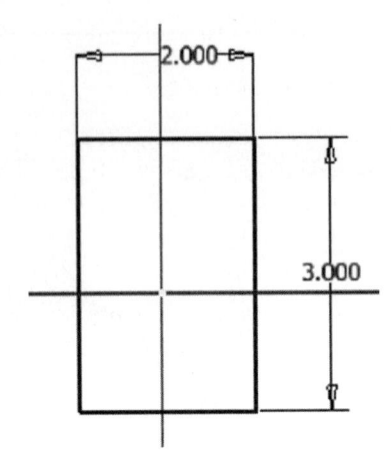

PROBLEM 12-2

Extrude the sketch to a distance of 96 inches.

PROBLEM 12-3

Use the Shell command to create a wall thickness of .10.

PROBLEM 12-4

Run a stress analysis on the part using Aluminum 1100-O for material. Use the left open edge for the Fixed edge. Use a load of 500 pounds along the top. The end result will be the creation of a Cantilever as shown.

Type: Von Mises Stress
Unit: ksi
3/3/2012, 2:44:37 PM

28.43 Max

22.75

17.07

11.39

5.71

0.02 Min

CHAPTER

13

INTRODUCTION TO THE DESIGN ACCELERATOR

OBJECTIVES

1. Create a Disc Cam
2. Edit the Disc Cam
3. Constrain the Disc Cam in an Assembly file
4. Animate the Disc Cam using the Drive Constraint command

Chapter 13 includes instruction on how to design the part shown above. This chapter contains a brief introduction to the Design Accelerator. The Design Accelerator contains numerous predefined parts. This chapter will cover one of these parts.

1.   Start Inventor by referring to Chapter 1.

2.   After Inventor is running, begin a New Sketch.

3.   Complete the sketch shown in Figure 13-1.

FIGURE 13-1

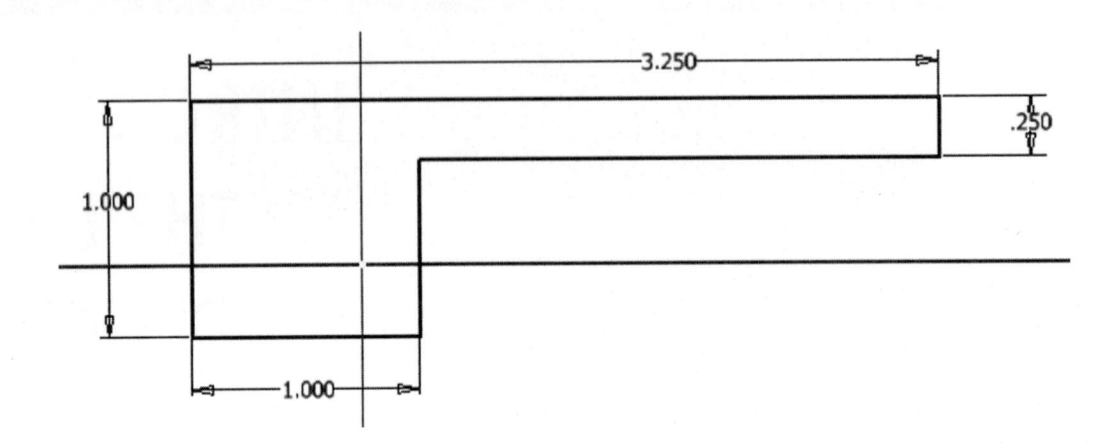

4.   Exit the Sketch Panel. Extrude the sketch to a thickness of **1.00** inch as shown in Figure 13-2.

FIGURE 13-2

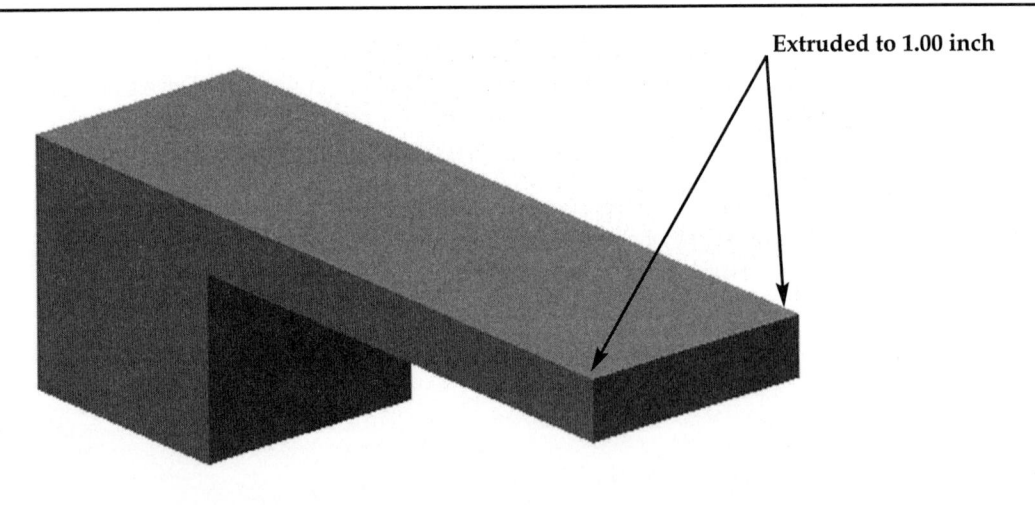

Extruded to 1.00 inch

5.   Use the **Fillet** command (.5 inch fillets) to radius the lower edge(s) as shown in Figure 13-3.

FIGURE 13-3

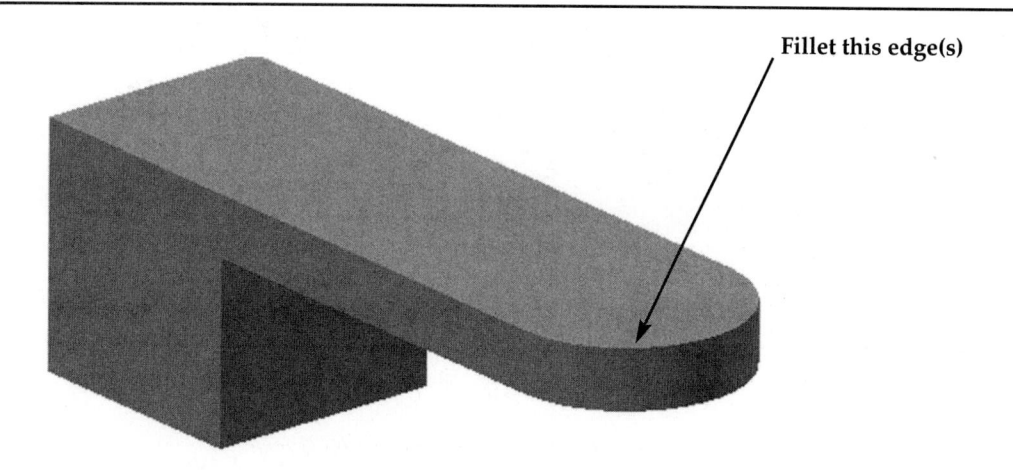

Fillet this edge(s)

6.   Use the **Rotate** command to rotate the part upward as shown in Figure 13-4.

FIGURE 13-4

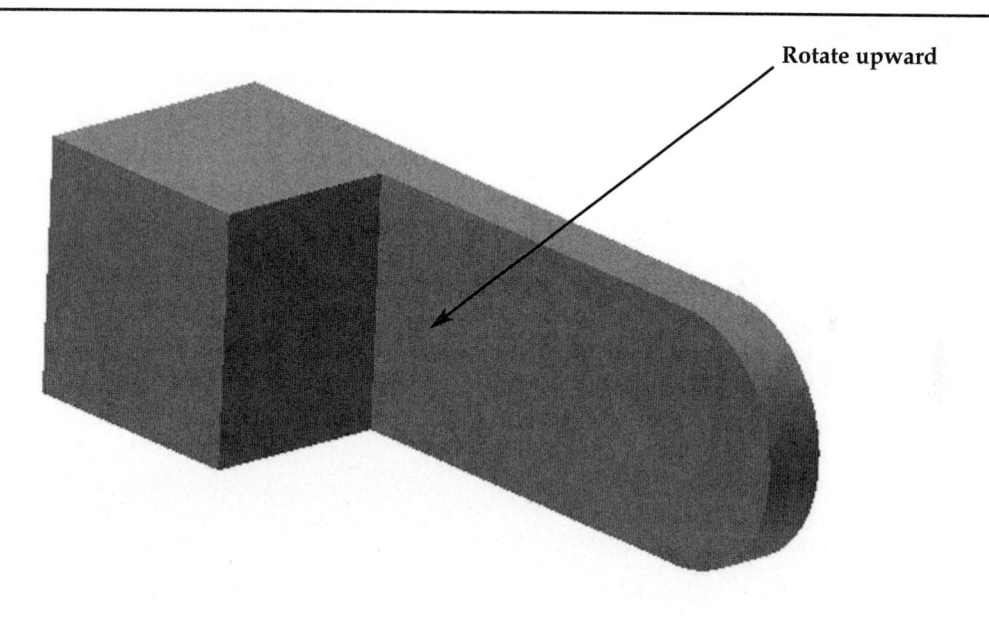

Rotate upward

7.   Complete the sketch as shown in Figure 13-5. The center of the circle is located on the center of the Fillet radius.

FIGURE 13-5

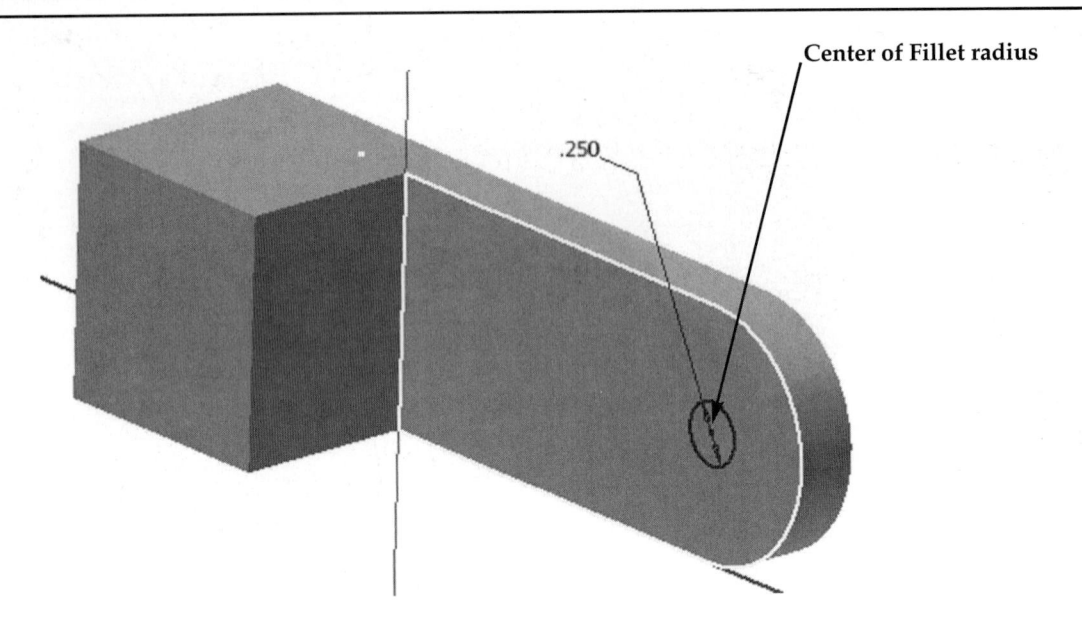

Center of Fillet radius

.250

8.   Extrude the circle to a distance of **.75** inches as shown in Figure 13-6.

FIGURE 13-6

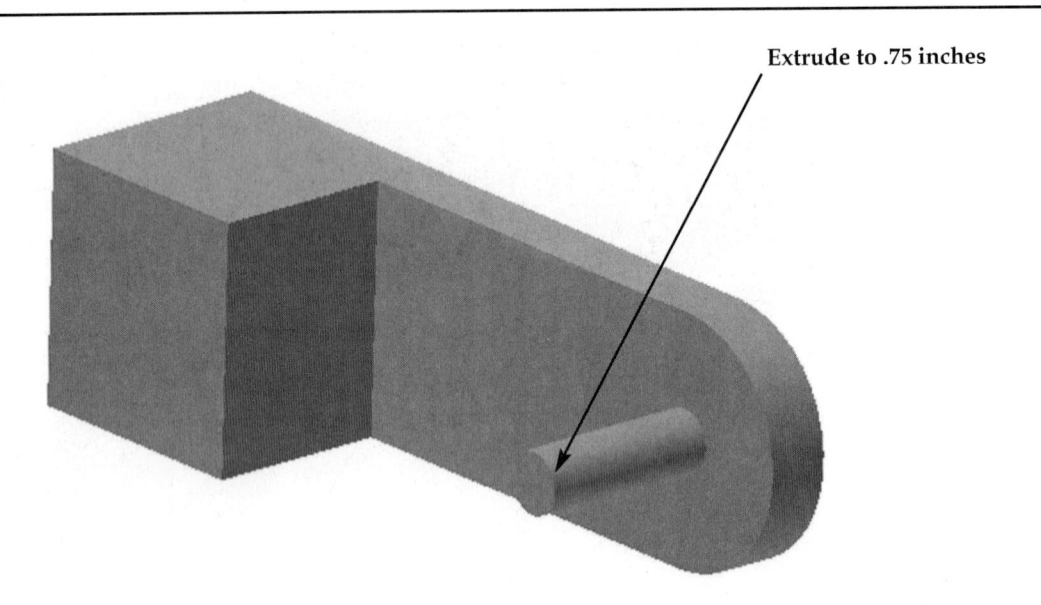

Extrude to .75 inches

9.   Complete the sketch as shown in Figure 13-7.

FIGURE 13-7

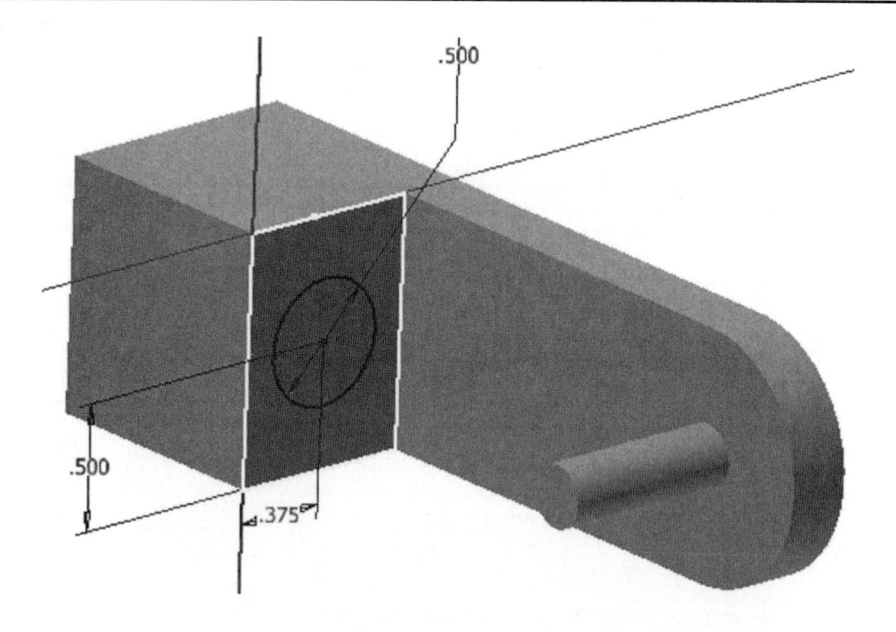

10.   Use the cut option in the **Extrude** command to cut a hole a distance of 1.00 inch as shown in Figure 13-8.

FIGURE 13-8

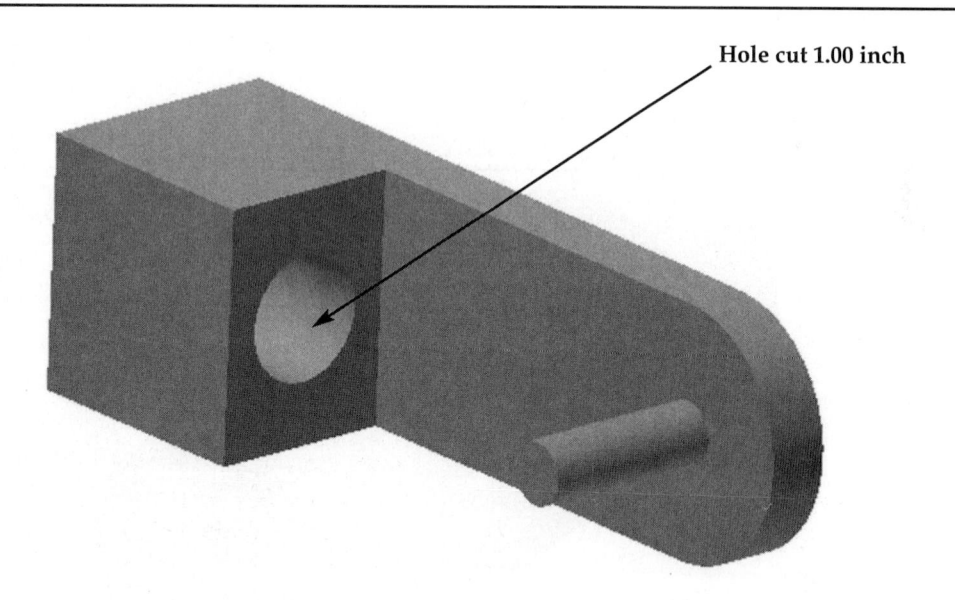

11.   Save the part as Camcase1.ipt where it can be easily retrieved later.

12.   Begin a new drawing as described in Chapter 1.

13.   Complete the sketch shown. Extrude the .500 inch diameter circle to a distance of 1 inch and the .625 inch diameter circle to a distance of .125 inches as shown in Figure 13-9.

FIGURE 13-9

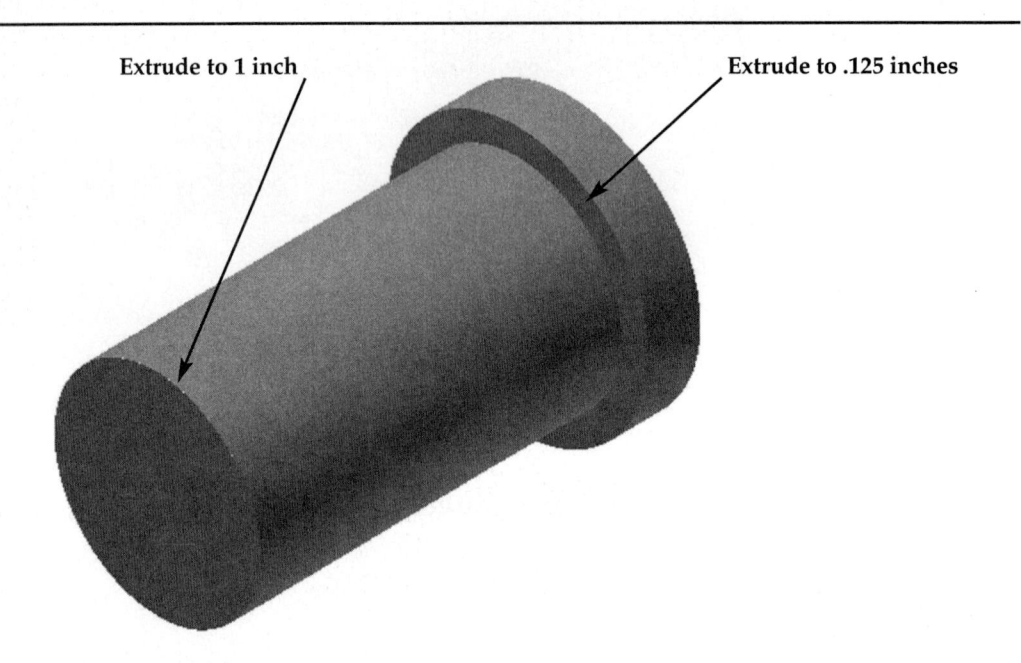

Extrude to 1 inch   Extrude to .125 inches

14.   The bottom of the part needs to be flat as shown in Figure 13-10.

FIGURE 13-10

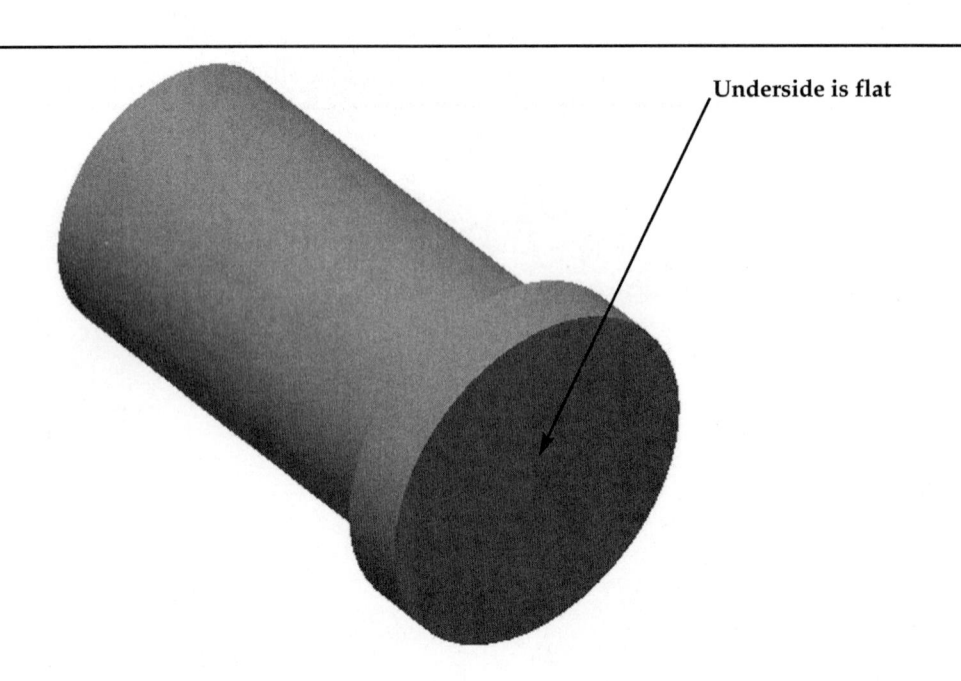

Underside is flat

15.   Save the part as Lifter1.ipt where it can be easily retrieved later.

16.   Begin a new Assemble drawing as described in Chapter 7 and shown in Figure 13-11.

FIGURE 13-11

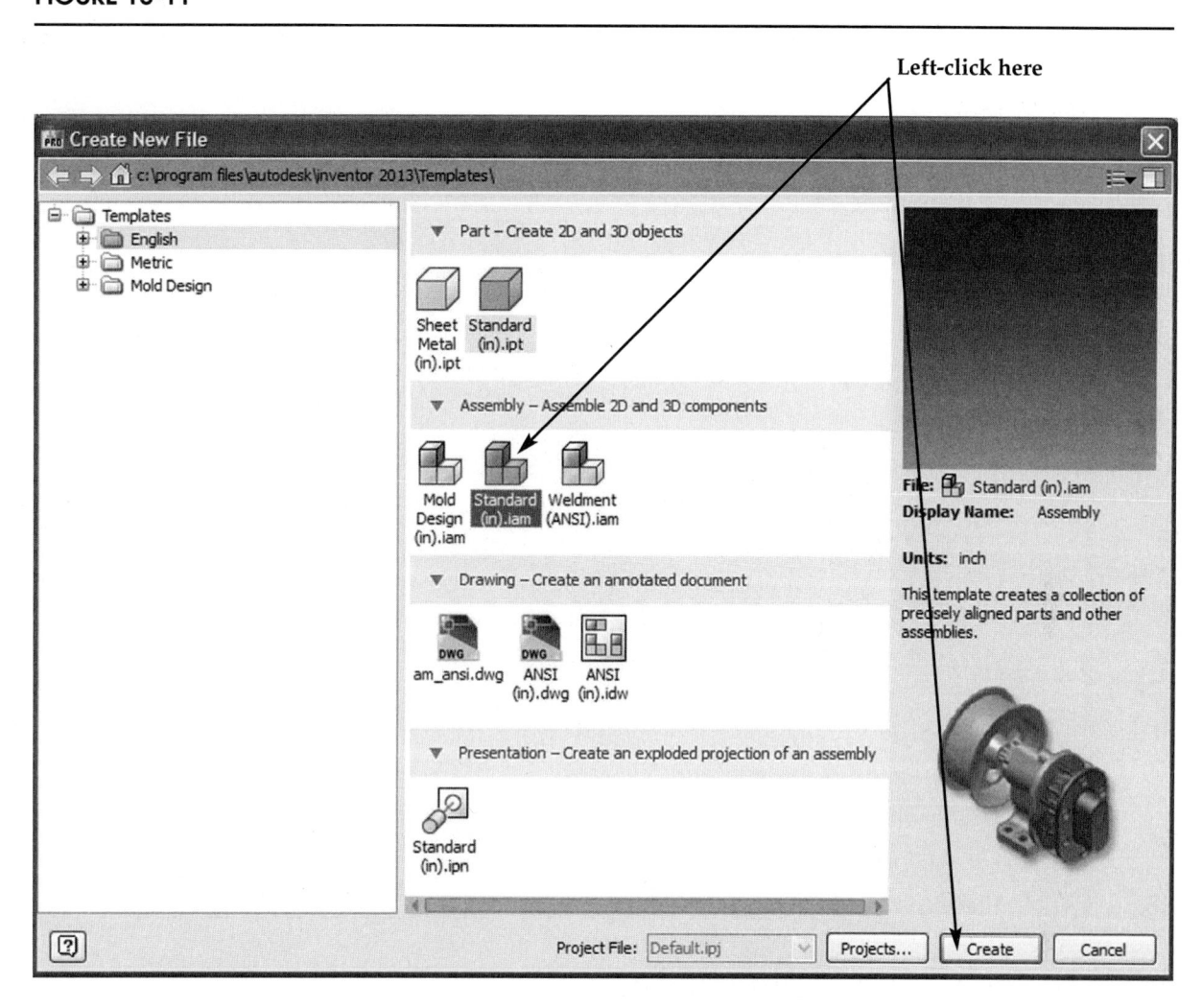

17.   The Assemble Panel will appear as shown in Figure 13-12.

FIGURE 13-12

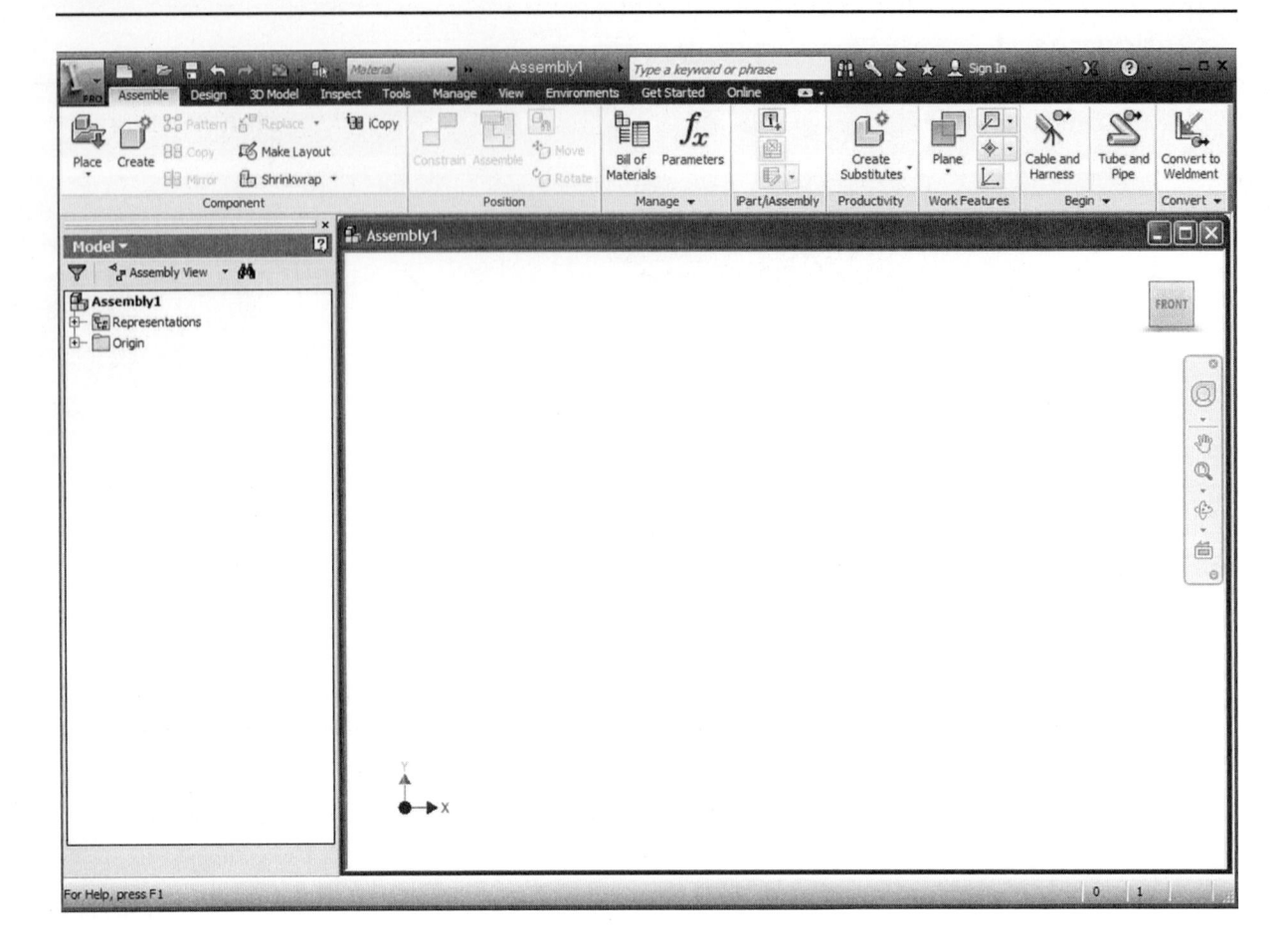

18.   Use the **Place** command to place the Camcase1.ipt file into the assembly as shown in Figure 13-13.

FIGURE 13-13

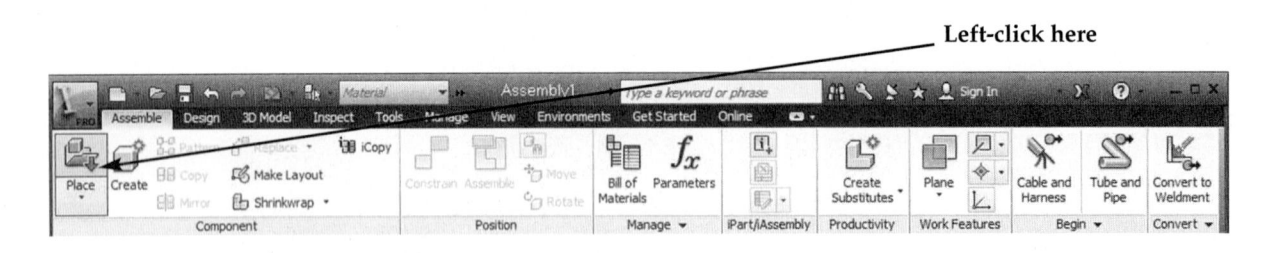

19. The Place Component dialog box will appear as shown in Figure 13-14.

FIGURE 13-14

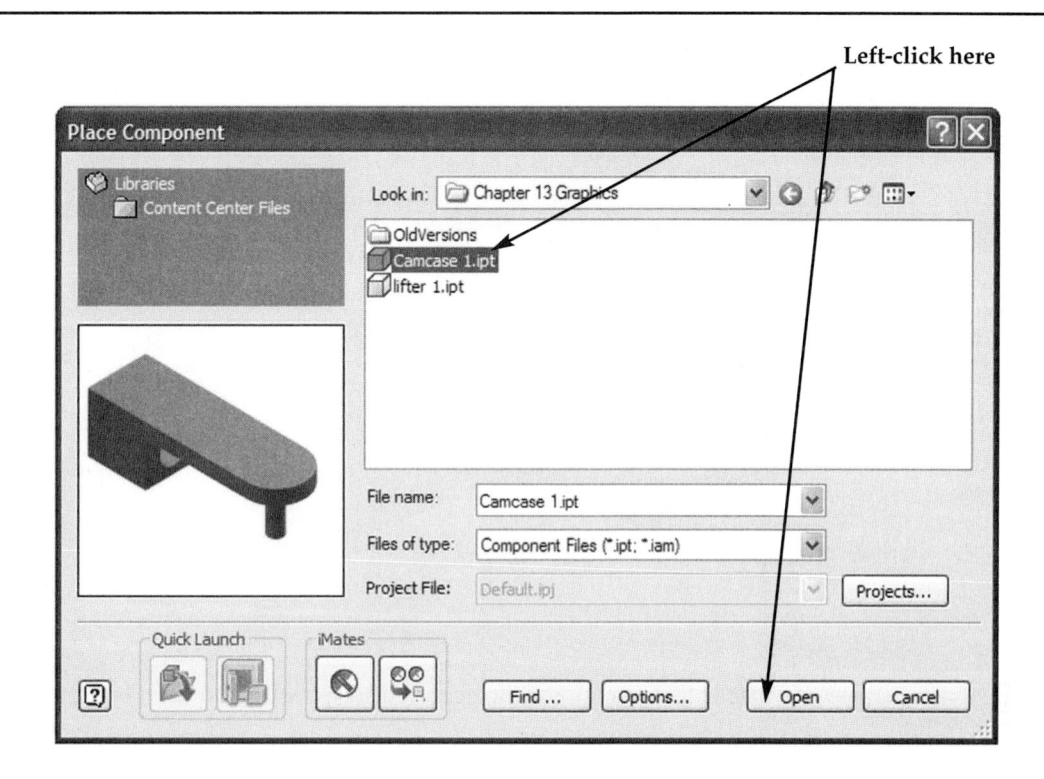

20. Repeat the same steps to place the Lifter1.ipt file into the assembly. Your screen should look similar to Figure 13-15.

FIGURE 13-15

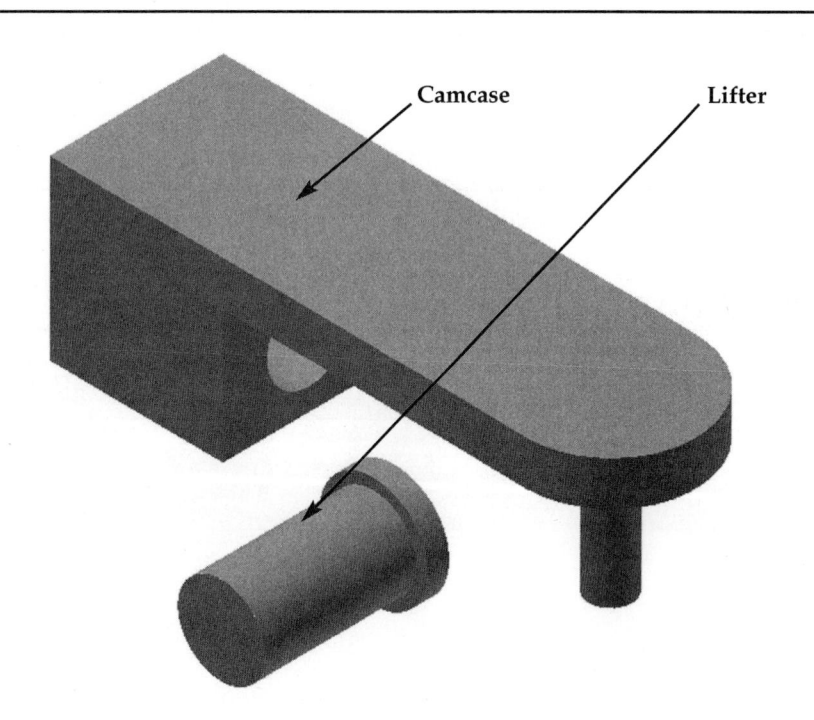

21. Use the **Rotate** command to rotate the case around for better access. Use the **Constraint** command to place the lifter into the lifter bore with the foot towards the bottom of the case as shown in Figure 13-16. Save the file before proceeding.

FIGURE 13-16

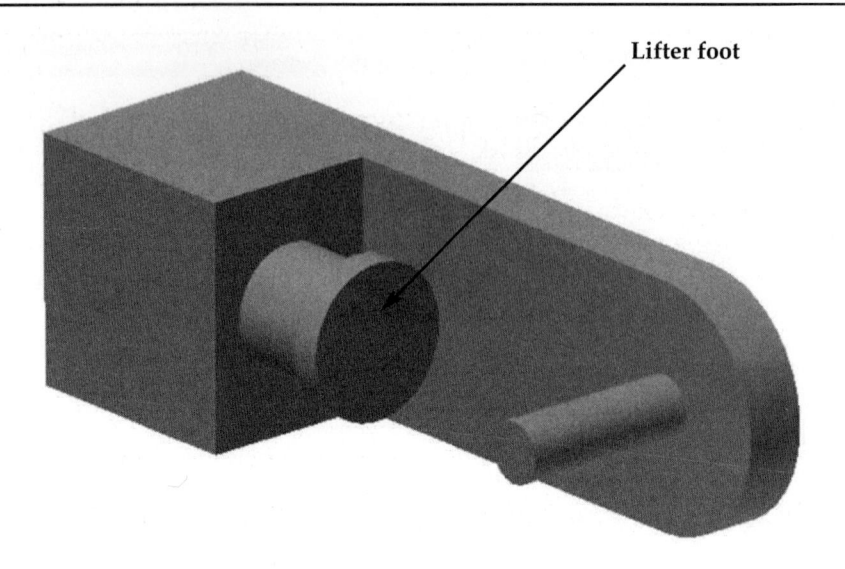

Lifter foot

22. Move the cursor to the upper left portion of the screen and left-click on the **Design** tab as shown in Figure 13-17.

FIGURE 13-17

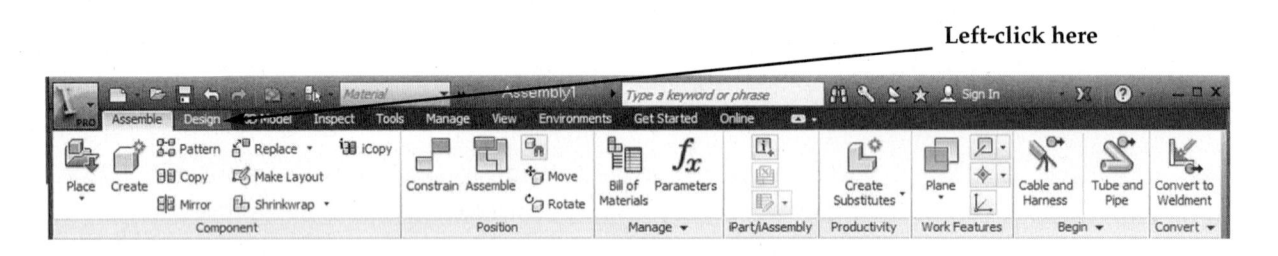

Left-click here

23. The Design Accelerator panel will open as shown in Figure 13-18.

FIGURE 13-18

CREATE A DISC CAM

24.   Move the cursor to the upper right portion of the screen and left-click on **Disc Cam** as shown in Figure 13-19.

FIGURE 13-19

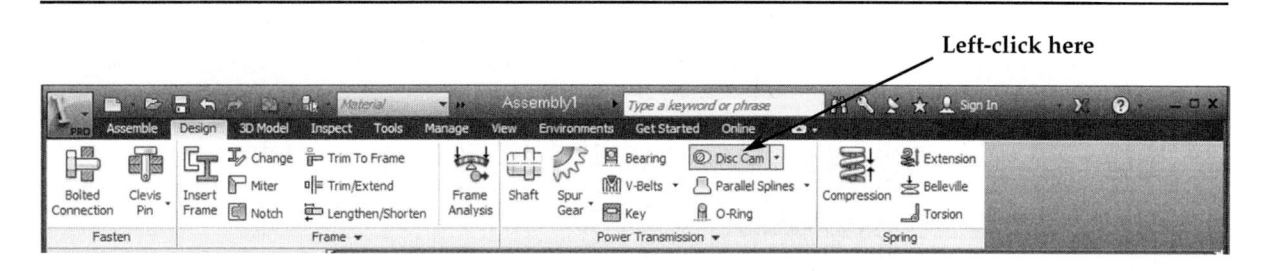

25.   The Disc Cam Component Generator dialog box will appear as shown in Figure 13-20.

FIGURE 13-20

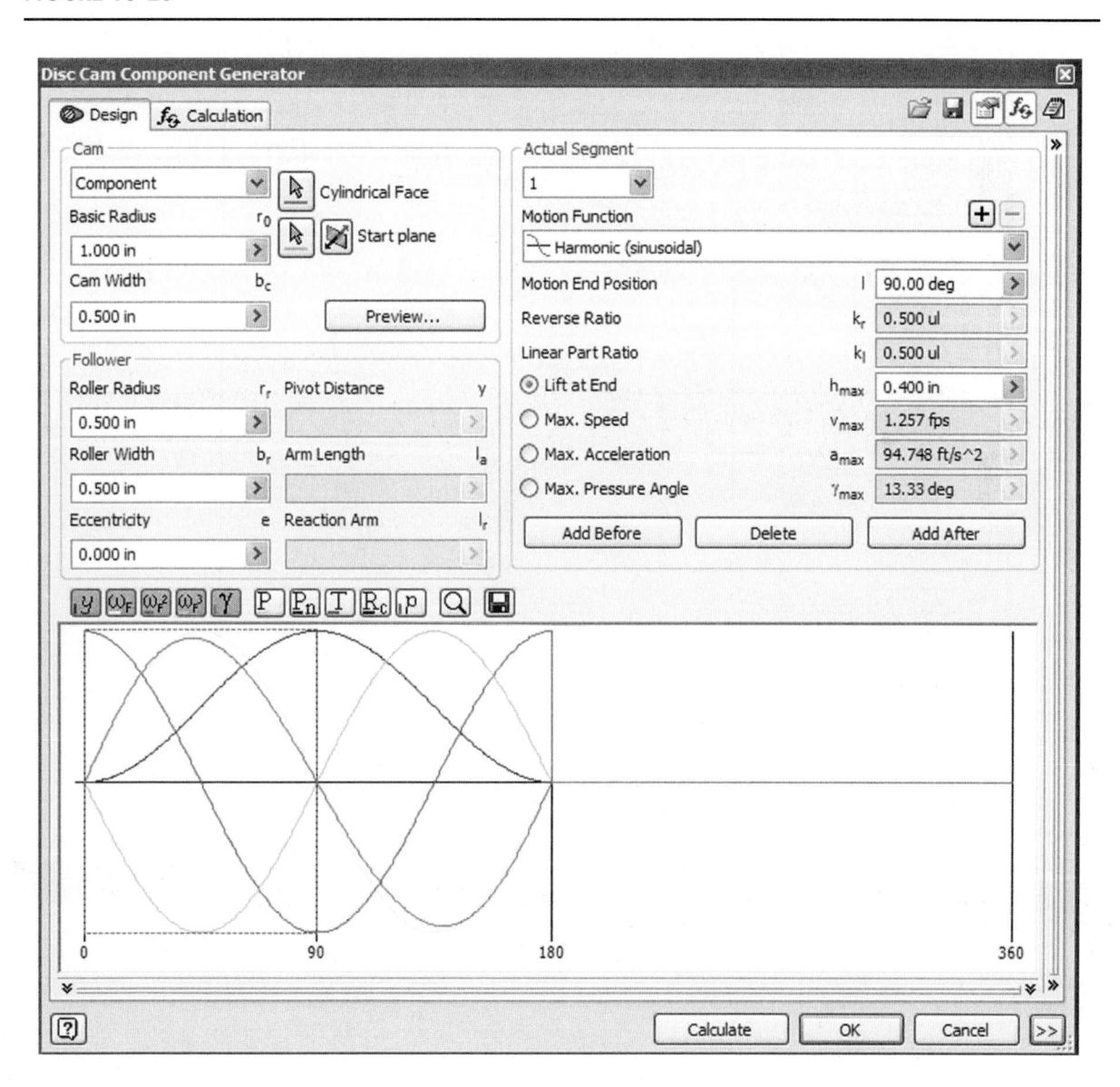

26. Highlight the text below Roller Radius and enter **2.000**. This will create a more pointed disc cam. Left-click on **Calculate** and **OK** as shown in Figure 13-21.

FIGURE 13-21

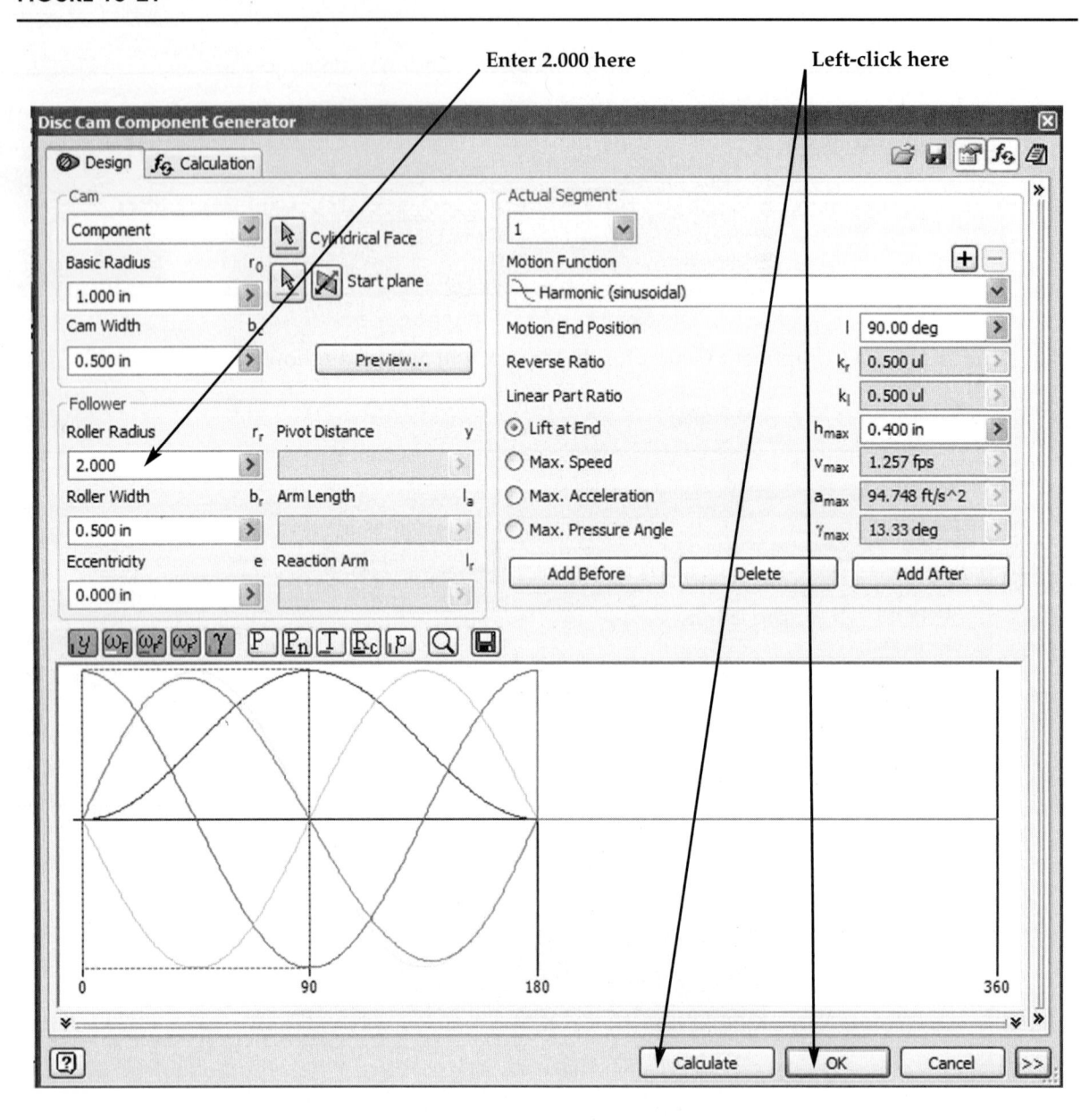

27. The disc cam will be attached to the cursor. If the File Naming dialog box (not shown) appears, left-click on **OK**. Left-click as shown in Figure 13-22.

FIGURE 13-22

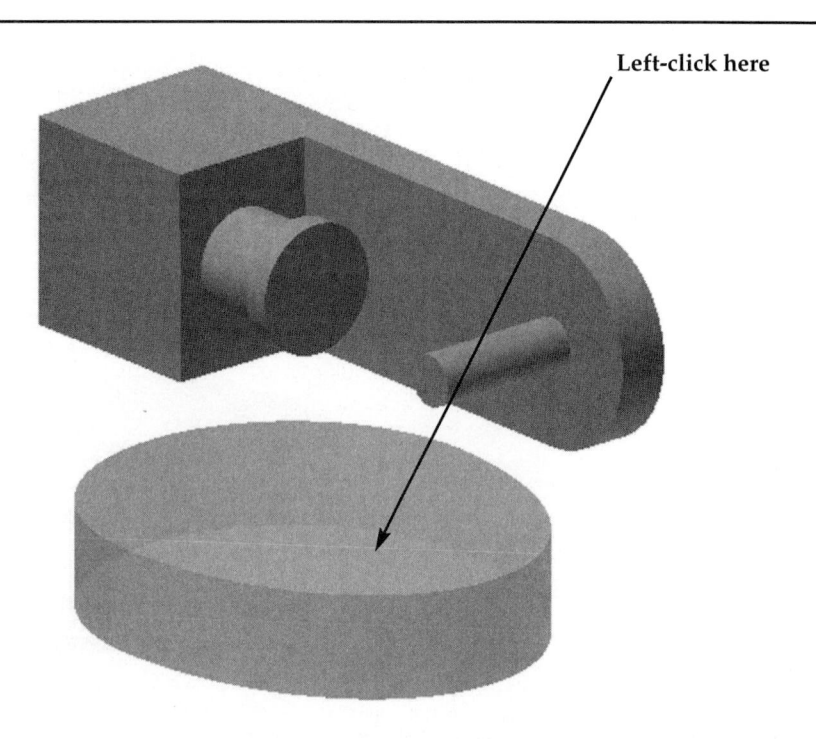

28. Your screen should look similar to Figure 13-23.

FIGURE 13-23

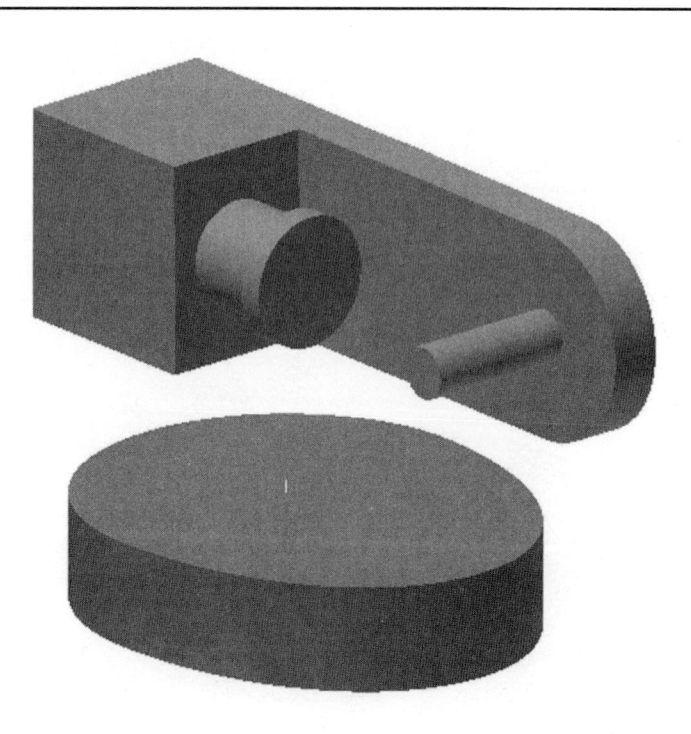

EDIT A DISC CAM

29. The disc cam will need the addition of a center hole and key slot. Move the cursor of the upper face of the disc cam, causing the edges to turn red. Double-click once as shown in Figure 13-24.

FIGURE 13-24

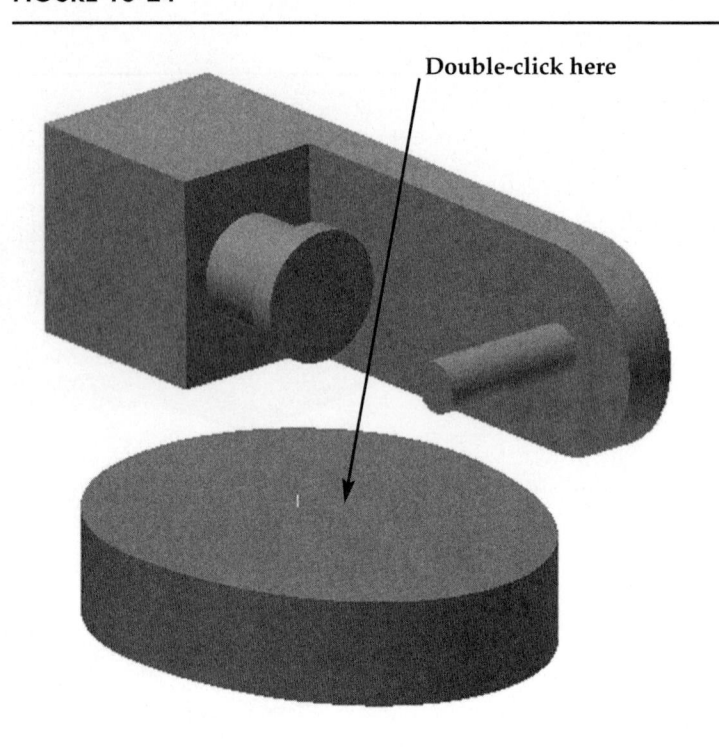

Double-click here

30. The rest of the parts in the assembly will become inactive as shown in Figure 13-25.

FIGURE 13-25

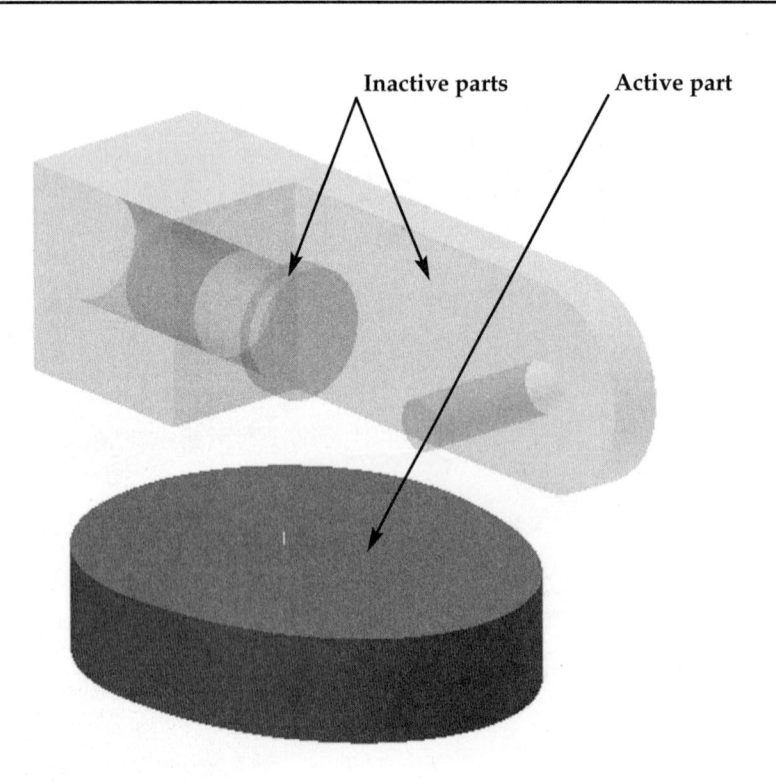

Inactive parts   Active part

31. Move the cursor to the upper face of the disc cam causing the edges to turn red and right-click once. A pop-up menu will appear. Left-click on **Edit** as shown in Figure 13-26.

FIGURE 13-26

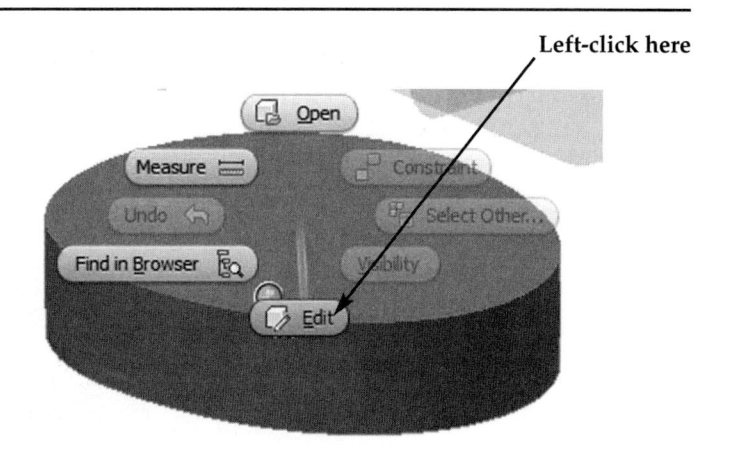

32. Move the cursor to the upper face of the cam disc, causing the edges to turn red and right-click once. A pop-up menu will appear. Left-click on **New Sketch** as shown in Figure 13-27.

FIGURE 13-27

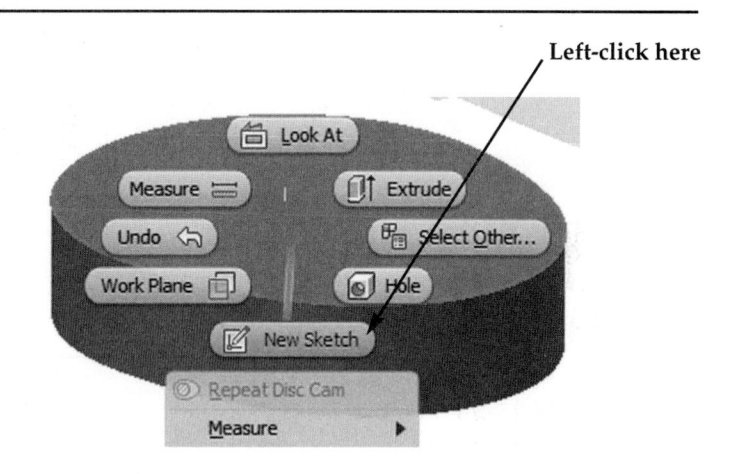

33. Inventor will return to the Sketch Panel as shown in Figure 13-28.

FIGURE 13-28

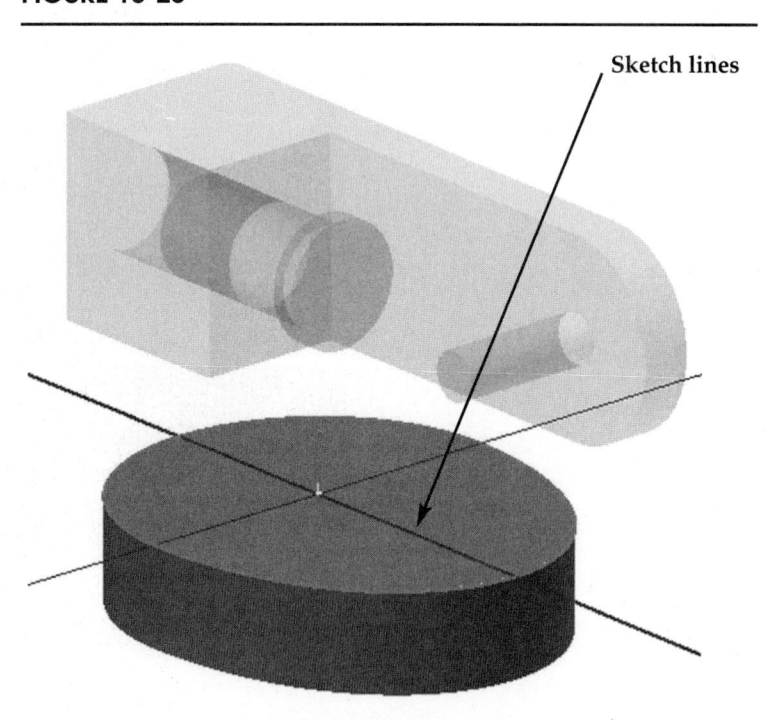

34. Use the **Look At** command to gain a perpendicular view of the upper face of the cam disc as shown in Figure 13-29.

FIGURE 13-29

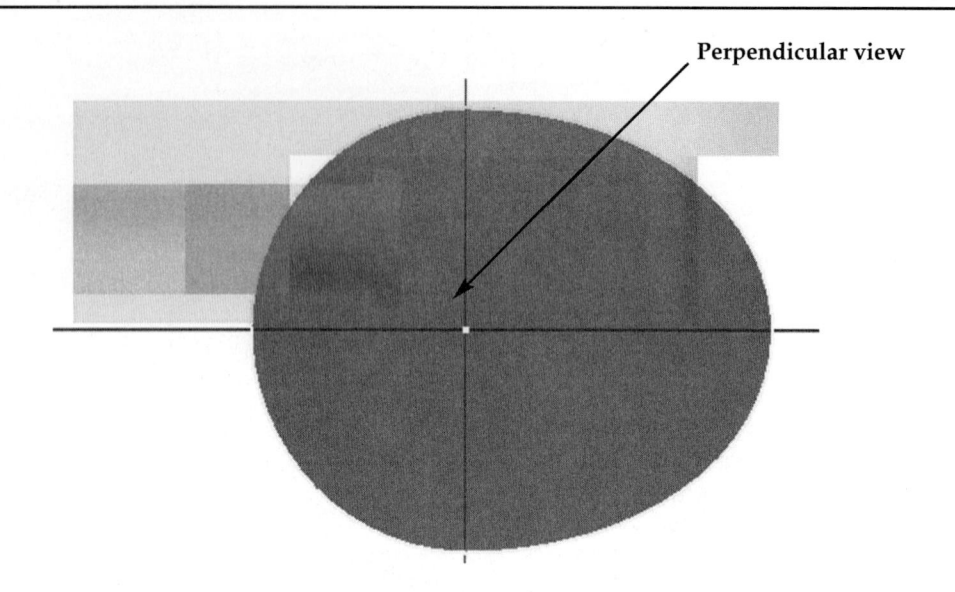

35. Complete the sketch as shown in Figure 13-30.

FIGURE 13-30

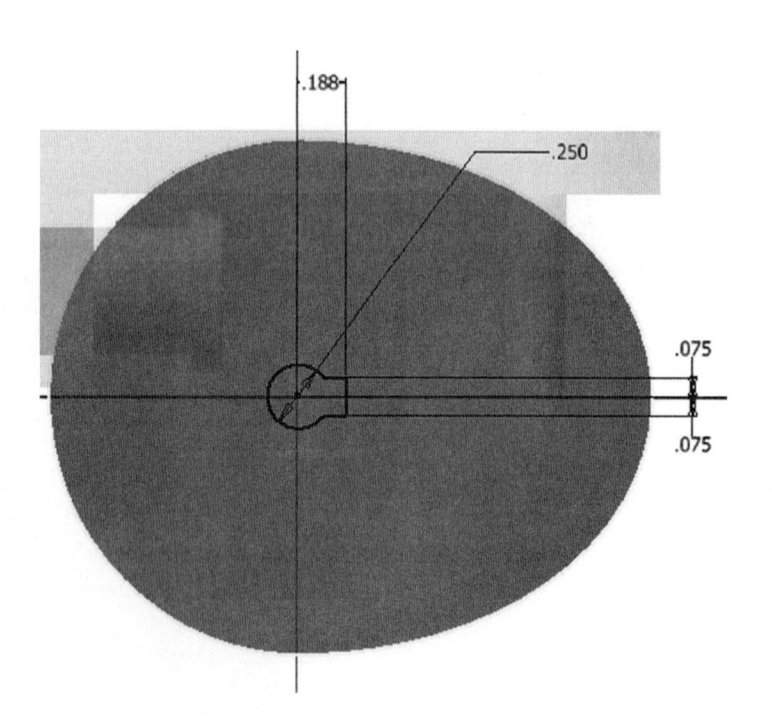

36.   Once the sketch is complete, exit out of the Sketch Panel as shown in Figure 13-31.

FIGURE 13-31

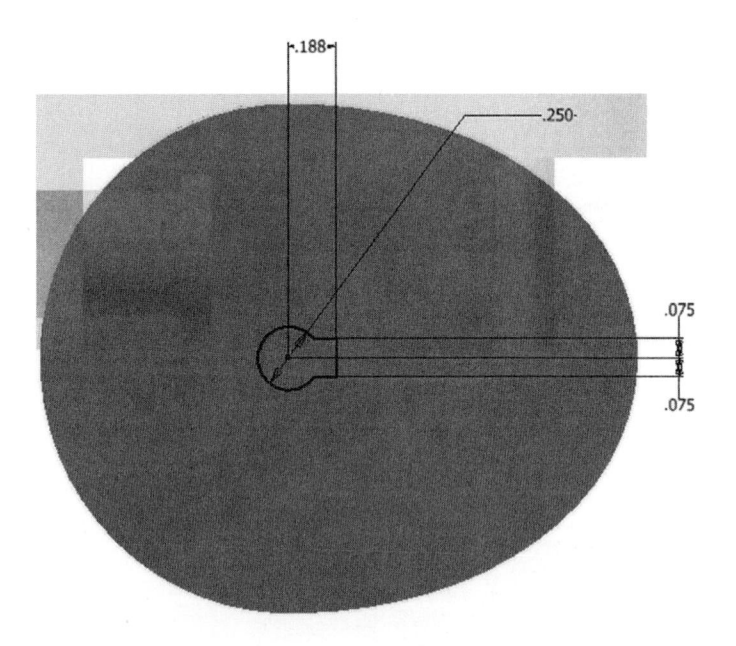

37.   Use the **Rotate** command to rotate the part around as shown in Figure 13-32. Use the **Extrude** command to cut a hole and key slot in the part as shown.

FIGURE 13-32

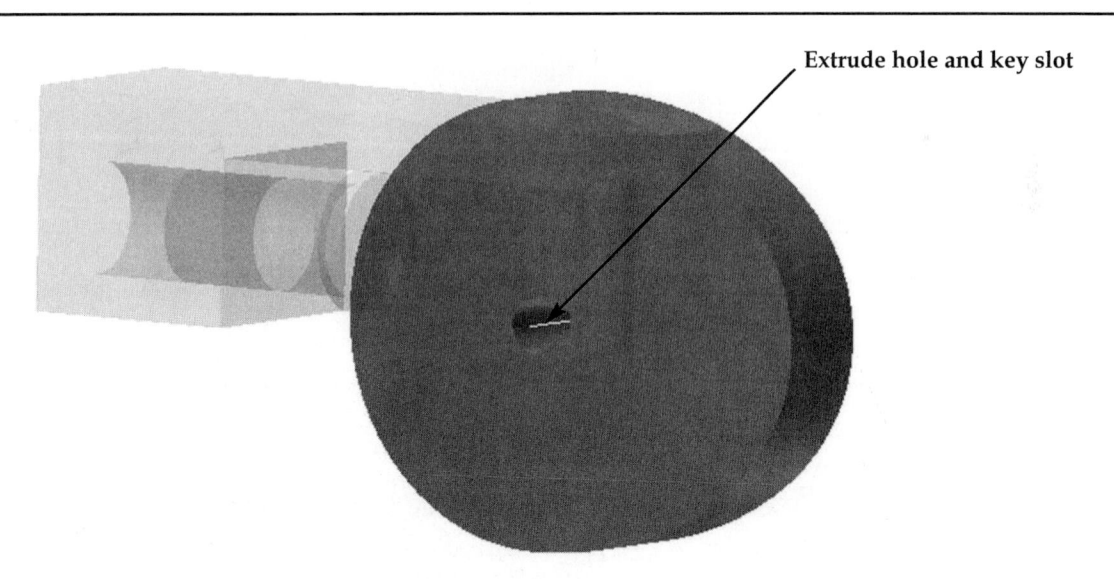

38. Move the cursor to the upper left portion of the screen and left-click on the **Manage** tab. Left-click on **Return** twice as shown in Figure 13-33.

FIGURE 13-33

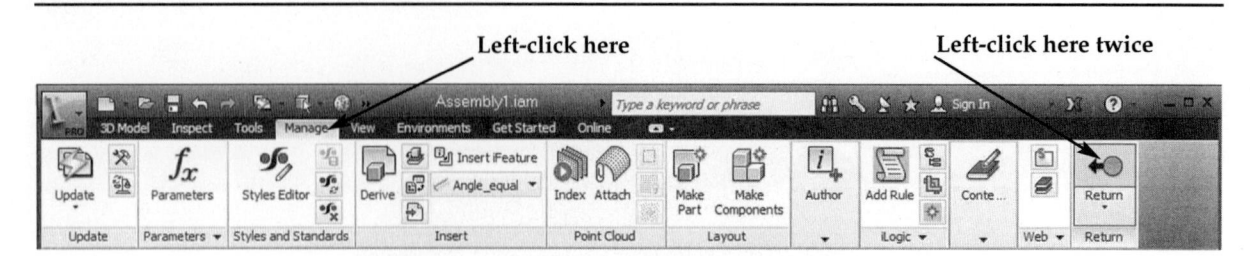

39. Your screen should look similar to Figure 13-34.

FIGURE 13-34

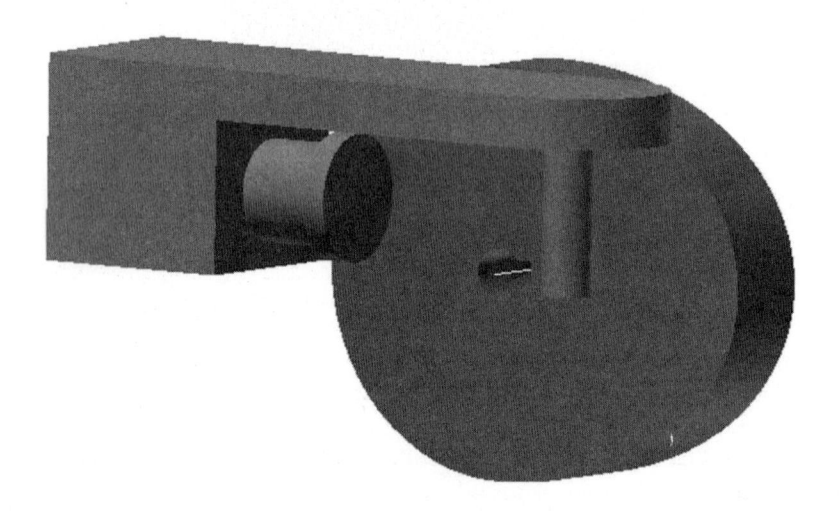

40. Move the cursor to the upper left portion of the screen and left-click on the **Assemble** tab. Inventor will return to the Assemble Panel if not already in the Assemble Panel.

FIGURE 13-35

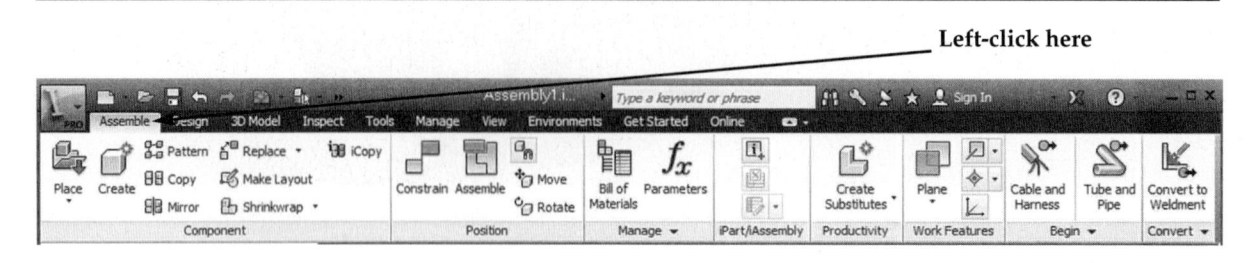

41.   Use the **Constraint** command to constrain the center of the disc cam to the center of the shaft as shown in Figure 13-36.

FIGURE 13-36

42.   Rotate the assembly around to gain access to the side of the disc cam. Use the **Constraint** command to constrain the disc cam an offset distance of **.125** inches from the inside of the cam case as shown in Figure 13-37.

FIGURE 13-37

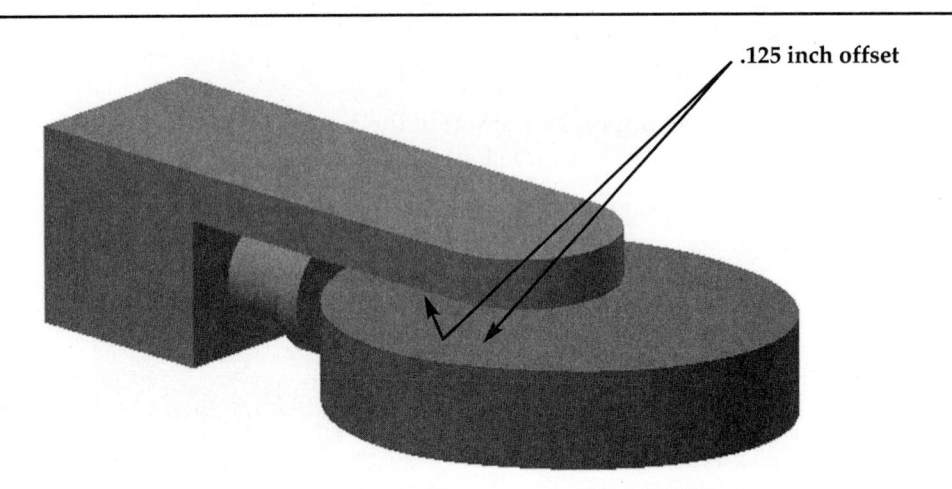

.125 inch offset

43. Use the **Rotate** command to rotate the part as shown. The disc cam's location on the shaft should be similar to Figure 13-38.

FIGURE 13-38

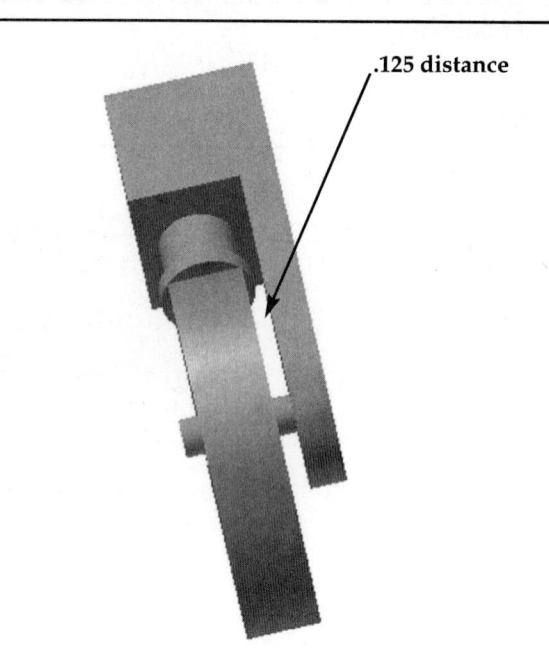

.125 distance

44. Use the **Rotate** command to rotate the assembly around as shown in Figure 13-39.

FIGURE 13-39

45. Move the cursor to the upper left portion of the screen and left-click on **Constrain** as shown in Figure 13-40.

FIGURE 13-40

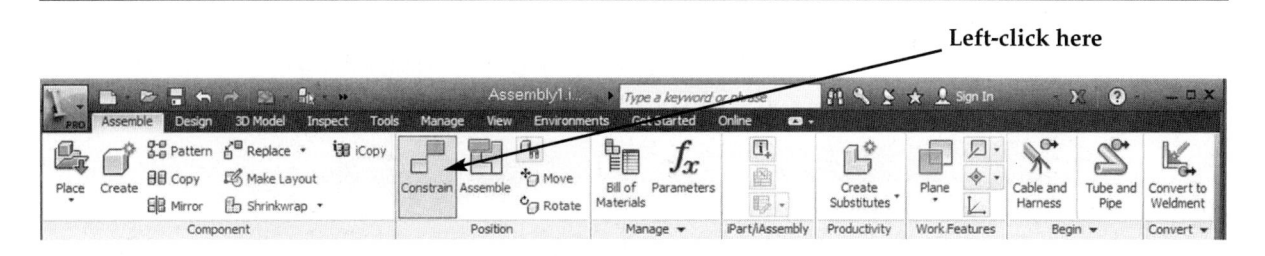

46. The Place Constraint dialog box will appear. Left-click on the "Tangent" icon. Left-click on the "Outside" solution as shown in Figure 13-41.

FIGURE 13-41

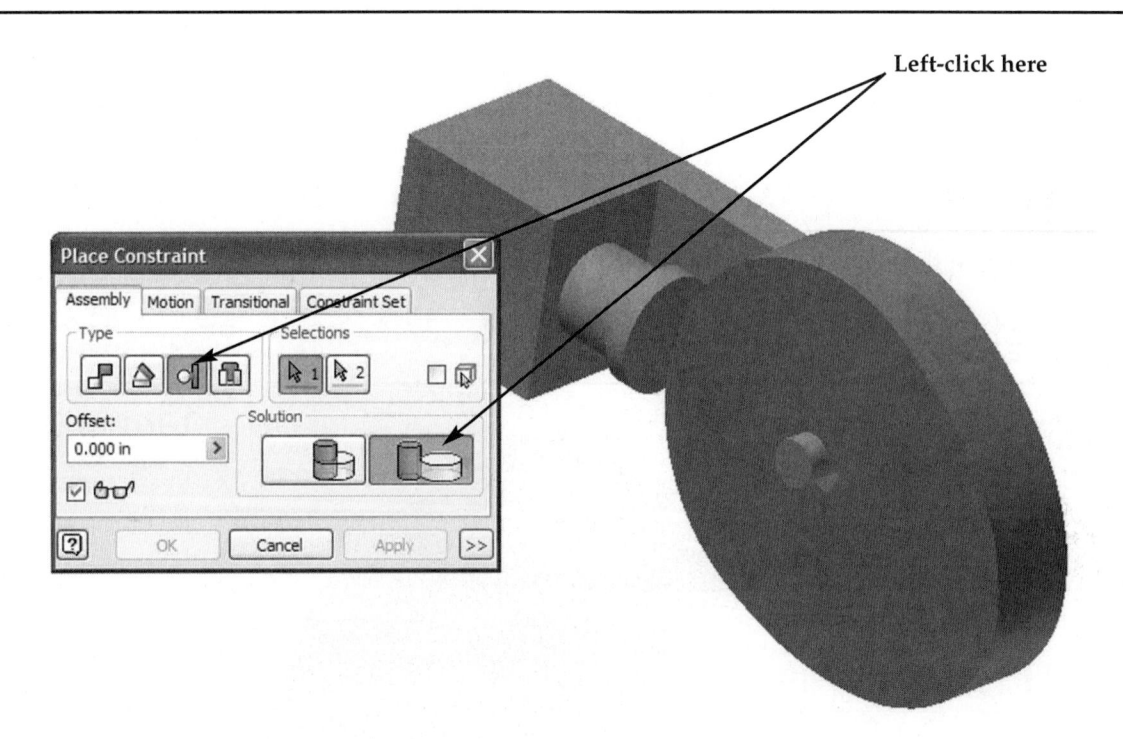

47. Left-click on the lifter foot as shown in Figure 13-42.

FIGURE 13-42

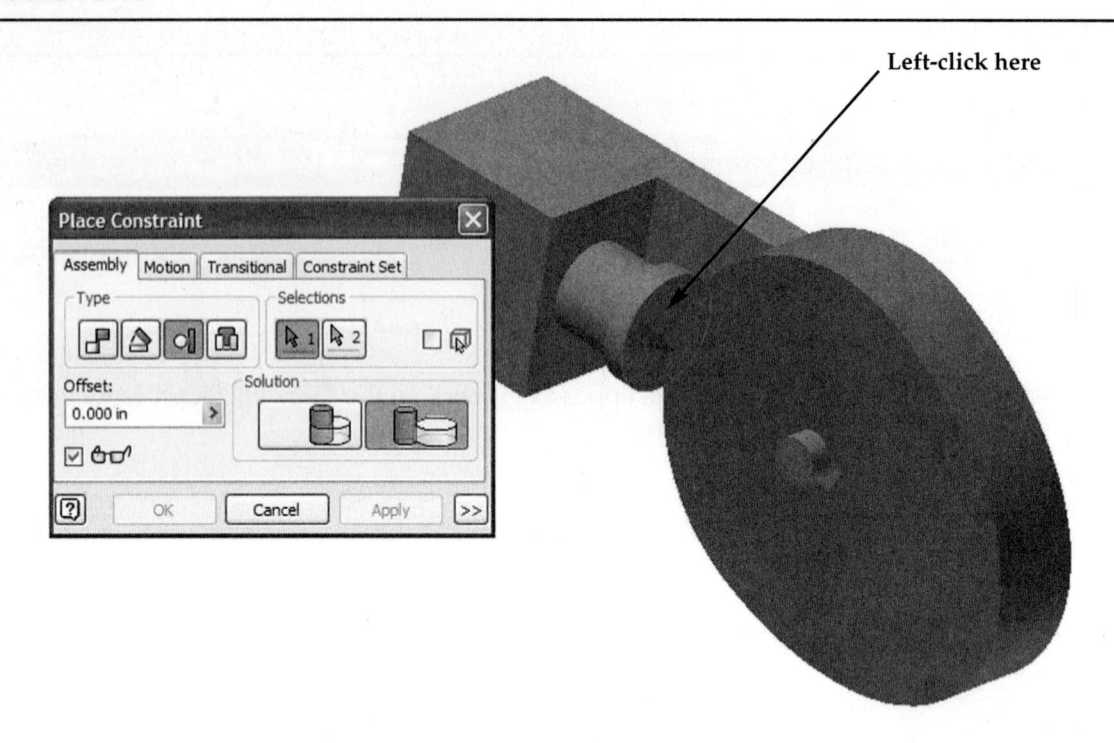

48. Left-click on the surface of the disc cam as shown in Figure 13-43.

FIGURE 13-43

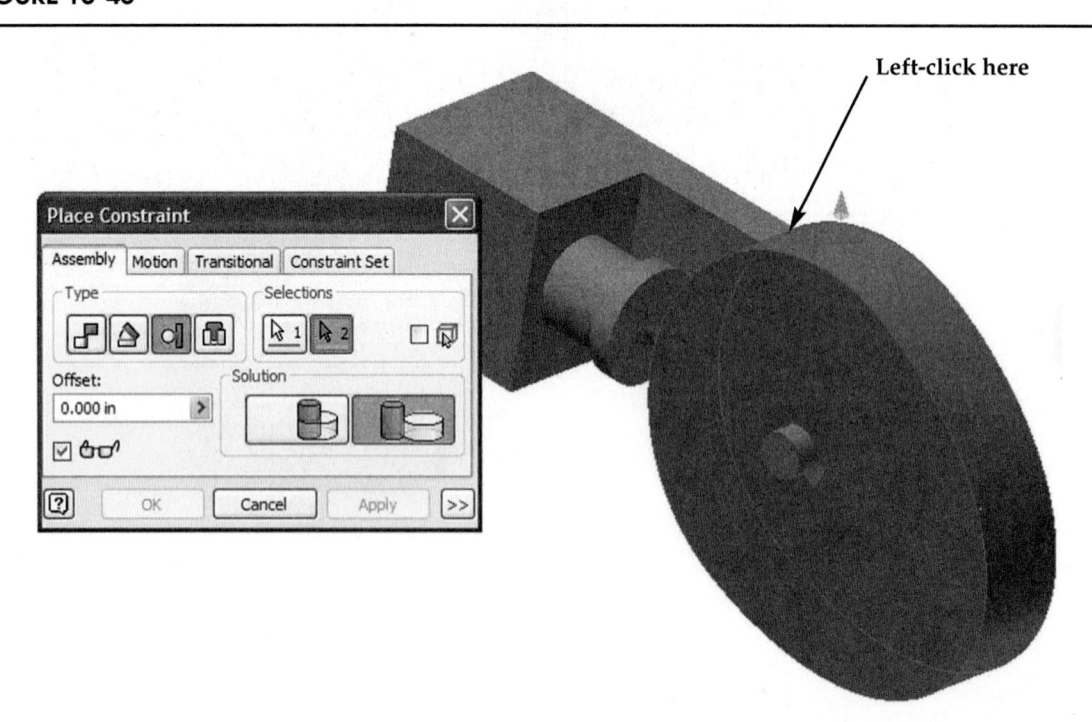

49.   Left-click on **OK** as shown in Figure 13-44.

FIGURE 13-44

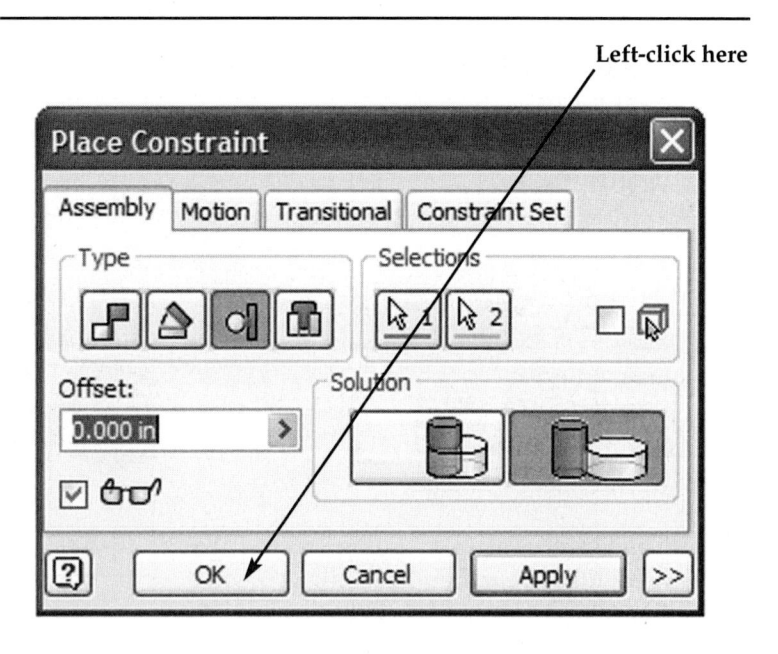

Left-click here

50.   Use the cursor to rotate the disc cam upward as shown in Figure 13-45.

FIGURE 13-45

ANIMATE THE ASSEMBLY

51.   Move the cursor to the upper left portion of the screen and left-click on **Constrain** as shown in Figure 13-46.

FIGURE 13-46

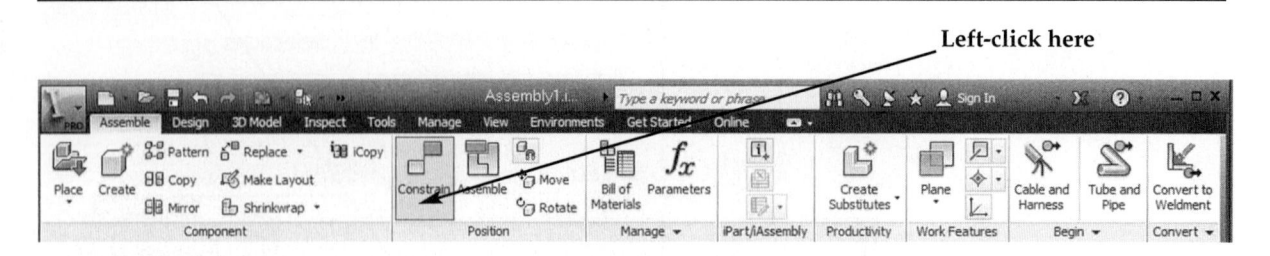

Left-click here

52.   The Place Constraint dialog box will appear. Left-click on the "Angle" icon. Left-click on the "Directed Angle" solution as shown in Figure 13-47.

FIGURE 13-47

Left-click here

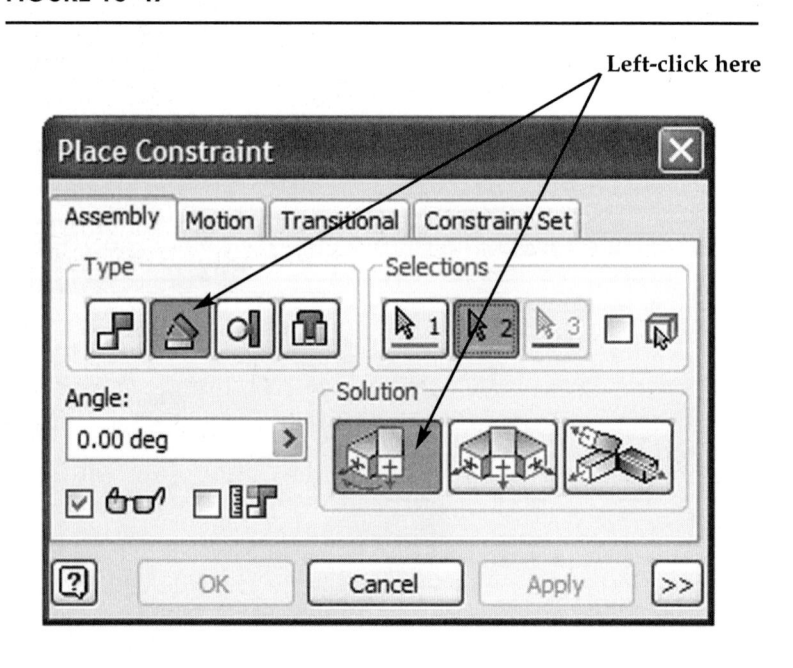

53.   Use the **Zoom** command to zoom in on the key slot. Left-click on the back (inside) of the key slot as shown in Figure 13-48.

FIGURE 13-48

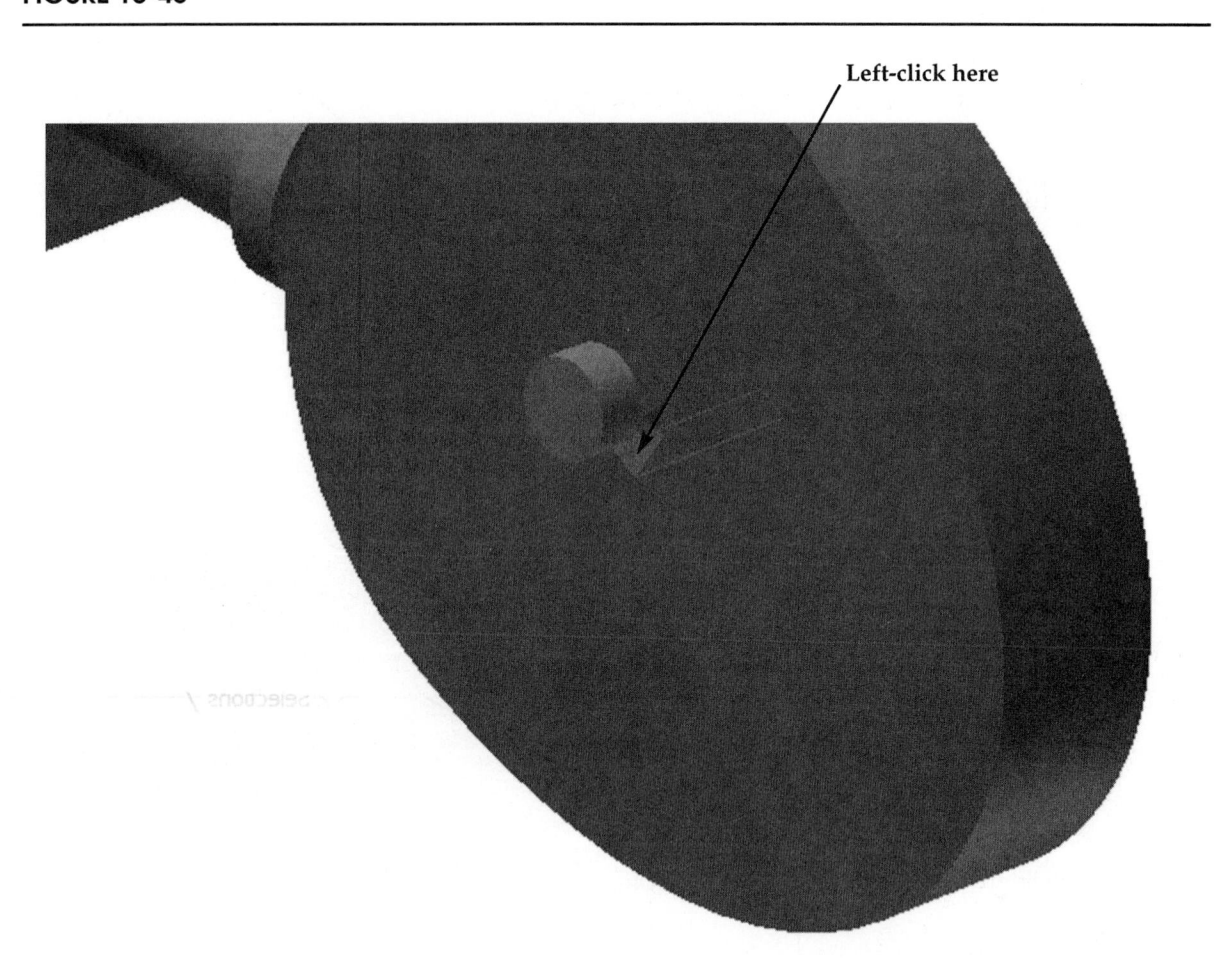

54. Left-click on the underside of the cam case as shown in Figure 13-49.

FIGURE 13-49

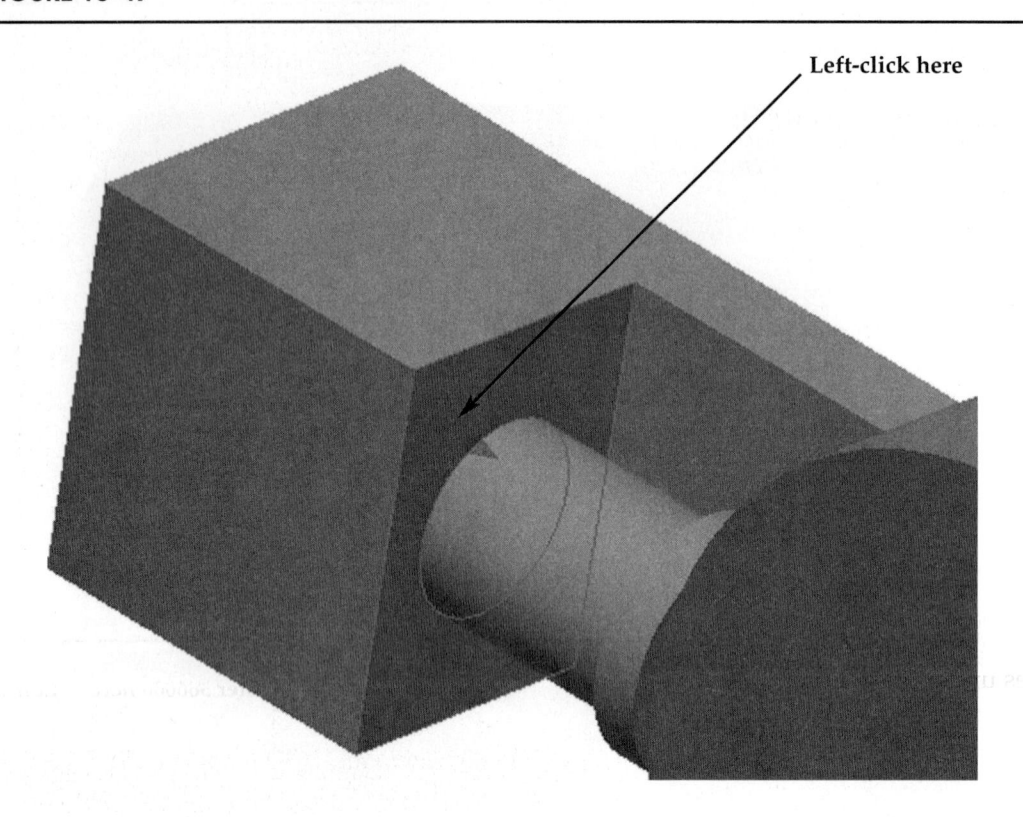

Left-click here

55. Left-click on **OK** as shown in Figure 13-50.

FIGURE 13-50

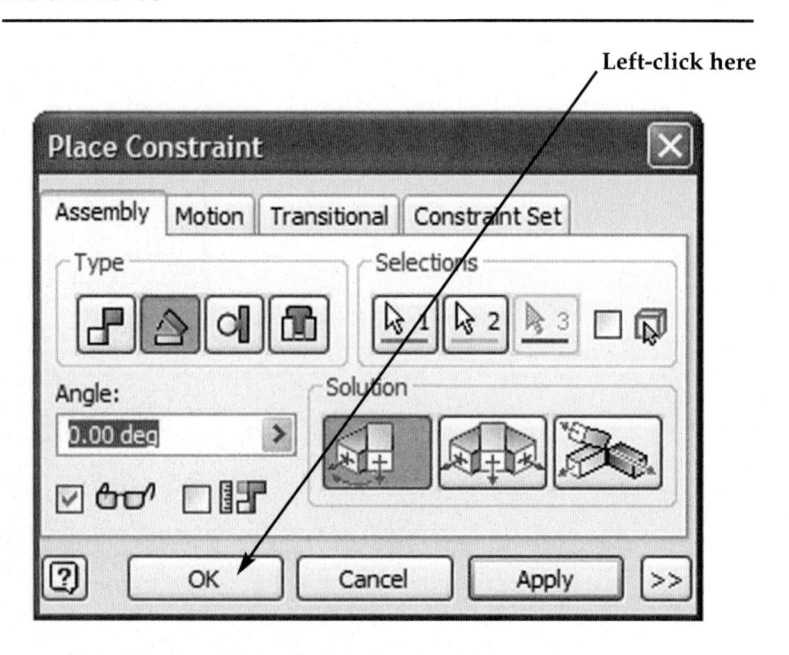

Left-click here

56. Refer back to Chapter 7 on Driving Constraints if needed. Move the cursor to the lower left portion of the screen to the part tree. Scroll down to **Angle:1** and right-click once. A pop-up menu will appear. Left-click on **Drive Constraint** as shown in Figure 13-51.

FIGURE 13-51

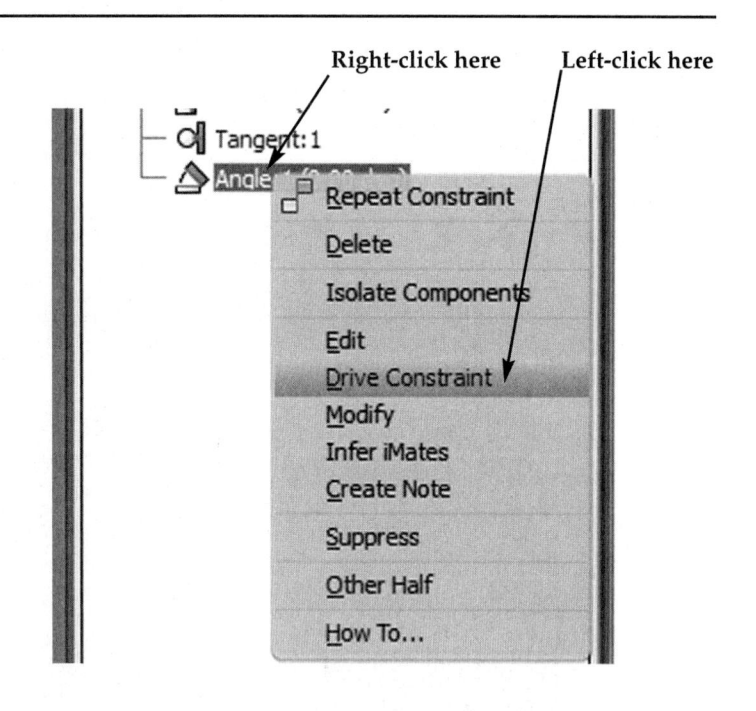

57. The Drive Constraint dialog box will appear. Enter **0** degrees under Start. Enter **360000** degrees under End. Left-click on the double arrows at the far lower right corner of the dialog box as shown in Figure 13-52.

FIGURE 13-52

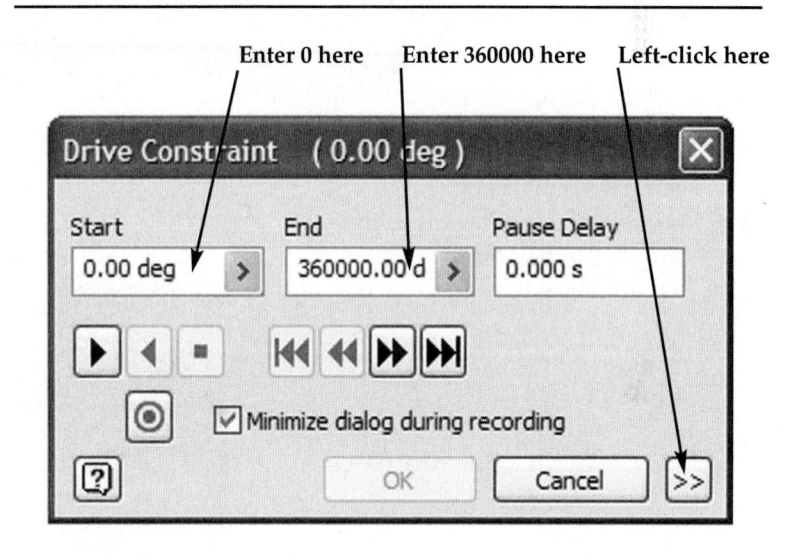

58. The Drive Constraint dialog box will expand, providing more options. Enter **10** for the number of degrees as shown in Figure 13-53.

FIGURE 13-53

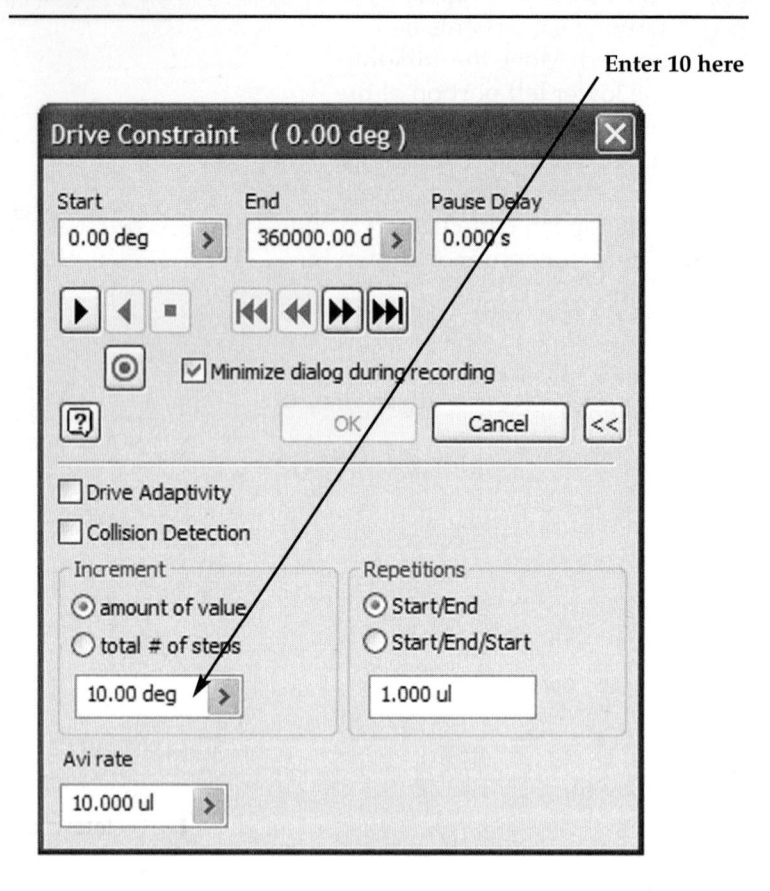

59. Use the **Zoom** option to zoom out. Use the **Pan** option to move the assembly off to the side. Left-click on the "Play" icon as shown in Figure 13-54.

FIGURE 13-54

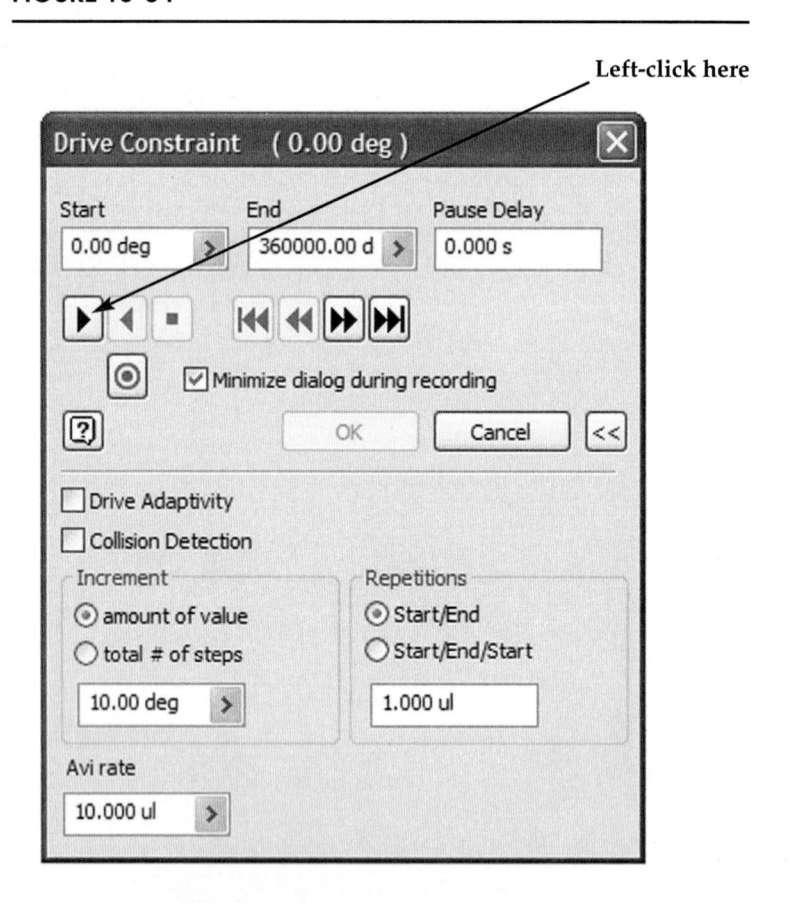

60.   Inventor will animate the assembly, causing the disc cam to rotate.

61.   Left-click on the "Stop" icon. The animation will stop. Left-click on the "Minimize" icon. The Drive Constraint dialog box will get smaller. Left-click on the "Rewind" icon. This will rewind the animation back to 0 degrees as shown in Figure 13-55. Refer to Chapter 7 for instructions on how to create an .avi or .wmv file.

FIGURE 13-55

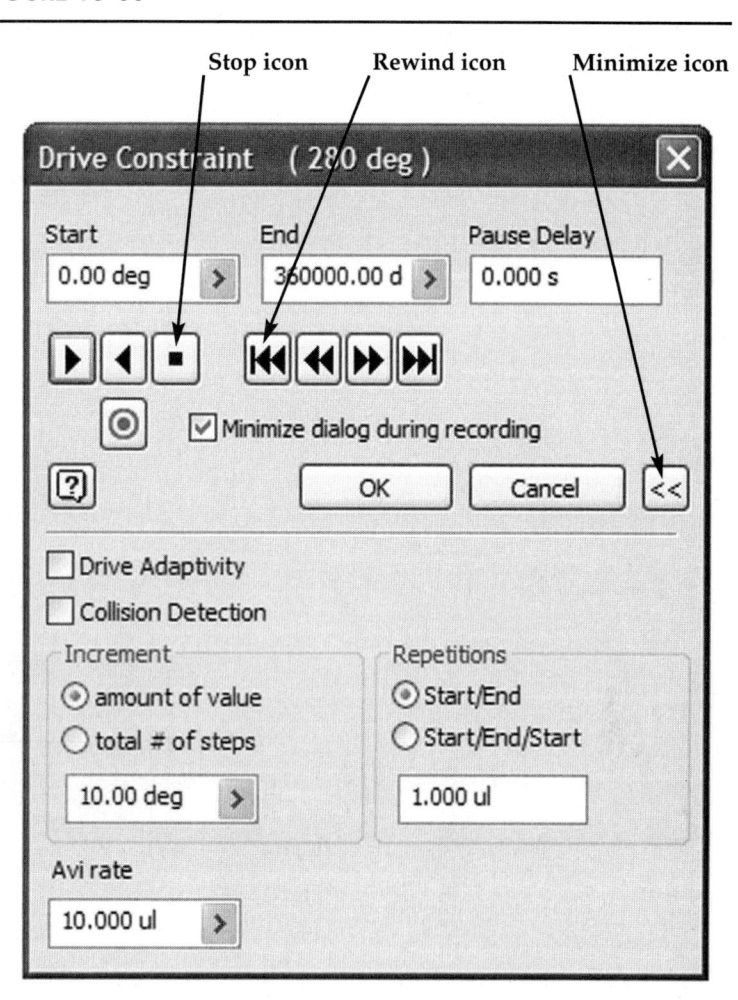

CHAPTER PROBLEMS

Create the following roller lifter and use it to replace the flat bottom lifter and constrain it to the Disc Cam as described in the chapter.

Using the existing flat bottom lifter, modify the base to utilize a small roller wheel. Use the Rectangle command to cut a .375 inch portion of the lifter as shown. Center the .063 hole on the lifter body.

PROBLEM 13-1

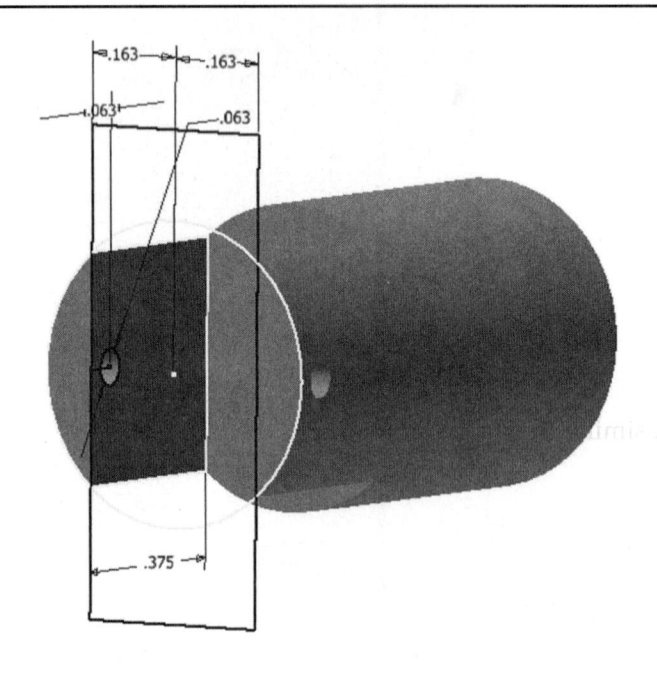

PROBLEM 13-2

Create the roller pin as shown.

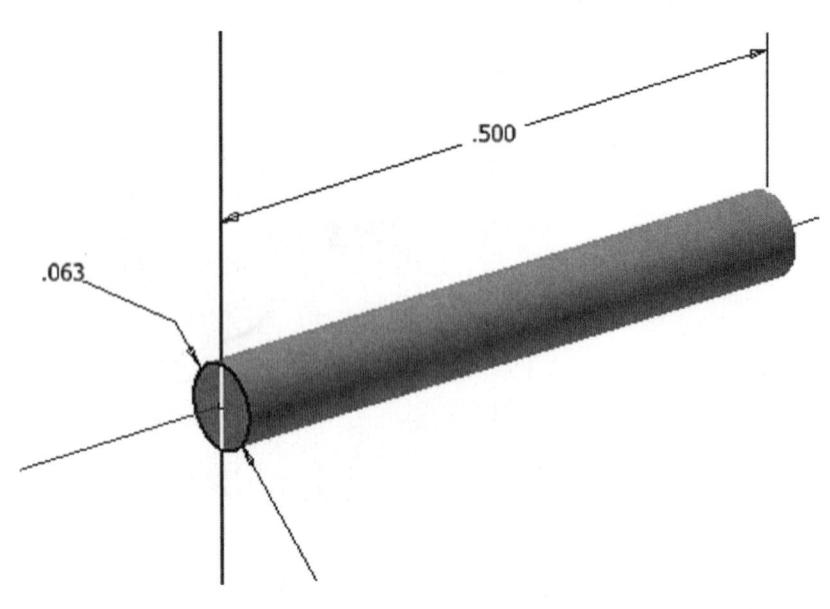

PROBLEM 13-3

Create the roller wheel as shown.

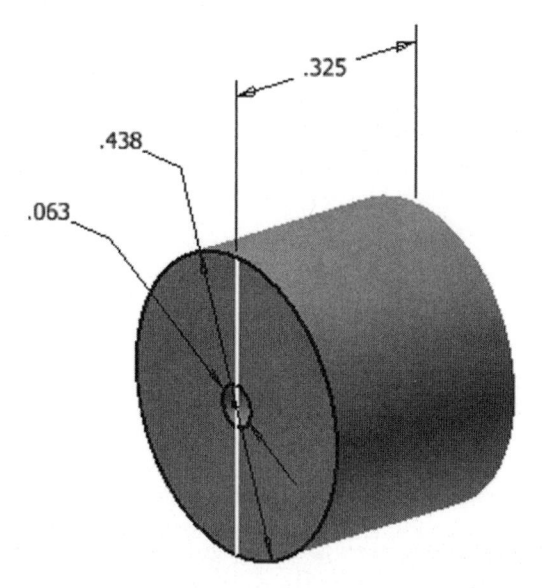

PROBLEM 13-4

Your screen should look similar to what is shown.

INDEX

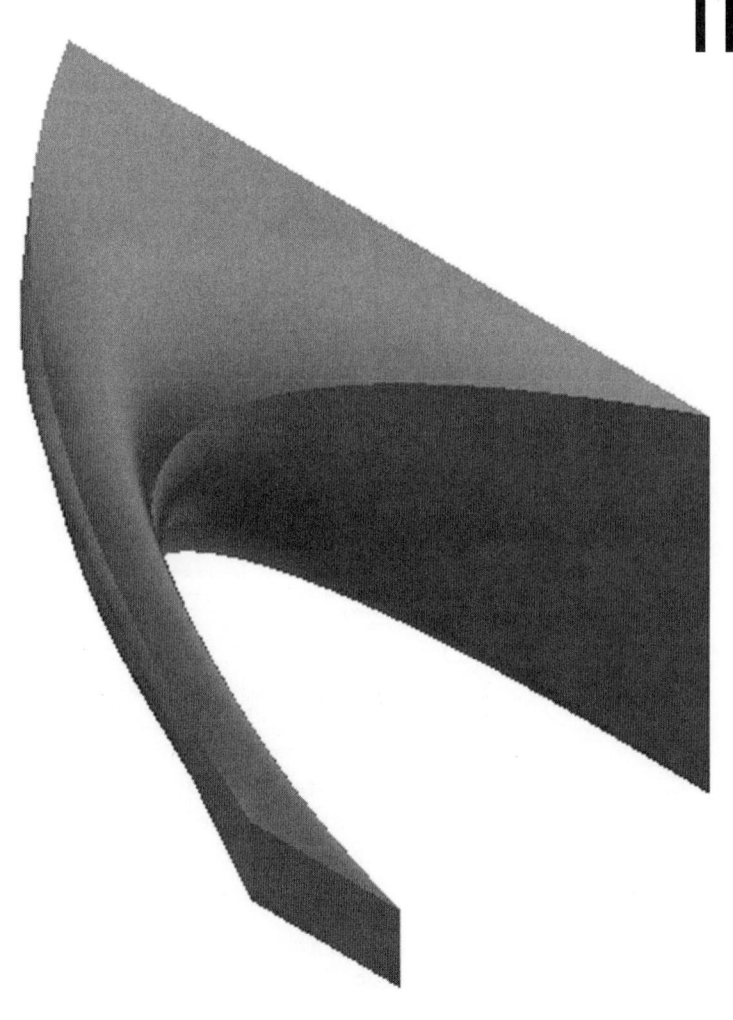